RADIATION EFFECTS IN MATERIALS

Edited by **Waldemar A. Monteiro**

Radiation Effects in Materials
http://dx.doi.org/10.5772/61498
Edited by Waldemar A. Monteiro

Contributors

Nelida Del Mastro, Maria Porubska, Mihai Oane, Rares Victor Medianu, Anca Buca, Markus Zehringer, Elahe Alizadeh, Sylwia Ptasinska, Leon Sanche, R Deepalaxmi, V Rajini, C Vaithilingam, Alexander Kir'Yanov, Jean Engohang-Ndong, Roberto Uribe, Farid Umarov, Dan Sporea, Adelina Sporea, Shahidi, Jakub Wiener, Mahmood Ghoranneviss, Taras Kavetskyy, Andrey Stepanov, Yanping Xu, Dejan Nikolic, Aleksandra Vasic-Milovanovic, Carlos Alberto Rinaldi, Maria Laura Azcarate, Cinthya Toro Salazar, Mohd Asyraf Kassim, Murthy Kolluri

CBS, Edition **2017**

Published by InTech
Janeza Trdine 9, 51000 Rijeka, Croatia

Publishing Process Manager Edi Lipovic

Technical Editor InTech DTP Team

Cover Designer InTech Design Team

Additional hard copies can be obtained from orders@intechopen.com

Radiation Effects in Materials, Edited by Waldemar A. Monteiro
p. cm.
Print ISBN 978-953-51-2417-7
Online ISBN 978-953-51-2418-4

Contents

Preface

This book shows us a special chapter concerning the effect of all radiation types in materials. The study of radiation effects has developed as a major field of materials science from the beginning, approximately 70 years ago. Its rapid development has been driven by two strong influences. The properties of the crystal defects and the materials containing them may then be studied. The types of radiation that can alter structural materials consist of neutrons, ions, electrons, gamma rays or other electromagnetic waves with different wavelengths. All of these forms of radiation have the capability to displace atoms/molecules from their lattice sites, which is the fundamental process that drives the changes in all materials. The effect of irradiation on materials is fixed in the initial event in which an energetic projectile strikes a target. The book is distributed in four sections: Ionic Materials; Biomaterials; Polymeric Materials and Metallic Materials.

The first section presents four specific chapters: the first chapter investigates optical loss property specifically of silica fibers, theoretically and experimentally, based on their Rayleigh scattering and absorption losses, which is very important and could help experts of this area. The second chapter presents the radiation effects on optical fibers and optical fiber–based devices as new materials, technologies and possible applications in radiation environments emerged. The third chapter presents the studies on radiation damage caused by neutrons, primarily related to the displacement of atoms from their positions in the lattice of silicon semiconductor. The fourth chapter shows the evaluation of the changes in dielectric parameters (breakdown voltage, dielectric strength, dielectric constant, dissipation factor) of virgin and gamma-irradiated SiR-EPDM blends (five different compositions).

The second section presents five chapters relating to biomaterials: the first chapter of this section refers to radiological event; there would be a major need to establish, within a few days, the radiation doses received by tens or hundreds of thousands of individuals. This chapter will be helpful in this area (radiation and environmental biophysics). A comparison of the biological effect of neutron and X-ray exposure on micronuclei yields in peripheral lymphocytes demonstrated that the IND-spectrum irradiator described above gives RBE values within the expected range; the second chapter of this section shows the importance in controlling the radioactivity and principally the environment. All forms of radiation have the capability to displace atoms from their lattice sites or human cells or plants in general; this third chapter involves radiation influence on edible materials. A polymeric material that can be easily consumed by human beings or lower animals in whole or part, via the oral cavity, and given harmless effect to the health, is fundamental to show a state of the art about the effects of ionizing radiation on edible polymers, that is, starch and vegetal proteins, and also on gelatin that come from animal origin. The fourth chapter concerns the experimental and theoretical results of LEE impact on single and double stranded DNA, its basic constituents, protein subunits, as well as radiosensitizers and chemotherapeutic agents alone or bound to DNA were reviewed. The fifth chapter refers to microbiological analyses on municipal sewage sludge treated in a pilot plant process utilizing an electron accelerator (3 MeV) that is very important due to the way to show the potential of this technology to decontaminate sludge with 15% solids and really open new opportunities for large urban agglomerations to save money on sewage sludge treatment.

The third section presents chapters covering polymeric materials: the first chapter shows the effects of some radiation types (electron beam; gamma beam; proton beam) on the chemical structure, crosslinking process, thermal and crystallinity characteristics, as well as mechanical properties of polyamide-6 are discussed depending on absorbed dose. The second chapter of this section investigates low-energy ion-induced processes in B: PMMA [Positron annihilation spectroscopy (2.15 keV), optical (UV-visible region and Raman spectroscopy), electrical measurements and nanoindentation test] is valuable and could help us understand other options for optoelectronic materials (polymers). The third chapter of this section involves the effect of plasma, laser, microwave, electron beam and ion beam on surface, and chemical, physical and mechanical properties of textile materials are fully discussed. The advantages of this technology are well known such as improvement in shades, enhancing colour fastness, color strength, low cost-effective and reduction of the concentration of the used chemicals. The fourth chapter discusses comprehensively the irradiation pretreatment of tropical biomass prior to the subsequent enzymatic saccharification and fermentation processes which can be applied as an alternative pretreatment approach for biofuel production.

The fourth section comprises metallic materials studies: the first chapter of this section concerns the experimental and computer simulation of low- and medium-energy ions collisions on the surface of a solid, and of the accompanying effects, namely, scattering, sputtering and surface implantation is treasured. The second chapter concerns the contribution of various irradiation damage mechanisms (thermal and fast neutrons) to the evolution of microstructure and mechanical properties in all four regimes for 5xxx and 6xxx series aluminum alloys to understand the expected changes in mechanical properties of HFR vessel material in relation to microstructural aspects beyond the current surveillance data to support the HFR SURP program. The third chapter of this section shows a parallel between laser irradiation and relativistic electrons irradiation of solids (graphite and tungsten). Simulations and experimental data are presented; the fourth chapter involves the nanostructuring of metal surfaces by laser ablation that is a very good technique which allows numerous technological developments ranging from Laser-Induced Breakdown Spectroscopy (LIBS), Pulsed Laser Deposition (PLD), laser propulsion, to surface modification and generation of nanoparticles, NPs. These effects or phenomena were found to occur during the implementation of the micromachining processes and have gained relevance and are a very important source for the modification of surfaces for technological uses.

Prof. Dr. Waldemar Alfredo Monteiro
Materials Science and Technology Center
Nuclear and Energy Research Institute
São Paulo, SP, Brazil.

Section 1

Ionic Materials

Chapter 1

Effects of Electron Irradiation Upon Absorptive and Fluorescent Properties of Some Doped Optical Fibers

Alexander V. Kir'yanov

Additional information is available at the end of the chapter

http://dx.doi.org/10.5772/63939

Abstract

A review of the recent studies of the effect of irradiating silica-based fibers doped with rare earths and metals by a beam of high-energy (β) electrons is presented. Of the review's main scope are the attenuation spectra' transformations occurring in optical fiber of such types under electron irradiation, allowing, from one side, to recover some general essence of the phenomena involved and, from the other side, to draw the features that would make such fibers useful for applications, for example, in dosimetry and space technologies. Among the fibers of the current review's choice, exemplifying the effect of electron irradiation most brightly, are ytterbium (Yb) and cerium (Ce) (the rare earths' representatives) and bismuth (Bi) (the post-transitional metals representative) doped fibers, where a diversity of the electron-irradiation-related effects is encouraged.

Keywords: electron irradiation, ytterbium-, cerium- and bismuth-doped silica fibers, photodarkening, optical bleaching

1. Introduction

In this chapter, a few examples are demonstrated of the impact of high-energy (β) electrons irradiation on the absorptive and fluorescence properties of silica-based optical fibers doped with rare earths and metals. The results presented hereafter seem to be useful for understanding the processes standing behind the highlighted phenomena and for possible applications of the fibers, say, in dosimetry and space technology.

In each case, we used for irradiating fiber samples a controllable linear accelerator of the LU type that emits β-electrons with a narrow-band energy spectrum (~6 MeV) in a short-pulse (~5 μs) mode. The samples with lengths of around 1–2 m were placed into the accelerator's chamber for various time intervals, which provided growing irradiation doses. The irradiated fibers were then left for 2 weeks prior to the main-course spectral measurements to avoid the role of short-living components in the decay of induced absorption (IA). The measurements were done during a limited time (viz., the following 2…3 weeks) for diminishing the effect of spontaneous IA recovering. Note that ionization, that is, the production of β-induced carriers by an electron beam (i.e., of secondary free holes and electrons), is the main cause of the spectral transformations in the fibers. This happens because high-energy primary β-electrons are virtually nondissipating at the propagation through a fiber sample; on the other hand, certain contribution in ionization of the fibers' core-glasses arising from γ-quanta born at inelastic scattering of the high-energy electrons cannot be disregarded.

We demonstrate below first a study of the resistance of a couple of cerium (Ce)-doped alumino-phospho-silicate fibers (one of them being codoped with gold (Au)), to β-electrons. The experimental data reveal a severe effect of β-irradiation upon the fibers' absorptive properties, given by noticeable susceptibility of Ce ions being in Ce^{3+}/Ce^{4+} states to the treatment, arising as growth followed by saturation of IA. We also report the essentials of posterior bleaching of β-darkened fibers, also in terms of attenuation spectra' transformations, at exposing them to low-power green (a He-Ne laser) and ultra violet (UV, a mercury lamp) light. It is shown that both phenomena are less expressed in Ce fiber codoped with Au than in Au-free one and that the spectral changes in the former are more regular versus dose and bleaching time.

Then, we provide a comparative experimental analysis of IA, induced by β-electrons, for a series of ytterbium (Yb)-doped alumino-germano-silicate fibers with different concentrations of Yb^{3+} ions and compare this effect with the photodarkening (PD) phenomenon in the same fibers, arising at resonant (into 977 nm absorption peak of Yb^{3+} ions) optical pumping. The experimental data obtained reveals that, in these two circumstances, substantial and complex but different in appearance changes affecting the resonant absorption band of Yb^{3+} ions and the off-resonance background loss are produced in the fibers.

Finally, we report a study of attenuation spectra' transformations in a set of bismuth (Bi)-doped silica fibers with various contents of emission-active Bi centers, which occur as the result of β-irradiation. Among the data obtained, notice a substantial decrease of concentration of Bi centers, associated with the presence of Germanium (Ge) in core-glass, with increasing irradiation dose (the "bleaching" effect), while, on the contrary, an opposite trend, that is, dose-dependent growth of resonant-absorption ascribed to Bi active centers, associated with the presence in core-glass of Aluminium (Al). These results are worth noticing for understanding the nature of Bi-related centers in silica fibers, yet uncovered.

2. The effects of electron irradiation and posterior optical bleaching in Ce-doped and Ce/Au-codoped alumino-phospho-silicate fibers

Development of suitable host glasses and fibers for dosimetry, which are based on formation of radiation-induced defects leading to glass coloration [1–6] or filling pre-existing traps, measured by means of thermally or optically stimulated fluorescence [7], became a hot task. Dosimetry systems can be used in high radiation fields, for example, in proximity to nuclear reactors, hazardous places, and in open space. Fiber-based dosimeters are being intensively investigated and recently a few systems have been proposed, based on versatile physical effects in radiation-sensitive silica fibers [8].

Cerium (Ce)-doped silica glass has interesting fluorescent properties [9], which makes it promising for utilizing as a scintillator for detecting X- and γ-rays, or neutrons [10, 11]. On the other hand, silica glass is known to suffer from the presence of point defects and OH groups, responsible for nonradiative recombination channels competing fluorescence. In turn, Au, when combined with cerium oxide (CeO_2) is known to be a promising catalyst for the reaction $CO + H_2O \rightarrow H_2 + CO_2$ [12, 13], giving a way to remove carbon-related impurities along with OH groups from silica matrix during synthesis. Thus, Ce/Au codoped glass is expected to enhance efficiency of energy transfer from the host matrix to emissive centers. The other motivation for Au codoping is to increase radiation resistance of Ce-doped fiber, as argued in more details below. The refereed properties of alumino-phospho-silicate glass doped with Ce and Ce/Au are also a concern of optical fibers made on its base.

Below, the results of experiments on irradiating Ce-doped alumino-phospho-silicate fibers by energetic β-electrons are highlighted, resulting in the fibers darkening. It is furthermore shown that the irradiated fibers are sensitive to weak light of a He-Ne laser (543 nm) and UV mercury lamp, both leading to partial recovery of their initial properties. The whole of experimental data evidences notable susceptibility of Ce-doped fibers to both kinds of treatment. As well, it is demonstrated that the spectral transformations occurring in Ce fiber codoped with Au are less expressed but more regular upon β-irradiation dose and exposure time when bleaching than those in Au-free fiber. A brief discussion in attempt of a reasonable explanation of the experimental laws completes the study, with the key point being a discussion about the species involved in the processes, which are associated with Ce.

The reported results may have value for using Ce-doped silica fibers for dosimetry in harmful environments [8, 14–20] and inscribing Bragg gratings [21–25]. As well, these results seem to be impactful, given by renewed interest to Ce codoping as a tool for diminishing PD in Yb-doped fibers (we inspect the last effect in detail in Paragraph 3).

2.1. Fiber samples and experimental arrangement

The sourcing Ce-doped and Ce/Au-codoped fiber preforms based on alumino-phospho-silicate glass have been made by means of modified chemical vapor deposition (MCVD) process employed in conjunction with solution doping (SD) technique; the final fibers have been drawn from the preforms using a drawing tower. Core diameters/numerical apertures of the two fibers were measured to be ~25 μm/0.15…0.16, respectively.

Estimated from EDX, average doping levels were found to be 5.0 wt.% Al_2O_3, 0.15 wt.% P_2O_5, 0.3 wt.% CeO_2 (in the Ce-doped fiber) and 5.1 wt.% Al_2O_3, 0.15 wt.% P_2O_5, 0.27 wt.% CeO_2, and 0.2 wt.% Au_2O_3 (in the Ce/Au-codoped fiber). Both fibers had multimode wave-guiding, which make them useful for sensor applications. A sample of standard multimode Al-doped (~6 wt.

% Al_2O_3) fiber was used in experiments for comparison. The β-irradiation dosage below corresponds to 1 × 10^{12} ("dose 1"), 5 × 10^{12} ("dose 2"), 1 × 10^{13} ("dose 3"), 5 × 10^{13} ("dose 4"), 1 × 10^{14} ("dose 5"), and 2.5 × 10^{15} ("dose 6") cm^{-2}.

Optical transmission spectra of fiber samples were measured (employing the cutback method), using a white light source and optical spectrum analyzer (OSA), turned to a 5 nm resolution. Such spectra were recorded before and after each stage of β-irradiation and at posterior exposure to light of a He-Ne laser (543 nm) or UV lamp (λ <450 nm). The attenuation spectra presented below were obtained after recalculating the measured transmissions into loss [dB/m]. In some of the figures below the difference spectra in terms of IA are provided, which were obtained after subtraction of the attenuation spectra of pristine samples from the ones taken after a certain dose of β-irradiation; this allows one straightforward view on the "net" spectral loss changes in the darkened fibers. The transmission dynamics at optical bleaching of β-darkened fibers by 543 nm light was inspected applying "frontal" detecting geometry where a beam of the He-Ne laser was coupled into a fiber sample, while the transmitted light was detected using a Si photodetector; this permitted detection of the changes in transmission *in situ*. The results of the measurements are given below in terms of absorption difference (AD) at bleaching with respect to the initial (β-darkened) state of the fiber. The experiments on optical bleaching of β-irradiated fibers by UV light were as well proceeding *in situ*, where transmission change at long-term exposure to UV light was analyzed. All experiments were made at room temperature.

2.2. Experimental

2.2.1. *IA as a result of β-irradiation*

In **Figure 1**, we demonstrate (a) attenuation spectra of the Ce-doped (black solid curve 1) and Ce/Au-codoped (grey dashed curve 2) fibers before irradiation, that is, in their "pristine" state, and (b and c) the fibers' cross-sections, obtained at white light illumination. Long (meters)

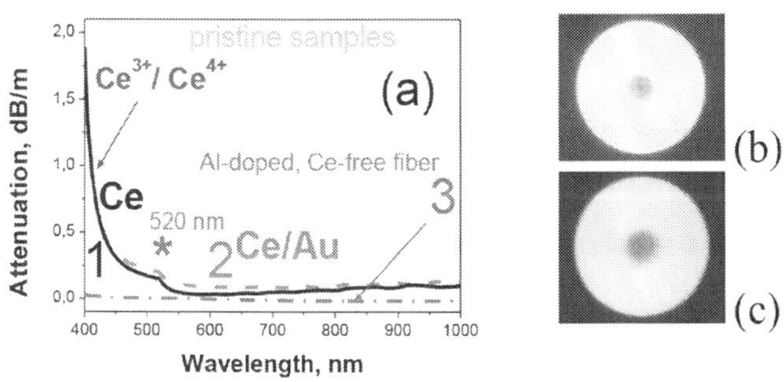

Figure 1. (a) Attenuation spectra of pristine Ce-doped (1), Ce/Au-codoped (2), and Al-doped Ce-free (3) fibers in a VIS-to-near-IR spectral range and micro-photographs of pristine Ce-doped (b) and Ce/Au-codoped (c) fibers. (Reproduced with permission from Kir'yanov et al. [75]. Copyright© 2014, Optical Society of America).

fibers were used in the measurements applying the cutback method, whereas short (10 centimeters) pieces of fibers—at microscopy. For comparison, spectral loss of "standard" Al-doped Ce-free fiber is presented in **Figure 1(a)**—see red dash-dotted curve 3.

We reveal from (a) that, in both Ce-doped and Ce/Au-codoped fibers, dramatic growth of absorption occurs toward UV, below ~550 nm, which is known to be a shoulder of the strong absorption bands adherent to Ce^{3+}/Ce^{4+} ions (mostly located in UV [23, 24]), and that no such feature is observed in the reference Ce-free fiber. Also notice steep loss rise in Ce-doped and Ce/Au-codoped fibers toward IR and a small peak at ~520 nm (asterisked), the features not observed in case of the Ce-free fiber.

Figure 2 shows the trends occurring in the fibers' attenuation spectra as the result of β-irradiation at moderate dose 4. Note that in this case, the measurements were proceeding with shorter fiber samples (~a few cm) in virtue of strong IA, established after β-irradiation.

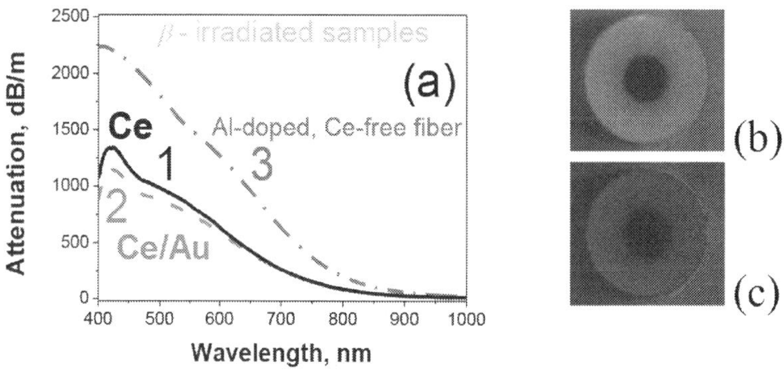

Figure 2. (a) Attenuation spectra of Ce-doped (1), Ce/Au-codoped (2), and Al-doped cerium-free (3) fibers, all measured after β-irradiation with dose 4 (5×10^{13} cm^{-2}) and micro-photographs of Ce-doped (b) and Ce/Au-codoped (c) fibers recorded after irradiation with this dose. (Reproduced with permission from Kir'yanov et al. [75], Copyright© 2014, Optical Society of America).

It is seen that IA in the Ce-free fiber is ~two times bigger than in the Ce-doped and Ce/Au-codoped ones. The other fact is that IA maxima are located near 400 and 500 nm in these two fibers, whereas the ones in the Ce-free one—at ~400 and ~600 nm, that is, in the range most probably attributing to well-known nonbridging oxygen-holes (NBOHCs) [26] (while the presence of other defect states in it—such as Si-/Al-defect centers cannot be excluded). Furthermore, it is seen from photos (b) and (c) that, in the Ce-doped and Ce/Au-codoped fibers, the core and adjacent core-cladding areas suffer darkening after β-irradiation, in the former, the effect being more pronounced.

Figure 3 demonstrates that IA in the Ce-doped (a) and Ce/Au-codoped (b) fibers increases monotonously with dose; this trend is noticeable for the 400–700 nm range, while for bigger wavelengths it fades. The other detail seen is that for moderate doses (1–4), IA is stronger in the Ce-doped fiber.

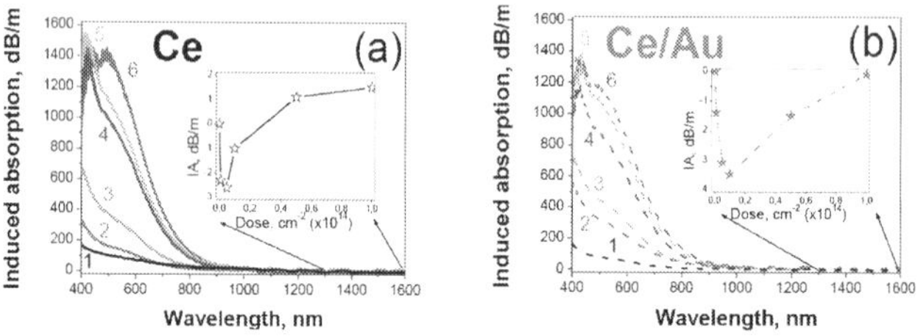

Figure 3. Main frames: IA spectra of Ce-doped (a) and Ce/Au-codoped (b) fibers; curves 1–6 correspond to doses of irradiation (in both figures) being: 1 × 10^{12} (1), 5 × 10^{12} (2), 1 × 10^{13} (3), 5 × 10^{13} (4), 1 × 10^{14} (5), and 2.5 × 10^{15} (6) cm^{-2}. Insets: average IA-losses measured within the 1300–1550 nm range *vs.* irradiation dose. (Reproduced with permission from Kir'yanov et al. [75]. Copyright© 2014, Optical Society of America).

The two-peaks structure of the IA spectra is apparent at higher irradiation doses for both fibers, with the first peak (bigger in magnitude) locating at ~415 ± 10 nm and the second one (lower in magnitude)—at ~520 ± 10 nm (compared to the ~520 nm peak asterisked in the attenuation spectra of pristine fibers in **Figure 1(a)**). To evaluate IA strength in the fibers in function of β-irradiation dose, let us compare the IA spectra with the attenuation spectra of the same fibers being in pristine state (refer to **Figure 1**). It is known that attenuation growth toward UV is common for Ce-doped glass, as stemming from the transitions inherent to Ce^{3+}/Ce^{4+} ions. (Unfortunately, IA arising in the UV-region, below 400 nm, was undetectable using our experimental equipment.) Regarding IA in the near-IR, note that the spectral transformations in this region are more complex (see insets to **Figure 3**) whose nature is unclear at the moment.

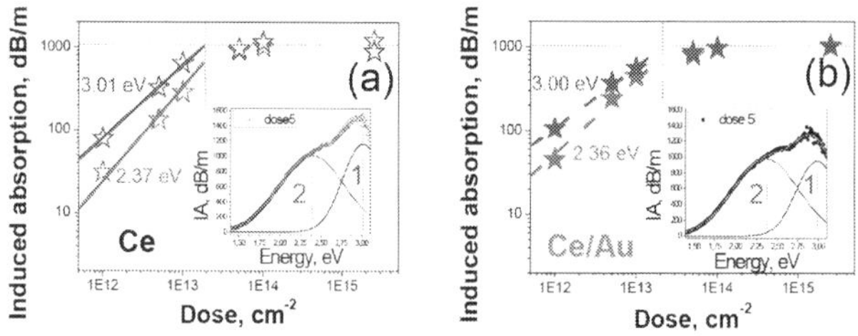

Figure 4. Main frames: dose dependences of IA for Ce-doped (a) and Ce/Au-codoped (b) fibers; blue and red symbols and lines show IA magnitudes of bands 1 and 2, obtained after deconvolution of the spectra shown in **Figure 3**. Insets: examples of deconvolution of the data obtained for the fibers, irradiated with dose 5 (spectra are plotted in eV-domain). (Reproduced with permission from Kir'yanov et al. [75]. Copyright© 2014, Optical Society of America).

Deconvolution of IA spectra (**Figure 3**) allows a closer view on their two-band structure (see insets in **Figure 4(a) and (b)**). Spectral locations of the bands (1 and 2) were found to be almost independent of irradiation dose, for both fibers: they are centered at ~3.0 and ~2.4 (±0.1) eV and are measured in half-widths at a 3 dB level by ~0.3 and ~0.5 (±0.05) eV, respectively. In main frames of **Figure 4**, IA—in terms of these two peaks' magnitudes—is plotted versus irradiation dose; these dependences are shown, respectively, by blue (band 1) and red (band

2) symbols. Fitting them within domain of smaller doses (up to ~2 × 10^{13} cm^{-2}), linear growth of IA in both bands versus dose is revealed (see the blue and red lines in the figure).

The slopes' values estimated as the result of fitting were found to be ~1.7 (~1.2 dB/m/cm^2) (Ce-doped fiber) and ~1.3 (~0.9 dB/m/cm^2) (Ce/Au-codoped fiber), correspondingly, for bands 2 and 1. It deserves mentioning that these ratios, on one hand, are almost equal for both fibers (~1.5) and, on the other hand, the slopes' ratios, when compared for bands 1 and 2, are vastly equal as well (~1.3). Furthermore, at bigger irradiation doses IA, in both bands and for either fiber, steadily approaches the "plateaus", marked by black dotted lines in the figures. It is interesting that IA in maxima of bands 1 and 2 at the plateaus (i.e. at doses exceeding 2 × 10^{14} cm^{-2}) has virtually the same magnitude, for both fibers.

2.2.2. Bleaching of IA as a result of posterior exposure to 543 nm/UV light

Hereafter, the featuring data on optical bleaching of β-irradiated Ce-doped and Ce/Au-codoped fibers by a low-power He-Ne 543 nm laser and UV mercury lamp are reported.

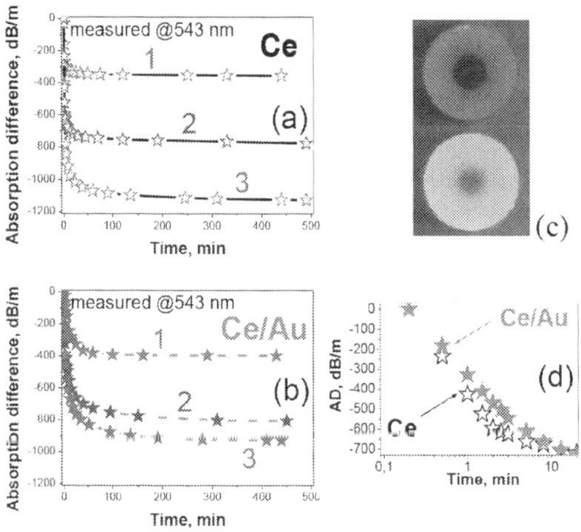

Figure 5. Dynamics of attenuation decay in terms of AD in Ce-doped (a) and Ce/Au-codoped (b) fibers under the action of 543 nm light (~0.5 mW); bleaching (negative AD) was realized after β-irradiation with doses 2 (curves 1), 4 (curves 2), and 6 (curves 3), for which AD is taken to be zero. (c) Micro-photographs of darkened (dose 5 of β-irradiation) Ce-doped fiber prior to (top) and after bleaching during 7.5 h (bottom). (d) Examples of the initial 543 nm bleaching stage, zooming the dependences shown by curves 2 in (a) and (b), respectively. (Reproduced with permission from Kir'yanov et al. [75]. Copyright© 2014, Optical Society of America).

Keeping in mind that, IA, a signature of color centers or defects in glass matrix produced at different kinds of irradiation, can be "bleached" by light (see e.g. [59–61]), we found reasonable to check whether such treatment has effect in our case.

First, we inspected the effect of weak 543 nm light delivered from a 1.5 mW He-Ne laser. In the experiments, power launched into both fibers was fixed (~0.5 mW; coupling efficiency

~30%). In this case, very short pieces (1…2 cm) of β-irradiated fibers were handled, given big IA, measured by hundreds of dB/m (refer to **Figures 3** and **4**), being established at β-darkening. The results are shown in **Figure 5**.

In the left part of **Figure 5**, we show the temporal dynamics of changes in attenuation of the Ce-doped (a) and Ce/Au-codoped (b) fibers under the action of 543 nm light, measured at the same wavelength. The effect of partial bleaching of β-induced loss (the negative AD) is apparent. Note that optical bleaching of both fibers demonstrates a saturating behavior and that the decay rate is bigger for the fiber codoped with Au than for Au-free one (compare curves 1–3 in (a) and (b)); also notice an almost exponential character of bleaching when the process gets starting (see (d) on the right side of **Figure 5**). The bleaching effect is clearly demonstrated by the photographs in **Figure 5(c)**, exemplifying the case of Ce-doped fiber. It is seen that its initial state (before β-irradiation) was almost restored under the action of 543 nm light: compare the photos in **Figure 1(b)** and **Figure 5(c)**.

One would speculate on whether bleaching of the Ce-doped and Ce/Au-codoped fibers arises solely due to laser-light-induced recombination or due to thermally assisted recombination, too, but as for us, the former appears to play a vital role.

Figure 6(a) and **(b)** shows how the bleached (main frames) and unbleached (insets) loss in the Ce-doped and Ce/Au-codoped fibers behave at 543 nm bleaching. Note that unbleached (remnant) loss is bigger in Ce/Au- than in Ce-doped fiber, that is, codoping of a Ce-doped fiber with Au results in a similar property of lesser susceptibility to exterior influence (compare with the results on β-irradiation); however, in the case of bleaching this feature appears to be a disadvantage.

The results of illuminating darkened Ce-doped and Ce/Au-codoped fibers with UV light are shown in **Figure 7**. In **Figure 7(a)** we exemplify the spectral dynamics of transmission of β-irradiated (at dose 5) Ce/Au-codoped fiber. The photographs in **Figure 7(b)** visualize the result of treatment, being almost a full fading of IA loss. This effect can be quantified by a shift of wavelength's transmission, measured at a 3 dB level (see gray line in **Figure 7(a)**), from near-

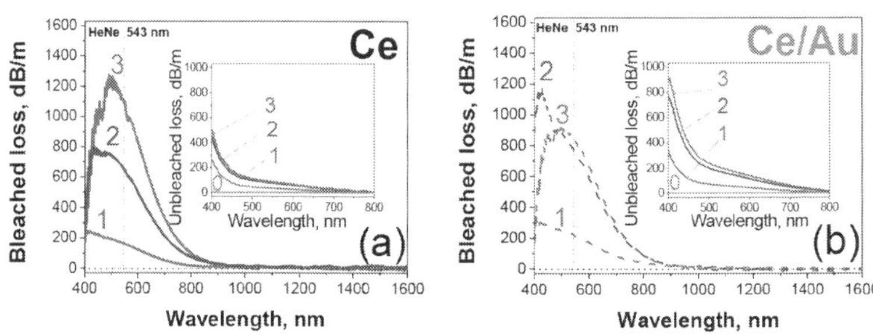

Figure 6. Bleached (main frames) and unbleached (insets) spectral loss in Ce-doped (a) and Ce/Au-codoped (b) fibers after ~0.5 mW 543 nm treatment, posterior to β-irradiation with doses 2 (curves 1), 4 (curves 2), and 6 (curves 3). For comparison, curves 0 demonstrate the attenuation spectra of pristine fibers. (Reproduced with permission from Kir'yanov et al. [75]. Copyright© 2014, Optical Society of America).

IR to VIS. It is seen from **Figure 7(c)**, where we demonstrate the results of experiments with Ce/Au-codoped (black open dots) and Ce-doped (gray open squares) fibers, that it has a similar character for both fibers.

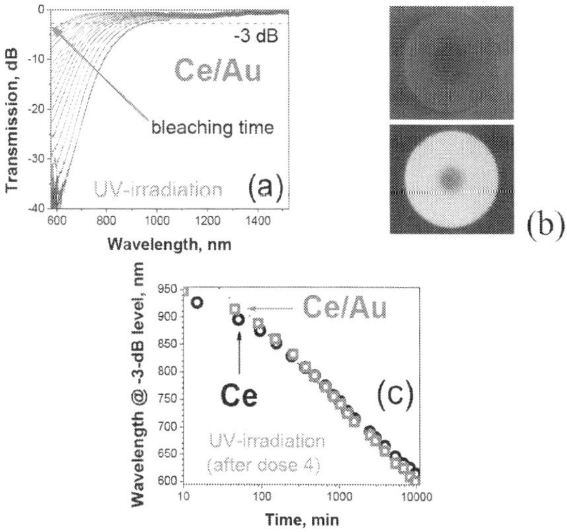

Figure 7. (a) Dynamics of attenuation decay (in terms of transmission of Ce/Au-codoped fiber under UV-lamp illumination with maximal spectral power @350 nm). Bleaching was realized in the darkened fiber, posterior to β-irradiation with dose 4. (b) Micro-photographs of darkened (dose 5 of β-irradiation) Ce/Au-codoped prior to optical bleaching with UV-lamp (top) and after continuous bleaching during 10,000 h (bottom). (c) Examples of the spectral transformations during UV-bleaching in terms of shifting of the fiber's transmission edge wavelength measured at -3 dB level; the data were obtained for Ce/Au-codoped (open black dots) and Ce-doped (open grey squares) fibers, preliminary β-irradiated with dose 4; both the data are fitted to the eye by dotted red lines. (Reproduced with permission from Kir'yanov et al. [75]. Copyright© 2014, Optical Society of America).

2.3. Discussion

2.3.1. Pristine fibers

Regarding pristine Ce-doped and Ce/Au-codoped fibers (**Figure 1**), apart from a strong growth of absorption seen at shorter (VIS to UV) wavelengths (apparently connected with the presence of Ce in valences Ce^{3+}/Ce^{4+} [23, 24]), the other two points deserve mentioning, being (*i*) monotonous growth of loss toward IR in both fibers (of not clear origin but inherent to Ce doping since such trend is absent in the reference Ce free fiber) and (*ii*) a distinct peak at ~520 nm (~2.4 eV) in the absorption spectra of both fibers (but absent in the Ce free one). We suppose that this peak has the same origin as band 2 risen at β-irradiation and located at ~2.4 eV (see **Figure 4**). This feature has not been reported for bulk Ce-doped silica but is frequently observed in Ce-doped fibers subjected to ionizing radiations [17, 24]. It can be related to quite stable $Ce^{3+}h^{+}$ centers, or alternatively while hypothetically, to $Ce^{4+}e^{-}$ centers (existence of which was not documented), but apparently not to sole Ce ions being in either trivalent or tetravalent state. As for us, a more realistic cause for the existence of $Ce^{3+}h^{+}$ or/and $Ce^{4+}e^{-}$ defect centers in pristine fibers, attributable by the 520 nm peak, can be ionization, that is, generation of electrons e^{-} and holes h^{+}, at the fiber preform's collapse stage [24] or during the fiber's drawing with posterior covering by acrylic outer cladding when—in both situations—strong UV light is produced, with a result being a trapping of free carriers by Ce^{3+}/Ce^{4+} species.

2.3.2. *β-irradiated fibers*

Consider in more details the results of β-irradiation of Ce-doped and Ce/Au-codoped fibers (**Figures 2–4**). The processes, involved at irradiating the fibers with the result being rise followed by saturation of IA, described by the "stretched-exponent" law [14, 24, 26]), comprise: (*i*) creating of secondary carriers (holes h^{+} and electrons e^{-}) in the core-glass matrix by β (primary) electrons and their trapping on such glass imperfections as Ce ions (Ce^{3+}/Ce^{4+}), nonbridging oxygen centers, other centers associated with Al and P, and oxygen vacancies; (*ii*) direct $h^{+}...e^{-}$ recombination (annihilation); (*iii*) thermally or/and radiatively activated recombination between the centers or defects that have arisen during and after β-irradiation. Concerning the role of Ce-doping, we assume that IA is produced via irradiation-induced reactions $Ce^{3+}+h^{+}\rightarrow Ce^{3+}h^{+}(\rightarrow ?Ce^{4+})$ and $Ce^{4+}+e^{-}\rightarrow(?Ce^{4+}e^{-})\rightarrow Ce^{3+}$ [26–28], implying Ce was in valences 3+/4+ in pristine fibers or/and being generated via irradiation. Note that determination of relative contents of Ce^{3+}/Ce^{4+} ions in the pristine state is hard and that at low Ce doping, mainly fluorescing Ce^{3+} are formed in the core-glass, while at higher overall Ce concentration both Ce^{3+} and Ce^{4+} (nonfluorescing) ions can be present. Unfortunately, the absorption spectra of glasses containing both Ce^{3+}/Ce^{4+} ions have the featuring bands within a 200–400 nm (UV) range (not detectable by our spectral equipment); so any arguing about $Ce^{3+}\leftrightarrow Ce^{4+}$ transformations for this range is impossible. In the meantime, the absorption bands of $Ce^{3+}h^{+}$ and $Ce^{4+}e^{-}$ defect centers are expectedly located in VIS (see above), on one hand, and, on the other hand, the detected spectral changes at β-irradiation occur in VIS, too (band 2); thus, formation of metastable centers $Ce^{3+}h^{+}/Ce^{4+}e^{-}$ as its result is a worthy proposal.

Furthermore, IA bands 1 (~3.0 eV) and 2 (~2.4 eV) (see **Figure 4**) have been undoubtedly separated; see above. The first of them, in Ce-doped and Ce/Au-codoped fibers, has seemingly the same origin as the one in the Ce-free fiber (see **Figure 2**), that is, it most probably belongs to one, most simply organized, type of the two NBOHCs centers, inherent to silica. The other would stem from Ce doping: it is seen from **Figure 2** that such band does not exist in the reference Ce-free fiber. However, the irradiated Ce-free fiber demonstrates ~600 nm band, probably attributing the other type of NBOHCs [26], absent in both Ce-doped fibers subjected to irradiation: compare spectra 1–3 shown in **Figure 2**. The fact that the dose dependences of IA (Figure 4) have different characters for the fibers points on different nature of the centers represented by bands 1 and 2. Therefore, our hypothesis that ~3.0 eV band stems from NBOHCs and that ~2.4 eV one is associated with a Ce-related center ($Ce^{3+}h^{+}/Ce^{3+}e^{-}$) seems to be relevant. In our case (alumino-silicate core glass of Ce-doped and Ce/Au-codoped fibers), such "point" defects as Al-E′ and Al-oxygen-deficient centers can be also created at trapping secondary electrons and holes born at β-irradiation (phosphorous (P) presented in small amount plays a little effect).

Thus, the processes roughly schematized as Ce3+↔Ce4+ seem to be a sole way to address the spectral transformations seen from Figures 2–4 to happen via the formation in the Ce- and Ce/Au-doped fibers of metastable states Ce3+h+/Ce4+e−. Furthermore, it deserves mentioning that overall susceptibility to β-irradiation (overall IA loss) of Ce-doped fiber is higher than of Ce/Au-codoped one. This may signify that the core glass containing gold is more stable than that solely doped with Ce. On the other hand, deviations in the experimental data for kinetics of IA versus β-irradiation dose for the fibers (refer to IA spectra in **Figure 3** and to dose dependences in **Figure 4**) are more pronounced in the former than in the latter fiber. A possible explanation for this can be that codoping with Au gives rise to the core-glass system more ordered. This property seems to have impact for establishing almost the same path kinetics of defect centers' formation as compared with other factors involved in such of type fibers.

2.3.3. *Optically bleached fibers*

Let us discuss now the effect of partial bleaching of β-irradiated Ce-doped and Ce/Au-codoped fibers under the action of low-power VIS/UV light (see Figures 5–7). Whereas a doubtless conclusion on its nature is hard, some discourse about the matters involved can be made. The processes responsible for recombination of radiation-induced defects or color centers, seen as IA fading (bleaching) of darkened fibers, can be of thermally and/or optically induced origin. Bleaching, with its result being decreasing IA versus time, seems to be an example of mainly optically induced recombination of both types of centers, NBOHCs and Ce-related $Ce^{3+}h^{+}$ (assumed to be represented by bands 1 and 2, respectively) ones.

As seen from **Figure 5(d)** and **Figure 7(c)**, IA decreases almost exponentially at the beginning of bleaching. However, within the whole interval of optical bleaching, IA in bands 1 and 2 is seen to fade (in terms of negative AD at 543 nm illumination, see **Figure 5(a, b)**, as well as in terms of shifting the transmission edge to shorter wavelength, at exposure the fibers to UV light, see **Figure 7(a, c)**), which obeys a "stretched exponent" law [26]. An explanation for this behavior can be not only complexity of the mechanisms involved at optical bleaching but also a fact of limited "penetration" of bleaching light into a fiber sample (especially in the case of 543 nm bleaching).

Concerning the essence of IA at optical bleaching, we can, at the current stage of our knowledge, propose them only tentatively. If our attribution of IA bands 1 and 2 as "signatures" of NBOHCs and Ce-related $Ce^{3+}h^{+}$ centers is correct, then these centers, formed at trapping free holes, should be breaking via the holes' detrapping and annihilating with free electrons born at interaction with VIS/UV light. Weak intensity of bleaching light is guessed to produce mainly extra electrons rather than holes in the core-glass, leading to dominance of the processes relating to the hole-trapped centers, such as NBOHC (~3.0 eV band) and $Ce^{3+}h^{+}$ ones (~2.4 eV band). Note that a strong candidate to be "in-charge" of production of e^{-} at the UV/VIS excitation may be Ce^{3+} ions themselves [15].

Comparison of the bleaching effect in Ce-doped (without Au codoping) and Ce/Au-codoped fibers show that it is less expressed in the latter than in the former, which is probably related to lower susceptibility to exterior influence of Ce/Au-codoped fiber (a consequence of its more ordered glass network, already noticed).

3. Electron irradiation versus PD of Yb-doped germano-alumino-silicate fibers: The effects comparison

Yb^{3+}-doped silica fibers (YFs) with different core-glass hosts codoped with Al, Ge, or P have been of considerable interest during the past decades as extremely effective media for fiber lasers for the spectral region 1.0–1.1 μm, when pumped at 0.9–1.0 μm wavelengths. A variety of diode-pumping configurations (core and cladding) and pump wavelengths were examined so far, resulting in recognition of optimal arrangements for multi-watt release from YF-based lasers with high optical efficiency ~70–75% and perfect beam quality [29, 30]. However, in spite of a remarkable progress in the field, there remain obstacles that limit the performance of YF-based lasers, one of them being PD [31], that is, long-term (minutes to hours) degradation of laser power, measured by units to tens %. This hardly mitigated disadvantage becomes notable when dealing with a laser based on heavily doped YF where a high Yb^{3+} population inversion is created, either at high-power continuous-wave or moderate-power pulsed lasing. A number of studies were aimed to understand the PD phenomenon which however remained unclear, although a few hypotheses have been proposed for its explanation [32–42].

On the other hand, a few studies aiming the characterization of susceptibility of YFs under such irradiations as X-rays, γ-quanta, and UV have been reported [43–45]. The main motivation was inspection of YF-resistance to harmful environments. In many cases, the excess-loss spectra induced in YFs resemble the ones, characteristic for PD at resonant pumping into Yb^{3+} resonant-absorption band, the fact undoubtedly deserving attention.

Here, the results of two sets of experiments, where susceptibility of YFs with similar germano-alumino-silicate glass-cores, doped with Yb in different concentrations to irradiation by a beam of β-electrons and to resonant (into Yb^{3+} resonant band) optical pumping, are presented. In both circumstances, qualitatively similar trends are revealed, being strong and monotonous change in attenuation in VIS (darkening), accompanied by more complex transformations within the resonant absorption band of Yb^{3+} ions, either upon dose (the case of β-electron irradiation), or exposing time (the case of optical pumping at 977 nm). Below, we compare and discuss the experimental results and attempt to explain them.

3.1. Fiber samples and experimental arrangement

The YFs inspected in these experiments were drawn from germano-alumino-silicate glass preforms fabricated using the "conventional" MCVD/SD route. The attenuation spectra of the fibers being in pristine (as-received) state are demonstrated in **Figure 8(a)**.

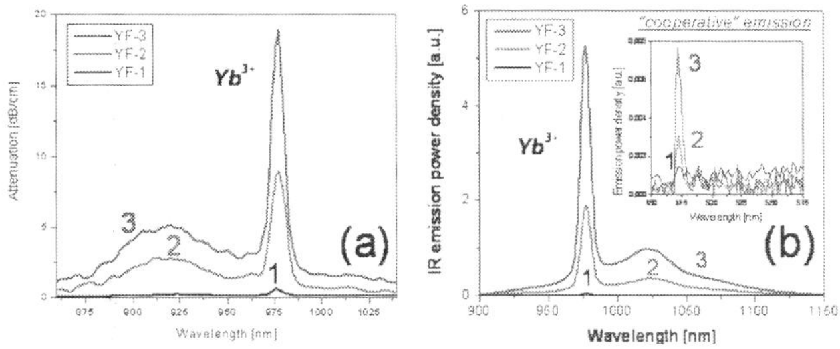

Figure 8. (a) Attenuation (small-signal absorption) spectra of fibers with low (YF-1), intermediate (YF-2), and high (YF-3) Yb^{3+} contents: curves 1, 2, and 3, respectively. (b) Fluorescence spectra of the fibers at resonant 977 nm excitation (pump power—300 mW). Labeling of curves 1, 2, and 3 is the same as in (a) and (b). Inset in (b) shows "cooperative" fluorescence in VIS. (Reproduced with permission from Kir'yanov [76]. Copyright© 2011, Scientific Research Publishing Inc).

The concentrations of Yb^{3+} ions in the fibers differed by more than an order of magnitude, so certain differences were expected after their exposure to β-electron irradiation (hereafter in this paragraph—*e-irradiation*) and to optical pumping (hereafter—*OP*) at 977 nm wavelength . The fibers, having the lowest, the intermediate, and the highest Yb^{3+} doping level, are referred further to as *YF-1*, *YF-2*, and *YF-3*, respectively.

The essences of experiments on e-irradiation of the YFs were completely the same as at irradiating the Ce-doped fibers (Paragraph 2). The indices "1," "2," and "3" label below the doses 2×10^{12}, 1×10^{13}, and 5×10^{13} cm^{-2}, respectively.

Experiments on OP at 977 nm were made in a similar way as described in Ref. [36]. YF samples were pumped using a standard 300 mW 977 nm laser diode (LD). The pump light was launched from LD to an YF sample under study through a splice. The end of the latter was spliced to a piece of SMF-28 fiber that was, in turn, connected to an OSA for the transmission spectra' measurements. In these experiments, we handled short (a few cm) pieces of YFs to ensure no-lasing conditions and negligible contribution of amplified spontaneous emission of Yb^{3+}.

The optical transmission spectra of the YF samples were obtained using a white light source with a fiber output and the OSA, turned to a 1 nm resolution. These spectra were recorded over the spectral range 400–1200 nm, where the most interesting spectral transformations occur as the result of e-irradiation/OP. The output of the white light source was connected to a fiber set containing an YF sample (pristine or subjected to e-irradiation/OP), while its attenuation was measured using the OSA. The attenuation spectra were recorded before and after each stage of e-irradiation (*doses*) or OP at 977 nm pumping (*times*). Lengths of the YFs were chosen to be short enough, from <1 cm (YF-3) to tens cm (YF-1), to avoid spectral noise artifacts. In some of the figures below, the difference (IA) spectra are demonstrated which were obtained after subtracting the attenuation spectra of pristine samples from the ones taken after certain dose/time of e-irradiation/OP. This allows insight to "net" spectral loss, established after darkening of either type. All the spectra presented beneath have been obtained after recalculating transmission coefficients in loss [dB/cm]. We also measured the fibers' fluorescence spectra and fluorescence kinetics of Yb^{3+} ions before and after e-irradiation/OP, applying the "lateral" geometry [46]. We used the same OSA for the fluorescence spectra measurements and a Ge photodetector (PD) and oscilloscope for the fluorescence-decay measurements. In the last case, LD power was modulated by a driver controlled by a function generator to achieve square-shaped pulses with sharp rise and fall edges. The time resolution of the setup was 8 μs. All the experiments were made at room temperature.

3.2. Experimental

3.2.1. E-irradiation

The attenuation spectra of samples YF-3 and YF-1, having correspondingly the highest and lowest Yb^{3+} concentrations, obtained after different doses of e-irradiation, along with the attenuation spectra of the samples in a pristine (dose "0") state are shown in **Figure 9(a, b)**.

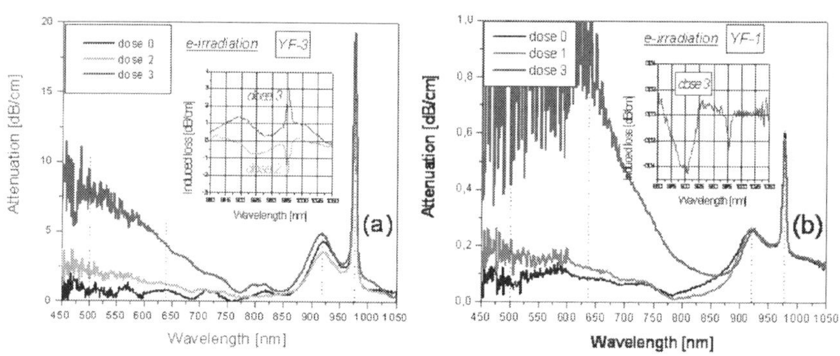

Figure 9. Attenuation spectra of samples YF-3 (a) and YF-1 (b). The data are for e-irradiation doses increased from "0" (pristine samples) through "1" and "2"–"3". Insets show the difference spectra obtained after subtraction of the spectra of pristine samples from the ones after e-irradiation of the samples. Dashed lines show the positions of wavelengths for which the data in **Figure 10** are built. (Reproduced with permission from Kir'yanov [76]. Copyright© 2011, Scientific Research Publishing Inc).

First, a notable increase of background loss in VIS with increasing e-irradiation dose is revealed (see main frames of **Figure 9**). Also notice a specific spectral character of this loss for both fibers, viz. a drastic rise of loss-magnitude toward shorter wavelengths. This is a well-known for Yb^{3+}-free silica fibers' trend in experiments on various kinds of irradiations. At the same time, apparent differences are seen in magnitude of e-irradiation-induced loss in these two fibers, that is, a higher degree of darkening in YF-3 than in YF-1. (For YF-2, intermediate in Yb^{3+} doping level, the effect of e-irradiation is intermediate, as compared with YF-3 and YF-1.]

Second, detectable but less pronounced spectral transformations are revealed for the resonant-absorption band of Yb^{3+} (850–1100 nm) (see insets to **Figure 9**), where the difference spectra are shown, obtained as explained above. Very weak in YF-1 (Figure 9(b)), the spectral transformations are noticeable in YF-3 (Figure 2(a)). These changes seem to be a result of some process, associated with e-irradiation of the fibers, which affects concentration of Yb^{3+} ions.

More details are seen in **Figure 10** where we plot the results for samples YF-1 (a) and YF-3 (b), taken for all doses. **Figure 10(a, b)** demonstrates how attenuation within the resonant-absorption of Yb^{3+} ions (peaks at 920 and 977 nm, see also **Figure 8(a)**) changes throughout e-irradiation: see curve 1 (for the 977 nm peak) and curve 2 (for the 920 nm peak), respectively. A decrease followed by an increase in the magnitude of small-signal absorption arises in both peaks with dose increasing in YF-3 (heavier doped with Yb^{3+}); this trend is, in contrast, less expressed in YF-1 (lower doped with Yb^{3+}).

For comparison, we plot in **Figure 10** the changes in attenuation of YF-3 (c) and YF-1 (d) fibers in VIS, where background (nonresonant) losses arise as the result of e-irradiation. Here we limit ourselves by the data, counted for 500 (curve 3) and 633 (curve 4) nm. It is seen that background loss steadily grows with dose, a common effect for silica fibers. Note that the rate of growth is higher in YF-3 than in YF-1. Furthermore, an initial level of background loss in pristine YF correlates with initial content of Yb^{3+} ions.

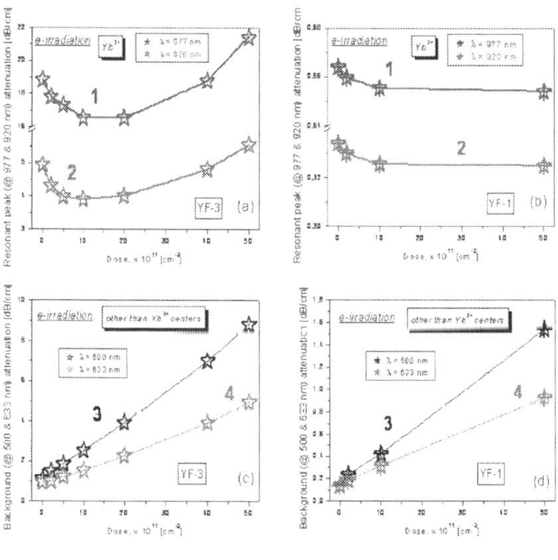

Figure 10. Dose dependences of attenuation in resonant-absorption Yb^{3+} peaks centered at 977 (curves 1) and 920 (curves 2) nm (top panels) and in VIS, for wavelengths 500 (curves 3) and 633 (curves 4) nm (bottom panels). The data are for samples YF-3 (a, c) and YF-1 (b, d). (Reproduced with permission from Kir'yanov [76]. Copyright© 2011, Scientific Research Publishing Inc).

Figure 11(a) and **(b)** gathers the experimental data obtained using all fibers, YF-1, YF2, and YF-3. From **Figure 11(a)**, it is seen that a monotonous increase of nonresonant loss in VIS (darkening), exampled by wavelengths 500 and 633 nm, with increasing Yb^{3+} concentration;

the latter is proportional to YF small-signal absorption at 977 nm. This demonstrates that the presence of Yb^{3+} dopants gain their degradation at e-irradiation. (Here we show the results obtained at dose "3" only, because for other doses the dependences are similar, given by a smooth dependence of induced loss in VIS versus e-irradiation dose (see **Figure 10(c, d)**). From **Figure 11(b)**, it is seen that the lowest levels to which the values of absorption in the 977 nm peak approach throughout e-irradiation (minima of curves 1 in **Figure 9**) decrease with increasing Yb^{3+} content (a similar trend is observed for the other peak of Yb^{3+}, at 920 nm). This fact seems to be in favor of that initial concentration of Yb^{3+} ions in pristine samples substantially decreases as a result of e-irradiation, at the primary stage. However, at the following stages, Yb^{3+} concentrations are re-established on levels comparable with those in pristine YFs (refer to **Figure 10(a)**). [The remainder of **Figure 11(c, d)** provides the data, obtained in the experiments on OP of the YFs, reported below.]

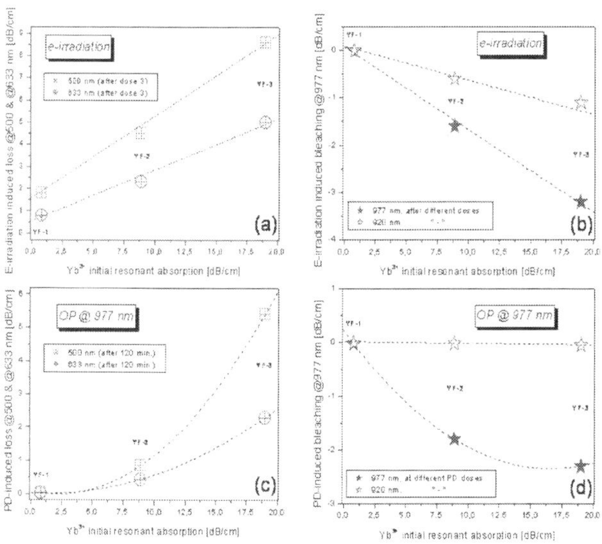

Figure 11. The results of experiments with fibers YF-1, YF-2, and YF-3, which were obtained for different e-irradiation doses (a, b) and OP times (c, d). The data are for the resonant-absorption peaks at 977 and 920 nm (filled and empty asterisks) (b, d)) and for the VIS region, exampled by wavelengths 500 nm (crossed squares) and 633 nm (crossed circles) (a, c). Dotted lines are for visual purposes only. (Reproduced with permission from Kir'yanov [76]. Copyright© 2011, Scientific Research Publishing Inc).

3.2.2. *PD at OP*

We report here the results of OP experiments for sample YF-3 mainly (see **Figures 12–14**), having the biggest content of Yb^{3+} ions. Then, we summarize all the results, obtained for YF-1, YF-2, and YF-3 fibers, in **Figure 11(c, d)**.

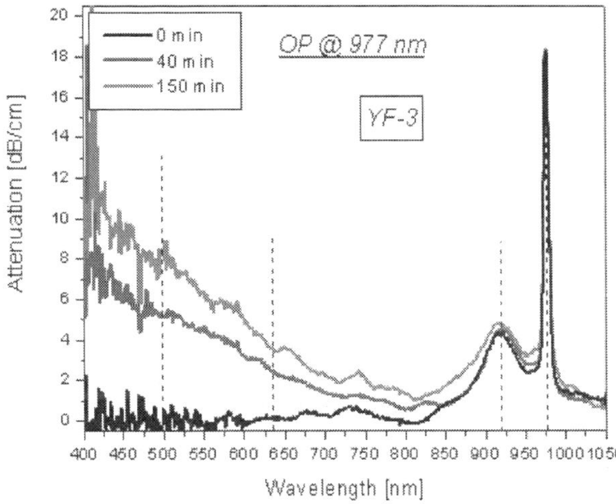

Figure 12. Attenuation (small-signal absorption) spectra of fiber sample YF-3 after OP @ 977 nm. The data are for a pristine sample (curve 1: "0 min") and for photo darkened samples (curves 2 and 3, obtained after 40 and 150 min of OP, respectively). Dashed lines show the positions of wavelengths for which the data in **Figure 13** are built. (Reproduced with permission from Kir'yanov [76]. Copyright© 2011, Scientific Research Publishing Inc).

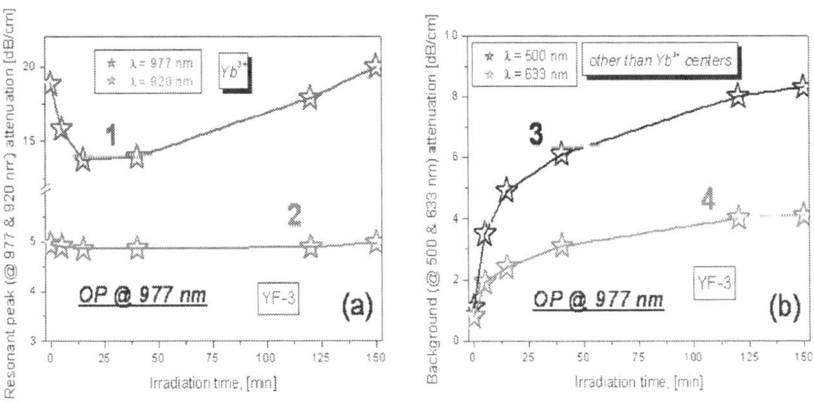

Figure 13. Dose dependences of attenuation in resonant-absorption (Yb^{3+}) peaks centered at 977 (curve 1) and 920 (curve 2) nm (a) and in VIS, for wavelengths 500 (curve 3) and 633 (curve 4) nm (b). The data are for sample YF-3. (Reproduced with permission from Kir'yanov [76]. Copyright© 2011, Scientific Research Publishing Inc).

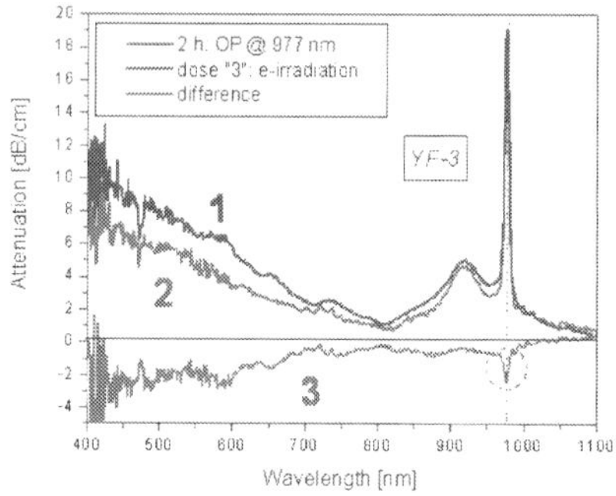

Figure 14. Difference attenuation spectra after dose "3" of e-irradiation (curve 1) and after 2 h of OP at 977 nm (pump power is 300 mW) (curve 2); curve 3 is the difference of spectra 1 and 2. The data are for sample YF-3. (Reproduced with permission from Kir'yanov [76]. Copyright© 2011, Scientific Research Publishing Inc).

Figure 12 shows the attenuation spectra of sample YF-3 (length, 0.8 cm) after 40 and 120 min. of OP. The LD power was fixed in these experiments at 300 mW, the highest in our circumstances level of Yb^{3+} ions inversion. For comparison, the attenuation spectrum of pristine (0 min) sample YF-3 is shown in **Figure 12**, too. Once compared with the attenuation spectra after e-irradiation (refer to **Figure 9(a)**), these spectra are seen to be similar. That is, a substantial increase of background loss is observed in VIS with increasing OP-time (the PD effect). Note that the spectral "signature" of PD resembles the one after e-darkening (see **Figure 9**).

In **Figure 13(a)**, we demonstrate the results of the experiments with sample YF-3, obtained at increasing OP time. Their representation is similar to the one used at the description of experiments on e-irradiation (see **Figure 10(a)**). From **Figure 13(a)**, it is seen how attenuations in the two absorption peaks of Yb^{3+} ions (at 977 and 920 nm) change throughout OP; see curves 1 and 2, respectively. The time dependence of OP-induced changes at 977 nm resembles the dose dependence at e-irradiation of sample YF-3. However, curve 1 in **Figure 13(a)** has "asymmetric" shape versus OP time, differing from "symmetric" shape of the dose dependence at e-irradiation given by curve 1 in **Figure 10(a)**. Furthermore, the time dependence of OP-induced changes at 920 nm, see curve 2 in **Figure 13(a)**, is very weak, being completely different from curve 2 in **Figure 10(a)** (e-irradiation). Therefore, we can propose that different mechanisms, responsible for the induced changes in the resonant-absorption band of Yb^{3+} at 977 and 920 nm, stand behind these two (e-irradiation and OP) treatments of the fibers.

In **Figure 13(b)**, we demonstrate the results of spectral transformations arising in YF-3 in VIS, at OP. Again, we provide in **Figure 13(b)**, the data for a couple of wavelengths, 500 (curves 3) and 633 (curves 4) nm, as most representative. In contrast to the dose dependences at e-irradiation, long-term OP at 977 nm results in completely different dynamics of background loss in time. Indeed, it is essentially nonlinear versus time: there is a short timing interval in the beginning (few minutes) where PD increases dramatically, while afterward (tens of minutes) it slows down and tends to saturate.

Figure 14 allows one to compare the attenuation spectra for YF-3 suffered dose "3" of e-irradiation (curve 1) and 2 h of OP (curve 2). The spectra look qualitatively similar, which may tell that the mechanisms involved are similar in these two circumstances. At the same time, if one spectrum is subtracted from another, the result (curve 3 in **Figure 14**) brings some news. That is, apart from the difference presented in VIS (in background loss), there is a feature in the Yb^{3+} resonant band: though no deviation from "plain" behavior of curve 3 is seen near 920 nm peak of Yb^{3+}, there is a well-defined (negative) 977 nm peak (it is marked by a dotted ring). This detail seems to be important as it lightens nonhomogeneity within the Yb^{3+} resonant-absorption band near 977 nm, present at OP but not—at e-irradiation.

This detail becomes expressed more when one analyzes the data obtained at PD of the other fibers, YF-1 and YF-2 (see **Figure 15**). In this figure, where we plot the difference spectra obtained for these fibers, analogous but clearer seen detail appears exactly within the 977 nm peak of Yb^{3+} ions (it is marked by a dotted ring in (a) and (b)).

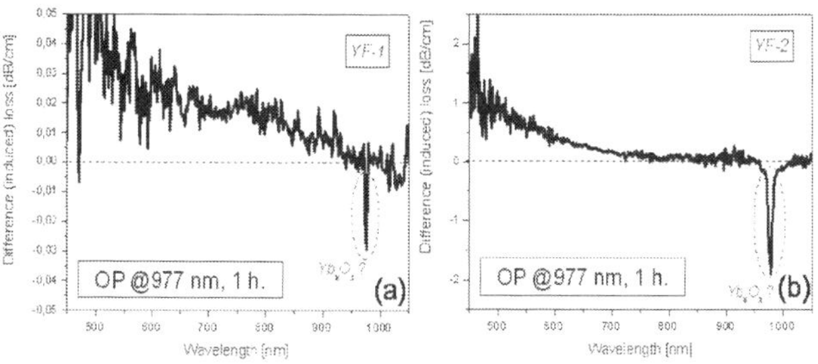

Figure 15. Difference loss spectra of YF-1 (a) and YF-2 (b), obtained after 1 h of OP at 977 nm. (Reproduced with permission from Kir'yanov [76]. Copyright© 2011, Scientific Research Publishing Inc).

Let us return to **Figure 11(c, d)**, where we gather the results on OP for all YF samples.

In contrast to the results on e-irradiation **(Figure 11(a, b))**, one can reveal first nonlinear growth of background loss at 500 and 633 nm with increasing Yb^{3+} concentration **(Figure 11(c))**. Apparently, this behavior is different from linear growth of background loss at e-irradiation **(Figure 11(a))**. Second, it is seen that, instead of a linear decrease of the resonant peaks at 977 and 920 nm with dose (occurring at primary stages of e-irradiation—see **Figure 11(b)**), a nonlinear law is obeyed by a decrease of the resonant peak at 977 nm while almost no change happens with the peak at 920 nm **(Figure 11(d))**.

Thus, the situation with OP-induced spectral transformations in the YFs is complex and curious at first glance. The 977 nm peak is strongly affected by OP, not the 920 nm one. This can be explained by the presence in the fibers of some other centers than Yb^{3+} dopants, but closely related to them and spectrally matching them near 977 nm. Moreover, partial weight of such centers in YF-core is expected to increase with increasing Yb^{3+} ions concentration. The nonlinear behavior of the nonresonant background loss versus OP time, discussed earlier (see **Figure 13(b)**), seems to be a related phenomenon.

3.2.3. *What's about fluorescence?*

The fluorescence spectra obtained using pristine YF-1, YF-2, and YF-3 fibers at 977 nm pumping are shown in **Figure 8(b)**. All these are similar in appearance and their intensities are proportional to Yb^{3+} ions concentrations in the fibers (The measurements were made at the same conditions and at the same pumps.)

We also measured the fluorescence spectra of the YFs after irradiation by an electron beam and after long-term OP at 977 nm, but almost no qualitative spectral changes were captured in the Yb^{3+} fluorescence band; so we don't provide them here. We could only notice a small decrease in the fluorescence power as the result of the treatment, but this trend could not be quantified. Furthermore, it was found that the characteristic Yb^{3+} fluorescence decay time slightly decreases in the set of pristine YFs. This is a result of the presence of two exponents in the fluorescence kinetics, measured by ~0.7 and ~0.2…0.3 ms. Note that insignificant growth of the latter contribution was detected for the fiber with the highest Yb^{3+} content (YF-3); see also Ref. [46–48]. However, the time constants obtained at fitting were nonaffected neither by e-irradiation nor by long-term OP. Concluding, we can reveal that none, or very insignificant, changes occurred in the YFs in the sense of Yb^{3+} fluorescence.

3.3. Discussion

Summarizing all the data, we notice that either at e-irradiation or at resonant OP substantial and complex but different in appearance changes arise within the resonant absorption band of Yb^{3+} ions ("reversible bleaching"), while monotonous growth of nonresonant background loss is observed in VIS ("darkening"). Furthermore, these trends are revealed to stem from the changes in concentrations of Yb^{3+} ions and, seemingly, of other centers, closely related to them and spectrally matching them near 977 nm. This is the main news of this study.

A general consequence of the experiments on e-irradiation, rise of background nonresonant loss in YFs in VIS (see **Figure 10 (c, d)**), is not surprising. This loss correlates spectrally with the excess loss arising in optical fibers at other types of irradiation (X-rays, γ-quanta, UV [33–35]). Some other aspects are as follows:

1. A monotonic increase of the background loss in VIS (darkening) with increasing Yb^{3+} content in the YFs, which demonstrates that the presence of Yb^{3+} dopants leads to a higher degree of the fibers' degradation at e-irradiation **(Figure 11(a))**.

2. A notable decrease followed by equally notable increase arising in the resonant-absorption peaks of Yb^{3+} (at 920 and 977 nm) with increasing e-irradiation dose **(Figure 10 (a, b))**, the effect also dependent on Yb^{3+} concentration **(Figure 11(b))**.

   Thus, the presence of Yb^{3+} dopants in the fibers results in a more pronounceable degradation at e-irradiation, with a probable reason being that Yb^{3+} ions are powerful sources of secondary carriers (electrons and holes) born at e-irradiation. That is, the changes within the resonant-absorption band of Yb^{3+} may stem from excitation of inner-shell (*f*) electrons of Yb^{3+} and their valence transformation through the charge-transfer (CT) processes (direct and reversed), sketched by the following reactions [36]: $e^- + Yb^{3+} \rightarrow Yb^{2+}$; $e^+ + Yb^{2+} \rightarrow Yb^{3+}$, where e^- and e^+ stand for secondary (irradiation induced) electrons and holes, and Yb^{2+} is the notation for Yb ions in 2+ valence state. In turn, the presence in the fibers of secondary carriers as the result of e-irradiation can produce such defects as oxygen-deficit center (ODC) and NBOH centers [47]. These centers are known to be responsible for the wide excess-loss spectral bands similar to the ones produced in the darkened fibers **(Figures 9** and **14)**.

Qualitatively similar observations can be made regarding the spectral transformations in the YFs as the result of OP at 977 nm (refer to **Figure 11(c, d)** and **Figures 12–15**). Analogously, the following trends are drawn:

3. Background loss in VIS substantially grows at long-term OP (see **Figures 12** and **13(b)**) while its character is typical for PD in YFs [31–41]. However, an increase of this loss in VIS with increasing small-signal absorption has, in contrast to e-irradiation, a nonlinear law **(Figure 11(c))**, thus revealing an almost quadratic dependence versus Yb^{3+} concentration.

4. The dependences of resonant absorption, measured in the peaks of Yb^{3+} at 977 and 920 nm upon OP time, have essentially different characters **(Figure 13(a))**. If the absorption coefficient in the 977 nm peak changes by a law similar to the one at e-irradiation, the absorption coefficient in the 920 nm peak is virtually constant throughout long-term OP. The concentration dependences shown in **Figure 11(d)** tell us more: the changes in these peaks with increasing content of Yb^{3+} ions are also different. We cannot interpret these details in terms of simple concentration dependences in regard to Yb^{3+} ions. Otherwise, an assumption can be made instead that the changes in the 977 nm peak are related to the changes in concentration of some others than Yb^{3+} ions centers but spectrally matching them in the 977 nm peak.

5. The spectral signature of the latter is seen from **Figures 14** and **15** where the difference attenuation spectra after OP are demonstrated. One can see from these figures that the PD effect (growth of nonresonant loss in VIS) is accompanied by bleaching of the resonant peak at 977 nm, whereas none happens with the peak at 920 nm. Note that a similar feature was reported earlier for the other type of YF, fabricated by the DND method [36].

The observations (3–5), when gathered together, tell us that PD in the YFs at high-power, long-term OP at 977 nm arises among the centers concentration of which is a nonlinear (almost quadratic) function of Yb^{3+} ions concentration. These are most probably the centers composed of couples of Yb^{3+} ions (*pairs*), or agglomerates of the latter. Furthermore, similar reactions: $e^- + Yb_p^{3+} \rightarrow Yb_p^{2+}$; $e^+ + Yb_p^{2+} \rightarrow Yb_p^{3+}$ (see above) can be proposed to address these transformations at OP, where index p stands to show that a pair of Yb^{3+} ions is involved in the processes and notations e^- and e^+ are used for an electron and a hole, free or trapped by the nearest ligand, say oxygen. Such reactions can go at the assistance of CT-processes between ion pairs where both constituents are in the excited state. Hence, the spectrally wide background loss (PD) in the fibers (see **Figures 12** and **14**) can be produced Yb_p^{2+} and of e^-/e^+-related centers (say, ODCs and NBOHCs) at OP, like this takes place at e-irradiation.

It is currently accepted that PD occurs among clusters of Yb^{3+} ions (obviously, pairs are their kind). However, a novelty found here is the spectral feature, occurring at OP (see dotted rings in **Figures 14** and **15**) but not—at e-irradiation.

There are evidences for that PD can be itself associated with nonbinding oxygen near surfaces of Yb/Al clusters that can be formed in alumino-silicate glass (our case). The nonbinding oxygen originates from Yb^{3+} substituting Si^{4+} sites. When subjecting a YF to 977 nm OP, the excess energy is radiated as phonons, causing a lone electron of a nonbinding oxygen atom to shift to a nearest neighbor nonbinding oxygen atom with creation of a hole and a pair of lone electrons, which results in a Coulomb field between the oxygen atoms to form an unstable "color" center. Conversion of such an unstable center to a semistable center requires shifting of one electron of the lone electron pair to a nearest neighbor site. As a result of this, the formation of Yb-related ODC can happen. On the other hand, PD in alumino-silicate YFs may take place via breaking of ODC, which gives rise to release of free electrons. The released electrons may be trapped at Al or Yb sites to form a color center resulting in PD. These hypotheses can serve as the arguments, bringing more clarity in understanding the similarity of the spectral transformations in YFs at e-irradiation (creation of "secondary" carriers by β-electrons) and at OP (creation of carriers and color centers by pump-light).

4. Effect of electron irradiation upon optical properties of Bi-doped silica fibers

Bi-doped silica fibers with core-glass codoped with Al, Ge, or P are currently of increasing interest, being a promising active medium for amplifying and lasing in the spectral range 1.1–1.6 μm (see e.g. [49–59]). In spite of remarkable success in the field, there remain certain obstacles for further improvements of Bi fiber lasers and amplifiers because the main problem is lack of clarity in the nature of Bi "active" centers (Further—BACs) in silica glass. Thus, any research aiming to recover the essences of BACs would have value.

Below we highlight the effect of irradiation of Bi-doped germano- and alumino-silicate fibers by a beam of free electrons of high energy. The main result of the treatment was found to be decrement ("bleaching")/increment (rise of resonant absorption) in the characteristic peaks, being ascribed to BACs in Bi-doped germano-silicate/alumino-silicate fibers. (Note that analogous trends were reported for similar fibers and glasses under the action of UV laser pulses and γ-quanta [60, 61]). Given that the other optical properties of the fibers under scope, such as BACs fluorescence spectra and lifetimes, were found to be weakly affected by electron irradiation, the changes in the absorption spectra should be associated with the changes in BACs concentration, as firmly justified in our study.

4.1. Fiber samples and experimental arrangement

The Bi-doped silica fibers were drawn from Ge and Al codoped silicate-glass preforms, fabricated applying the MCVD/SD technique. Core radii of the fibers were measured to be in the 2…3 μm range. The representative attenuation spectra of pristine (as-received) Bi-doped germano- and alumino-silicate fibers, having comparable contents of BACs, are shown in **Figure 16**. Hereafter, the emission-active BACs are referred to as Bi(Ge,Si) and Bi(Al), respectively, in these two types of fiber. Impact of electron irradiation on the basic characteristics of the fibers, referred to further as *Bi-1, Bi-2,* and *Bi-3* (germano-silicate) and *Bi-4* (alumino-silicate), is addressed subsequently.

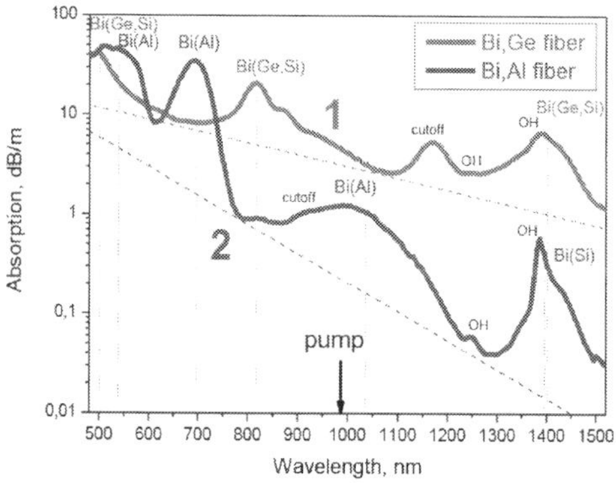

Figure 16. Attenuation spectra of typical Bi-doped germano-silicate (curve 1) and alumino-silicate (curve 2) silica fibers. Arrow shows the pump wavelength (977 nm) used in the experiments on fluorescence spectra and lifetimes measurements. Dashed lines show schematically a trend of the background loss to grow toward shorter wavelengths. (Reproduced with permission from Kir'yanov [67]. Copyright© 2011, Optical Society of America).

Electron irradiation of Bi-1…Bi-4 fibers was proceeding in the conditions, described in Introduction; the indices "1," "2," and "3" label to doses 2×10^{12}, 1×10^{13}, and 5×10^{13} cm^{-2}, respectively. The technique applied to reveal the spectral transformations in attenuation of the Bi-doped fibers as the result of irradiation was completely the same as described in Paragraphs 2 and 3 and is not repeated here. When measuring Bi-related fluorescence, we utilized the same LD (pump wavelength, 977 nm) for excitation. As seen from **Figure 16**, the pump wavelength was on the Stokes tail of the 750–950 nm absorption band of BACs in Bi-doped germano-silicate fiber and, correspondingly, on the anti-Stokes slope of the absorption band (centered at 1050 nm) of BACs in Bi-doped alumino-silicate fiber. We applied in the BACs fluorescence measurements the lateral detecting geometry, when it was collected from the surface of a Bi-doped fiber sample; the same OSA and a Ge PD were handled to proceed the fluorescence measurements.

4.2. Experimental

The experimental results are presented by **Figures 17–21**. The attenuation spectra of Bi-doped germano-silicate fiber sample Bi-2 subjected to electron irradiation with doses "2" and "3" are shown in **Figure 17(a)** along with the attenuation spectrum of a pristine (dose "0") fiber of the same type. A strong irradiation-induced bleaching effect can be revealed from the figure, seen as drop of magnitude of the absorption peaks labeled "1" (the 750–950 nm band) and "2" (the

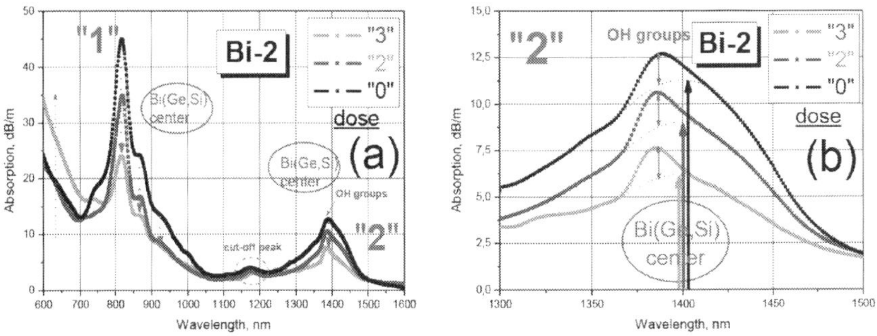

Figure 17. (a) Attenuation spectra of Bi-doped germano-silicate fiber sample Bi-2 obtained before (dose "0") and after (doses "2" and "3") irradiation. The spectral area, comprising the resonant-absorption peaks "1" and "2" which attribute Bi(Ge,Si) centers in the host glass, is shown (b) Insight to the spectral area of peak "2" in a vaster scale. (Reproduced with permission from Kir'yanov [67]. Copyright© 2011, Optical Society of America).

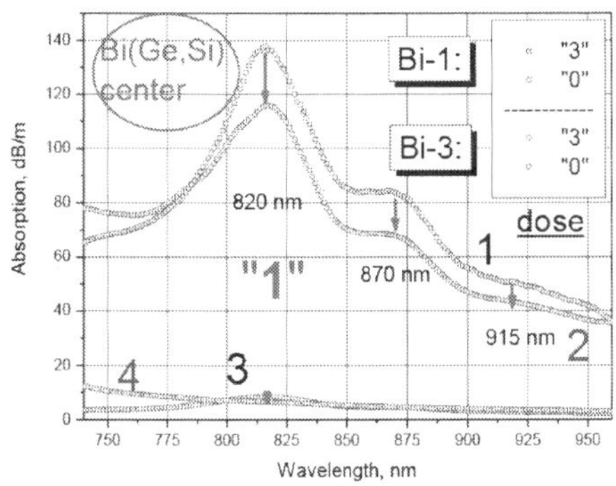

Figure 18. Attenuation spectra of Bi-doped germano-silicate fibers Bi-1 (curves 1 and 2) and Bi-3 (curves 3 and 4), obtained before (dose "0") and after (dose "3") electron irradiation. [The spectral area for the peak "1" is zoomed.] (Reproduced with permission from Kir'yanov [67]. Copyright© 2011, Optical Society of America).

1250–1450 nm band). It is accompanied by an increase of background loss at shorter wavelengths (refer to the left side of **Figure 17(a)**), a well-known feature in experiments on influence of various type of irradiations on optical properties of Ge-doped silica fibers (see e.g. Refs. [62–66]). Unfortunately, such a drastic growth of background loss did not allow us to make well-resolved measurements of the irradiation-induced transformations of BACs band peaked at ~500 nm (see **Figure 16**), so we inspected mostly the changes in peaks "1" and "2". Also notice that almost no changes arise in the attenuation peak at 1180 nm, which corresponds to the cutoff wavelength: this and other Bi-doped germano-silicate fiber samples were drawn to provide single-mode propagation for wavelengths >1200 nm.

http://dx.doi.org/10.5772/63939

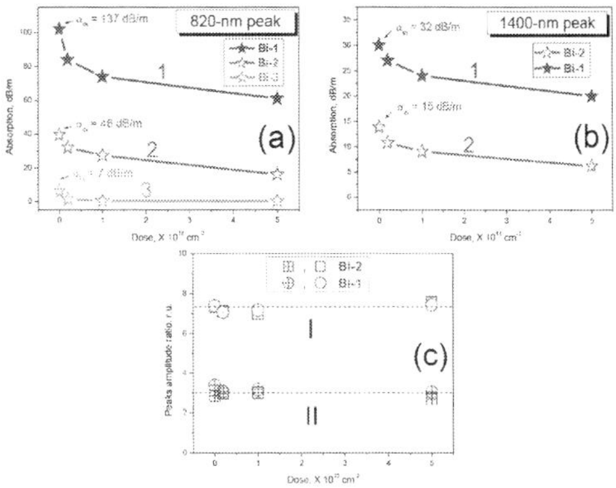

Figure 19. Dose dependences of attenuation of the resonant-absorption peaks "1" (~820 nm) (a) and "2" (~1400 nm) (b): The data for Bi-1 (curves 1), Bi-2 (curves 2), and Bi-3 (curve 3) are shown. (c) insights dose dependences of the peaks magnitudes' ratios (820…1400 nm – curve I and 500…1400 nm – curve II), for fibers Bi-1 (circles) and Bi-2 (squares). (Reproduced with permission from Kir'yanov [67]. Copyright© 2011, Optical Society of America).

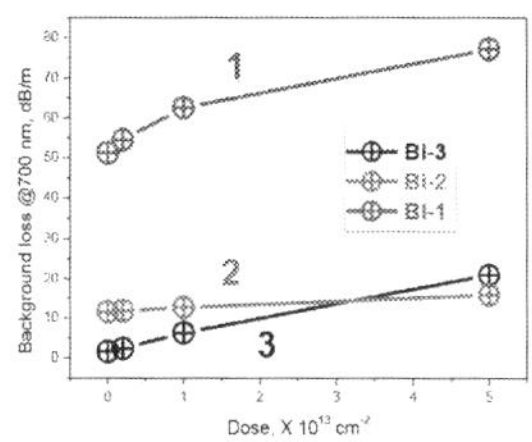

Figure 20. Dose dependences of background loss measured at 700 nm for fibers Bi-1 (1), Bi-2 (2), and Bi-3 (3). (Reproduced with permission from Kir'yanov [67]. Copyright© 2011, Optical Society of America).

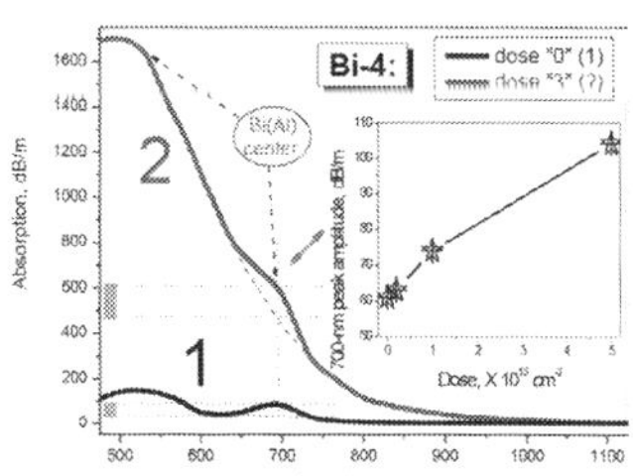

Figure 21. Attenuation spectra of Bi-doped alumino-silicate fiber Bi-4, obtained before (curve 1, dose "0") and after (curve 2, dose "3") electron irradiation. A part of the spectra is shown where the main resonant-absorption peaks of Bi(Al) centers are observed. Inset highlights the behavior of one of the peaks (at ~700 nm) against the irradiation dose. (Reproduced with permission from Kir'yanov [67]. Copyright© 2011, Optical Society of America).

Of separate interest is the behavior of absorption peak "2." Since absorption of BACs in this spectral area is covered by an absorption peak of OH groups (1385 nm), we found reasonable to zoom the spectral transformations for this range (see **Figure 17(b)**). From the figure, it is seen that the contribution in attenuation which comes from contaminating by OH groups is unchanged after irradiation, while the one stemming from the presence of the Bi-dopants is substantially reduced.

One more example of the irradiation-induced bleaching effect is given in **Figure 18** where we make insight to the spectral transformations in the absorption peaks within the 750–950 nm band ("1") after electron irradiation of the rest of Bi-doped germano-silicate fibers, Bi-1 and Bi-3. These two have, in pristine state, a higher and lower than Bi-2 concentration of BACs, correspondingly (see **Figure 17**). The spectra shown in **Figure 18** have been obtained before (dose "0": black curves 1 and 3) and after (dose "3": blue curves 2 and 4) electron irradiation. Qualitatively, the same law, viz., bleaching of the resonant-absorption peaks through the interval 750–950 nm as the result of electron irradiation, is revealed, now for fiber samples Bi-1 and Bi-3. Hence, the bleaching effect is found to be a general essence of the Bi-doped germano-silicate fibers.

The next graphs plotted in **Figure 19** (a, b) demonstrate how absorption peaks "1" (namely, its main subpeak centered at 820 nm) and "2" (the one centered at ~1400 nm) are reduced via electron irradiation (these dose dependences are shown for all fibers: Bi-1, Bi-2, and Bi-3). The initial absorption values (in peaks; these are given near each curve in **Figure 19(a, b)**) were taken from the attenuation spectra of pristine (dose "0") samples. Curves 1–3 for resonant-absorption peaks "1" (Figure 19(a)) and "2" (Figure 19(b)) were obtained from the spectra shown in Figures 16 and 17 after subtracting the background loss, which grows at irradiation (refer to **Figure 16** and also to **Figure 20**). Note that, for fiber Bi-3 characterized by the lowest content of Bi centers, the data are provided for peak "1" only because the measurements for peak "2" were below the resolution limit. It is seen from **Figure 19(a, b)** that bleaching of the resonant-absorption bands after electron irradiation is a characteristic feature of the Bi-doped germano-silicate fibers. Furthermore, resonant absorption bleaching in peaks "1" and "2" has almost the same character, which is evident from **Figure 19(c)** where we plot the ratio of absorption coefficients in peaks "1" and "2" in function of irradiation dose for Bi-1 and Bi-2 samples; see curve I. As seen, this quantity is kept virtually unchanged via irradiation, being equal to its initial value measured in pristine state. The same conclusion can be made for the ratio of absorption coefficients in peaks at ~500 nm and ~1400 nm ("2"), see curve II in **Figure 19(c)**. This is a justification of that resonant-absorption bands peaked at ~500, ~820, and ~1400 nm (and accordingly emission-active BACs attributed by these peaks, see Figures 16–18) are affected by the same or by a very similar manner by electron irradiation.

Then, as seen from **Figure 20**, the background loss (measured in the dip at 700 nm, between the absorption peaks ascribed to BACs in germano-silicate fiber; see **Figure 17**) monotonously increases with dose, a common effect for all kinds of Ge-doped silica materials. (Growth of the background loss is even more pronounceable in the UV.)

The results of electron irradiation of the Bi-doped alumino-silicate fibers (exemplified for fiber Bi-4) deserve a separate attention. **Figure 21** shows how the attenuation spectra of this fiber are changed after a maximal dose of electron irradiation. It is seen from a direct comparison of curves 1 and 2 (obtained before and after irradiation) that in the Bi-doped aluminate fiber an opposite (to the case of the Bi doped germanate fiber) trend exists, viz. instead of resonant-absorption bleaching (see **Figures 17–19**), weaker but detectable extra absorption arises in the peaks centered at ~520 and ~700 nm. Inset to **Figure 21** examples the dynamics of the absorption peak at ~700 nm upon dose; note that almost the same dose behavior is observed for the peaks at ~520 and ~1050 nm.

We do not present here the results of measuring fluorescence spectra and fluorescence lifetimes adherent to BACs, obtained before and after irradiation; the reader is advised to refer to [67] for details. The only thing to mention in this regard is that the fluorescence spectra of both types of the Bi-doped fibers (germano- and alumino-silicate) were not affected qualitatively by electron irradiation, with a sole result of the latter being a decrease/increase of integrated fluorescence power emitted by the germano-/alumino-silicate fibers. Also note that almost no change was detected in the fluorescence kinetics for pristine and irradiated fibers of these two types ($0.38 \pm 0.03/0.89 \pm 0.04$ ms). Thus, the changes in the resonant-absorption peaks, detected above, should be related to a decrement/increment of the BACs concentration in the germanate/aluminate Bi-doped fibers.

4.3. Discussion

First of all, the attenuation spectra of typical pristine Bi-doped germano- and alumino-silicate fibers **(Figure 16)** need examination. From these spectra that cover an extended wavelengths interval (400–1600 nm), one can recognize the "fingerprints" of Bi dopants in the fibers, appearing through the correspondent resonant-absorption bands: these were referred to as Bi(Ge,Si) and Bi(Al) centers. Specifically, the main absorption peaks at 520, 700, and 1050 nm (the Bi-doped alumino-silicate fiber) seem to belong to the center Bi(Al), whereas the ones at 500, 820 (910), and 1400 nm (the Bi-doped germano-silicate fiber)—to physically similar Bi(Ge) and Bi(Si) BACs. Note that the peaks at 1400 nm look indistinguishable for both fiber types; so they can be related to Si forming host of both the glasses (see e.g. Refs. [68, 69]). (Other spectral features not linked to the presence in the fibers of Ge, Al, and Si originate either from contaminating by water (OH peaks at 1385 and 1240 nm) or from special design of the fibers (the cutoff peaks). Regarding the experimental results on electron irradiation, they are remarkable but not enough to make a definite conclusion on real processes involved. The only thing to propose is possible correlation of the rise and decrease of IR emission-active BACs concentrations after electron irradiation in alumino- and germano-silicate fibers, respectively, with known facts that substitutional four-coordinated Al in alumino-silicate glass is a hole trap, whereas substitutional Ge in germanate glass is an electron trap [64, 66, 70–72]. This difference can strongly affect the residuary charge state of the Bi specie after the electron irradiation. The process of radiation-induced charge trapping of both electrons and holes can be accompanied by the formation of different point defects (say, Ge(1), Ge(2), GeE′, Al-E′, and Al-ODCs [73, 74]), detectable in ESR and optical spectra′ measurements.

5. Concluding remarks

Resuming, we have shown in this chapter that a diversity of effects can be encouraged at irradiating optical silica-based fibers with dopants of different kinds by high-energy β-electrons. This has been demonstrated on the examples of Ce- and Ce(Au)-doped alumino-phospho-silicate fibers, Yb-doped germano-alumino-silicate fibers, and Bi-doped germano- and alumino-silicate fibers. The data presented in this Chapter is a collection of our recent results, published in and in part reproduced from Refs [67, 75, 76]. In each case, unavoidable "darkening" of the fibers in VIS arises as the main feature of electron-irradiation. Meanwhile, such treatment allows one to detect interesting laws in transformations that "active" dopants presented in the fibers suffer as well as to propose mechanisms responsible for the phenomena involved. Also note that the new knowledge arising as the result of considering these transformations can be helpful for some applications of these or other doped fibers in such areas as dosimetry (on nuclear plants) and space technology and can be as well valuable when designing fiber devices (lasers and amplifiers) for the next-day telecom systems.

Acknowledgements

The author sincerely acknowledges the following people, contributing in the researches highlighted above: Dr. N.S. Kozlova (National University of Science and Technology (MISIS), Moscow, Russia)—for assistance in electron irradiation of all fibers; Drs. M.C. Paul and S. Ghosh (Central Glass & Ceramic Research Institute, Kolkata, India)—for providing samples of Ce and Ce(Au)-doped fibers and for discussions; Drs. V.V. Dvoyrin, V.M. Mashinsky, and E.M. Dianov (Fiber Optics Research Center of the Russian Academy of Sciences, Moscow, Russia)—for providing samples of Bi-doped fibers and valuable discussions; Dr. Yu.O. Barmenkov (Centro de Investigaciones en Optica, Leon, Mexico)—for participating in experiments with Ce and Ce(Au)-doped fibers and useful discussions. The author also thanks support from the Ministry of Education and Science of the Russian Federation under the "Increase Competitiveness Program of NUST «MISiS»" (Grant K3-2015-056).

Author details

Alexander V. Kir'yanov

Address all correspondence to: kiryanov@cio.mx

Centro de Investigaciones en Optica (Center for Optical Researches), Leon, Guanajuato, Mexico

References

[1] Berthold III JW. Overview of prototype fiber optic sensors for future application in nuclear environments. Proc. SPIE. 1994; 2425: 74–83.

[2] Berghmans F, Deparis O, Coenen S, Decréton M, Jucker P. Optical fibres in nuclear radiation environments: potential applications radiation effects need for standards. Soares ODD, editor. Trends in Optical Fibre Metrology and Standards ...(NATO ASI Series E: Applied Sciences). Springer (Netherlands); 1995; 285: 131–156. DOI: 10.1007/978-94-011-0035-9

[3] Partin JK. Radiation response of optical fibers in a nuclear reactor. Proc. SPIE. 1984; 506: 46–49.

[4] Liu P, Bao X, Brown K, Kulkarni N. Gamma-induced attenuation in normal single- and multi-mode, Ge-doped and P-doped optical fibers: A fiber optic dosimeter for low dose levels. Can. J. Phys. 2000; 78: 89–97.

[5] Paul MC, Bohra D, Dhar A, Sen R, Bhatnagar PK, Dasgupta K. Radiation response behavior of high phosphorous doped step-index multimode optical fibers under low dose gamma irradiation. J. Non-Crystal. Sol. 2009; 355: 1496–1507.

[6] Ghosh S, Das S, Paul MC, Dasgupta K, Bohra D, Chaudhary HS, Panwar L, Bhatnagar PK, Vaijapurkar SG. Evaluation of the performance of high phosphorous with germanium codoped multimode optical fiber for use as a radiation sensor at low dose rates. Appl. Opt. 2011; 50: E80-E85.

[7] Jones SC, Sweet JA, Braunlich P, Hoffman JM, Hegland JE. A remote fibre optic laser TLD system. Rad. Prot. Dos. 1993; 47: 525–528.

[8] Houston AL, Justus BL, Falkenstein PL, Miller RW, Ning H, Altemus R. Remote optical fiber dosimetry. Nuclear Instr. Meth. Phys. Res. B 2011; 184: 55–67.

[9] Canevali C, Mattoni M, Morazzoni F, Scotti R, Casu M, Musinu A, Krsmanovic R, Polizzi S, Speghini A, Bettinelli M. Stability of luminescent trivalent cerium in silica host glasses modified by Boron and Phosphorus. J. Am. Chem. Soc. 2005; 127: 42–50.

[10] Nikl M, Nitsch K, Mihokova E, Solovieva N, Mares JA, Fabeni P, Pazzi GP, Martini M, Vedda A, Baccaro S. Efficient radioluminescence of the Ce^{3+}-doped Na-Gd phosphate glasses. Appl. Phys. Lett. 2000; 77: 2159–2161.

[11] Baccaro S, Dall'Igna R, Fabeni P, Martini M, Mares JA, Meinardi F, Nikl M, Nitsch K, Pazzi GP, Polato P, Susini C, Vedda A, Zanella G, Zannori R. Ce^{3+} or Tb^{3+} -doped phosphate and silicate scintillating glasses. J. Lumin. 2000; 87: 673–675.

[12] Fu Q, Saltsburg H, Flytzani-Stephanopoulos M. Active-non-metallic Au and Pt species on Ceria-based water-gas shift catalysts. Science. 2003; 301: 935–938.

[13] Wang X, Rodriguez JA, Hanson JC. In situ time-resolved characterization of Au–CeO_2 and AuO_x–CeO_2 catalysts during the water-gas shift reaction: Presence of Au and O vacancies in the active phase. J. Chem. Phys. 2005; 123: 221101.

[14] Friebele EJ. Radiation protection of fiber optics materials: Effect of cerium doping on the radiation-induced absorption. Appl. Phys. Lett. 1975; 27: 210–212.

[15] Anoikin EV, Guryanov AN, Gusovsky DD, Dianov EM, Mashinsky VM, Miroshnichenko SI, Neustruev VB, Tikhomirov VA, Zverev YB. UV and gamma radiation damage in silica glass and fibres doped with germanium and cerium. Nuclear Instr. Meth. Phys. Res. B. 1992; 65: 392–396.

[16] Vedda A, Chiodini N, Di Martino D, Fasoli M, Keffer S, Lauria A, Martini M, Moretti F, Spinolo G, Nikl M, Solovieva N, Brambilla G. Ce^{3+}–doped optical fibres for remote radiation dosimetry. Appl. Phys. Lett. 2004; 85: 6536–6538.

[17] Chiodini N, Brambilla G, Vedda A, Di Martino D, Fasoli M, Lauria A, Redaelli M, Rosetta E. SiO_2 – based scincillating fibres for X-ray detection. Proc. SPIE. 2004; 5198: 298–305.

[18] Cruz JL, Lliso-Valverde F, Andres MV, Perez-Calatayud J. Induced attenuation in Ce and Nd doped fibers irradiated with electron beams under low dose regime. Opt. Commun. 2005; 252: 286–291.

[19] Mones E, Veronese I, Moretti F, Fasoli M, Loi G, Negri E, Brambilla M, Chiodini N, Vedda A. Feasibilty study for the use of Ce^{3+}–doped optical fibers in radiotherapy. Nuclear Instr. Meth. Phys. Res. A. 2006; 562: 449–455.

[20] Mones E, Veronese I, Vedda A, Loi G, Fazoli M, Moretti F, Chiodini N, Canillo B, Brambilla M. Ce-doped optical fibre as radioluminescent dosimeter in radiotherapy. Radiat. Measur. 2008; 43: 888–892.

[21] Broer MM, Cone RL, Simpson JR. Ultraviolet-induced distributed-feedback gratings in Ce^{3+}–doped silica optical fibers. Opt. Lett. 1991; 16: 1391–1393.

[22] Dong L, Archambault JL, Reekie L, Russel PSJ, Payne D. Bragg gratings in Ce^{3+}–doped fibers written by a single excimer pulse. Opt. Lett. 1993; 18: 861–863.

[23] Poignant H, Boj S, Delavaque E, Monerie M, Taunay T, Niay P, Bernaje P, Xie WX. Ultraviolet-induced permanent Bragg gratings in Ce-doped fluorozirconate glasses or optical fibers. J. Non-Crystal. Sol. 1995; 184: 282–285.

[24] Taunay T, Bernage P, Douay M, Xie WX, Martinelli G, Niay P, Bayon JF, Delavaque E, Poignant H. Ultraviolet-enhanced photosensitivity in cerium-doped aluminosilicate fibers and glasses through high-pressure hydrogen loading. J. Opt. Soc. Am. B. 1997; 14: 912–925.

[25] Saad M, Chen LR, Gu X. Highly reflective fiber Bragg gratings inscribed in Ce/Tm co-doped ZBLAN fibers. IEEE Photon. Technol. Lett. 2013; 25: 1066–1068.

[26] Farah K, Mejri A, Hosni F, Hamzaoui AH, Boizot B. Formation and decay of colour centres in a silicate glasses exposed to gamma radiation: Application to high-dose dosimetry. Current Topics in Ionizing Radiation Research, Nenoi M, editor. InTech (Croatia); 2012. DOI: 10.5772/32765

[27] Engholm M, Jelger P, Laurell F, Norin L. Improved photodarkening resistivity in ytterbium-doped fiber lasers by cerium cooping. Opt. Lett. 2009; 34: 1285–1287.

[28] Vivona M, Girard S, Marcandella C, Robin T, Cadier B, Cannas M, Boukenter A, Ouerdane Y. Influence of Ce codoping and H pre-loading on Er/Yb-doped fiber: radiation response characterized by Confocal Micro-Luminescence. J. Non-Crystal. Solid 2011; 357: 1963–1965.

[29] Pask HM, Carman RJ, Hanna DC, Tropper AC, Mackechnie CJ, Barber PR, Dawes JM. Ytterbium-doped silica fiber lasers: versatile sources for the 1–1.2 μm region. IEEE J. Quant. Electron. 1995; 1: 2–13.

[30] Richardson DJ, Nilsson J, Clarkson WA. High-power fiber lasers: current status and future perspectives. J. Opt. Soc. Am. B. 2010; 27: B63–B92.

[31] Koponen JJ, Soderlund MJ, Hoffman HJ, Tammela SKT. Measuring photodarkening from single-mode ytterbium doped silica fibers. Opt. Expr. 2006; 14: 11539–11544.

[32] Kirchhof J, Unger S, Schwuchow A, Grimm S, Reichel V. Materials for high-power fiber lasers. J. Non-Crystal. Sol. 2006; 352: 2399–2403.

[33] Manek-Honninger I, Boullet J, Cardinal T, Guillen F, Podgorski M, Bello Doua R, Salin F. Photodarkening and photobleaching of an ytterbium-doped silica double-clad LMA fiber. Opt. Expr. 2007; 15: 1606–1611.

[34] Jetschke S, Unger S, Ropke U, Kirchhof J. Photodarkening in Yb doped fibers: experimental evidence of equilibrium states depending on the pump power. Opt. Expr. 2007; 15. 14838–14843.

[35] Kitabayashi T, Ikeda M, Nakai M, Sakai T, Himeno K, Ohashi K. Population inversion factor dependence of photodarkening of Yb-doped fibers and its suppression by highly aluminum doping. In: Optical Fiber Communications Conference; Anaheim, USA; 2006. paper OThC5.

[36] Guzman Chavez AD, Kir'yanov AV, Barmenkov YO, Il'ichev NN. Reversible photodarkening and resonant photo-bleaching of ytterbium-doped silica fiber at in-core 977 nm and 543 nm irradiation. Laser Phys. Lett. 2007; 4: 734–739.

[37] Koponen J, Soderlund M, Hoffman HJ, Kliner DAV, Koplow JP, Hotoleanu M. Photodarkening rate in Yb doped silica fibers. Appl. Opt. 2008; 47: 1247–1256.

[38] Koponen J, Laurila M, Hotoleanu M. Inversion behavior in core- and cladding-pumped Yb-doped fiber photodarkening measurements. Appl. Opt. 2008; 47: 4522–4528.

[39] Yoo S, Basu C, Boyland AJ, Sones C, Nilsson J, Sahu JK, Payne D. Photodarkening in Yb-doped aluminosilicate fibers induced by 488 nm irradiation. Opt. Lett. 2007; 32: 1626–1628.

[40] Engholm M, Norin L, Aberg D. Improved photodarkening resistivity in ytterbium-doped fiber lasers by cerium cooping. Opt. Lett. 2009; 34: 1285–1287.

[41] Engholm M, Jelger P, Laurell F, Norin L. Strong UV absorption and visible luminescence in ytterbium-doped aluminosilicate glass under UV excitation. Opt. Lett. 2007; 32: 3352–3354.

[42] Suzuki S, McKay HA, Peng X, Fu L, Dong L. Highly ytterbium-doped silica fibers with low photo-darkening. Opt. Expr. 2009; 17: 9924–993.

[43] Fox BP, Schneider ZV, Simmons-Potter K, Thomes Jr. WJ, Meister DC, Bambha RP, Kliner DAV, Soderlund MJ. Gamma radiation effects in Yb-doped optical fiber. Proc. SPIE. 2007; 6453: paper 645328.

[44] Arai T, Ichii K, Tanigawa S, Fujimaki M. Gamma-irradiation-induced photodarkening in ytterbium-doped silica glasses. Proc. SPIE. 2011; 7914: paper 7914OK.

[45] Groothoff N, Canning J, Aslund M, Jackson S. 193 nm photodarkening of ytterbium-doped optical fibre. In: *OSA/BGPP Meeting*; 2007; paper BTuC2.

[46] Kir'yanov AV, Barmenkov YO, Lucio Martinez I, Kurkov AS, Dianov EM. Cooperative luminescence and absorption in ytterbium-doped silica fiber and the fiber nonlinear transmission coefficient at λ=980 nm with a regard to the ytterbium ion-pairs' effect. Opt. Expr. 2006; 14: 3981–3992.

[47] Carlson CG, Keister KE, Dragic PD, Croteau A, Eden JG. Photoexcitation of Yb-doped aluminosilicate fibers at 250 nm: evidence for excitation transfer from oxygen deficiency centers to Yb^{3+}. J. Opt. Soc. Am. B. 2010; 27: 2087–2094.

[48] Burshtein Z, Kalisky Y, Levy SZ, Le Boulanger P, Rotman S. Impurity local phonon nonradiative quenching of Yb fluorescence in ytterbium-doped silicate glasses. IEEE J. Quant. Electron. 2000; 36: 1000–1007.

[49] Peng M, Qiu J, Chen D, Meng X, Yang I, Jiang X, Zhu C. Bismuth- and aluminum-codoped germanium oxide glasses for super-broadband optical amplification. Opt. Lett. 2004; 29: 1998–2000.

[50] Dianov EM, Dvoyrin VV, Mashinsky VM, Umnikov AA, Yashkov MV, Guryanov AN. CW bismuth fibre laser. Quantum Electron. 2005; 35: 1083–1084.

[51] Peng M, Qiu J, Chen D, Meng X, Zhu C. Superbroadband 1310 nm emission from bismuth and tantalum codoped germanium oxide glasses. Opt. Lett. 2005; 30: 2433–2435.

[52] Dvoyrin VV, Mashinsky VM, Bulatov LI, Bufetov IA, Shubin AV, Melkumov MA, Kustov EF, Dianov EM, Umnikov AA, Khopin VF, Yashkov MV, and Guryanov AN.

Bismuth-doped-glass optical fibers – a new active medium for lasers and amplifiers. Opt. Lett. 2006; 31: 2966–2968.

[53] Dianov EM, Shubin AV, Mel'kumov MA, Medvedkov OI, Bufetov IA. High-power CW bismuth fiber lasers. J. Opt. Soc. Am. B. 2007; 24: 1749–1755.

[54] Rulkov AB, Ferin AA, Popov SV, Taylor JR, Razdobreev I, Bigot L, Bouwmans G. Narrow-line, 1178 nm CW bismuth-doped fiber laser with 6.4 W output for direct frequency doubling. Opt. Expr. 2007; 15: 5473–5476.

[55] Razdobreev I, Bigot L, Pureur V, Favre A, Bouwmans G, Douay M. Efficient all-fiber bismuth-doped laser. Appl. Phys. Lett. 2007; 90: 031103.

[56] Bufetov IA, Dianov EM. Bi-doped fiber lasers. Laser Phys. Lett. 2009; 7: 487–504.

[57] Fujimoto Y, Nakatsuka M. Infrared luminescence from bismuth-doped silica glass. Jpn. J. Appl. Phys. 2001; 40: L279–L281.

[58] Meng XG, Qiu JR, Peng MY, Chen DP, Zhao QZ, Jiang XW, Zhu CS. Near infrared broadband emission of bismuth-doped aluminophosphate glass. Opt. Expr. 2005; 13: 1628–1634.

[59] Dvoyrin VV, Kir'yanov AV, Mashinsky VM, Medvedkov OI, Umnikov AA, Guryanov AN, Dianov EM. Absorption, gain, and laser action in bismuth-doped aluminosilicate optical fibers. IEEE J. Quant. Electron. 2010; 46: 182–190.

[60] Peng M, Zhao Q, Qiu J, Wondraczek L. Generation of emission centers for broadband NIR luminescence in bismuthate glass by femtosecond laser irradiation. J. Am. Ceram. Soc. 2009; 92: 542–544.

[61] Ou Y, Baccaro S, Zhang Y, Yang Y, Chen G. Effect of gamma-ray irradiation on the optical properties of $PbO\text{-}B_2O_3\text{-}SiO_2$ and $Bi_2O_3\text{-}B_2O_3\text{-}SiO_2$ glasses. J. Am. Ceram. Soc. 2010; 93: 338–341.

[62] Griscom DL, Gingerich ME, Friebele EJ. Radiation-induced defects in glasses: Origin of power-law dependence of concentration on dose. Phys. Rev. Lett. 1993; 71: 1019–1022.

[63] Médjahdi K, Boukenter A, Ouerdane Y, Messina F, Cannas M. Ultraviolet-induced paramagnetic centers and absorption changes in singlemode Ge-doped optical fibers. Opt. Expr. 2006; 14: 5885–5894.

[64] Alessi A, Agnello S, Gelardi FM, Grandi S, Magistris A, Boscaino R. Twofold co-ordinated Ge defects induced by gamma-ray irradiation in Ge-doped SiO_2. Opt. Expr. 2008; 16: 4895–4900.

[65] Girard S, Ouerdane Y, Origlio G, Marcandella C, Boukenter A, Richard N, Baggio J, Paillet P, Cannas M, Bisutti J, Meunier JP, Boscaino R. Radiation effects on silica-based preforms and optical fibers—I: Experimental study with canonical samples. IEEE Trans. Nucl. Sci. 2008; 55: 3473–3482.

[66] Dianov EM, Mashinsky VM, Neustruev VB, Sazhin OD, Brazhkin VV, Sidorov VA. Optical absorption and luminescence of germanium oxygen-deficient centers in densified germanosilicate glass. Opt. Lett. 1997; 22: 1089–1091.

[67] Kir'yanov AV, Dvoyrin VV, Mashinsky VM, Il'ichev NN, Kozlova NS, Dianov EM. Influence of electron irradiation on optical properties of bismuth doped silica fibers. Opt. Expr. 2011; 19: 6599-6608.

[68] Razdobreev I, El Hamzaoui H, Ivanov VY, Kustov EF, Capoen B, Bouazaoui M. Optical spectroscopy of bismuth-doped pure silica fiber preform. Opt. Lett. 2010; 35: 1341–1343.

[69] Bufetov IA, Semenov SL, Vel'miskin VV, Firstov SV, Bufetova GA, Dianov EM. Optical properties of active bismuth centres in silica fibres containing no other dopants. Quantum Electron. 2010; 40: 639–641.

[70] Skuja LN, Trukhin AN, Plaudis AE. Luminescence in germanium-doped glassy SiO_2. Phys. Status Solidi A. 1984; 84: K153–K157.

[71] Griscom DL. Self-trapped holes in pure-silica glass: A history of their discovery and characterization and an example of their critical significance to industry. J. Non-Cryst. Sol. 2006; 352: 2601–2617.

[72] Watanabe Y, Kawazoe H, Shibuya K, Muta K. Structure and mechanism of formation of drawing- or radiation-induced defects in SiO_2:GeO_2 optical fiber. Jpn. J. Appl. Phys. 1986; 25: 425–431.

[73] Friebele EJ, Griscom DL, Sigel, Jr. GH. Defect centers in a germanium-doped silica-core optical fiber. J. Appl. Phys. 1974; 45: 3424–3428.

[74] Trukhin AN, Sharakovski A, Grube J, Griscom DL. Sub-band-gap-excited luminescence of localized states in SiO_2–Si and SiO_2–Al glasses. J. Non-Cryst. Sol. 2010; 356: 982–986.

[75] Kir'yanov AV, Ghosh S, Paul MC, Barmenkov YO, Aboites V, Kozlova NS. Ce-doped and Ce/Au-codoped alumino-phospho-silicate fibers: Spectral attenuation trends at high-energy electron irradiation and posterior low-power optical bleaching. Opt. Mater. Expr. 2014; 4: 434–448.

[76] Kir'yanov AV. Electron-irradiation and photo-excitation darkening and bleaching of Yb doped silica fibers: comparison. Opt. Photon. J. 2011; 1: 155–166.

Chapter 2

Radiation Effects in Optical Materials and Photonic Devices

Dan Sporea and Adelina Sporea

Additional information is available at the end of the chapter

http://dx.doi.org/10.5772/62547

Abstract

The chapter continues previous reviews on radiation effects in optical fibers and on the use of optical fibers/optical fiber sensors in radiation monitoring, published by InTech in 2010 and 2012, by referring to radiation effects in optical materials, with an emphasis on those operating from visible to mid-IR, and on some photonic devices such as optical fibers for amplifiers, fiber Bragg gratings and long period gratings. The focus is on optical materials and fiber-based devices designed for both terrestrial and space-borne applications. For the presented subjects, an overview of available data on X-rays or gamma rays, electron beams, alpha particles, neutrons, and protons effects is provided. In addition, comments on dose rate, dose, and/or temperature effects on materials and devices degradation under irradiation are mentioned, where appropriate. The optical materials and photonic devices reliability under ionizing radiation exposure is discussed as well, as the opportunities to use them in developing radiation sensors or dosimeters. The chapter includes an extensive bibliography and references to last published results in the field. Novel proposed applications of photonic devices in charged particle beam diagnostics, quasi-distributed radiation field mapping and the evaluation of radiation effects in materials for mid-IR spectroscopy are briefly introduced to the reader.

Keywords: radiation effects, optical materials, optical fibers, fiber Bragg grating, long period grating

1. Introduction

The goal of this chapter is to continue previous reviews on radiation effects in optical fibers [1] and on the use of optical fibers and optical fiber sensors in radiation related measurements [2], by referring to radiation effects in optical materials and some photonic devices. According to this vision, the chapter is organized to cover the interaction of ionizing radiation with some optical materials and optical fibers, followed by a reference to radiation effects on some photonic devices based on optical fibers. The discussion addresses radiation effects produced by both energetic photons and charged particles, as appropriate [3]. In this context, an overview of some recently published results in the field is included, with a focus on original authors' contributions.

The terrestrial radiation environments where optical and photonics components can be found include, but are not limited to, high energy physics experiments, nuclear power plants [3], fusion installations as the International Thermonuclear Experimental Reactor – ITER, or the Laser Mégajoule – LMJ [4–8], high power laser installations [9], nuclear waste repositories [10], high energy physics [11, 12], medical equipments for diagnostics or treatment [13]. On the other side, applications of optical components or photonic devices can be found in spaceborne instrumentation [14–16]. These environments involve various types of ionizing radiations, depending on the application considered: X-rays or gamma rays, electron beams, alpha particles, neutrons, protons, and Bremsstrahlung [3, 14, 17].

2. Radiation effects in optical materials

2.1. Optical materials

Extensive research was involved in the elucidation of defects formation in glasses, as investigations were performed in glasses with various compositions under ionizing and non-ionizing radiation exposure. The studies were focused either on the materials degradation upon irradiation or on the possible use of such materials in radiation dosimetry [18, 19]. The radiation induced changes depend on the glass composition, total dose, dose rate, temperature and humidity during exposure, and post irradiation heating of the sample [19]. The operation of a glass-based dosimeter can be decided as function of radiation sensitivity, linearity of the response, stability of the radiation produced effect, and possibility to re-use the material.

Besides glass-based optical materials, radiation hardening tests were performed on various other optical materials. More than 20 years ago radiation induced defects were studied in BaF_2 crystals by exposing them to gamma rays (from 10 Gy to 47 kGy) and observing the optical attenuation recovery (between 300 and 700 nm) under UV radiation and the scintillating signal [20]. Samples from different manufacturers exhibited radiation induced attenuation (RIA) saturation starting from 10^2 Gy. Crystal impurities and defects are the primary source of the optical attenuation increase in the 190–250 nm and 500–600 nm spectral bands induced by gamma rays [21]. BaF_2 and LaF_3 were subjected to Ne and U ions (at energies from 1.4 to 13.3 MeV/u) bombardment, and their degradation was investigated by scanning force microscopy (SFM), optical spectroscopy and surface profilometry. RIA for BaF_2 shown an increase at λ = 240, 420, 550 and 750 nm, while LiF_3 crystals remained almost unchanged spectrally. Surface topography studies indicated the presence of hillock in the irradiated zone [22].

Neutron irradiation was done on $Y_3Al_5O_{12}$, CaF_2 and LiF and RIA was monitored for UV-visible spectra. For $Y_3Al_5O_{12}$ samples an increase of the optical attenuation was present for wavelength lower than 350–400 nm. CaF_2 and LiF single crystals degrade their optical transmission after neutron exposure mostly in the 400–500 nm region. When heated after the irradiation RIA for the three crystals recovers according to different patterns [23].

Gamma irradiation (dose rate 110 Gy/h, total doses of 500 Gy, 2 kGy, 8 kGy, 20 kGy) was conducted on CaF_2, Fused Silica and Clearceram in order to evaluate their qualification for space applications. RIA modification was measured over the 350–800 nm spectral interval at normal incidence. The optical investigations were completed by ellipsometry tests before and after the irradiation, from 200 nm to 1 μm. In the case of Clearceram, for example, three absorption bands located at 3.20, 2.20, and 1.81 eV are present. Over a quite long period these peaks decrease exponentially [24].

The scintillation properties of different optical materials were studied for their possible use in the development of radiation detectors. Two radiation induced luminescence (RIL) bands were observed in polycrystalline BaF_2 irradiated by X-rays at 295 K, for the wavelengths intervals 380–600 nm and around 659 nm [25]. Under X-ray irradiation, ZnSe crystals present a degradation of the optical transmission at 475–575 nm, and four luminescence spectra at $\lambda = 460$ nm, 610 nm, 645 nm and 970 nm [26]. An X-ray induced RIL peak was reported in CaF_2 crystals at $\lambda = 420$ nm, accompanied by a small one at about 350 nm [27].

Sapphire is one of the most intensively studied optical material under different ionizing radiation: X-ray (40 kV, 15 mA) and β from a ^{90}Sr source at 1.5 Gy/min [28]; 8 MeV proton (flux 2×10^{12} p cm^{-2} s^{-1} and 150 MeV argon (4.3×10^8 ion cm^{-2} s^{-1}) and 253 MeV krypton (4.3×10^8 ion cm^{-2} s^{-1}) [29]; gamma rays (dose rate 6.7 Gy/s, total dose 10^8 Gy) combined with neutrons (energy 2.4 MeV and fluences of 10^{17}–10^{20} n cm^{-2}) [30]; fast neutrons (energy 1.2 MeV and fluence of 1.4×10^{18} n cm^{-2}) [31], neutron irradiation followed by heating (energy 10 MeV, flux of 6.6×10^{12} n cm^{-2} s^{-1} and fluence up to 10^{19} n cm^{-2}), maximum temperature 1000°C) [32]. Sapphire proved to be radiation hardened under gamma exposure up to high doses (10^8 Gy), but is more susceptible to optical transmission degradation under gamma-neutron irradiation as the optical attenuation increases with the fluence for wavelengths below 600 nm. RIL spectra change under gamma irradiation in relation to RIA modification [30]. For fast neutron irradiation, absorption bands develop at $\lambda = 203$, 255, 300, 357 and 450 nm, while post irradiation excitation produce photoluminescence signals at $\lambda = 320$, 377 and 551 nm [31]. High purity Al_2O_3 crystals present absorption peaks at $\lambda = 206$, 230, 258, 305, 358 and 452 nm, when irradiated by fast neutrons. Post irradiation annealing (up to 1000°C heating) contributes to partial recover of RIA. Excitation at $\lambda = 302$ nm induces luminescence at $\lambda = 325$, 482, 543 nm [32]. The RIL associated to the F- and F^+- bands was found to be dependent on the proton dose [29].

The invention of quantum cascade laser (QCL) [33] and subsequent research in the field made possible the development of compact, very accurate, and portable spectroscopic instruments for the mid-IR spectral range of interest for organic compounds identification (3–12 μm). Tunable QCLs or arrays of QCLs operating at different wavelengths proved to be affordable substitutes for Fourier transform IR (FTIR) systems to be used in astronomy, astrophysics, astrochemistry, and space missions [34, 35]. In order to operate reliably under extreme conditions as components to be included in spaceborne equipment, the composing parts of such equipment have to be tested under specific radiation exposure. Within this context, a program to evaluate for the Romanian Space Agency passive and active mid-IR components is under way. In this chapter, reference will be made to tests carried out on mid-IR windows materials (CaF_2, BaF_2, ZnSe, and sapphire—Al_2O_3). The subjects of these investigations are COTS (components-on-the-shelf) products, manufactured in Europe and China. The windows have 10–25 mm diameter and a thickness of 2–3 mm. Considering the complex space radiation environments which can be encountered during extra planetary missions, irradiation tests were run under various irradiation conditions: gamma rays (dose rate of 5.7 kGy/h +/–1.8%, four irradiation steps at total doses of 0.1 kGy; 1 kGy; 10 kGy; 20 kGy), alpha particles (doses from 1.46×10^6 kGy to 5.79×10^6 kGy, depending on the window material, at three beam currents of 100, 200 and 300 μC, beam diameter 3–4 mm), protons (at 10^{15} p cm^{-2}, 10^{16} p cm^{-2}, 10^{17} p cm^{-2} fluences), electron beam (dose rate of 4 kGy/min, total dose – several kGy). Gamma and electron beam irradiation were performed in air at room temperature, while proton and alpha particle irradiation were done in vacuum [36–38]. Prior and after each irradiation step the samples were measured in relation to their optical transmittance and reflectance over the spectral range from 250 nm to 18 μm. **Figure 1** illustrates the setup for spectral reflectance evaluation, which was carried out with the Gooch and Housego OL Series 750 Automated Spectroradiometric Measurement System and the accessories (tungsten lamp and IR glower, diffracting gratings, integrating spheres, reflectance standards, optical detectors - Si & Ge, PbS, InSb and HgCdTe) appropriate to each spectral interval over which the measurements were done. Optical transmittance of the irradiated samples was used to assess the color centers generation (RIA) in mid-IR optical materials under various irradiation conditions (type of radiation, dose, and dose rate). In the mean time, during alpha particle irradiation the RIL was monitored on-line in order to associate this signal with the presence of some dopants/impurities. The peak wavelengths of the detected radioluminescence are: $\lambda = 285$ nm (Caf_2), $\lambda = 405$ nm (BaF_2), $\lambda = 465$ nm (ZnSe), and $\lambda = 700$ nm (Al_2O_3). Significant decrease of the optical transmission was obtained for CaF_2 and BaF_2, below 1 μm, while a drop of the optical transmission over the investigated spectral range (0.3–16 μm) was noticed for all the materials except for sapphire, case when the transmission change was smaller as compared to other mid-IR materials. For the same purpose, RBS (Rutherford backscattering spectrometry) measurements were done on the investigated samples.

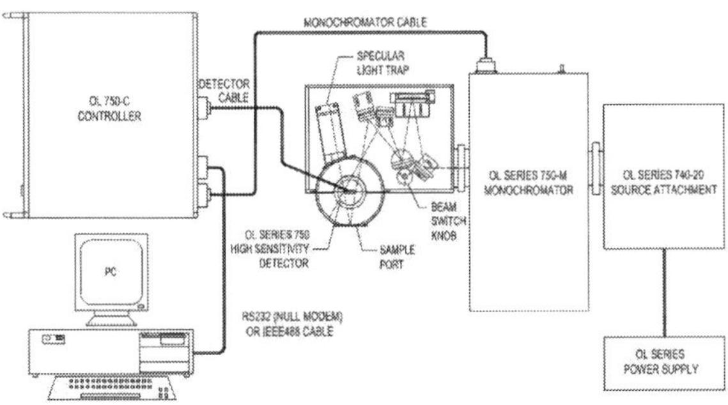

http://dx.doi.org/10.5772/62547

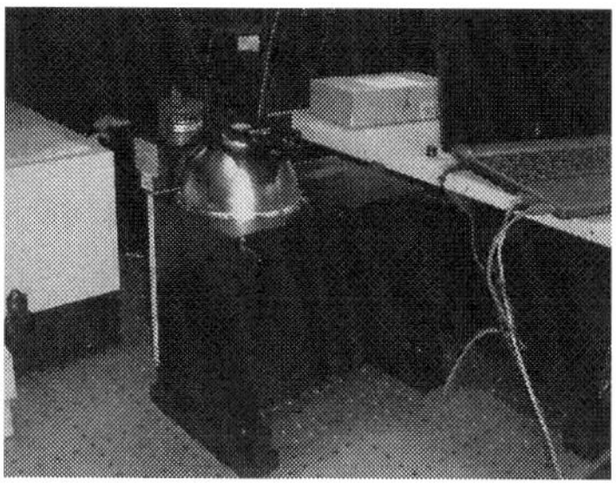

Figure 1. The sketch (a) and the picture (b) of the setup for spectral reflectance measurements over the UV to mid-IR (Reproduced with permission and courtesy of Gooch and Housego).

Spectral reflectance measurements were associated to optical microscopy tests as charged particles impinging on windows surface affect its quality (**Figure 2**).

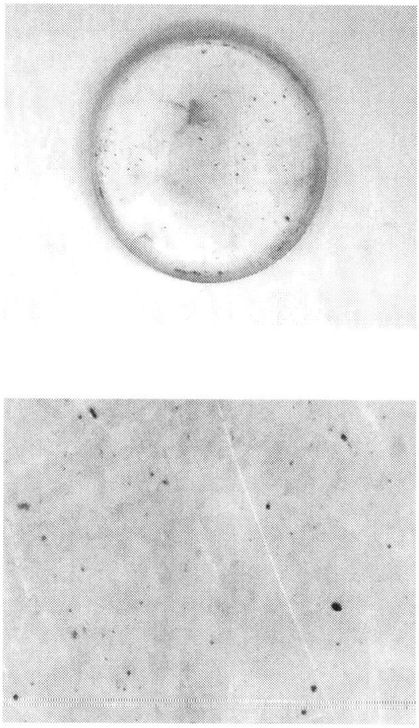

Figure 2. The degradation of windows surface quality after alpha particle irradiation: (a) BaF2; (b) ZnSe (Courtesy of Laura Mihai).

For the first time, THz (Terahertz) spectroscopy and imaging were introduced (**Figure 3**) in the analysis of irradiated mid-IR optical materials.

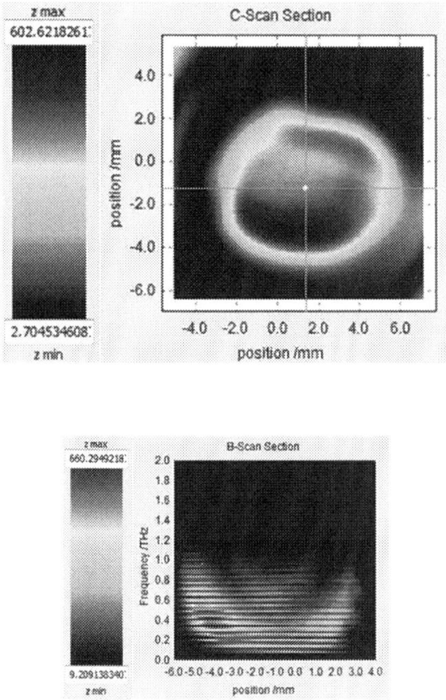

Figure 3. 2D plot of the THz reflected signal for BaF_2 window as the sample was scanned in the XY plane, for 0–2 THz (a); the frequency signal along the X axis (b), at a resolution of 50 μm (Courtesy of Laura Mihai).

2.2. Optical fibers

Optical fibers were suggested to be possible candidates for radiation dosimetry through RIA, RIL or thermoluminescence monitoring [2]. Such applications were considered in the case of glass optical fibers and plastic optical fibers as well. The radiation induced effects depends strongly on the optical fiber composition of the irradiated optical fibers, a detailed discussion on radiation induced defects in Pure Silica Core (PSC) and doped optical fibers can be found in [39, 40].

Optical fibers based on different dopants were proposed for radiation dosimetry using the optical attenuation change under gamma irradiation: TiO_2 and $GeO_2 + TiO_2$ in the silica core SM optical fibers [41]; Ge/Al co-doped SM fibers [42]; P_2O_5 doped step-index MM [43]. Depending on the experiment, low doses (dose rates of 0.01–1 Gy/h, for total doses up to 1 Gy) [43], moderate (dose rates 5–10 Gy/h, total doses of 10–100 Gy) [41] or high doses (dose rates of 2, 4 or 6 Gy/min, and the maximum total dose of 13 kGy) [42] were used. RIA was measured either at λ = 980, 1310 and 1530 nm [41, 42] or at λ = 502, 540 and 560 nm [43]. The recovery of the irradiation induced attenuation was studied at room temperature, through photobleaching (Ar laser radiation at λ = 514 nm) or by heating the samples (1000°C in O_2 atmosphere at the pressure of 10 psi). Some of the studied samples (depending of the dopants concentration, wavelength considered, dose used, and radiation sensitivity) indicate a linear dependence of RIA with the dose, its independence with the dose rate and low recovery, which provide the ground for applications in radiation dosimetry.

Step index multimode, 16.0 mole% P_2O_5 co-doped (with and without 6.0 mole% GeO_2 in the core) optical fibers were tested under gamma irradiation and shown a linear response at 505 nm for dose rates of 0.1 to 1 Gy/h, and a radiation sensitivity of 0.69–0.97 dB m^{-1} Gy, up to a maximum dose 1 Gy [44]. As the samples present an independence of the radiation response on the dose rate and a low recovery they seem to be suitable for medical dosimetry.

An investigation of gamma rays on polarization maintaining (PM) optical fibers to be used in interferometric fiber optic gyroscope (IFOG) for spaceborne assembles was carried out on three types of fibers: (i) pure-silica-core, (ii) P-doped (1 mol%), and germanium Ge-doped (15 mol %), as they were exposed to total doses of 100 Gy and 1 kGy. RIA was monitored over the spectral range 1100–1700 nm, with a focus on 1310 and 1550 nm. All the three fibers are more sensitive to radiation at 1550 nm than at 1310 nm, and the most degradation as it concerns RIA was observed in the case of the P-doped optical fiber [45].

On-line evaluation of gamma radiation produced RIA was investigated in 40 cm long Co/Fe co-doped alumino-silicate optical fibers fabricated by modified chemical vapor deposition (MCVD), and exposed to dose rates of 6.7, 18.4, 37.0, and 78.3 Gy/min, for 30 min to total doses of 201, 551, 1110, and 2348 Gy [46]. The measurements were done at 1310 nm with a Yokogawa AQ6370C optical spectrum analyzer. Following the irradiation, the attenuation in the doped fiber increases 258 times as compared to a reference optical fiber (SMF-28™), and a small decay of the RIA signal was noticed after the irradiation ended. An almost linear relationship was present between RIA and the dose up to a total dose of 2.4 kGy at 78.3 Gy/min, which recommend this optical fiber for radiation dosimetry, as RIA dose rate independence was observed.

The performances of Ce-doped alumino-phospho-silicate optical fibers without Au (5.0 wt.% Al_2O_3, 0.15 wt.% P_2O_5, 0.3 wt.% CeO_2) and with Au co-doping (5.1 wt.% Al_2O_3, 0.15 wt.% P_2O_5, 0.27 wt.% CeO_2, and 0.2 wt.% Au_2O_3), produced by MCVD, were evaluated under electron beam irradiation at energy of ~6 MeV up to the total doses of 10^{12} e/cm², 5 × 10^{12} e/cm², 10^{13} e/cm², 5 × 10^{13} e/cm², 10^{14} e/cm², and 2.5 × 10^{15} e/cm². RIA was monitored between 400 nm and 1100 nm, and post irradiation photobleaching under He-Ne laser radiation laser (λ = 543 nm) and UV. The results of the study indicated that the Ce/Au-doped optical fibers are more suitable for fiber-based dosimetry [47].

In the last years, the research focused also on the X-ray irradiation (10 keV energy) effects on various optical fibers: PSC, Ge-doped, P-doped, Fluorine-doped/ co-doped, in some cases the irradiation being combined with sample heating to 300°C [48, 49].

The irradiation outcomes in changing the optical fiber characteristics were investigated by on-line RIA measurements, confocal micro-luminescence (CML) and electron paramagnetic resonance (EPR), for total doses up to 3 MGy (dose rate of 50 Gy/s). RIA modifications in the UV-visible spectral range for Ge/F co-doped optical fibers produced under different conditions (draw speed and tension) were monitored with a mini optical fiber spectrometer (Ocean Optics HR4000) and deuterium-halogen DH2000 light source. Almost complete recovery of the irradiation induced attenuation was observed at room temperature for λ = 360 nm, corresponding to Ge(1) defects, at room temperature during the day following the exposure [49]. For various production conditions and different chemical composition the RIA vs. total irradiation dose graph indicates a linear dependency up to the dose of 300 kGy.

An interesting aspect in the research of optical fiber behaviour exposed to radiation is represented by the evaluation of radiation induced changes as heating is applied to the investigated sample. Different types of multimode UV optical fibers (deep UV enhanced, high OH step-index; solarization resistant, high OH step-index; UV enhanced extended spectral response, H_2-loaded, step-index) were tested under gamma irradiation. In some circumstances, the samples were heated to 573 K while irradiated, as the variation of the optical attenuation at specific wavelengths in the UV spectral range (λ = 248 nm, λ = 265 nm, λ = 320 nm and λ = 330 nm) was monitored. For some wavelengths (λ = 248 nm and λ = 265 nm) H_2-loading improves radiation resistance, while heating during the irradiation of the same optical fiber increases their radiation sensitivity [50].

A complex research was carried out to evaluate the combined effect of radiation (1keV X-ray, 50 Gy/s dose rate and total dose from 1.5 KGy to 1 MGy) and heating up to 300°C, for different types of optical fibers [48]. The results of this investigation indicated that no unitary behavior with temperature during radiation exposure can be predicted in the case of PSC, Ge-doped, P-doped, Fluorine-doped optical fibers, as in some situations (fiber compositing, temperature range, total dose) the sample heating has no effect in compensating the defects generation, while in some others RIA increases or decreases with the temperature increase. In any case, some of the study conclusions refer (i) to the possible use of such fibers in radiation dosimetry, especially at low doses, and (ii) to the limited length of fiber which can be incorporated into distributed radiation sensing systems.

A recent contribution to the use of optical fibers in radiation dosimetry refers to some studies carried out in cooperation with a team working at the European Synchrotron Radiation Facility, in Grenoble on UV multimode optical fibers subjected to synchrotron radiation [51]. In this experiment were tested five types of commercially available UV optical fibers and off-line measurements concerning the degradation of the optical fibers transmission in the UV-visible spectral range were done, as color centers developed upon exposure to synchrotron radiation. The measurements were performed for different doses: 5, 10, 30, 60, 200, 400, 1000 and 2000 Gy. The radiation induced optical attenuation was monitored in relation to recovery phenomenon both at room temperature and after samples heating to 560 K. The change of optical transmission and the increase of the attenuation at specific wavelengths (λ = 215 nm; λ = 229 nm; λ = 248 nm; λ = 265 nm; λ – 330 nm) were performed using the setup presented in **Figure 4**. The measurements were done with a dedicated software control developed under LabVIEW, by using a broadband UV-visible light source, a sensitive optical fiber spectrometer, and an optical fiber multiplexer.

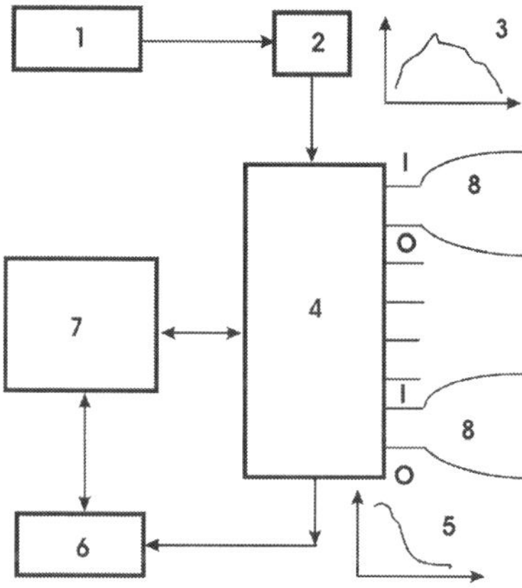

Figure 4. The sketch of the measuring setup: (1) light source; (2) optical fiber attenuator; (3) spectrum of the light source; (4) optical fiber multiplexer; (5) optical fiber sample absorption spectrum; (6) optical fiber mini spectrometer; (7) laptop; (8) connecting optical fibers; (9) irradiated optical fiber samples. I, multiplexer input; O, multiplexer output.

The graphical user's interface is illustrated in **Figure 5**. The multiplexer is employed to connect different tested samples to the light source and the spectrometer and help for the determination of the optical absorption from 215 nm to 600 nm. The operator can preset the number of averaging cycles, the box car value and can select the wavelengths of interest to be monitored. Of interest for these tests was to monitor specific wavelength associated to the presence of color centers in glass fibers. By observing the dependency of the optical absorption on the total irradiation dose at the specified wavelengths (**Figure 6**) can be estimated the linearity of this dependency and the spectral interval over which a saturation effect occurs. This can help to evaluate the possible use of these optical fibers in radiation dosimetry, at specified dose rates.

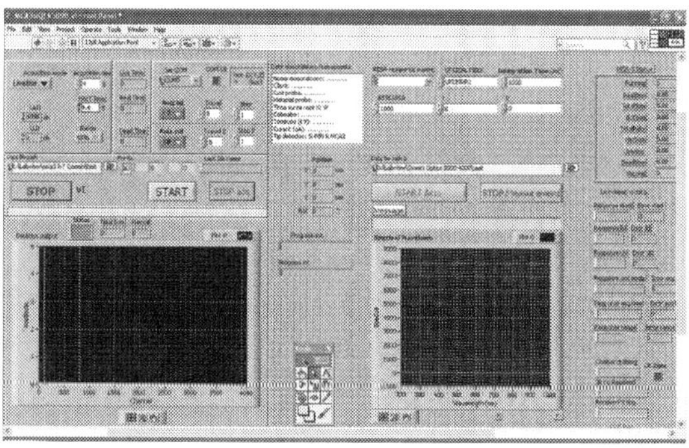

Figure 5. The LabVIEW graphical user's interface for the experimental setup (Courtesy of Laura Mihai).

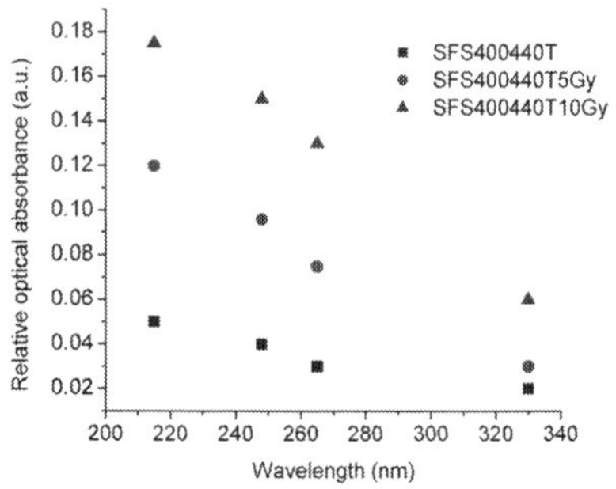
SFS400440T
SFS400440T5Gy
SFS400440T10Gy
Relative optical absorbance (a.u.)
Wavelength (nm)

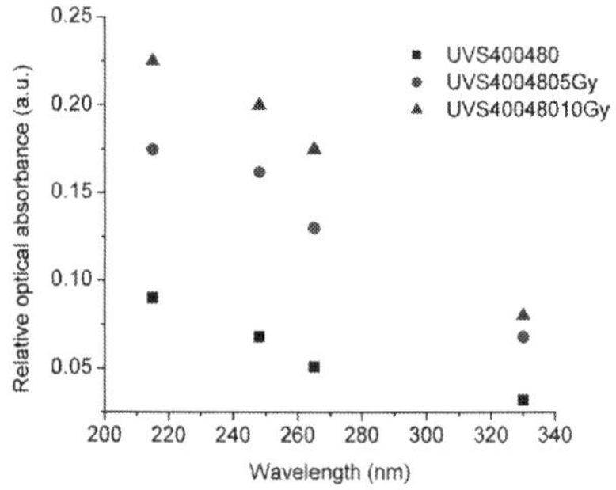
UVS400480
UVS4004805Gy
UVS40048010Gy
Relative optical absorbance (a.u.)
Wavelength (nm)

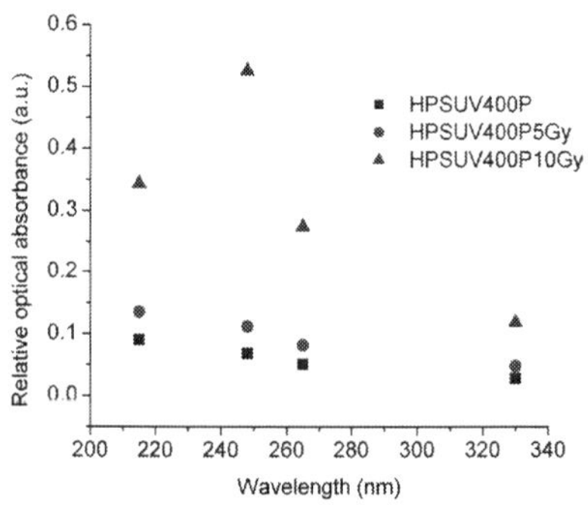
HPSUV400P
HPSUV400P5Gy
HPSUV400P10Gy
Relative optical absorbance (a.u.)
Wavelength (nm)

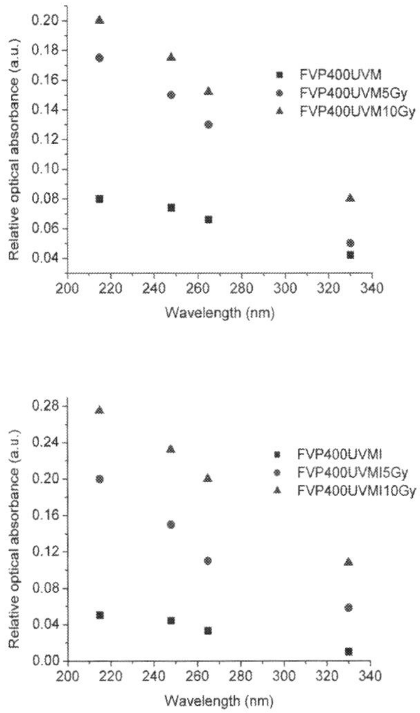

Figure 6. The dynamics of the color center formation as function of the dose (i.e. 5 Gy and 2000 Gy in this case) for the five tested optical fibers, at λ = 215 nm, λ = 248 nm, λ = 265 nm, λ = 330 nm: (a) SFS400/440T (Fiberguide Industries Inc.); (b) UVS400/480 (Fiberguide Industries Inc.); (c) HPSUV400P (Oxford Electronics); (d) FVP400/UVM (Polymicro Technologies); (e) FVP400/UVMI (Polymicro Technologies).

As a premiere, the use of THz spectroscopy and imaging to assess the synchrotron radiation induced changes in the core and coating of the investigated optical fibers (**Figure 7**) was reported. This approach makes possible the "visualization" of the dielectric constants change of the irradiated materials.

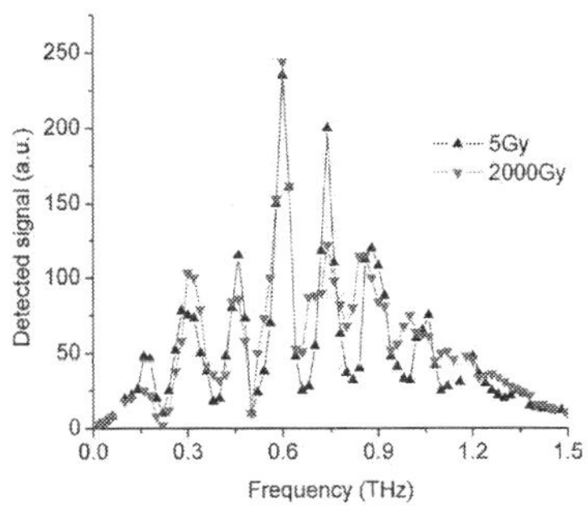

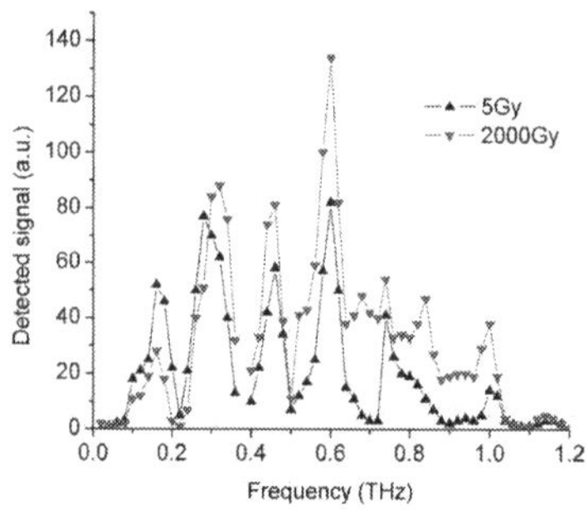

Figure 7. The THz spectral signal of the core (a) and coating (b) of an irradiated UV optical fiber, for two synchrotron radiation doses: 5 Gy and 2000 Gy (Courtesy of Laura Mihai).

The potential to employ P-doped core multimode optical fibers as X-ray (1 keV) radiation detectors for doses up to 3 kGy, under different dose rates (1, 10, 50 Gy/s) was studied by monitoring in real time the RIA in the spectral range from 200 to 900 nm. The results indicated a sub linear dependence of the irradiation induced attenuation with the dose and dose rate. The temperature did not change RIA during the irradiation for values between 5 and 50°C. The highest radiation sensitivity was observed at 300 nm, ~0.60 $dB/m^{-1}\ Gy^{-1}$ [52].

For the case RIA values during the irradiation is dependent on the radiation dose and exhibit a recovery after the exposure is interrupted, a differential scheme can be applied when the measurand is represented by the difference of RIA at two wavelengths, $\lambda = 413$ nm and $\lambda = 470$ nm, respectively [53].

Another novelty recently introduced refers to the test of gamma radiation on perfluorinated polymer optical fibers (pPOF). These results complement the investigations previously reported in literature on the spectral characteristics of gamma irradiated PMMA optical fibers. In this case, commercially available MM perfluorinated fiber GigaPOF-50SR from Chromis Fiberoptics were exposed to gamma ray at a dose rate of 5.7 kGy/h for total doses of 1, 5, 20, 50 and 100 kGy. The irradiation took place at maximum temperature of 32°C, and the samples were measured before and after the irradiation using a ν-OTDR from Luciol Instruments, operating at 1310 and 1550 nm [54]. In addition, strain tests were also performed to evaluate the mechanical degradation of the polymer fiber under gamma irradiation. For doses up to 5 kGy the attenuation vs. the dose presents a linear dependency with a radiation sensitivity of 0.096 ± 0.006 dB m-1 kGy^{-1} at 1310 nm, and 0.25 ± 0.05 dB m-1 kGy^{-1} at 1550 nm. The obtained results suggest the possible use of these optical fibers in radiation dosimetry. The dynamics of the color centers formation under gamma irradiation was investigated by on-line measurements and by monitoring RIA change at several wavelengths ($\lambda = 420$, 525, and 750 nm). The highest radiation sensitivity was noticed at $\lambda = 420$ nm (up to 0.5 kGy) dose, while the best linearity response of RIA vs. dose was obtained at $\lambda = 750$ nm for a dose reaching 2.5 kGy [55].

The last decade recorded as a breakthrough the use of sapphire optical fibers in radiation environments, as sapphire supports very high operating temperatures and presents minor optical transmission changes when exposed to ionizing radiation over a wide spectral range [56]. Recently, sapphire optical fibers were subjected to neutron flux of 1.5×10^{12} n cm^{-2} s^{-1} and a gamma dose rate of 84 kGy/hr (dose in sapphire) for a total neutron fluence of 1.1×10^{17} n cm $^{-2}$ and total gamma dose of 1.8 MGy [57]. The optical attenuation in the sapphire fiber was measured on-line over the 500–2200 nm spectral range. The major RIA occurs below the 500 nm measuring limit. A change of RIA was noticed as the sample was heated to 1000°C, under gamma irradiation [58].

The use of optical fibers for distributed measurements in ionizing radiation environments became a hot topic in the last years as several tests were run and some possible applications were suggested: Raman and Brillouin strain/ temperature sensing [10, 59]; Rayleigh temperature sensing [60]. Tests were performed on standard communication, highly doped GeO_2, F-doped optical fibers [10, 59] and (PSC), Ge-doped, radiation hardened and Al-doped optical fibers [60], at dose rates of 700 Gy/h [60], 1.5 kGy/h [10], 28 kGy/h [59], and total doses of 110 kGy [60], 618 kGy [10], 10 MGy [59]. RIA proved to be the major challenge in using optical fibers distributed Brillouin temperature sensors, as the operating distance is reduced to several hundreds of meters due to irradiation [59]. Limitations in temperature monitoring are present also in the case of Rayleigh-based setup [60]. Temperature measurements using Raman scattering encounters some difficulties as the temperature evaluation uncertainty is quite high upon irradiation [10].

The complementary aspect of interest is represented by the possible use of radiation sensitive optical fibers for distributed dosimetry. Two approaches were proposed to be used: (i) the optical time domain reflectometry – OTDR [61], (ii) the optical frequency domain reflectometery – OFDR [62]. In both experiments the fiber samples were subjected to gamma irradiation: P-doped optical fibers [61] and standard communication, Al-doped, and P-doped optical fibers [62], at dose rates 590 Gy/h at 30°C and 605 Gy/h at 80°C, total dose of about 71 kGy [62], and dose rate of 23 mGy/s, for a total dose of 300 Gy [61]. The OTDR approach exhibits a limited spatial resolution for distributed dosimetry. Depending on the optical fiber type, the interrogating instrument dynamic range and the dose, the usable length of the dosimetric fiber is limited to several centimeters [62].

2.3. Rare earth-doped optical fibers

Generally, radiation hardening tests of rare earth-doped optical fibers were performed in relation to their possible use in space applications such as inter-satellite communications and fiber optic gyroscopes (FOGs) applications [63–65]. The investigations focused on the radiation induced absorption in the optical fiber and on changes of the emission efficiency. Tests were carried out under gamma ray [63, 65, 66], neutrons [65], protons [65–67], or X-ray [66] exposure. The reported studies indicate that:

- the irradiation results are primarily dependent on the dose and not on the type of irradiation involved [65],
- the device degradation is function more on the co-dopant concentration than on the rare-earth elements [65],

- annealing and photobleaching can contribute to partial recovery of the irradiation induced effects [65, 67, 68].

Tests on photobleaching of two Erbium-doped fibers (EDFs) subjected to 1.3 and 2.24 kGy gamma irradiation were carried out at two laser wavelengths (532 and 976 nm). A higher efficiency was achieved with the shorter wavelength excitation. The annealing effect induced by the 976 nm laser is comparable to that present in the case of thermal annealing at 500 K [69].

In another experiment, an EDF was tested under electron irradiation (beam energy 1.2 MeV, total dose 10 kGy), and the output power, noise figure and the central wavelength were measured after the irradiation, with the pump at 980 nm and detection at 1550 nm. The central wavelength did not change after irradiation, while the output power and noise figure were deteriorated (i.e. the output power dropped from 10 to –60 dBm for the same pump level). Within 2 weeks, the two parameters partially recovered [70].

Single mode and multi-mode Yb-doped optical fibers, acting as amplifiers at 1064 nm, were subjected to gamma rays and mixed gamma-neutron irradiation, up to a total dose of 1 Gy(Si) [71]. It was noticed a linear decrease of the output power with the irradiation dose. A slight recovery was observed at room temperature, after 20 hours.

In his PhD thesis, Fox [72] concluded that EDFs are more radiation sensitive than Yb^{3+}-doped fibers, and the most radiation hardened to gamma ray are optical fibers of Er^{3+}/Yb^{3+} co-doped type. In this investigation, the radiation induced degradation increases with the increase of the dose rate.

Optical fiber preforms of Yb-free aluminosilicate core, Yb-doped Al-free silicate core, Yb-doped alumino-silicate cores and Yb-free germanosilicate core produced by MCVD and solution-doping techniques were irradiated by 45 kV X-rays, at ~2.5 Gy(SiO_2)/ min, for doses up to 0.3–0.5 kGy [73]. The effects of the irradiation were studied by thermally stimulated luminescence (TSL) and optical absorption measurements.

Investigations were carried out to evaluate the influence of H_2-loading of EDFs, under pumping at 980 and 1480 nm, as the samples were subjected to gamma irradiation, having doses from 0.1 to 10 kGy. One such sample was a H_2-loaded EDF, while the other was H_2-free carbon-coated [74]. At 980 nm pumping a photobleaching effect was observed which increases the efficiency of the process especially in the H_2-loaded optical fiber. No such effect was present with the pump radiation of 1480 nm. H_2-loading and the use of a hermetical coating of the optical fiber, which prevents H_2 diffusion, produce a radiation resistant EDF [75].

One approach to enhance the radiation resistance of the optical amplifier consists in the use of Er-doped-nanoparticles optical fibers [76]. Another proposed method is based on the doping of the optical fiber core with Ce, reducing in this way the radiation sensitivity [77].

Tests on a thulium doped optical fiber amplifier were performed under neutron irradiation and a 17.1 dB was observed for a dose of 720 Gy, which is close to the computed values [78].

The development of Bi co-doped silica optical fibers [79] having an extensive emission operation range (from 1100 to 1800 nm) called researchers attention on their investigation under ionizing radiation. Tests run under gamma irradiation, dose rate – 5.5 kGy/h and maximum total dose – 50 kGy, on Bismuth/Erbium/Ytterbium co-Doped Fiber – BEYDF ([Bi] ~ 0.07, [Er] < 0.1, [Al] ~ 0.35, [Ge] ~ 1.18, [Yb] ~0.02 atom%, respectively, indicated the formation of bismuth related active center (BAC), as the absorption at 830nm is increased significantly, while the absorption at longer wavelength diminishes [80].

In the case of Bi/Al-co-doped silica optical fibers the fluorescence peak increases with the increase of the gamma dose (dose rate, 800 Gy/h; maximum total dose, 3 kGy), when the sample is pumped at 980 nm. The preform of the investigated materials had the following concentrations: Si—28 mol/% (core), Ge— 6 mol/% (core), O—66 mol/% (core and inner layer), Bi—0.016 mol/% (core and inner layer), and Al—0.04 mol/% (core and inner layer). It is presumed that this enhancement of the photoluminescence signal upon irradiation is due to creation of subvalence Bi ions [81].

Another investigation was focused on the effect of gamma irradiation (dose rate – 5.5 kGY/h, total doses of 1, 5, 15, 30 and 50 kGy) on samples of Bismuth/Erbium co-doped fiber (BEDF). The irradiation has significantly increased unsaturable absorption by about 8 dB. This corresponds to a very significant unsaturable attenuation coefficient change of ~32 dB/cm. The results indicate that the saturable absorption of the BEDF at 830 nm is also increased by about 4 dB. This corresponds to a significant saturable attenuation coefficient change of 16 dB/cm. The increase in saturable pump absorption implies that the irradiation has significantly increased the number of bismuth related active centers (BAC-Si). In this case the overall emission at 1430 nm (which is related to BAC-Si) is only slightly decreased while the unsaturable absorption is significantly increased. A good radiation survivability of the BEDF for emission or amplification was noticed [82].

3. Radiation effects in photonic devices

3.1. Fiber Bragg gratings

Fiber Bragg gratings were excessively studied under various type of ionizing radiation: gamma ray [12, 83, 84], neutron [85], mixed gamma-neutron [86–88] or 13.5 MeV protons [89]. In most cases, the changes of the FBG Bragg wavelength are relatively low. Tests were carried out using Type I, Type II, Type IA, Type IIA, chemical composition or fs laser engraved gratings, and were focused on the effect of fiber composition, dose rate, total dose, heating during the irradiation or possible photobleaching during the exposure to ionizing radiation [90]. For example, Type I gratings produced in H_2-loaded standard communication optical fibers (SMF28™) or B/Ge co-doped optical fibers present an increase of the central wavelength followed by a plateau, with an overall change between 10 pm to 34 pm, under 0.54 MGy exposure. In the case of Type II gratings written in B/Ge co-doped optical fibers exhibit a decrease of the central wavelength between -30 and -60 pm. Degradation (30–50%) of gratings reflectivity was produced after irradiation, while no recovery was observed in the studied sensors [91]. A quite significant impact of gamma radiation on Ge-doped core optical fibers was observed as such uncoated fibers are immersed in water during gamma exposure [83], for a total dose of up to 5 MGy. Generally, all the investigations carried out on radiation effects on FBGs focused on the possible use of such devices in radiation environments for temperature or humidity [12] monitoring. Under some design circumstances (optical fiber type, technology used to write the grating), these modifications are significant, making possible the use of FBGs as radiation detectors [92, 93].

More recently, the use of FBGs developed in sapphire optical fiber was proposed for monitoring very high temperature in nuclear reactors [94].

Through these studies, the paradigm was changed as [95]:

- FBGs were produced in radiation hardened optical fibers by two-beam interferometer and deep ultraviolet fs laser radiation;
- for the first time FBGs written in both standard commercial optical fibers and radiation hardener optical fibers were tested, exposed to the electron beam from a linear accelerator;
- a mash of FBGs was used for beam diagnostics of charged particle beams, as a novelty.

This research indicated a linear dependence of the Bragg wavelength shift with the irradiation dose for commercially available FBGs and grating produced in radiation hardened optical fibers, and this shift was monitored by measuring simultaneously with a thermocouple the temperature change in the irradiation plane. This aspect is important because no saturation or permanent modification of FBG wavelength was observed; hence these sensors can be used for on-line measurements. The principle of the charged particle beam diagnostics is illustrated in **Figure 8**, for the case of an electron beam having a diameter of about 100 mm. The spatial resolution is limited by the number of FBGs engraved in an optical fiber, their individual length and the distance between two adjacent FBGs. In the described proof-of-concept design, gratings having 12 mm and 4 mm length were employed. As compared to other solutions for charge particle beam monitoring (arrays of Faraday cups; ionization chambers; micro strip metal detector; pepper-pot device, slit-grid or rotating slits; scintillating screen or gas detector; flat panel detectors, arrays of p-i-n diodes; moving wire or vibrating wire scanners) there are several advantage of this approach: its immunity to electro-magnetic noise, remote monitoring capability and the possibility to multiplex the acquired signals by using few connecting lines.

The operation of the proposed instrument is similar to that employed for laser beam diagnostics. In the evaluation of a laser beam quality an image sensor is used to acquire the transversal distribution of the laser beam's intensity. Periodically, the electric charge generated inside the image sensors is removed upon reading and new acquisition starts. In this implementation the instrument operation is based on the proved reliability of the tested FBGs under electron beam exposure and on the linear shift of FBGs Bragg wavelength with the temperature in the detection plan, and hence with the deposed energy by the charge particle beam. Prior to the use, the FBGs were calibrated as it concerns their wavelength change vs. temperature.

When the charge particle beam (1) is propagating from the linear accelerator output (2) its diameter increases (4) in the detection plan (5) due to its divergence (**Figure 8a**). In the detecting plan (item 1 in **Figure 8b**) a mesh composed of FBGs (3) is placed, the sensors being embedded into a thermally insulating material to prevent the lateral dissipation of the heat. As the individual gratings are exposed to different beam energies their temperature increases according to the energy deposed on each detecting site. In this way, in time, a map of the transferred energy at each location is obtained.

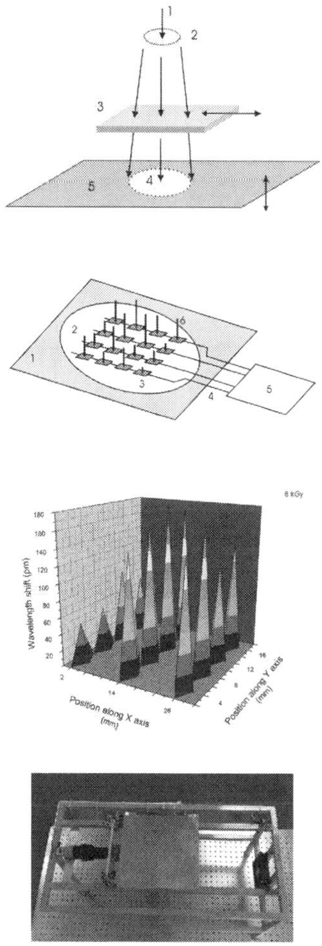

Figure 8. The operation principle of the electron beam analyzer: (a) the moving shutter used to interrupt FBGs exposure to the electron beam: (1) the incident electron beam; (2) beam diameter at the exit of the linear accelerator focusing system; (3) moving shutter; (4) the electron beam diameter at its incidence on the detection plan; (5) detecting plan. (b) The position of the FBGs mash: (1) the detecting plan; (2) electron beam pattern on the detection plan; (3) FBGs; (4) connecting optical fibers; (5) optical fiber interrogator; (6) bars symbolizing the integral energy deposited on a particular FBG at a specified moment. (c) Example of the data acquired for the 8 kGy dose [95]. (d) Top view of the shutter and cooling system.

The detector operates as an energy integrator (dose related measurements). Permanently, the central wavelength of all sensors is acquired by a Micron Optics optical fiber interrogator, which makes possible the real time mapping of the electron beam cross section energy distribution. In order to avoid saturation, periodically the FBGs signal has to be "reset" by blocking the electron beam and forced cooling the FBGs array. For this purpose a thick (20 mm) Al shutter restricts under the software control (**Figure 8a** and **8d**) the exposure of the FBGs by interrupting the electron beam. During the interval the beam is blocked, a cooler pushes air over the sensors' matrix. The Al shutter goes back and forth as the array has to be exposed or cooled. The flow chart for the instrument operation is presented in **Figure 9**.

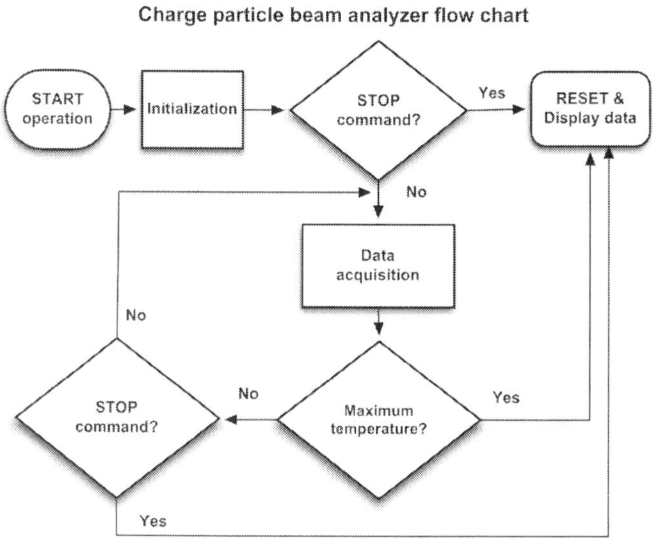

Figure 9. The flow chart of the electron beam analyzer.

During the "Initialization" step, the shutter is closed, the FBGs are cooled, and the functioning parameters set by the operator are introduced in the system. If a STOP command was issued by the operator the system stops and displays in 3D format the integrated energy to which each sensor was subjected, measured as Bragg wavelength shift in response to the local temperature increase. If no STOP command was given the system periodically acquires on the central wavelength shift for each sensor and checks if the maximum set temperature was reached by any of the FBGs. If this temperature was reached the system stops and displays the results as a 3D representation. If no such signal was received the software commands during the "Data acquisition" step the closing of the shutter and the cooling of the FBGs array. The periodicity of the cooling cycles, and the upper limit of the temperature to which any of the sensors can be subjected are set by the operator. The Bragg wavelength shift with the dose increase is given in **Figure 10**. In this case, during the irradiation pause, the grating's cooling was obtained by convection in air and not by forced cooling. The proposed instrument can be used during the adjusting process of the charged particle accelerator or to check the stability in time of the output beam.

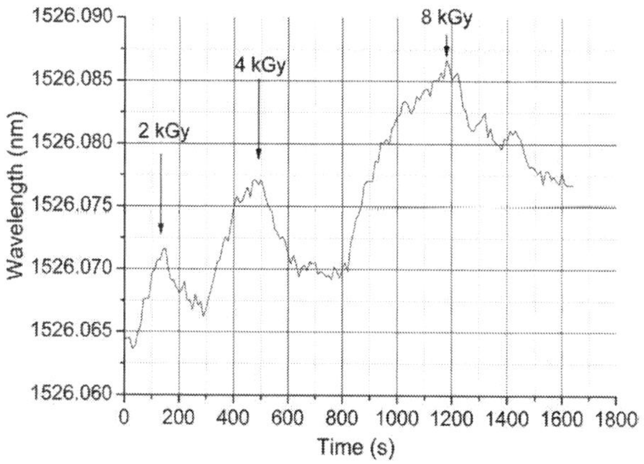

Figure 10. The change of the Bragg wavelength with the dose increase. Periodically, between two exposures the grating was cooled by natural convection [95].

A novelty in the field can be considered the investigation of FBGs performances written in polymer fibers under fast neutrons irradiation. In the paper it is suggested that neutrons produce a degradation of the fiber structure, which in turn causes a shift of the Bragg wavelength up to 14 pm. The wavelength change with neutron dose can be exploited in radiation dosimetry [96].

3.2. Long period gratings

Long period gratings (LPGs) were tested only under gamma ray exposure [90]. Most of the measurements were performed off-line [97–99], with some exceptions when the device behavior was monitored during the irradiation [100, 101]. The LPGs were obtained by different techniques: CO laser or UV engraving [99], electric arc-discharge (EAD) technique [101], CO_2 laser point-by-point writing [97], as chiral gratings [100], having a turnaround point (TAP) design produced by a CO_2 laser [98]. Various optical fibers were employed: N-doped and Ge-doped optical fibers [99], pure-silica-core/ F-doped silica cladding fibers [101], SMF-28™ [97], core doped with P, Ge, F, rare earth elements [100], photosensitive fibers (B/Ge co-doped) [98]. The irradiation conditions are modified from 5 kGy [97] up to 1.47 MGy [99]. Depending on the optical fiber type, the writing technique, the dose rate/total dose used, the reported results led to different conclusions: (i) no effect was observed on the LPG transmission spectrum for doses lower than 5 kGy [97], lower than 500 kGy [101] or as high as 1.47 MGy [99], (ii) a 10 nm shift of the transmission deep was noticed for doses of 100 kGy [100], (iii) very high shift of the dual resonance deep (35 nm) is present for a dose of 6 kGy [98]. This variety of investigations outcomes suggest that, by an appropriate selection of the above mentioned parameters, LPGs can be produced either to be radiation hardened or with pronounced radiation sensitivity, appropriate to be used in radiation dosimetry. No change in the temperature sensitivity of LPGs upon irradiation was observed [101].

A new proposed approach in studying gamma irradiation effects on LGPs Includes several novelties [102, 103]:

- the use of an OFDR model LUNA OBR 4600, operating in the transmission mode or, for comparative purpose, of an optical fiber interrogator model sm125;
- LPGs on-line monitoring when the irradiation is on and off, which makes possible the observation of room temperature recovery;
- the comparative evaluation of the radiation induced LPG changes for grating produced in standard communication and specially designed radiation hardened optical fibers.

The use of the OFDR improved drastically the detection S/N as compared to classical reading with an optical spectrum analyzer. During the irradiation the sensors were encapsulated into ceramic radiation transparent cases to avoid any strain induced changes in the LPG spectrum, and in the mean time, the gratings were placed into a thermally insulated box. The temperature was permanently monitored both inside this box and in the irradiation chamber and temperature related corrections were applied to the LPG characteristics. Based on the referred papers, **Figure 11** illustrates comparatively the behaviour of gamma irradiation on the LPG developed into a standard communication optical fiber (LPGsc) and one written in a radiation hardened optical fiber (LPGrh). The wavelength deep of the two samples move in opposite directions, the wavelength of the grating produced in the standard communication fiber increases with the dose increase, while for the other one the wavelength decreases with the dose increase. Besides that, it can be noticed the magnitude of the two changes, LPGsc being by far more

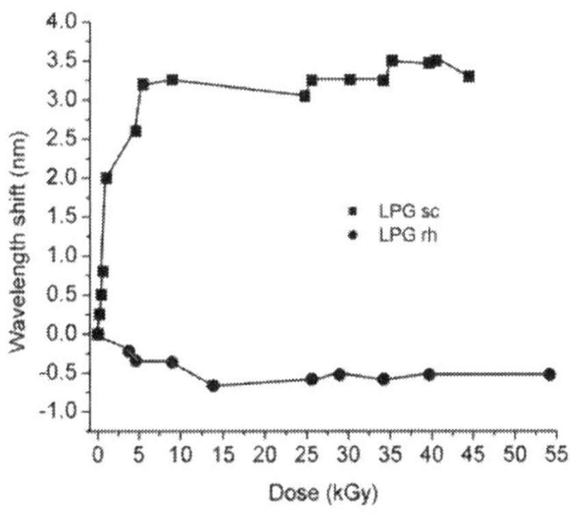

Figure 11. The change of the wavelength deep for two LPGs written in standard communication optical fibers LPGsc and in F-doped core optical fiber LPGrh.

http://dx.doi.org/10.5772/62547

sensitive. The fluctuations of the wavelength deep values during gamma exposure as well as the partial saturation trend are associated to the on-off operation of the irradiation facility. In this way, for the first time the recovery effect during the irradiation pauses was observed.

Because for the LPGrh the wavelength change is by far less significant than that in the case of LPGsc, it can be expected that the first one can be used in radiation environments as a sensor for temperature/humidity, while the second LPG can be employed within specific dose range as an optical fiber-based dosimeter.

4. Conclusions

The last four to five years brought additional data concerning radiation effects in optical fibers and optical fiber-based devices as new materials, technologies and possible applications in radiation environments emerged. In the frame of this chapter, the major trends were presented and some novel techniques and applications promoted recently were addressed, in relation to different types of irradiations (i.e. electron beam, synchrotron radiation). In this way, the data reviewed in previous publications are updated and the extensive possibilities offered by new materials and experimental setups in using radiation effects on optical fibers were demonstrated. By focusing such type of research on newly developed fibers (plastic, micro-structured, multi-core) and by using novel photonic devices based on optical fibers, new classes of equipment will emerge for terrestrial applications (dosimetry, medicine, remote and distributed monitoring) and for space missions.

Acknowledgements

The authors acknowledge the financial support of the Romanian Executive Agency for Higher Education, Research, Development and Innovation Funding (UEFISCDI), under grant 8/2012, project "Sensor Systems for Secure Operation of Critical Installations - SOCI" and of the Romanian Space Agency (ROSA) through the project "Evaluation of Components for Space Applications - ECSA", grant 67/ 2013.

Author details

Dan Sporea* and Adelina Sporea

*Address all correspondence to: dan.sporea@inflpr.ro

Photonics Investigations Laboratory, National Institute for Laser, Plasma and Radiation Physics, Magurele, Romania

References

[1] Sporea D, Agnello S, Gelardi FM. Irradiation effects in optical fibers. In: Pal B, editor. Frontiers in Guided Wave Optics and Optoelectronics. InTech; 2010. p. 46–66. DOI: 10.5772/303

[2] Sporea D, Sporea A, O'Keeffe S, McCarthy D, Lewis E, Optical fibers and optical fiber sensors used in radiation monitoring. In: Yasin M, Harun SW, Arof H, editors. Selected Topics on Optical Fiber Technology. InTech; 2012. p. 607-652. DOI: 10.5772/2429

[3] Berghmans F, Brichard B, Fernandez FA, Gusarov M, Van Uffelen M, Girard S. An introduction to radiation effects on optical components and fiber optic sensors. In: Bock WJ, Gannot I, Tanev S, editors. Optical Waveguide Sensing and Imaging, NATO Science for Peace and Security Series B: Physics and Biophysics. Springer; 2008. p. 127-165.

[4] Bourgade JL, Allouche A, Baggio J, Bayer C, Bonneau F, Chollet C, et al. New constraints for plasma diagnostics development due to the harsh environment of MJ class lasers. Rev. Sci. Instrum. 2004;75(10):4204-4212. DOI: http://dx.doi.org/10.1063/1.1789610

[5] Bourgade JL, Costley A.E, Reichle R, Hodgson ER, Hsing W, Glebov V, et al. Diagnostic components in harsh radiation environments: Possible overlap in R&D requirements of inertial confinement and magnetic fusion systems. Rev. Sci. Instrum. 2008;79:10F304. DOI: http://dx.doi.org/10.1063/1.2972024

[6] ITER. Available from: https://www.iter.org/mach [Accessed: January 29, 2016]

[7] Girard S, Baggio J, Leray J-L, Meunier J-P, Boukenter A, Ouerdane Y. Vulnerability analysis of optical fibers for Laser Megajoule Facility: Preliminary studies. IEEE Trans. Nucl. Sci. 2005;52(5):1497-1503. DOI: 10.1109/TNS.2005.855813

[8] Yamamoto S, Shikama T, Belyakov V, Farnum E, Hodgson E, Nishitani T,et al. Impact of irradiation effects on design solutions for ITER diagnostics. J. Nucl. Mater. 2000;283–287(1):60–69. DOI: 10.1016/S0022-3115(00)00157-4

[9] ELI-Nuclear Physics Working Groups. The White Book of ELI Nuclear Physics Bucharest-Magurele, Romania [Internet]. Available from: http://www.eli-np.ro/documents/ELI-NP-WhiteBook.pdf [Accessed: January 29, 2016]

[10] Delepine-Lesoille S, Pheron X, Bertrand J, Pilorget G, Hermand G, Farhoud R, et al. Industrial qualification process for optical fibers distributed strain and temperature sensing in nuclear waste repositories. J. Sensors. 2012;369375. DOI: 10.1155/2012/369375

[11] Wijnands TJ, De Jonge LK, Kuhnhenn J, Hoeffgen SK, Weinand U. Optical absorption in commercial single mode optical fibres for the LHC machine. Proceedings of the Topical Workshop on Electronics for Particle Physics [Internet], Prague, 3–7 September 2007, pp. 121-124. Available from: http://cds.cern.ch/record/1089249/files/p121.pdf [Accessed: January 29, 2016]

[12] Makovec A, Berruti G, Consales M, Giordano M, Petagna P, Buontempo S, et al. Radiation hard polyimide-coated FBG optical sensors for relative humidity monitoring in the CMS experiment at CERN. J. Instrum. 2014;9(03):C03040. DOI: 10.1088/1748-0221/9/03/C03040

[13] O'Keeffe S, McCarthy D, Woulfe P, Grattan MW, Hounsell AR, Sporea D, et al. A review of recent advances in optical fibre sensors for in-vivo dosimetry during radiotherapy. Br. J. Radiol. 2015;88:1050. DOI: http://dx.doi.org/10.1259/bjr.20140702

[14] Barth JL, Dyer CS, Stassinopoulos E.G. Space, atmospheric, and terrestrial radiation environments. IEEE Trans. Nucl. Sci. 2003;50(3):466-482. DOI: 10.1109/TNS.2003.813131

[15] Barth JL, LaBel KA, Poivey, C. Radiation assurance for the space environment. In: International Conference on Integrated Circuit Design and Technology ICICDT '04; Austin. 2004. p. 323-333. DOI: 10.1109/ICICDT.2004.1309976

[16] ECSS Secretariat. Space environment, ECSS-E-10-04A, ESA Publications Division [Internet], 2000. Available from: www.spacewx.com/Docs/ECSS-E-ST-10-04C_15Nov2008.pdf [Accessed: January 29, 2016]

[17] Schwank JR, Shaneyfelt MR, Dodd PE. Radiation hardness assurance testing of microelectronic devices and integrated circuits: Radiation environments, physical mechanisms, and foundations for hardness assurance. IEEE Trans. Nucl. Sci. 2013;60(3): 2074-2100. DOI: 10.1109/TNS.2013.2254722

[18] El-Kheshen AA. Glass as radiation sensor. In: Nenoi M, editors. Current Topics in Ionizing Radiation Research. InTech; 2012. p. 579-602. DOI: 10.5772/34612

[19] Farah K, Mejri A, Hosni F, Hamzaoui AH, Boizot B. Formation and decay of colour centres in a silicate glasses exposed to gamma radiation: Application to high-dose dosimetry. In: Nenoi M., editors. Current Topics in Ionizing Radiation Research. InTech; 2012. p. 603-624. DOI: 10.5772/32765

[20] Woody CL, Kierstead JA, Levy PW, Stoll S. Radiation damage in BaF2 crystals. IEEE Trans. Nucl. Sci. 1992;39:515-523. DOI: 10.1109/23.159658

[21] Li PJ, Xie YY, Zhao YL, Yin ZW. Radiation damage barium fluoride in large crystals [Internet], 1992. Available from: http://lss.fnal.gov/archive/other/ssc/ssc-gem-tn-92-000077.pdf [Accessed: January 29, 2016]

[22] El-Said AS, Cranney M, Ishikawa N, Iwase A, Neumann R, Schwartz K, et al. Study of heavy-ion induced modifications in BaF2 and LaF3 single crystals. Nucl. Instrum. Methods B. 2004;218:492–497. DOI: 10.1016/j.nimb.2003.12.057

[23] Izerrouken M, Meftah A, Nekka M. Color centers in neutron-irradiated Y3Al5O12, CAF2 and LiF single crystals. J. Lumin. 2007;127:696–702. DOI: 10.1016/j.jlumin.2007.04.005

[24] Fernández-Rodríguez M, Alvarado CG, Núñez A, Álvarez-Herrero A. Analysis of optical properties behaviour of Clearceram, fused silica and CaF2 glasses exposed to

simulated space conditions. In: Proceedings of the International Conference on Space Optics (ICSO 2010), 4–8 October 2010. Available from: http://www.congrexprojects.com/custom/ icso/Papers/Session%2011a/FCXNL-10A02-2018721-1-FERNANDEZ-RODRIGUEZ_ICSO_ PAPER.pdf

[25] Brovetto B, Maxia V, Salis M. On the radioluminescent emission of BaF2. Il Nuovo Cimento D. 1996;18(9):1107-1110. DOI: 10.1007/BF02457676

[26] Dafinei I, Fasoli M, Ferroni F, Mihokova E, Orio F, Pirro S, et al. Low temperature scintillation in ZnSe crystals. IEEE Trans. Nucl.Sci. 2010;57(3):1470-1474. DOI: 10.1109/TNS.2009.2035914

[27] Yanagida T, Fujimoto Y, Koshimizu M, Fukuda K. Scintillation properties of CdF2 crystal. J. Lumin. 2015;157:293–296. DOI: http://dx.doi.org/10.1016/j.jlumin.2014.09.013

[28] Kristianpoller N, Rehavi A, Shmilevich A, Weiss D, Chen R.Radiation effects in pure and doped Al2O3 crystals. Nucl. Instrum. Meth B. 1998;141:343-346. DOI: 10.1016/S0168-583X(98)00096-2

[29] Plaksin OA, Stepanov VA, Demenkov PV, Stepanov PA, Skuratov VA, Kishimoto N. Radioluminescence of alumina during proton and heavy ion irradiation. Nucl. Instrum. Methods B. 2003;206:1083–1087. DOI: 10.1016/S0168-583X(03)00915-7

[30] Islamov AKh, Ibragimova EM, Nuritdinov I. Radiation-optical characteristics of quartz glass and sapphire. J. Nucl. Mater. 2007;362:222–226. DOI: 10.1016/j.jnucmat.2007.01.047

[31] Izerrouken M, Benyahia T. Absorption and photoluminescence study of Al2O3 single crystal irradiated with fast neutrons. Nucl. Instrum. Methods B. 2010;268:2987-2990. DOI: 10.1016/j.nimb.2010.05.024

[32] Zhang MF, Zhang HL, Han JC, Guo HX, Xu CH, Ying GB. Effects of neutron irradiation and subsequent annealing on the optical characteristics of sapphire. Effects of neutron irradiation and subsequent annealing on the optical characteristics of sapphire. Physica B. 2011;406:494–497. DOI: 10.1016/j.physb.2010.11.021

[33] Faist J, Capasso F, Sivco DL, Sirtori C, Hutchinson AL, Cho AY. Quantum cascade laser. Science. 1994;264(5158):553-556. DOI: 10.1126/science.264.5158.553

[34] Wilson CF, Chassefière E, Hinglais E, Baines KH, Balint TS, Berthelier J-J, et al. The 2010 European Venus Explorer (EVE) mission proposal. Exp. Astron. 2012;33(2):305-335. DOI: 10.1007/s10686-011-9259-9

[35] Forouhar S, Borgentun C, Frez C, Briggs R, Bagheri M, Webster CR, et al. Mid-IR semiconductor lasers for space. In: Mid-Infrared Optoelectronics: Materials and Devices (MIOMD-XII); 5–9 October 2014; Montpellier, France.

[36] Mihai L, Sporea D, Sporea A. Vâţă I. Testing of ZnSe, CaF2, BaF2, and sapphire windows under alpha particles irradiation. In: 3rd International Conference and Exhibition on Lasers, Optics and Photonics; 1–3 September 2015; Valencia, Spain.

[37] Mihai L, Sporea D, Sporea A, Ighigeanu D, Neguţ D. Optical and THz evaluation of components for gas sensing spectroscopy in hazardous environments. In: NATO ARW on THz Diagnostics of CBRN effects and Detection of Explosives & CBRN; 3–6 November 2015; Izmir, Turkey.

[38] Mihai L, Sporea D, Sporea A, Craciun G, Manaila E. Electron beam irradiation of materials and components to be used in mid-IR spectroscopy. In: Photonics Europe 2016; 3–7 April, 2016; Brussels, Belgium. paper EPE117-144.

[39] Girard S, Kuhnhenn J, Gusarov A, Brichard B, Van Uffelen M, Ouerdane Y, et al. Radiation effects on silica-based optical fibers: Recent advances and future challenges. IEEE Trans. Nucl. Sci. 2013;60(3):2015-2036. DOI: 10.1109/TNS.2012.2235464

[40] Griscom DL. A minireview of the natures of radiation-induced point defects in pure and doped silica glasses and their visible/near-IR absorption bands, with emphasis on self-trapped holes and how they can be controlled. Phys Res Int. 2013;ID 379041. DOI: http://dx.doi.org/10.1155/2013/379041

[41] Paul MC, Sen R, Bhadra SK, Pal M, Giri PP, Dasgupta K, et al. Gamma ray radiation induced absorption in Ti doped single mode optical fibres at low dose levels. Opt. Mater. 2007;29:738–745. DOI: 10.1016/j.optmat.2005.12.004

[42] Paul MC, Sen R, Bhadra SK, Dasgupta K. Radiation response behaviour of Al codoped germano-silicate SM fiber at high radiation dose. Opt. Commun.. 2009;282:872–878. DOI: 10.1016/j.optcom.2008.11.052

[43] Paul MC, Bohra D, Dhar A, Sen R, Bhatnagar PK, Dasgupta K. Radiation response behavior of high phosphorous doped step-index multimode optical fibers under low dose gamma irradiation. J. Non-Cryst. Solids. 2009;355:1496-1507. DOI: 10.1016/j.jnoncrysol.2009.05.017

[44] Ghosh S, Das S, Paul MC, Dasgupta K, Bohra D, Chaudhary HS,et al. Evaluation of the performance of high phosphorous with germanium codoped multimode optical fiber for use as a radiation sensor at low dose rates. Appl. Opt. 2011;50(25):E80-E85. DOI: 10.1364/AO.50.000E80

[45] Jin J, Liu J, Xu J. Effect of spectral dependencies of radiation-induced attenuation in polarization maintaining fibers on interferometric fiber optic gyroscope at near-infrared wavelengths. Optik. 2013;124:5679-5682. DOI: http://dx.doi.org/10.1016/j.ijleo.2013.04. 057

[46] Kim Y, Ju S, Jeong S, Kim JY, Lee NH, Jung HK, et al. Gamma-ray irradiation-induced optical attenuation in Co/Fe co-doped alumino-silicate optical fiber for dosimeter

application. J. Lightwave Technol. . 2014;32(22):3791-3797. DOI: 10.1109/JLT. 2014.2357798

[47] Kir'yanov AV, Ghosh S, Paul MC, Barmenkov YO, Aboites V, Kozlova NS. Ce-doped and Ce/Au-codoped alumino-phosphosilicate fibers: Spectral attenuation trends at high-energy electron irradiation and posterior low-power optical bleaching. Opt. Mater. Express. 2014;4(3):434-448. DOI: 10.1364/OME.4.000434

[48] Girard S, Marcandella C, Morana A, Perisse J, Di Francesca D, Paillet P, et al. Combined high dose and temperature radiation effects on multimode silica-based optical fibers. IEEE Trans. Nucl. Sci. 2013;6(6):4305-4313. DOI: 10.1109/TNS.2013.2281832

[49] Di Francesca D, Boukenter A, Agnello S, Girard S Alessi A, Paillet P, et al. X-ray irradiation effects on fluorine-doped germanosilicate optical fibers. Opt. Mater. Express. 2014;4(8):1683-1695. DOI: 10.1364/OME.4.001683

[50] Sporea D, Sporea A, Oproiu C. Effects of hydrogen loading on optical attenuation of gamma-irradiated UV fibers. J. Nucl. Mater. 2012;423(1-3):142-148. DOI: 10.1016/ j.jnucmat.2012.01.023

[51] Sporea D, Mihai L, Sporea A, Lixandru A, Bräuer-Krisch E. Investigation of UV optical fibers under synchrotron irradiation. Opt. Express. 2014;22(25):31473-31485. DOI: 10.1364/OE.22.031473

[52] Girard S, Ouerdane Y, Marcandella C, Boukenter A, Quenard S, Authier N. Feasibility of radiation dosimetry with phosphorus-doped optical fibers in the ultraviolet and visible domain. J. Non-Cryst. Solids. 2011;357:1871-1874. DOI: 10.1016/j.jnoncrysol. 2010.11.113

[53] Tomashuk AL, Grekov MV, Vasiliev SA, Svetukhin VV. Fiber-optic dosimeter based on radiation-induced attenuation in P-doped fiber: suppression of post-irradiation fading by using two working wavelengths in visible range. Opt Express. 2014;22(14): 16779-16783. DOI: 10.1364/OE.22.016778

[54] Stajanca P, Sporea D, Mihai L, Negut D, Schukar M, Krebber K. Gamma radiation induced effects on perfluorinated polymer optical fibers for sensing applications. In: Bunge C-A, Kruglov R, editors. Perfluorinated Polymer Optical Fibers for Sensing Applications. BoD - Books on Demand: Norderstedt; 2015. ISBN 978-3-7392-1499-3.

[55] Mihai L, Sporea D, Negut D, Stajanca P, Krebber K. On-line monitoring of gamma irradiated perfluorinated polymer optical fiber. In: Photonics Europe 2016; 3–7 April, 2016; paper EPE104-58.

[56] Sporea D, Sporea A. Radiation effects in sapphire optical fibers. Physica Status Solidi (c). 2007;4(3):1356–1359. DOI: 10.1002/pssc.200673709

[57] Petrie ChM, Windl W, Blue ThE. In-situ reactor radiation-induced attenuation in sapphire optical fibers. J. Am. Ceram. Soc. 2014;97:3883-3889. DOI: 10.1111/jace.13211

[58] Petrie ChM, Wilson B, Blue ThE. In situ gamma radiation-induced attenuation in sapphire optical fibers heated to 1000°C. J. Am. Ceram. Soc. 2014;97:3150-3156. DOI: 10.1111/jace.13089

[59] Phéron X, Girard S, Boukenter A, Brichard B, Delepine-Lesoille S, Bertrand J,et al. High γ-ray dose radiation effects on the performances of Brillouin scattering based optical fiber sensors. Opt. Express. 2012;20(24):26978-26985. DOI: 10.1364/OE.20.026978

[60] Faustov A, Gusarov A, Wuilpart M, Fotiadi AA, Liokumovich LB, Kotov OI, et al. Distributed optical fibre temperature measurements in a low dose rate radiation environment based on Rayleigh backscattering. In: Berghmans F, Mignani AG, De Moor P, editors. Proceedings of Optical Sensing and Detection II; SPIE 8439: 84390C; 2012. DOI: 10.1117/12.922082

[61] Toccafondo I, Brugger M, Di Pasquale F, Guillermain E, Kuhnhenn J. First steps towards a distributed optical fiber radiation sensing system. International Conference on Space Optics, Tenerife, Canary Islands, Spain [Internet], 7–10 October 2014. Available from: http://congrexprojects.com/Custom/ICSO/2014/Papers/5.%20Posters/Poster%20Session%202/2.10.66978_Toccafondo.pdf [Accessed: January 29, 2016]

[62] Faustov AV, Gusarova A, Wuilpart M, Fotiadi A, Liokumovichc LB, Zolotovskiyf IO, at al. Remote distributed optical fibre dose measuring of high gamma-irradiation with highly sensitive Al- and P-doped fibres. In: Baldini F, Homola J, Lieberman RA, editors. Proceedings of Optical Sensors 2013; SIPE 8774:877404; 2013. DOI: 10.1117/12.2017331

[63] Brichard B, Fernandez, AF, Ooms H, Berghmans F. Gamma dose rate effect in erbium-doped fibers for space gyroscopes. In: Proceedings of the 16th International Conference on Optical Fiber Sensors OFS16, Nara, 13–17 October 2003, pp. 336-339. Available from: https://www.researchgate.net/profile/Francis_Berghmans/publication/228759091_Gamma_Dose_Rate_Effect_in_Erbium-Doped_Fibers_for_Space_Gyroscopes/links/09e4150b625689e300000000.pdf [Accessed: January 29, 2016]

[64] Brichard B, Fernandez AF, Ooms H, van Uffelen M, Berghmans F. Study of the radiation-induced optical sensitivity in erbium and aluminium doped fibres. In: Proceedings of the 7th European Conference on Radiation and Its Effects on Components and Systems RADECS2003, Noordwijk, 15–19 September 2003, pp. 35-38. Available from: http://citeseerx.ist.psu.edu/viewdoc/download?doi=10.1.1.59.2175&rep=rep1&type=pdf [Accessed: January 29, 2016]

[65] Lezius M, Predehl K, Stower W, Turler A, Greiter M, Hoeschen C, et al. Radiation induced absorption in rare earth doped optical fibers. IEEE Trans. Nucl. Sci. 2012;59(2): 425-433. DOI: 10.1109/TNS.2011.2178862

[66] Girard S, Ouerdane Y, Tortech B, Marcandella C, Robin T, Cadier B, et al. Radiation effects on Ytterbium- and Ytterbium/Erbium-doped double-clad optical fibers. IEEE Trans. Nucl. Sci. 2009;56(6):3293-3399. DOI: 10.1109/TNS.2009.2033999

[67] Rose TS, Gunn D, Valley GC. Gamma and proton radiation effects in Erbium-doped fiber amplifiers: active and passive measurements. J. Lightwave Technol. 2001;19(12): 1918-1923. DOI: 10.1109/50.971685

[68] Chang SH, Liu R-Y, Lin C-E, Chou F-I, Tai C-Y, Chen CC. Photo-annealing effect of gamma-irradiated erbium-doped fibre by femtosecond pulsed laser. J. Phys. D: Appl. Phys. 2013;46:495113-495119. DOI: 10.1088/0022-3727/46/49/495113

[69] Peng T-S, Huang Y-W, Wang LA, Liu R-Y, Chou F-I. Photo-annealing effects for Erbium doped fiber sources after gamma irradiation tests by using 532 nm and 976 nm lasers. In: Jones J, Culshaw B, Ecke W, López-Higuera JM, Willsch R, editors. Proceedings of the 20th International Conference on Optical Fibre Sensors; SPIE 7503: 750375; 2009. DOI: 10.1117/12.835304

[70] Ma J, Li M, Tan L-Y, Zhou Y-P, Yu S-Y, Che C. Space radiation effect on EDFA for inter-satellite optical communication. Optik - Int. J. Light Electron Opt. 2010;121(6):535-538. DOI: 10.1016/j.ijleo.2008.09.009

[71] Singleton BJ. Radiation Effects on Ytterbium-doped Optical Fibers. No. AFIT-ENP-DS-14-J-15, Air Force Institute of Technology Wright-Patterson AFB OH Graduate School of Engineering and Management [Internet], 2014. Available from: http://www.dtic.mil/cgi-bin/GetTRDoc?Location=U2&doc=GetTRDoc.pdf&AD=ADA602890 [Accessed: January 29, 2016]

[72] Fox BP. [thesis]. Investigation of ionizing-radiation-induced photodarkening in rare-earth-doped optical fiber amplifier materials, The University of Arizona; 2013.

[73] Mady F, Benabdesselam M, Mebrouk Y, Dussardier B. Radiation effects in ytterbium-doped silica optical fibers: traps and color centers related to the radiation-induced optical losses. In: Proceedings of RADECS 2010, Paper LN2; 2010. Available from: https://hal.archives-ouvertes.fr/hal-00559422/file/Paper_LN2.pdf [Accessed: January 29, 2016]

[74] Zotov KV, Likhachev ME, Tomashuk AL, Kosolapov AF, Bubnov MM, Yashkov MV, et al. Radiation resistant Er-doped fibers: optimization of pump wavelength. IEEE Photonics Technol. Lett. 2008;20(17):1476–1478. DOI: 10.1109/LPT.2008.927909

[75] Zotov KV, Likhachev ME, Tomashuk AL, Bubnov MM, Yashkov MV, Guryanov AN, et al. Radiation-resistant erbium-doped fiber for spacecraft applications. IEEE Trans. Nucl. Sci. 2008;55(4):2213-2215. DOI: 10.1109/TNS.2008.2001834

[76] Thomas J, Myara M, Troussellier L, Burov E, Pastouret A, Boivin D, et al. Radiation-resistant erbium-doped-nanoparticles optical fiber for space applications. Opt. Express. 2012;20(3):2435-2444. DOI: 10.1364/OE.20.002435

[77] Girard S, Mescia L, Vivona M, Laurent A, Ouerdane Y, Marcandella C, et al. Design of radiation-hardened rare-earth doped amplifiers through a coupled experiment/

simulation approach. J. Lightwave Technol. 2013;31(8):1247-1254. DOI: 10.1109/JLT. 2013.2245304

[78] Hamdalla, TA, Nafee SS. Radiation effects on the gain of thulium doped fiber amplifier: Experiment and modeling. Opt. Laser Technol. 2014;55:46-49. DOI: 10.1016/j.optlastec. 2013.06.024

[79] Luo Y, Wen J, Zhang J, Canning J, Peng G-D. Bismuth and erbium codoped optical fiber with ultrabroadband luminescence across O-, E-, S-, C-, and L-bands. Opt. Lett. 2012;37(16):3447-3449. DOI: 10.1364/OL.37.003447

[80] Yan B, Luo Y, Sporea D, Mihai L, Neguţ D, Sang X, et al. Gamma radiation-induced formation of bismuth related active centre in Bi/Er/Yb co-doped fibre. In: Asia Communications and Photonics Conference (ACP); 19–23 November 2015; Hong Kong.

[81] Wen J, Liu W, Dong Y, Luo Y, Peng G-D, Chen N, et al. Radiation-induced photoluminescence enhancement of Bi/Al-codoped silica optical fibers via atomic layer deposition. Opt. Express. 2015;23(22):29004-29013. DOI: 10.1364/OE.23.029004

[82] Sporea D, Mihai L, Negut D, Luo Y, Yan B, Ding M, et al. Effects of gamma rays on rare Bi/Er co-doped optical fibers. In: ROMOPTO2015 Conference; 1–4 September 2015; Bucharest, Romania.

[83] Butov OV, Golant KM, Shevtsov IA, Fedorov AN. Fiber Bragg gratings in the radiation environment: Change under the influence of radiolytic hydrogen. J. Appl. Phys. 2015;118:074502. DOI: 10.1063/1.4928966

[84] Fernandez AF, Brichard B, Berghmans F, Decréton M. Dose-rate dependencies in gamma-irradiated fiber Bragg grating filters. IEEE Trans. Nucl.Sci. 2002;49(6): 2874-2878. DOI: 10.1109/TNS.2002.805985

[85] Fielder RS, Klemer D, Stinson-Bagby KL. High neutron fluence survivability testing of advanced fiber Bragg grating sensors. In: AIP Conference Proceedings on Thermophysics in Microgravity; Commercial/Civil Next Generation Space Transportation; 21st Symposium Space Nuclear Power & Propulsion; Human Space Explorer; Space Colonization; New Frontiers & Future Concepts; 2004, Vol. 699, pp. 650-657. DOI: http://dx.doi.org/10.1063/1.1649627

[86] Fernandez AF, Brichard B, Berghmans F, Rabii H, Fokine M, Popov M. Chemical composition fiber gratings in a high mixed gamma neutron radiation field. IEEE Trans. Nucl. Sci. 2006;53(3):1607-1613. DOI: 10.1109/TNS.2005.863273

[87] Gusarov A, Fernandez AF, Vasiliev S, Medvedkov O, Blondel M, Berghmans F. Effect of gamma–neutron nuclear reactor radiation on the properties of Bragg gratings written in photosensitive Ge-doped optical fiber. Nucl. Instrum. Methods B. 2002;187(1):79-86. DOI: 10.1016/S0168-583X(01)00829-1

[88] Perry M, Niewczas P, Johnston M. Effects of neutron-gamma radiation on fiber Bragg grating sensors: a review. IEEE Sensors. 2012;12(11):3248-3257. DOI: 10.1109/JSEN.2012.2214030

[89] Curras E, Virto AL, Moya D, Vila I, Carrión JG, Frövel M, Palomo FR. Influence of the fiber coating type on the strain response of proton-irradiated fiber Bragg gratings. IEEE Trans. Nucl.Sci. 2012;59(4):937-942. DOI: 10.1109/TNS.2012.2206049

[90] Gusarov A, Hoeffgen SK. Radiation effects on fiber gratings. IEEE Trans. Nucl. Sci. 2013;60(3):2037-2053. DOI: 10.1109/TNS.2013.2252366

[91] Maier RRJ, MacPherson WN, Barton JS, Jones JDC, McCulloch S, Fernandez AF, et al. Fibre Bragg gratings of type I in SMF-28 and B/Ge fibre and type IIA B/Ge fibre under gamma radiation up to 0.54 MGy. In: Proceedings of SPIE 17th International Conference on Optical Fibre Sensors, 2005, Vol. 5855, pp. 511-514. DOI: 10.1117/12.624037

[92] Faustov A, Saffari P, Koutsides C, Gusarov A, Wuilpart M, Mégret P, et al. Highly radiation sensitive type IA FBGs for future dosimetry applications. IEEE Trans. Nucl. Sci. 2012;59(4):1180-1185. DOI: 10.1109/TNS.2012.2202247

[93] Krebber K, Henschel H, Weinand U. Fibre Bragg gratings as high dose radiation sensors? Meas. Sci. Technol. 2006;17:1095-1102. DOI: 10.1088/0957-0233/17/5/S26

[94] de Villiers GJ, Treurnicht J, Dobson RT. In-core high temperature measurement using fiber-Bragg gratings for nuclear reactors. Appl. Therm. Eng. 2012;38:143-150. DOI: 10.1016/j.applthermaleng.2012.01.024

[95] Sporea D, Stancalie A, Becherescu N, Becker M, Rothhardt M. An electron beam profile instrument based on FBGs. Sensors 2014. 2014;14(9):15786-15801. DOI: 10.3390/s140915786

[96] Hamdalla TA, Nafee SS. Bragg wavelength shift for irradiated polymer fiber Bragg grating. Opt. Laser Technol. 2015;74:167–172. DOI: 10.1016/j.optlastec.2015.06.008

[97] Chaubey S, Joshi P, Kumar M, Arya R, Nath AK, Kher S. Design and development of long-period grating sensors for temperature monitoring. Sādhanā. 32(5):513-519.

[98] Kher S, Chaubey S, Kashyap R, Oak SM. Turnaround-Point Long period fiber gratings (TAP-LPGS) as high radiation dose sensors. IEEE Photonics Technol. Lett. 2012;24(9):742-744. DOI: 10.1109/LPT.2012.2187637

[99] Vasiliev SA, Dianov EM, Golant KM, Medvedkov OI, Tomashuk AL, Karpov VI,et al. Performance of Bragg and long-period gratings written in N- and Ge-doped silica fibers under γ-radiation. IEEE Trans. Nucl. Sci. 1998;45(3):1580-1583. DOI: 10.1109/23.685243

[100] Henschel H, Hoeffgen SK, Kuhnhenn J, Weinand U. High radiation sensitivity of chiral long period gratings. IEEE Trans. Nucl. Sci. 2010;57(5):2915-2922. DOI: 10.1109/TNS.2010.2059043

[101] Rego G, Fernandez AF, Gusarov G, Brichard B, Berghmans F, Santos JL. Effect of ionizing radiation on the properties of arc-induced long-period fiber gratings. App. Opt. 2005;44(29):6258-6263. DOI: 10.1364/AO.44.006258

[102] Sporea D, Stancălie A, Negut D, Pilorget G, Delepine-Lesoille S, Lablonde L. On-line tests of an optical fiber long-period grating subjected to gamma irradiation. IEEE Photonics J. 2014;6(6):1-9. DOI: 10.1109/JPHOT.2014.2337877

[103] Sporea D, Stăncalie A, Neguţ D, Pilorget G, Delepine-Lesoilled S, Lablonde L. Comparative study of long period and fiber Bragg gratings under gamma irradiation. Sensor Actuator A-Phys. 2015;233:295-301. DOI: http://dx.doi.org/10.1016/j.sna.2015.07.007

Chapter 3

The Impact of Successive Gamma and Neutron Irradiation on Characteristics of PIN Photodiodes and Phototransistors

Dejan Nikolić and Aleksandra Vasić-Milovanović

Additional information is available at the end of the chapter

http://dx.doi.org/10.5772/62756

Abstract

The aim of this paper is to explore the impact of increased gamma and neutron radiation on the PIN photodiodes and phototransistors and their output characteristics. Special attention was paid to the successive impact of gamma and neutron radiation when the components were located in the field of gamma radiation and after that in the field of neutron radiation. The impact of successive irradiation was compared with the influence of gamma and neutron radiation when they appear individually. An important result of this research is the observation that neutron irradiation of photovoltaic detectors, applied after gamma irradiation, leading to partial reparations of distorted semiconductor structure and increasing disrupted output characteristics (photocurrent, spectral response). Monte Carlo simulation of gamma photons transfer through the crystal lattice of the semiconductor has been shown that the cause of such effect of neutron radiation is a large number of divacancies caused by successive operation of the previous gamma radiation and the neutron radiation itself. Divacancies have created the basis for increased generation of charge carriers by direct transfer (tunneling) of carriers through the traps (recombination centers). This is so called intercenter charge transfer.

Keywords: photovoltaic detectors, gamma and neutron radiation, divacancies, intercenter charge transfer, Monte Carlo simulation

1. Introduction

Science and technology that deals with photovoltaic semiconductor detectors is an area with an extremely rapid development in the last 20 years. The reasons for this are, on the one hand, practically countless possibilities of application of these detectors (optical communication systems, medical devices, military equipment, automatic control systems, various electronic devices), and, on the other hand, miniaturization of electronic components and development of these devices mass production allowed them to have relatively low cost and to be accessible to the wide population. Particularly interesting applications of semiconductor photovoltaic detectors are in military systems, medical devices and equipment, and cosmic systems. These are areas where the probability for photovoltaic detectors to be in increased radiation field is very large.

The area of photovoltaic detectors and radiation type which they can be exposed is very large. This work is limited to the observation of the PIN photodiodes and phototransistors and their behavior in terms of gamma and neutron radiation considering that with particle emission from the core, as a rule, there have been a simultaneous de-excitations descendant core by a discrete gamma-ray emission. Semiconductor devices, therefore, are exposed to summary effect of neutron and gamma radiation.

The aim of this paper is to explore the impact of increased gamma and neutron radiation on the PIN photodiodes and phototransistors and their output characteristics. Special attention was paid to the observation of semiconductor devices' behavior when they have been exposed to the field of gamma radiation and after that to the field of neutron radiation (successive gamma and neutron radiation).

2. Experimental

Experimental measurement in this paper was carried out on the commercially available photovoltaic detectors. In this experiment, the following were used:

1. four types of silicon PIN photodiodes (BP104, BPW41N, BPW34 all manufactured by Vishay, and SFH203FA by Osram),
2. two types of silicon NPN phototransistors (BPW40 manufactured by Telefunken electronic and LTR4206 by LITEON).

Devices were first exposed to gamma radiation from Co^{60} source and then, after 30 days, to ^{241}Am-Be neutron and gamma source. Both sources were housed in Institute of Nuclear Sciences "Vinča" in Belgrade, Serbia.

The dose of Co^{60} gamma source is 2000 Gy, the energy of 1.25 MeV, and half-life time of 5.27 years. The samples were placed in controlled environment at a distance of 150 mm away from the radioactive source with a glass between them. The dose rate was 100 Gy/hr which was measured by electrometer with ionization chamber TW 30012-0172 produced by PTW, Germany. Measurement uncertainty of the system is less than 1.2%.

^{241}Am-Be source emits gamma photons of low energy (60 and 14 keV) with the activity of 3.7×10^{10} Bq, the intensity of the neutron emission of 2.7×10^{6} neutrons s^{-1} and the mean energy of the neutrons E_{nav} = 5.5 MeV. The panels were at a distance of 5 cm from the source, so the photon equivalent dose rate is $\dot{H}_\gamma$ = 12 mSv/hr, and the photon absorbed dose rate is $\dot{D}_\gamma$ = 12 mGy/hr. Calculated neutron absorbed dose rate is $\dot{D}_n$ = 1.714 mGy/hr and the equivalent dose rate of neutrons is $\dot{H}_n$ = 12 mSv/hr with the quality factor Q_n = 7. In this experiment, the semiconductor devices were placed at a distance of 5 cm from the ^{241}Am-Be source, and the exposure period was 16.75 hr. Since the total absorbed dose, for that distance, is $\dot{D}_{tot}$ = 13.714 mGy/hr and the total equivalent dose is $\dot{H}_{tot}$ = 24 mSv/hr, the total absorbed dose for material components is D_{tot} = 229.71 mGy and the total equivalent dose is H_{tot} = 402 mSv.

Both irradiation and those from Co^{60} gamma source and those from ^{241}Am-Be source were performed in the air at a temperature of 21°C and relative humidity of 40–70%.

Before and after every step of irradiation spectral response and photocurrent have been measured. The measurement were performed on the photodiodes and phototransistors, in highly controlled conditions at room temperature, which have previously been removed from the irradiation room. Samples have been divided in two groups. First group was irradiated only with neutron radiation and the second one with successive gamma and neutron radiation.

For the first group, there have been performed three measurements of the photodiodes and phototransistors parameters:

1. first measurement: just before neutron irradiation,
2. second measurement: just after neutron irradiation,
3. third measurement: 30 days after neutron irradiation.

For the second group there have been performed five measurements of the photodiodes and phototransistors parameters:

1. first measurement: just before gamma irradiation,
2. second measurement: just after gamma irradiation,
3. third measurement: 30 days after gamma irradiation (just before neutron irradiation),
4. fourth measurement: just after neutron irradiation,
5. fifth measurement: 30 days after neutron irradiation.

In order to perform the long-term isothermal annealing i.e. to give detectors enough time to recovery, the third and fifth measurement have been undertaken 30 days after the irradiation. Because of that, the changes occurring in the samples of the second group after the first irradiation (gamma) can be considered as a permanent. Standard measurement equipment (the professional digital multimeter AMPROBE 33XR) was used for measurement. Combined measurement uncertainty for all measurements was less than 1.2% [1, 2].

In order to understand the state of the semiconductor crystal lattice after exposure to gamma radiation and before neutron irradiation, a Monte Carlo transfer simulation of gamma photons through the photodiode and phototransistor have been performed.

3. Results and discussion

3.1. Photodiodes and phototransistors response to neutron radiation

Free neutrons, in small amounts, are everywhere in nature. The main source of neutrons is cosmic radiation. They also occur in nuclear reactions of natural α radiation and spontaneous fission of heavy nuclei. Neutron is a unique particle, it is uncharged, has a relatively large mass, and leads to radioactive disintegrations. From the point of nuclear reaction, neutrons are much more important than any other particles. As neutral particles, neutrons do not have the ability for direct ionization of materials. The basic mechanism of neutron interaction with the matter is via elastic collisions with atomic nuclei of environment. The interaction with electrons, although it exist, is negligible. Thereby, neutron loses some of its energy and slows down, while the environment can suffer different types of transformation.

Neutrons interact with the material in two different ways:

- through collisions with other particles,
- through the process of absorption.

In the case of high-energy neutrons (fast one), the dominant process is elastic scattering, while with low-energy neutrons, the absorption process is more likely [3].

Displacement of atoms can be compared with a collision between two solid spheres. If the transferred energy higher than the energy required for displacement (*displacement energy Ed*) atom will be shifted from their original positions in the lattice, and there will be defect (*PKA —primary knock-on atom*). Assuming that there is enough energy, displaced atom could be able to move other atoms or to produce electron-hole pairs. In the case of very high energy particles, a cascade distortion can be formed.

Different types of displacement defects could occur due to neutron irradiation (**Figure 1**):

- vacancies,
- divacancies,
- interstitials,
- Schottky defects,
- Frenkel defects.

Figures 2–4 show the results of measurements of PIN photodiodes and phototransistors spectral response before and after neutron irradiation and after a period of 30 days recovery are presented [4]. As can be seen from Figures 2 to 4, neutron radiation caused the deterioration of photodiodes and phototransistors characteristics.

High-energy particles like neutrons create much more displacement damages than gamma radiation. When an atom is ejected from its position, it creates a vacancy in the lattice. The ejected atom may recombine with a vacancy or stay in an interstitial position in the lattice.The

http://dx.doi.org/10.5772/62756

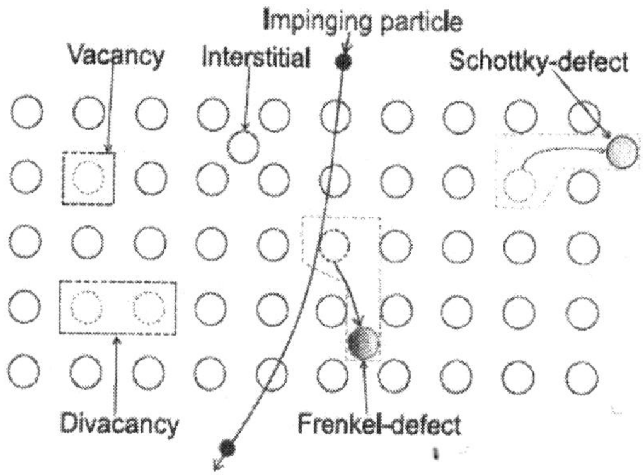

Figure 1. Displacement defects [3].

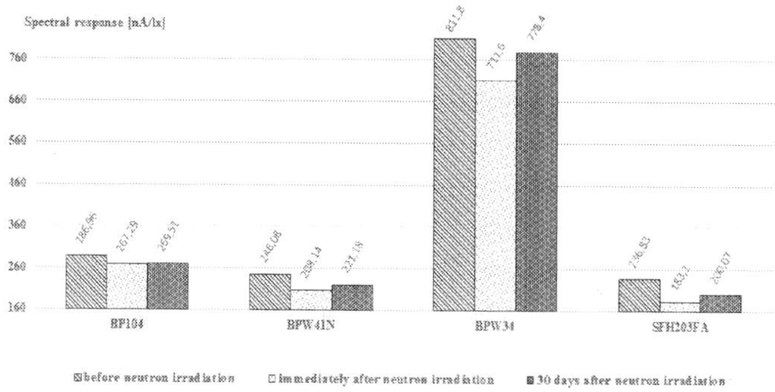

Figure 2. Spectral response of photodiodes before and after neutron irradiation.

vacancies are mobile and combine with other vacancies or with impurities of the semiconductor [5, 6], thus creating recombination centers that cause the reduction of charge carrier lifetime. Axness *et al.* [7] showed that the damage to the crystal lattice and reduction of the charge carriers lifetime are spatially dependent. Sporea *et al.* [8] have calculated that the major degradation of the photodiode responsivity, for the total gamma dose of 1.23 MGy and to the neutron fluence of 1.2×10^{13} n/cm^2, occurs in the case of neutron irradiation (37.5%) as compared to the gamma irradiation (7.2%).

Steady defects act as recombination centers and traps for charge carriers and because of that the resistance of the material could be increased [6]. Mobile vacancies represent a strong recombination instrument for capturing of minority charge carriers and thus reduce their lifetime. Defects responsible for the capture of electrons called E-defects while the H-defects actually traps holes [3]. Displacement defects mainly affect the electrical characteristics of the semiconductor substrate and thus the electrical characteristics of the whole electronic components. As a result, there have been the reduction of the spectral response and lower photocurrent photodiode (**Figure 2**).

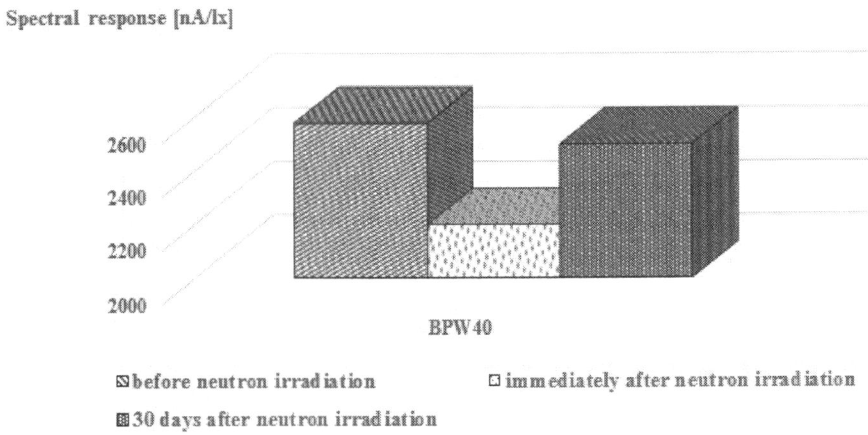

Figure 3. Spectral response of phototransistor BPW40 before and after neutron irradiation.

Phototransistors are very susceptible to neutron radiation. Neutron radiation affects the characteristics of phototransistors primarily by creating defects in the crystal lattice which can dramatically increase the level of charge carriers recombination. On the other hand, the increment of the recombination rate reduces the current gain. Many studies of the damage relocation mechanism in bipolar transistors have shown that the current gain of the transistor with a common emitter decreases with increasing of recombination centers number. The measurement data of phototransistors before and after irradiation showed that the adverse effects of neutron radiation are the most pronounced on transistors base current. Phototransistor is light controlled device where the output current is controled by the base current and brightness. Cluster defects caused by fast neutrons are the dominant mechanism for damaging of phototransistors exposed to neutron radiation. Number of displaced atoms caused by neutrons is very large. The result is forming of recombination-generation centers. Electron-hole recombination causes a decrease of current gain. Generation of electron-hole pairs cause an increase in leakage current. Removing the majority charge carriers and the reduction of carrier mobility causing an increase in voltage between the collector and emitter. Current gain is determined by the number of majority carriers emitted from the emitter which are passing through the base as minority carriers and are collected by collectors as the major carriers. Increasing of density of recombination-generation centers due to defects created by radiation causes a reduction of minority carrier lifetime, and because of that, the rate of electron-hole recombination in the base increases. Accordingly, the current gain decreases as a result of reduced injection of charge carriers from the emitter to the collector and, as a result, the photocurrent and spectral response decreases (Figures 3 and 4) [9].

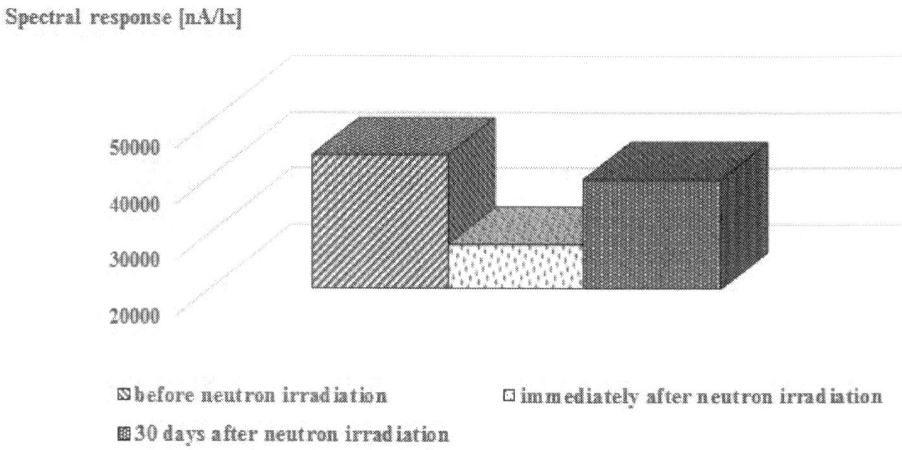

Figure 4. Spectral response of phototransistor LTR4206 before and after neutron irradiation.

In this experiment, a long-term isothermal annealing at room temperature was applied. The recovery period, labelled as a short-term annealing, begins immediately after the occurrence of damage and fully complete within a few minutes to 1 hour after irradiation. Damage, remaining after, that are often referred as a permanent damage. However, relatively slow process of recovery or long-term annealing, continues even after the short-term annealing is completed [10]. Recovery causes partially increasing of spectral response and the photocurrent.

Figures 2–4 show that the response of new, unused photodiode and phototransistor to neutron irradiation is in accordance with the theoretical principles described in the literature.

3.2. Photodiodes and phototransistors response to successive gamma and neutron irradiation

In recent few years there have been carried out a number of studies with the aim of observing the behavior of different photovoltaic detectors in terms of gamma and neutron radiation [11–17]. Most common topics were photodiode, as one of the most used and simplest types of optical sensors. The effect of gamma and neutron radiation on semiconductors is well known and described in the available literature. This chapter will present the results of research of behavior of photovoltaic detector due to successive gamma and neutron radiation. The samples were first exposed to gamma radiation and after 30 days to neutron radiation.

Figure 5 shows the results of the photodiodes and the phototransistors spectral response measurements before and after gamma and neutron radiation [18].

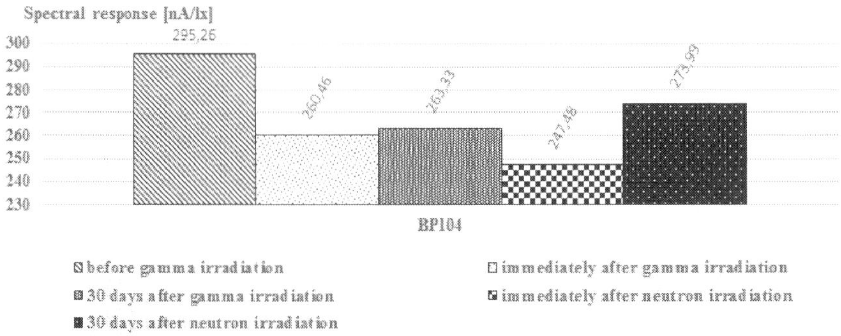

Figure 5. Spectral response of the reverse biased photodiode BP104 before and after gamma and neutron irradiation.

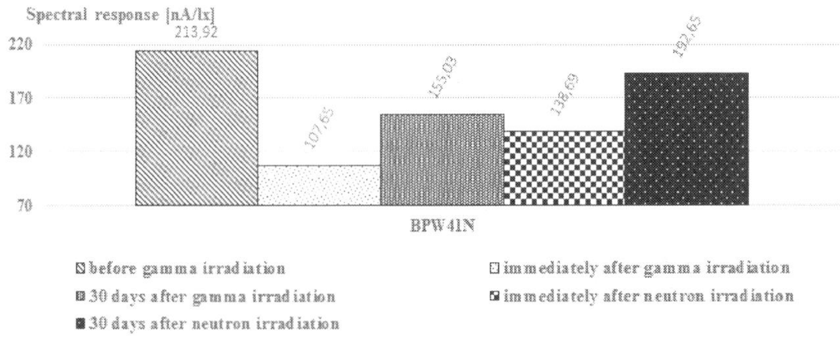

Figure 6. Spectral response of the reverse biased photodiode BPW41N before and after gamma and neutron irradiation.

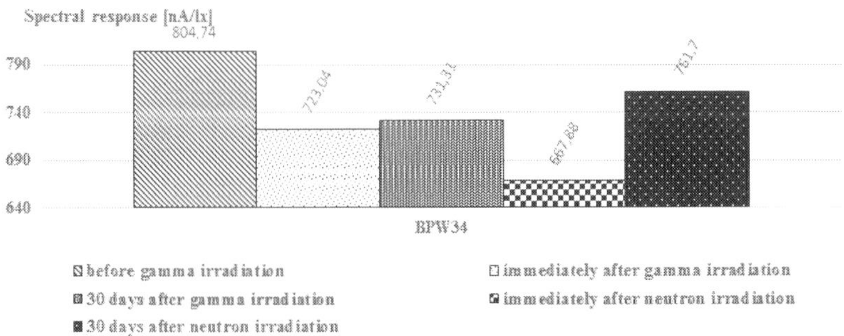

Figure 7. Spectral response of the reverse biased photodiodeBPW34 before and after gamma and neutron irradiation.

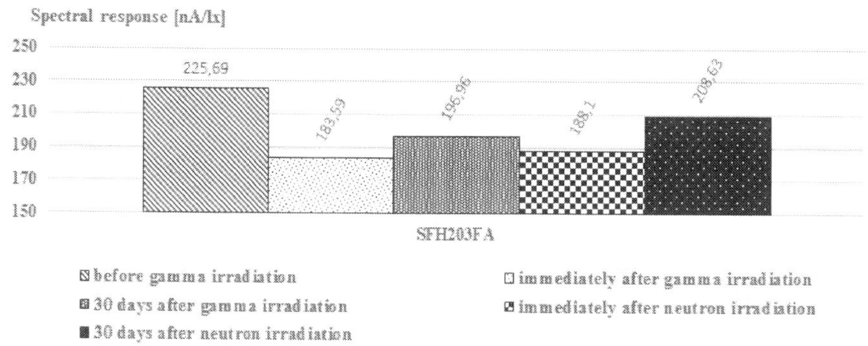

Figure 8. Spectral response of the reverse biased photodiode SFH203FA before and after gamma and neutron irradiation.

As it can be seen from Figures 5 to 10, neutron irradiation, applied 30 days after gamma irradiation, at first was deteriorate response and characteristics of photodetectors. However, after 30 days of recovery, there was a partial improvement of the spectral photodetector response and the increasing of photocurrent. The degree of improvement is different for each type of photodetector.

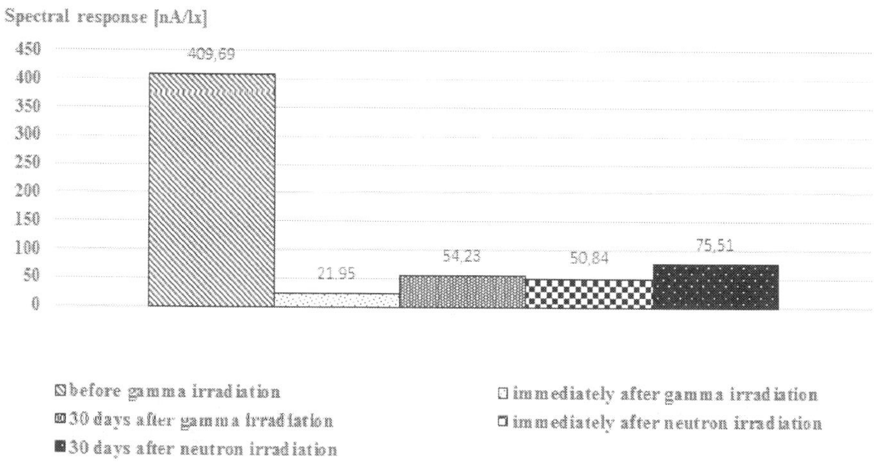

Figure 9. Spectral response of phototransistor BPW40 before and after gamma and neutron irradiation.

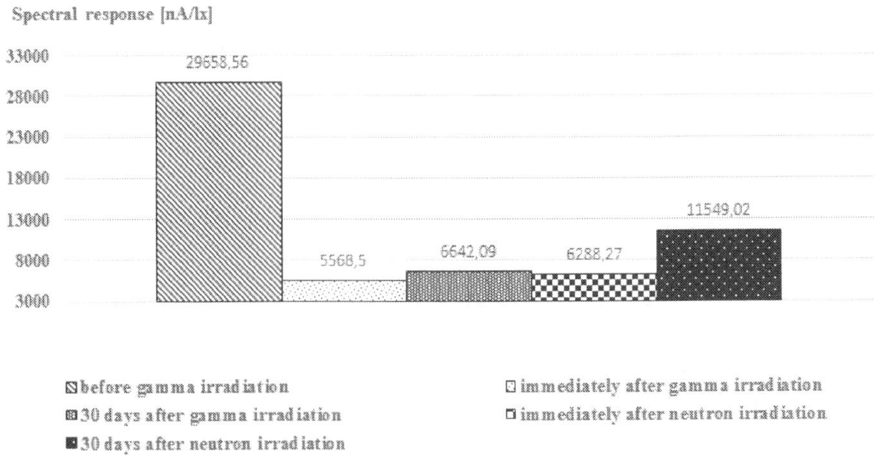

Figure 10. Spectral response of phototransistor LTR4206 before and after gamma and neutron irradiation.

Neutron irradiation, by itself, causes the formation of displacement damage in photodiodes and phototransistors, which leads to the degradation of their electrical characteristics, as is shown in Chapter 3.1 (Figures 2–4). However, if it is applied after the gamma radiation, neutron radiation makes such changes which increasing the efficiency of the recovery process and, as a result, we have improved electrical characteristics. To achieve these effects to be occurred, the concentration of charge carriers must be increased in semiconductor material. Taking previous studies into account [12, 19–21], it can be concluded that the possible cause is tunneling of charge carriers supported by traps and increased generation.

Defects in the material represent traps for the free charge carriers and that can lead to tunneling supported by traps, and this increases the tunneling current at low voltages which are commonly attributed to SILC (*Stress-Induced Leakage Current*) [22–24]. Tunneling supported by traps is a process where particle spend some time trapped in the defect (trap) before tunneling through the barrier (*Trap Asissted Tunneling—TAT*) [20, 21]. This process is caused by inelastic transfer of charge carriers with the help of emission of phonons [21].

Let the electron from the field 1 in **Figure 11** receive enough energy to cross the barrier and came to the area 2. This process undermines the law of conservation of energy for a short period of time determined by Heisenberg's uncertainty principle. Now, if some other electron from the field 2 tunneled in a similar way in a similar time in the area 3, then total number of electrons that are passed from area 1 to area 3 is one. This tunneling is called inelastic tunneling because the excited electron-hole pair occurs, which dissipates after a short time through the interaction carrier-carrier [25].

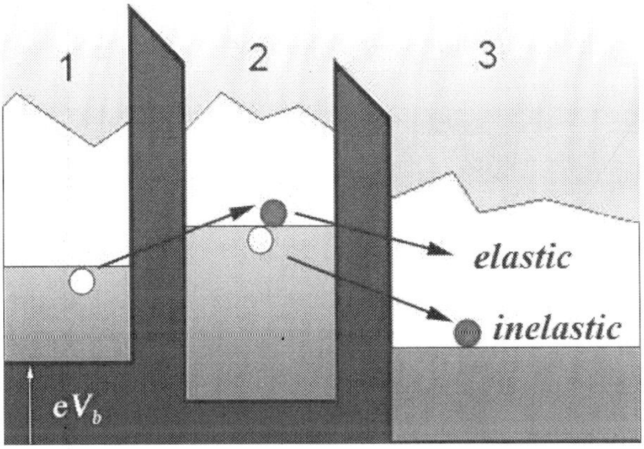

Figure 11. Elastic and inelastic tunnelling through a double barrier [25].

Elastic tunneling is the process of tunneling of same electrons from and in the region 2 with preserving of the phase which is why this process is coherent. Elastic tunneling depends on the internal structure of the area between the barrier and the amount and polarization of the applied voltage. Inelastic tunneling is the dominant process in comparison with the elastic tunneling except in the case of low voltage, low temperature, or low density of quantum dot states [26, 27]. Area 2 is also called a virtual state and simultaneous tunneling through this state, co-tunneling.

Shockley–Read–Hall model describes the process of recombination and the generation of charge carriers in a semiconductor with the help of quantum tunneling mechanism [19, 28, 29]. The transition of an electron from the valence band to the conductive band represents the generation of electron-hole pair, because in the valence band, hole remains in the place of electrons which contribute to the current. The reverse process is recombination. In order to be transferred from the valence band into the conductive band electron must have greater energy than the energy gap. As it may be defects in the crystal structure of the semiconductor caused by impurities (or other causes, eg. radiation), it may appear within the bandgap. Such defects are called traps and they represent energy levels that can trap electrons ejected from the valence band [30]. According to Dharival-Rajvanshi's model, traps can be near the edge of the valence and conduction band (*Tail State*) and near the Fermi level on both sides (*Dangling Bond*). In order to move electron from the valence band into the trap, it requires much less energy than for the transition to the conductive band, so the traps actually facilitate the process of generation of free carriers. The probability that an electron will fall into the trap and spend some time in it depends on the material, the density of defects in the energy gap, the present electric field, temperature, concentration of electrons in the conduction band, and the concentration of holes in the valence band. Schokley–Read–Hall model assumes one level within a gap where electrons or holes can come, which dynamic is quasi-stationary [25, 31, 32].

When gamma radiation and neutron radiation are acting individually on a photodiode there is, as the final result, an increase in the concentration of recombination centers which, according to Schokley-Read formula [33], result in a reduction of minority charge carriers lifetime:

$$\tau = \frac{1}{\langle c_n \rangle \cdot N_t} \frac{n_0 + \delta n + n_1}{n_0 + p_0 + \delta n} + \frac{1}{\langle c_p \rangle \cdot N_t} \frac{n_0 + \delta n + p_1}{n_0 + p_0 + \delta p} \tag{1}$$

where $\tau = \tau_p = \tau_n$ is the life time of electrons and holes and N_t concentration of R-centers (recombination centers which can accept both electrons and holes). The reduction of minority carrier lifetime causes photocurrent decreasing. Previously stated explanation is related to the influence of neutron irradiation on the new, previously non-irradiated photodiodes. However, if we change the initial conditions, i.e. if the photodiode previously has been exposed to gamma radiation, the effects of neutron irradiation will be different. One of the results of gamma radiation are interstitial (PKA), vacancies, and their complexes [34, 35]. Vacancies are also one of the main products of neutron irradiation of the material. When the material, which already contains a number of vacancies, is exposed to the effects of neutron radiation, there is high probability that the defects such as vacancies would be found physically close to each other. When the two vacancies occur next to each other within the grid, they form defective complex called divacancies complex. This complex captures electrons and also can stress the homopolar bonds, which can lead to the termination of the connection. Straining of homopolar connections and its termination can lead to the release of one or two electrons from the defective complex in the conductive band, which results in increased generation.

In some previous studies, increased generation [12] and increased recombination [36, 37] have been observed through the process of electron transfer directly between the defects located close to each other without passing through the conductive belt. This process can be very fast and therefore dominant compared to the Shockley–Read–Hall process. In order to occur the intercenter charge transfer, defects must be physically close to one another. Two irradiation of the same material, such as gamma and neutron, allowing some defects to be close to one another.

The divacancy has three energy levels in the bandgap: a hole trap and two aceptor states. In standard Shockley–Read–Hall theory, current generation in silicon depletion regions is mediated by isolated defect levels in the forbidden bandgap. Generations occurs when a hole is emitted from the defect level into the valence band (i.e. electron captured from it) and an electron is emitted into the conduction band. Each transition occurs with a rate, e_n or e_p, and is governed by the time constant τ_{ne} ili τ_{pe}.[1] If several defect levels exist, they are regarded as the sum of the individual components. In coupled defect generation, illustrated in **Figure 12**, an electron is first captured by the donor state in the bottom half of the bandgap. This is an efficient process with time constant τ_{pe1} being very short hence the fractional occupation of this level is ≈1. The electron can then transfer directly to a higher state in a nearby defect without going

[1] $e_n = 1/\tau_{ne}$ and $e_p = 1/\tau_{pe}$.

via the conduction band. The time constant for this step is denoted $\tau_{1\to2}$. The final transition to the conduction band then occurs as normal with a time constant τ_{ne2}. The enhancement of the generation rate arises because the large transition from the valence band to the above midgap level is mediated by the donor level. This shortens the time taken for the upper state to become filled and hence increases its fractional occupancy [12, 38].

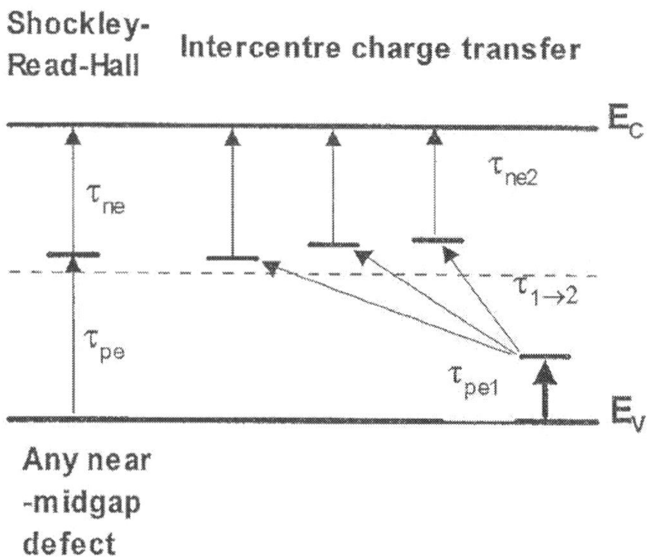

Figure 12. Schematic diagram of Schokley–Read–Hall theory and intercenter charge transfer generation processes [12].

The enhancement of the fractional occupancy increases the number of electrons generated per unit of time from a defect state and hence increases the photocurrent [33].

3.3. Monte Carlo simulation of radiation transfer through photovoltaic detectors

In order to understand the state of the semiconductor after irradiation, Monte Carlo simulations of radiation particles transfer through the material were performed. Monte Carlo simulation gets the answers by simulation of each individual particle and memorizing of certain aspects of their middle behavior. For simulation, FOTELP-2K10 and MCNP programs were used. FOTELP-2K10 is a program that gives the Monte Carlo simulation of the transport of photons, electrons, and positrons [39], while the MCNP (*Monte Carlo N-Particle*) is a general purpose software that can simulate the transport of neutrons, photons, electrons, or a combination of neutron/photon/electron through arbitrary geometric configurations [40].

For this experiment, two Monte Carlo simulation were made, γ-photon transfer through the PIN photodiode and through the phototransistor. The simulations were done with the aim of understanding the processes occurring in the photodiode and phototransistor between gamma and neutron irradiation, i.e. to provide a review process, which is gamma radiation caused in a semiconductor since the final result of these processes represents the initial conditions for neutron irradiation that followed.

3.3.1. *Monte Carlo simulation of gamma photon transport through a pin photodiode*

Figure 13 presents a cross-section of a PIN photodiode used for the simulation.

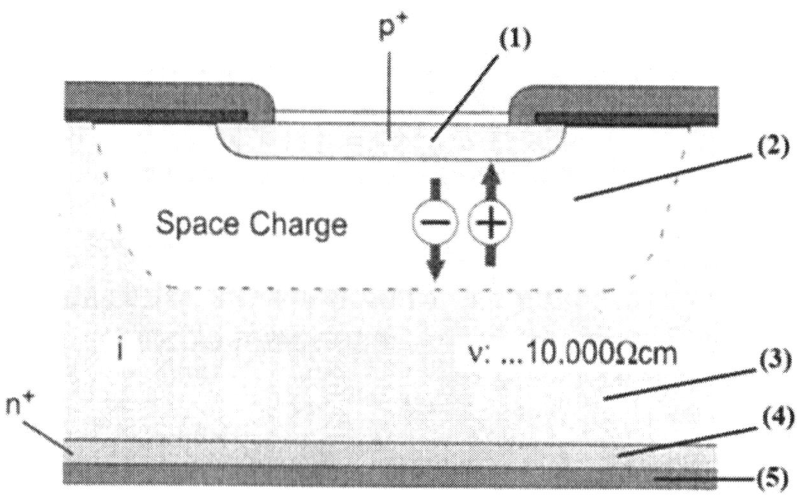

Figure 13. Cross-section of a PIN photodiode [41].

The results of Monte Carlo simulations are shown in **Tables 1–5** and in **Figure 14**. **Table 1** shows the deposed energy per input particle in each zone of photodiode, where the zones are semiconductors area: p^+ (zone 1), p (zone 2), and n^+ (zone 4), and the pure semiconductor (zone 3). **Figure 14** shows the ratio of energy absorbed during each interaction in different layers per depth of semiconductors, i.e. each zone.

Zone	Deposed energy (eV)	Relative error (%)
1	556 .77	0.165
2	257.78	0.255
3	293.31	0.239
4	1386.9	0.112

Table 1. Deposed energy per input particle obtained by Monte Carlo simulation using FOTELP-2K10.

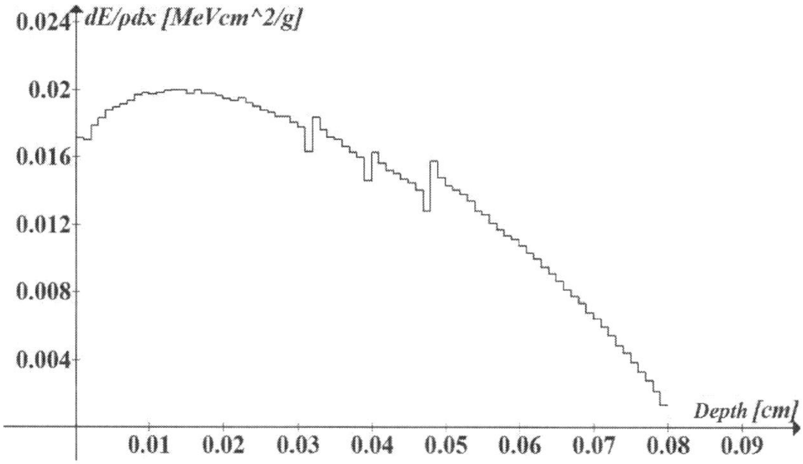

Figure 14. Depth dose distribution in PIN photodiode obtained using FOTELP-2K10.

In order for a lattice atom to be displaced, a minimum amount of energy must be transferred to the target atom. This threshold energy is called the displacement energy E_d (*threshold displacement energy—TDE*) [42, 43]. By using molecular dynamics (*MD*) simulations, Perlado *et al.* [44] predicted TDE values, at 300 K, ranging from 42 to 112 eV for Si. Average TDE values of 93 eV for Si are suggested by El-Azab and Ghoniem from MD simulations [45].

In each zone of photodiode (**Table 1**), and in almost every layer (**Figure 14**) deposed energy per incident particle is high enough to move the atom, i.e. to create vacancy.

Tables 2 and **3** show the probability of creating new photons and electrons per incident particle (photon) through individual interactions.

Probability of creating new photons	Interaction that creates photons
8.4333E-06	Bremsstrahlung
1.0000E-06	Positron-electron annihilation
4.0000E-07	Electron x-rays

Table 2. Probability of creating new photons per incident particle (photon) obtained using MCNP.

Probability of creating new electrons	Interaction that creates electrons
1.3333E-06	Pair production
2.1085E-06	Compton effect
3.3667E-07	Photoelectric effect

Probability of creating new electrons	Interaction that creates electrons
3.3333E-06	Auger photons
4.1667E-06	Auger electrons
5.3281E-07	Electrons knocked out in a collision with impact electrons

Table 3. Probability of creating new electrons per incident particle (photon) obtained using MCNP.

Bremsstrahlung is the interaction that has the highest probability to generate new photons (**Table 2**), while the highest probability for creation have Auger electrons (**Table 3**).

Tables 4 and **5** show the number of physical interactions in which are created or disappeared photons and electrons per input particle (per cell).

Physical interaction	Area 1—p^+	Area 2—p	Area 3—i *(intrinsic)*	Area 4—n^+	Area 5—*Al contact*
From neutrons	0	0	0	0	0
Bremsstrahlung	9.3333E-07	7.3333E-07	6.3333E-07	2.1333E-06	4.0000E-07
Capture of photons	−2.233E-06	−1.433E-06	−1.867E-06	−1.500E-06	−1.133E-06
P-annihilation	6.6667E-08	0	0	6.6667E-08	0
Pair production	−3.333E-08	0	0	−3.333E-08	0
Photonuclear effect	0	0	0	0	0
Electron x-rays	1.4333E-06	9.0000E-07	1.4667E-06	0	9.6667E-07
Total	**1.6667E-07**	**2.0000E-07**	**2.3333E-07**	**6.6667E-07**	**2,3333E-07**

Table 4. Number of physical interactions in which are created or disappeared photons per incident particle (per cell) obtained using MCNP.

Physical interaction	Area 1—p^+	Area 2—p	Area 3—i *(intrinsic)*	Area 4—n^+	Area 5—*Al contact*
Pair production	6.6667E-08	0	0	6.6667E-08	0
Compton recoil	1.7967E-05	9.1000E-06	9.1667E-06	2.3833E-05	7.7333E-06
Photoelectric effect	2.2333E-06	1.4333E-06	1.8667E-06	1.5000E-06	1.1333E-06
Photon Auger	0	3.3333E-08	3.3333E-08	1.3333E-07	0
Electron Auger	4.7667E-05	3.4133E-05	3.7300E-05	0	5.5800E-05
PKA	2.5647E-04	1.8663E-04	2.0543E-04	7.2073E-04	2.1520E-04
p-annihilation	−3,333E-08	0	0	−3.333E-08	0
Total	**3.2437E-04**	**2.3133E-04**	**2.5380E-04**	**7.4623E-04**	**2.7987E-04**

Table 5. Number of physical interactions in which are created or disappeared electrons per incident particle (per cell) obtained using MCNP.

Simulation results show that the number of interactions that result in a vacancy i.e. PKA (*primary knock-on atom*) are 10 to 1000 times higher than all other possible types of interaction (**Table 5**—shaded part). Among the total number of electrons caused by gamma radiation in all areas of photodiode, 78–80% is produced by PKA (in area 4 even up to 97%). This is an unequivocal sign that the gamma radiation caused a very large number of vacancies.

In order for neutron irradiation of photodiodes (applied after gamma radiation) to cause intercenter charge transfer and tunneling supported by traps, as already mentioned, it is necessary for neutron radiation to form defects in a semiconductor (vacancies), which are close to each other, and to create a sufficient number of divacancies. As a relatively heavy and uncharged particles, neutrons, in a collision with the atoms of the crystal lattice, lead to the displacement of entire atoms from the lattice. This naturally causes the breaking and destruction of the local lattice structure by displacing atoms and creating vacancies. Displaced atom is called interstitial because it takes place in the space between knots, and a pair of interstitial atom and vacancy is called Frenkel defect. If the energy of incident neutron is high enough, it can give sufficient energy to displaced atom, which can displace other atoms in the lattice. In the case of high-energy incident neutrons, this process has a cascade (avalanche) character. This requires quick energy neutrons from 10 keV to 10 MeV. At the end, all displaced atoms lose their excess energy and the heat balance in the grid established. Some of the atoms return to vacancies and reconstruct the structure of the local grid. Some of these atoms come together with dopants or impurity atoms and form stable electrically inactive defects, which do not contain recombination centers and trap. On the other hand, moving vacancies associate with impurity atoms, vacancies, and other donors forming temperature stable defects (complex defects) that represent recombination centers and trap centers. Since the mean energy of neutrons from a source in the experiment was 5.5 MeV, it follows that neutrons have sufficient energy to cause a cascading process of creating vacancies. Previously, gamma irradiation created a large number of vacancies, increasing the probability for vacancies, created by neutron irradiation, to be physically close to the preformed vacancy. Divacancies, formed like this, facilitate intercenter charge transfer supported by traps and provide increased generation of charge carriers and this, as already mentioned in Section 3.1., leads to partial reparation of semiconductor structure and increase the spectral response and the photocurrent of the photodiode.

3.3.2. *Monte Carlo simulation of gamma photon transport through a phototransistor*

Figure 15 presents a cross-section of a phototransistor used for the simulation.

The results of Monte Carlo simulations are given in **Tables 6–8**. **Tables 6** and **7** show the probability of creating new photons and electrons per incident particle (photon) through individual interactions.

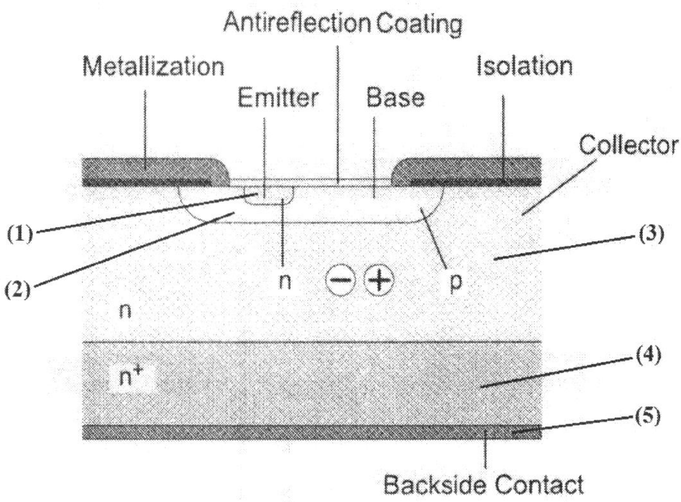

Figure 15. Cross-section of a phototransistor [41].

Probability of creating new photons	Interaction that creates photons
1.3333E-05	Bremsstrahlung
1.1333E-06	Positron-electron annihilation
5.1667E-06	Electron x-rays

Table 6. Probability of creating new photons per incident particle (photon).

Probability of creating new electrons	Interaction that creates electrons
1.4667E-06	Pair production
2.1764E-03	Compton effect
1.1633E-05	Photoelectric effect
2.3333E-07	Auger photons
1.7910E-04	Auger electrons
7.0135E-03	Electrons knocked out in a collision with impact electrons

Table 7. Probability of creating new electrons per incident particle (photon).

According to the simulation results in any semiconductor field within the phototransistor, there was no interaction in which are created or disappeared photons. **Table 8** show the number of physical interactions in which are created or disappeared electrons per input particle (per cell).

Physical interaction	Area 1—*emitter n*	Area 2—*base p*	Area 3—*colector n*	Area 4—n^+	Area 5—*Al contact*
Pair production	0	0	0	0	0
Compton recoil	0	3.3333E-08	1.6667E-07	1.6667E-07	0
Photoelectric effect	0	0	0	0	0
Photon Auger	0	0	0	0	0
Electron Auger	0	3.3333E-08	0	0	0
PKA	0	0	4.3333E-07	7.0000E-07	6.6667E-08
p-annihilation	0	0	0	0	0
Total	**0**	**6.6667E-08**	**6.0000E-07**	**8.6667E-07**	**6.6667E-08**

Table 8. Number of physical interactions in which are created or disappeared electrons per incident particle (per cell).

In phototransistor, as in the photodiode, the largest number of integration caused by gamma radiation are vacancies (PKA) (**Table 8**—shaded part). When the semiconductor material, with structure like this, is exposed to neutron radiation, due to the nature of radiation, a number of new vacancies will be created together with those previously formed. The final result of both types of radiation action are divacancies. As already mentioned in Section 3.2, divacancies cause the increased generation of charge carriers through the two dominant effects [19]:

1. divacancies strain homopolar bond and break it, causing a release of one or two electrons;
2. divacancies allow direct transfer of electrons between the defects located close to each other without passing through the conduction band (intercentre charge transfer) [12].

According to the results in **Table 8** the largest number of divacancies have creating in collector and the n^+ area of phototransistor, increasing the concentration of electrons in these areas. On the other hand, due to Compton scattering and Auger electron and increases the concentration of electrons in the base. The final result of these effects is increasing the transistors photocurrent after neutron irradiation (compared to its value after gamma irradiation), which is consistent with the results of the experiment presented in chapter 3.2.

4. Conclusion

Gamma and neutron radiation, applied individually, affect the semiconductor material creating defects and changing the existing structure, which results in a change in the output characteristics of the device and reducing their functionality.

Gamma irradiation of silicon semiconductor causing numerous defects of the crystal lattice. Monte Carlo simulations showed that in this experiment were represented almost all of the effects described in the literature: displacement of atoms (PKA), Auger electrons, Compton scattering, photoelectric effect, pair production. The impact of all these effects are manifested in the generation of energy levels in the energy gap of crystal lattice which decreases the the minority charge carriers lifetime resulting in a decrease in the photocurrent and spectral response. The big change of phototransistors output characteristics can be explained by the influence of radiation on the current gain. The current gain is proportional to the minority charge carriers lifetime so the degradation of their lifetime directly affects the degradation of current gain. This degradation is caused by a displacement of atoms in the semiconductor bulk which affects the increase in the number of recombination centers and also oxidation of the oxide pasivisation layer especially over the emitter-base junction.

Neutron irradiation causes damage in the photovoltaic detector which is primarily related to the displacement of silicon atoms from their positions in a grid and creating vacancies. Together with the vacancies, other effects appeared. Monte Carlo simulations showed that after the vacancies, the most frequent are Auger electrons, Compton scattering, pair production, and the photoelectric effect. Because of the combination of complex defects, defects that act as recombination centers are created and reduce the minority charge carriers' lifetime which can lead to the degradation of electrical parameters of photovoltaic detectors.

When the semiconductor photovoltaic detectors are first exposed to gamma radiation and after a month to neutron, one can see that neutron radiation, applied after gamma radiation, partially corrects the characteristics of semiconductor devices which are exacerbated by gamma radiation, and that is manifested through increased spectral response and output photocurrent. This behavior of photodiodes and phototransistors can be explained by the increased generation of charge carriers as a result of direct transfer (tunneling) of the charge through the traps (recombination centers). Direct (intercenter) charge transfer is a process where charge carriers spend some time trapped in the defect of material (traps) before tunneling through the barrier. To become free (transferred from the valence to the conductive band), an electron must have enough energy to overcome the energy gap. However, if the traps, that represent energy levels, are located near the edge of the conduction and valence band and near the Fermi level on both sides (according to Dharival-Rajvanshi model), then moving electrons from the valence band into the trap require notably less energy than for direct transit to the conductive band, which means that the traps actually facilitate the process of generation of free carriers. Also, according to the Shockley–Read–Hall model, there is one quasi-stationary energy level within the gap where the electron or hole could come. The probability that an electron will fall into the trap and spend some time in it depends, among other causes, on the density of defects in the energy gap. One of the ways to increase the density of defects in the energy gap is creating a large number of vacancies located physically close to each other in semiconductor material. Monte Carlo simulation of γ-photons transfer through the photovoltaic detectors showed that gamma radiation leaves behind itself a number of displaced atoms (vacancies). Since the radiation damage caused by neutrons primarily related to the displacement of atoms from their positions in the lattice of silicon semiconductor, i.e. forming of vacancies, so neutron irradiation of photovoltaic detectors applied after gamma irradiation gives a possibility for the creation of a sufficient number of divacancies which can

http://dx.doi.org/10.5772/62756

cause intercenter transfer and increased generation of charges and thereby increasing the photocurrent and other parameters. The requirement for creation of divacancies by neutron irradiation is the existence of vacancies in a semiconductor caused by previous gamma radiation.

Author details

Dejan Nikolić[1*] and Aleksandra Vasić-Milovanović[2]

*Address all correspondence to: nikolcorp@gmail.com

1 Brcko District Government, Brcko, Bosnia and Herzegovina

2 Faculty of Mechanical Engineering, University of Belgrade, Belgrade, Serbia

References

[1] Stanković K., Vujisić M., Kovačević D., Osmokrović P. Statistical analysis of the characteristics of the some basic mass-produced passive electrical circuits used in measurements. Measurement. 2011;44(9):1713–1722.

[2] Vujisić, M., Stanković, K., Osmokrović, P. A statistical analysis of measurement results obtained from nonlinear physical laws. Applied Mathematical Modeling. 2011;35(7): 3128–3135.

[3] G. Barbottin and A. Vapaille. Instabilities in Silicon Devices, New Insulators, Devices and Radiation Effects. Elsevier; 1999.

[4] B. Simić, D. Nikolić, K. Stanković, Lj. Timotijević, S. Stanković. Damage induced by neutron radiation on output characteristics of solar cells, photodiodes and phototransistors. International Journal of Photoenergy. 2013;2013:6. ID 582819.

[5] A. Holmes-Siedle and L. Adams. Handbook of Radiation Effects. Oxford University Press; 2004.

[6] G.C. Messenger and M.S. Ash. The Effects of Radiation on Electronic Systems, 2nd edition. Van Nostrand Reinhold Company: New York; 1992.

[7] C.L. Axness, B. Kerr, E.R. Keiter. Analytic 1-D pn junction diode photocurrent solutions following ionizing radiation and including time-dependant changes in the carrier lifetime from a nonconcurrent neutron pulse. IEEE Transactions on Nuclear Science. 2010; 57(6):3314–3321.

[8] D.G. Sporea, R.A. Sporea, C. Oproiu, I. Vat®. Comparative study of gamma-ray, neutron and electron beam irradiated index-guided laser diodes. in Proceedings of the

8th European Conference on Radiation and Its Effects on Components and Systems (RADECS '05). 2005; PA31–PA39.

[9] G.E. Schwarze, A.J. Frasca. Neutron and gamma irradiation effects on power semiconductor switches. In Proceedings of the 25th Intersociety Energy Conversion Engineering Conference (IECEC '90). 1990; 6: 30–35.

[10] J.R. Srour, C.J. Marshall, P.W. Marshall. Review of displacement damage effects in silicon devices. IEEE Transactions on Nuclear Science. 2003; 50(3): 653–670.

[11] M.B. El Mashade, M. Ashry, Sh.M. Eladl, M.S. Rageh. Experimental measurements of some optoelectronic devices before and after gamma irradiation. Journal of Microwaves and Optoelectronic. 2004; 3(4): 1–12.

[12] S.J. Watts, J. Matheson, I.H. Hopkins-Bond, A. Holmes-Siedle, A. Mohammadzadeh, R. Pace. A new model for generation-recombination in silicon depletion regions after neutron irradiation. IEEE Transaction on Nuclear Science. 1996; 43(6):2587-2594.

[13] Z.D. Kovalyuk, V.N. Katerynchuk, O.A. Politanska, O.N. Sydor, V.V. Khomyak. Effect of gamma radiation on the properties of InSe photodiodes. Technical Physics Letters. 2005; 31(5):359–360.

[14] M. Vujisić. Simulated effects of proton and ion beam irradiation on titanium dioxide memristors. IEEE Transactions on Nuclear Science. 2010; 57(4):1798–1804.

[15] M. Vujisić, K. Stanković, E. Dolićanin, P. Osmokrović. Radiation hardness of COTS EPROMs and EEPROMs. Radiation Effects and Defects in Solids: Incorporating Plasma Science and Plasma Technology. 2010; 165(5):362–369.

[16] D.R. Alexander. Transient ionizing radiation effects in devices and circuits. IEEE Transaction on Nuclear Sciences. 2003; 50(3):565–582.

[17] D. Nikolić, A. Vasić-Milovanović, M. Obrenović, E. Dolićanin. Effects of successive gamma and neutron irradiation on solar cells. Journal of Optoelectronics and Advanced Materials. 2015; 17(3–4): 351–356.

[18] D. Nikolić, A. Vasić, Đ. Lazarević, M. Obrenović. Improvement possibilities of the I-V characteristics of PIN photodiodes damaged by gamma irradiation. Nuclear Technology & Radiation Protection. 2013; 28(1): 84–91.

[19] G.A.M. Hurkx, D.B.M. Klaassen, M.P.G. Knuvers. A new recombination model for device simulation including tunnelling. IEEE Transactions on Electron Devices. 1992; 39:331–338.

[20] S. Palpacelli, M. Mendoza, H. J. Herrmann, S. Succi. Klein tunneling in the presence of random impurities. International Journal of Modern Physics C. 2012; 23:12.

[21] T.K. Kang, M.J. Chen, C.H. Liu, Y.J. Chang, S.K. Fan. Numerical confirmation of inelastic trap-assisted tunneling (ITAT) as SILC mechanism. IEEE Transactions on Electron Devices. 2001; 48:2317–2322.

[22] S. Aritome, R. Shirota, G. Hemink, T. Endoh, F. Masuoka. Reliability issues of flash memory cells. Proceedings of the IEEE. 1993; 81:776–788.

[23] E. Rosenbaum, L. F. Register. Mechanism of stress-induced leakage current in MOS capacitors. IEEE Transaction on Electron Devices. 1997; 44:317–323.

[24] C.D. Young, G. Bersuker, M. Jo, K. Matthews, J. Huang, S. Deora, K. Ang, T. Ngai, C. Hobbs, P.D. Kirsch, A. Padovani, L. Larcher. New insights into SILC-based life time extraction. IEEE International Reliability Physics Symposium (IRPS). 2012.

[25] M. Razavy. Quantum Theory of Tunneling. World Scientific Publishing Co. Pte. Ltd: Singapore; 2003.

[26] R. T. Collins, J. Lambe, T. C. McGill. Elastic and inelastic tunneling characteristics of AlAs/GaAs heterojunctions . Journal of Vacuum Science and Technology B. 1984; 2.

[27] D.V. Averin, Y.V. Nazarov. Virtual electron diffusion during quantum tunneling of the electric charge. Physical Review Letters. 1990; 65.

[28] L. Pelaz, J.L. Orantes, J. Vincente, L.A. Bailon, J. Barbolla. The Poole-Frenkel effect in 6H-SiC diode characteristics. IEEE Transactions on Electron Devices. 1994; 41:587–591.

[29] T. Goudon, V. Miljanović, C. Schmeiser. On the Shockley-Read-Hall Model: Generation-Recombination in Semiconductors. 2007 .

[30] S.M. Sze, K.K. Ng. Physics of Semiconductor Devices. Wiley Interscience: New Jersey. 2007.

[31] R.K. Srivastava. Quantum Mechanics. Prentice-Hall of India Private Limited: India; 2007.

[32] M. Vukadinović, F. Smole, M. Topić, R.E.I. Schropp, F.A. Rubinelli. Transport in tunneling recombination junctions: a combined computer simulation study. Journal of Applied Physics. 2004; 96:7289–7299.

[33] W. Shockley, W.T. Read. Statistics of the recombination of holes and electrons. Physical Review. 1952; 87(5):835–842.

[34] D. Nikolić, K. Stanković, Lj. Timotijević, Z. Rajović, M. Vujisić. Comparative study of gamma radiation effects on solar cells, photodiodes and phototransistors. International Journal of Photoenergy. 2013; 2013:6; ID 843174.

[35] D. Nikolić, A. Vasić, I. Fetahović, K. Stanković, P. Osmokrović. Photodiode behavior in radiation environment. Scientific Publications of the State University of Novi Pazar Series A. 2011; 3(1): 27–34.

[36] W.M. Chen, B. Monemar, E. Janzén, J.L. Lindström. Direct observation of intercenter charge transfer in dominant nonradiative recombination channels in silicon. Physical Reviews Letters. 1991; 67(14):1914–1917.

[37] A.M. Frens, M.T. Bennebroek, A. Zakrzewski, J. Schmidt, W.M. Chen, E. Janzén, J.L. Lindström, B. Monemar. Observation of rapid direct transfer between deep defects in silicon. Physical Reviews Letters. 1994; 72(18):2939–2942.

[38] J. Racko, M. Mikolášek, P. Benko, O. Gallo, L. Harmatha, R. Granzner, F. Schwierz. Coupled defect level recombination in the P-N Junction. Journal of Electrical Engineering. 2011; 62(6):355–358.

[39] Radovan D. Ilić. FOTELP-2K10, Photons, electrons and positrons Monte Karlo transport simulation. Institute of Nuclear Sciences Vinča: Belgrade; 2009.

[40] X-5 Monte Carlo Team. MCNP—a general Monte Carlo N-Particle Transport Cod. Los Alamos National Laboratory. Version 5, 2008.

[41] Physics and Technology. Vishay Semiconductors, www.vishay.com. 2008.

[42] R. Smith (ed.). Atomic & Ion Collisions in Solids and at Surfaces: Theory, Simulation and Applications. Cambridge University Press: Cambridge, UK; 1997.

[43] M. Nastasi, J. Mayer, and J. Hirvonen. Ion-Solid Interactions—Fundamentals and Applications. Cambridge University Press: Cambridge, Great Britain; 1996.

[44] J.M. Perlado, L. Malerba, A. Sánchez-Rubio, T. Díaz de la Rubiab. Analysis of displacement cascades and threshold displacement energies in beta-SiC. Journal of Nuclear Materials. 2000; 276:235–242.

[45] A. El-Azab, N. Ghoniem. Molecular dynamics study of the displacement threshold surfaces and the stability of frenkel pairs in beta-sic. Journal of Nuclear Materials. 1992; 191:1110–1114.

Chapter 4

Electron Beam Irradiation Effects on Dielectric Parameters of SiR–EPDM Blends

R. Deepalaxmi, V. Rajini and C. Vaithilingam

Additional information is available at the end of the chapter

http://dx.doi.org/10.5772/62624

Abstract

The survival of an electrical system is mostly governed by the endurance limit of the dielectric material employed in it. The five different compositions of SiR–EPDM blends were prepared. Electron beam radiation has been widely used in the cable manufacturing industries in order to increase the life of the cable. Hence, the five blends were irradiated to 5, 15 and 25 Mrad dose levels by electron beam accelerator. The dielectric parameters such as breakdown voltage (BDV), dielectric strength (DS), dielectric constant (DC), and dissipation factor (DF) were measured as per ASTM/IEC standards. This chapter evaluates the effect of electron beam irradiation on dielectric parameters of SiR–EPDM blends.

Keywords: silicone rubber (SiR), ethylene propylene diene monomer (EPDM), breakdown voltage (BDV), dielectric strength (DS), dielectric constant (DC), dissipation factor (DF), electron beam irradiation

1. Introduction

Electron beam irradiation has been effectively utilized in power cable industry and identified as one of the most advanced processing techniques. The products processed with electron beam radiation, experience shorter exposure time, which could result in less oxidative effects on certain materials. It is essential to investigate the effect of electron beam irradiation upon the dielectric parameters of the five different compositions of SiR–EPDM blends [1–4, 9]. Hence, the samples of SiR–EPDM were irradiated to 5, 15 and 25 Mrad dose levels by electron beam accelerator. The new functional groups formed during blending and after the electron beam irradiation were investigated through physicochemical investigation techniques like Fourier transform infrared spectroscopy (FTIR). To observe the morphological changes and also to identify the elemental composition, scanning electron microscope (SEM) analysis and energy dispersive X-ray analysis (EDXA) were performed on SiR–EPDM blends. The effects of electron beam irradiation on the dielectric parameters of various compositions of SiR–EPDM blends were reported.

2. Experimental

2.1. Preparation of SiR–EPDM blends

Commercially, available SiR and EPDM were used. Type of SiR-polydimethyl siloxane (PDMS). The composition of EPDM is ethylene-65%; propylene-25%; diene monomer-10%. Diene type is ethylidene norbornene (ENB). They are supplied by M/S Joy Rubbers, India. The five different compositions of SiR–EPDM blends were prepared [1–4, 9].

1. Blend A—SiR 90%/EPDM 10%.
2. Blend B—SiR 70%/EPDM 30%.
3. Blend C—SiR 50%/EPDM 50%.
4. Blend D—SiR 30%/EPDM 70%.
5. Blend E—SiR 10%/EPDM 90%.

2.2. Electron beam irradiation

The five different compositions of SiR–EPDM blends were irradiated up to 5, 15 and 25 Mrad doses using an electron beam accelerator of 1.5 MeV rating at M/S Siechem Industries, Pondicherry, India.

3. Characterization of SiR–EPDM blends

3.1. Dielectric characterization

In order to analyze the dielectric behavior of SiR–EPDM blends in harmful environments, blends have been tested as per ASTM/IEC standards.

3.1.1. Breakdown voltage and dielectric strength

As per standard ASTM D 149 (IEC 60243), BDV and DS were measured. The sample dimensions of 5 × 5 × 0.3 cm were placed between two electrodes, and the voltage was increased at a fixed rate of 500 V/s. The voltage at which dielectric breakdown occurs was measured as BDV. DS was calculated.

3.1.2. Dielectric constant and dissipation factor

DC and DF were measured as per ASTM D 150 (IEC 60250) at 1 MHz. The sample dimensions were 5 × 5 × 0.3 cm.

3.2. Physicochemical investigations

Various physicochemical techniques such as FTIR, EDXA and SEM were used to identify the nature of changes in the electron beam irradiated samples of SiR–EPDM blends.

3.2.1. Fourier transform infrared spectroscopy (FTIR) analysis

FTIR spectra of electron beam irradiated samples were taken using Perkin Elmer spectrophotometer, in the wave number ranging from 500 cm^{-1} to 4000 cm^{-1}. The number of scans for each IR spectrum was 4.

3.2.2. Energy dispersive X-ray (EDXA) analysis

EDXA analysis has been performed using EDXA analysis setup (Make HITACHI), in order to determine the elemental composition of the materials at the surface of the electron beam irradiated samples of SiR–EPDM blends.

3.2.3. Scanning electron microscopy (SEM) analysis

SEM analysis was performed using a scanning electron microscope (Make HITACHI) with a magnification of 5–300,000, in order to study the morphology of the surface of electron beam irradiated samples of SiR–EPDM blends.

4. Results

4.1. Dielectric characterization of virgin SiR–EPDM blends

The virgin SiR rich blends (A and B) have higher breakdown voltage (BDV) and dielectric strength (DS), when compared to remaining blends. This may be due to the occurrence of maximum self cross-linking during the blending process itself. During the blending process, the cross-linking reaction has taken place between the side chains of SiR and EPDM. The blend C and EPDM rich blends (D and E) were found to have lesser values of BDV, DS, and higher values of DC and DF in comparison with SiR rich blends (A and B).

4.2. Effect of electron beam irradiation on dielectric behavior of SiR–EPDM blends

4.2.1. Effect on breakdown voltage and dielectric strength

Figures 1 and **2** depict the variations in breakdown voltage and dielectric strength of five different compositions of SiR–EPDM for various doses of electron beam irradiation.

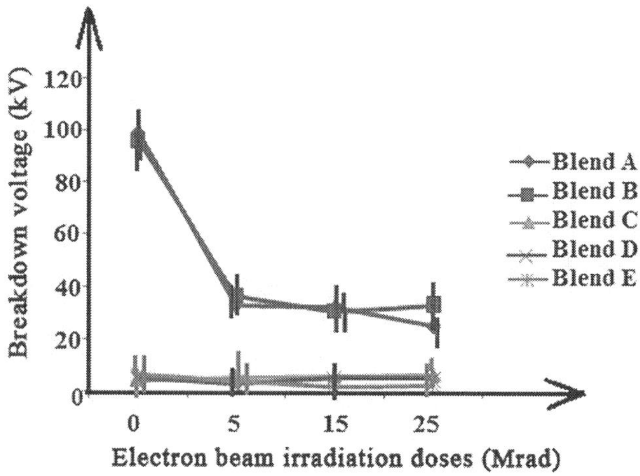

Figure 1. Variations in breakdown voltage of SiR–EPDM blends for various doses of electron beam irradiation.

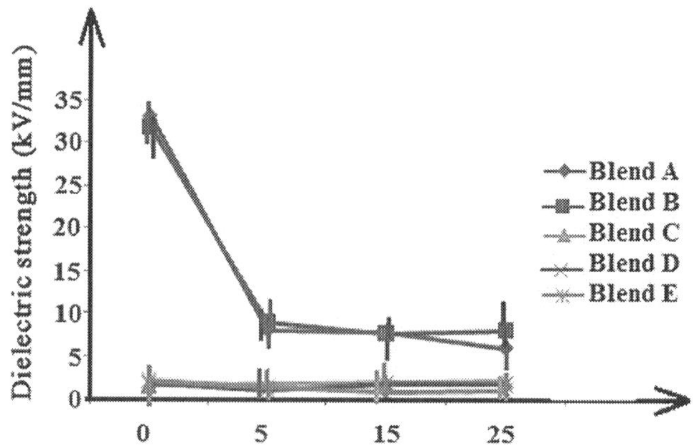

Figure 2. Variations in dielectric strength of SiR–EPDM blends for various doses of electron beam irradiation.

The BDV and DS of SiR-rich blends (A and B) and EPDM-rich blends (D and E) reduced for all doses of electron beam. The BDV and DS of the blend C improved for all doses of electron beam. The DC of the blend D and E has been improved at 5 and 5/25 Mrad respectively.

4.2.2. Effect on dielectric constant and dissipation factor measurement

Figures 3 and **4** depict the variations in dielectric constant and dissipation factor of five different compositions of SiR–EPDM blends for various doses of electron beam irradiation.

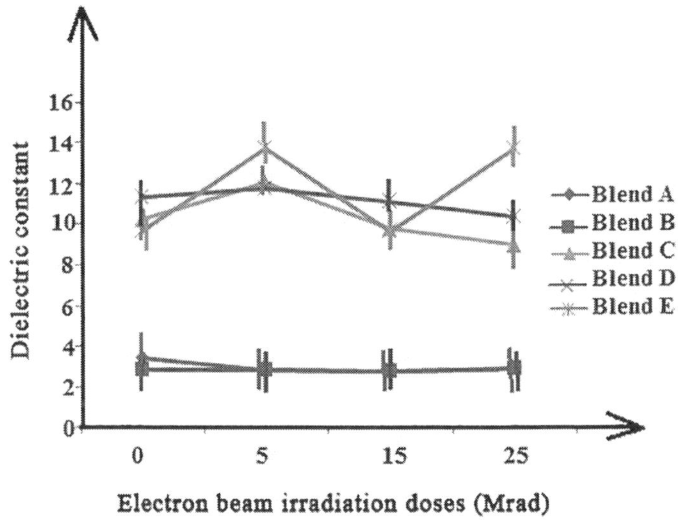

Figure 3. Variations in dielectric constant of SiR–EPDM blends for various doses of electron beam irradiation.

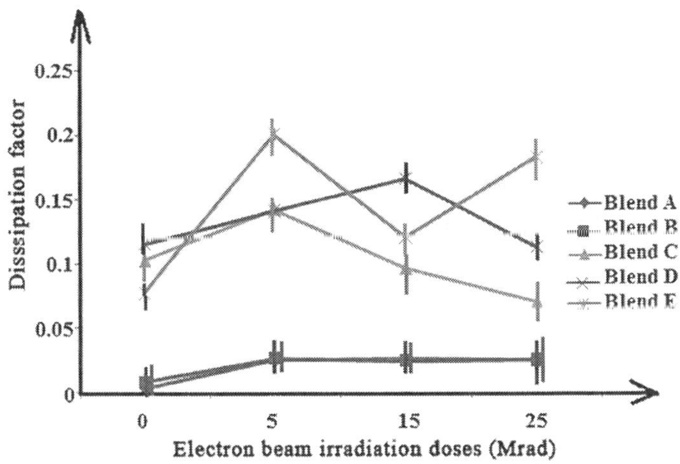

Figure 4. Variations in dissipation factor of SiR–EPDM blends for various doses of electron beam irradiation.

The DC of the blend A reduced for all doses of electron beam. The DC of blend B reduced at 5 and 15 Mrad. The DF of SiR-rich blends (A and B) reduced for all doses of electron beam. The DC and DF of the blend C improved at 5 and 15 Mrad respectively. The blend D has the improved DF at 25 Mrad.

5. Discussion

5.1. Dielectric performance of virgin and electron beam irradiated SiR–EPDM blends

For SiR–EPDM blends, it has been observed that cross-linking and chain scission may modify the macromolecular chains of the material. The consequence is the change in the dielectric parameters of the material. The effect of dominant mechanism can be noted from the changes in dielectric parameters.

5.2. FTIR analysis

FTIR spectra of electron beam irradiated samples of SiR–EPDM blends were obtained to identify the mechanism for the change in dielectric parameters after the electron beam irradiation. FTIR spectra of the virgin and electron beam irradiated samples of three compositions of SiR–EPDM blends are depicted in **Figures 5–7** respectively.

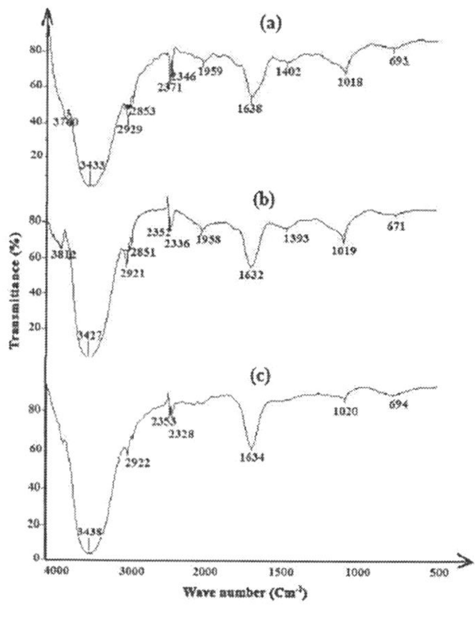

Figure 5. FTIR spectra of electron beam irradiated samples of blend A. (a) 5 Mrad, (b) 15 Mrad and (c) 25 Mrad.

The FTIR investigations on electron beam radiated samples revealed that the radiation has induced the chemical and morphological changes. The variation in dielectric parameters was validated through FTIR spectra. It depicts the occurrence of new functional groups along with the % absorbance and the corresponding wave number. **Tables 1** and **2** list the correlation of variation in dielectric parameters of electron beam irradiated samples of SiR rich blends and EPDM rich blends and blend C using FTIR respectively.

http://dx.doi.org/10.5772/62624

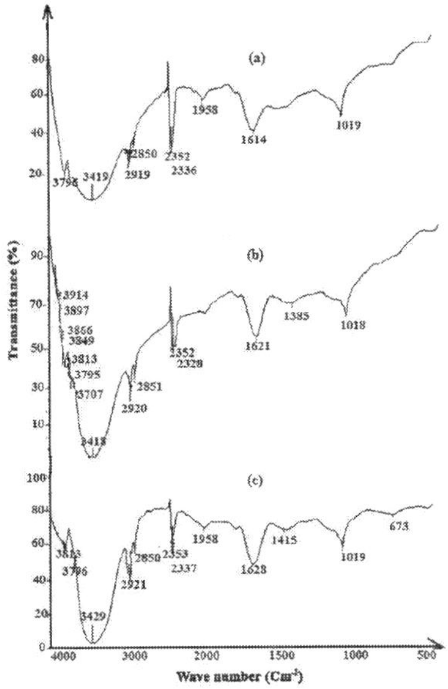

Figure 6. FTIR spectra of electron beam irradiated samples of blend C. (a) 5 Mrad, (b) 15 Mrad and (c) 25 Mrad.

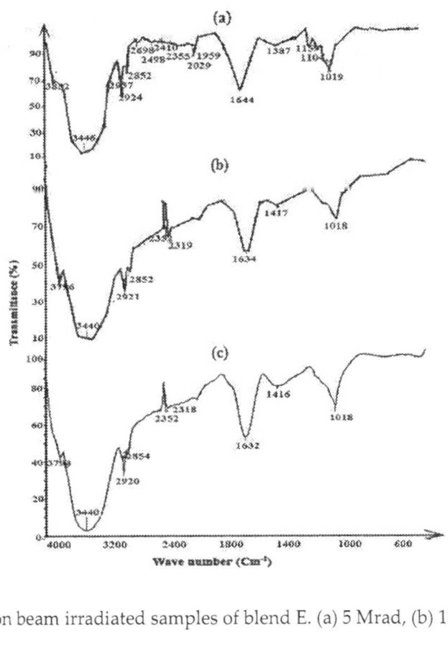

Figure 7. FTIR spectra of electron beam irradiated samples of blend E. (a) 5 Mrad, (b) 15 Mrad and (c) 25 Mrad.

Behavior of dielectric parameter/doses	5 Mrad	15 Mrad	25 Mrad
B → Improvement in DC at 25 Mrad	Alcohol(OH) bonded, strong, broad A → 3433 cm^{-1} with 150% B → 3427 cm^{-1} with 116% Si–O–Si A →1018 cm^{-1} with 18% B →1018 cm^{-1} with 10%	Alcohol(OH) bonded, strong, broad A → 3427 cm^{-1} with 150% B → 3446 cm^{-1} with 150% Si–o–Si A → 1019 cm^{-1} with 15% B → 1018 cm^{-1} with 2%	Alcohol(OH) bonded, strong, broad A → 3438 cm^{-1} with 150% B → 3429 cm^{-1} with 150% Si–O–Si A → 1020 cm^{-1} with 4% B → 1018 cm^{-1} with 49%
Reduction in BDV, DS and DF A → Reduction in DC	Absence of alcohol (OH)-free, strong, sharp group and acid (COOH) group		

Table 1. Correlation of variation of dielectric parameters of electron beam irradiated samples of SiR rich blends using FTIR.

The BDV, DS, and DF of the SiR rich blends (A and B) found to reduce for all doses of electron beam irradiation. This is due to the disappearance of acid (COOH) group in them. The BDV and DS of the blend C is improved for all doses of electron beam irradiation. This is due to the appearance of Si–O–Si group at 1019, 1018, and 1019 cm^{-1} with 29, 20, and 23% absorbance in it. The dielectric constant is improved at 5 Mrad. This may be due to the appearance of =C–H (Alkene, bending, strong). The DF has been reduced at 15 Mrad. This may be due to the disappearance of =C–H (Alkene, bending, strong). The maximum improvement in DC of blend D occurred at 5 Mrad. This may be due to the increase in Si–O–Si group at 1018 cm^{-1} with 34% absorbance and also due to the shifting of alcohol (OH)-free group to higher wave number [16–18]. The maximum improvement in DC of blend E has occurred at 25 Mrad. This may be due to the increase in alcohol (OH)-free group at 3795 cm^{-1} with 138% absorbance.

Behavior of dielectric parameter/ doses	5 Mrad	15 Mrad	25 Mrad
C (50:50) Improvement in BDV, DS Improvement in DF except at 15 Mrad D → 3445 cm^{-1} Improvement in DC at 5 Mrad	Absence of =C–H (Alkene), bending strong	Improvement in	=C–H (Alkene), bending strong at 673 cm^{-1} with 11% Si–O–Si group at 1019 cm^{-1} with 23% (C–H) alkane group at 1416 cm^{-1} with 16%
EPDM rich blends (D and E)	Absence of =C–H	Alcohol(OH) free,	Alcohol(OH) bonded, strong,

http://dx.doi.org/10.5772/62624

Behavior of dielectric parameter/ doses	5 Mrad	15 Mrad	25 Mrad
D →	(Alkene), bending	strong, sharp	broad with
DC except at 15/25	strong	D → 3813 cm^{-1}	150% E → 3440 cm^{-1} with 150%
Mrad Improvement in	Increase in	with 15%	
DF except at 5/15 Mrad	Si–O–Si	E → 3796 cm^{-1}	
E → Improvement in	group at 1018	with 45%	
DC except at 15 Mrad	cm^{-1} with 34%	Alcohol(OH) bonded,	
	absorbance	strong, broad	
	*Shifting of	D → 3434 cm^{-1}	
	alcohol (O–H)	with 150%	
	free group to	E → 3440 cm^{-1}	
	higher wave	with 150%	
	number		

Table 2. Correlation of variation of dielectric parameters of electron beam irradiated samples of EPDM rich blends and blend C using FTIR.

5.3. EDXA analysis

Figures 8–10 show the EDXA curves of the electron beam irradiated samples of blends A, C, and E respectively. The inferences from EDXA curves of all the irradiated samples of SiR–EPDM blends are listed in **Tables 3**and **4**.

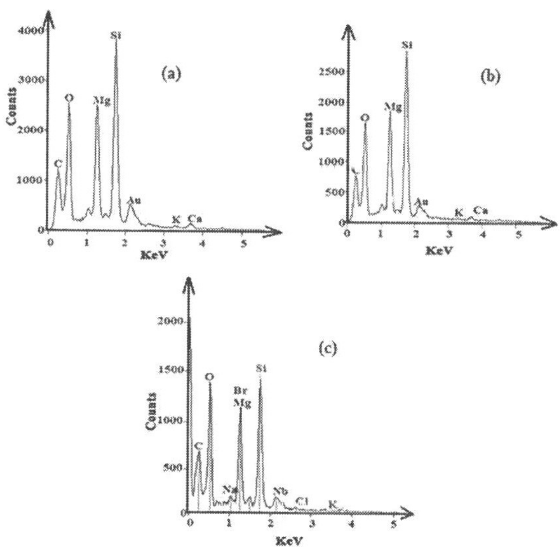

Figure 8. EDXA curves of electron beam irradiated samples of blend A. (a) 5 Mrad, (b) 15 Mrad and (c) 25 Mrad.

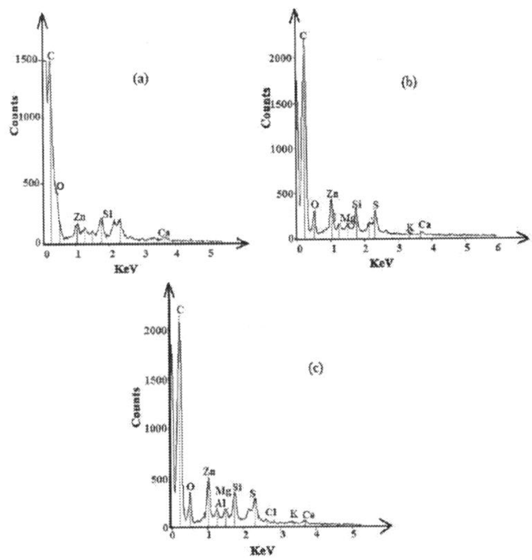

Figure 9. EDXA curves of electron beam irradiated samples of blend C. (a) 5 Mrad, (b) 15 Mrad and (c) 25 Mrad.

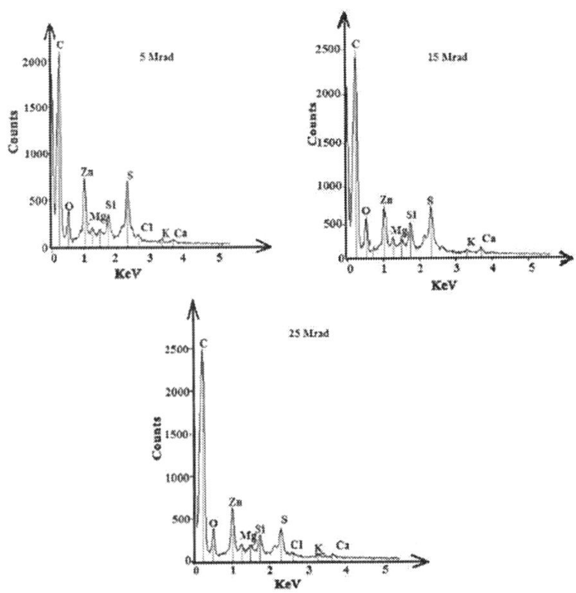

Figure 10. EDXA curves of electron beam irradiated samples of blend E. (a) 5 Mrad, (b) 15 Mrad and (c) 25 Mrad.

Doses/elements	Carbon wt (%)		Silicon wt (%)		Oxygen wt (%)	
	A	B	A	B	A	B
0 Mrad	38.82	54.2	13.17	6.51	28.69	19.26
5 Mrad	39.35	40.21	14.91	11.32	27.6	30.3
15 Mrad	46.01	40.85	10.7	12.08	36.38	31.41
25 Mrad	39.92	38.87	11.17	11	36.72	32.07

Table 3. Inferences from EDXA curves of electron beam irradiated samples of SiR rich blends (A and B).

Doses/elements	Carbon wt (%)			Silicon wt (%)			Oxygen wt (%)		
	C	D	E	C	D	E	C	D	E
0 Mrad	71.42	84.99	89.13	2.11	0.76	0.8	11.84	3.78	3.14
5 Mrad	78.47	77.56	73.78	3.83	1.83	1.29	0.77	8.26	8.73
15 Mrad	82.43	81.66	79	1.26	1.62	1.63	5.21	6.86	6.69
25 Mrad	77.09	77.63	78.36	2.06	1.73	1.34	9.23	7.9	7.96

Table 4. Inferences from EDXA curves of electron beam irradiated samples of EPDM rich blends (D and E) and blend C.

5.4. Correlation of EDXA results with FTIR

The interpretations between EDXA and FTIR of the electron beam irradiated samples of SiR–EPDM blends are listed in **Tables 5–7**.

Inference from EDXA	5 Mrad	15 Mrad	25 Mrad
Increase in carbon content	C=C (alkene, stretch variable)/C=C (asymmetric, stretch, strong)/C–H (alkane)/=C–H (alkene) bending strong	C=C (alkene, stretch variable)/C=C (asymmetric, stretch, strong)/C–H (alkane)	C=C (alkene, stretch variable)/)/=C–H (alkene) bending strong
Decrease in silicon content except at 5 Mrad	Presence of Si–O–Si/Si–CH_3–CH_2 and Si–H		
Increase in oxygen content except 5 Mrad	Increase in Alcohol (–OH) free group content Occurrence of Si–O–Si group		

Table 5. Interpretation between EDXA and FTIR of the electron beam irradiated samples of blend A.

Inference from EDXA	5 Mrad	15 Mrad	25 Mrad
Increase in carbon content	Increase in CH_3–CH_2–CH		
	Appearance of C=C (asymmetric, stretch, strong)	Appearance of C–H (alkane), bending ,strong	Appearance of C=C (asymmetric, stretch, strong)/C–H (alkane), bending, strong/=C–H (alkene), bending, strong
Decrease in silicon content except at 5 Mrad	Absence of Si–H (amorphous Si)		
Decrease in oxygen content	Absence of acid COOH group		

Table 6. Interpretation between EDXA and FTIR of the electron beam irradiated samples of blend C.

Inference from EDXA	5 Mrad	15 Mrad	25 Mrad
Decrease in carbon content	Absence of acid (–COOH) group		
Increase in silicon content	Appearance of Si–H (amorphous Si) Increase in Si–CH_3–CH_2/Si–H content	Increase in Si–CH_3–CH_2/Si–H content	
Increase in oxygen content	Increase in alcohol (–OH)-free group content		

Table 7. Interpretation between EDXA and FTIR of the electron beam irradiated samples of blend E.

5.5. SEM analysis

Figure 11 (a1, b1, c1, d1, e1) and (a2, b2, c2, d2 , e2) are the SEM micrographs of the electron beam irradiated samples of SiR–EPDM blends exposed to 15 Mrad dose of electron beam irradiation for a magnification of 500 and 4000 respectively.

Figure 12 (a1, b1, c1, d1, e1) and (a2, b2, c2, d2, e2) are the SEM micrographs of the electron beam irradiated samples of SiR–EPDM blends exposed to 25 Mrad dose of electron beam irradiation for a magnification of 500 and 4000 respectively.

It is observed from **Figure** 11(a1, a2) that the surface of SiR rich blend (A) has larger number of cracks. This may be due to the decrease in silicon content for 15 Mrad dose of electron beam irradiation (inferred from EDXA analysis), but the surface of blend B and EPDM rich blends (D and E) has smaller number of cracks. This is validated through the increase in oxygen and silicon concentrations (inferred from EDXA analysis). The surface of blend C has smaller cracks. This may be due to the reduction in oxygen and silicon concentrations. The availability of white particles on the surface of the blends B, D, and E may be due to the decrease in carbon content in them (inferred from EDXA curves).

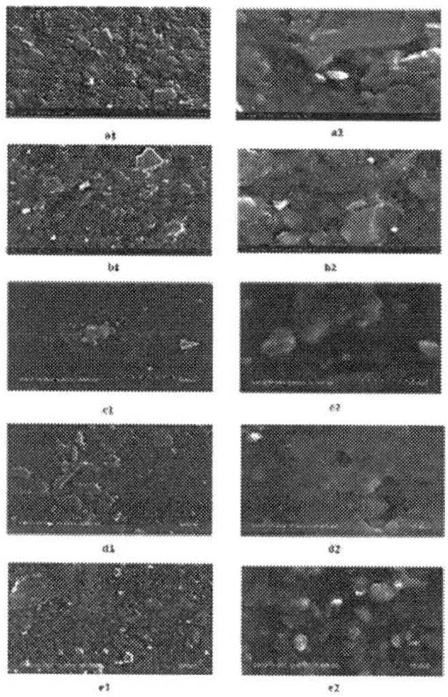

Figure 11. SEM micrographs of electron beam irradiated (15 Mrad) samples of SiR–EPDM blends. 11 (a1), 11 (b1), 11 (c1), 11 (d1) and 11 (e1) -500 magnification; 11 (a2), 11 (b2), 11 (c2), 11 (d2) and 11 (e2) -4000 magnification.

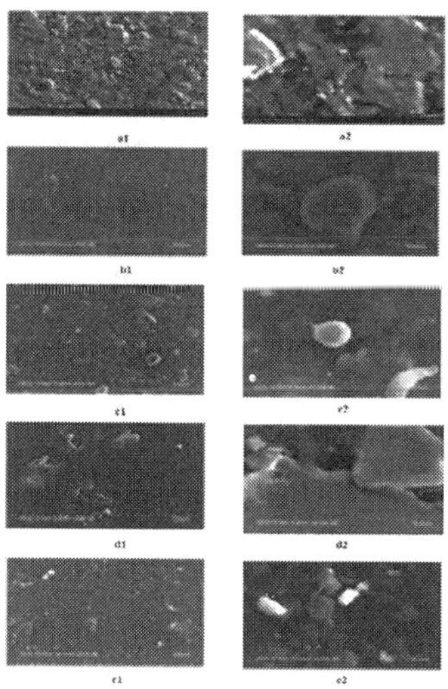

Figure 12. SEM micrographs of electron beam irradiated (25 Mrad) samples of SiR–EPDM blends. 12 (a1), 12 (b1), 12 (c1), 12 (d1) and 12 (e1) – 500 magnification; 12 (a2), 12 (b2), 12 (c2), 12 (d2) and 12 (e2) – 4000 magnification.

It is observed from **Figure 12**(a1, b1) that the surface of SiR rich blend (A) has large number of cracks. This may be due to the decrease in silicon content for 25 Mrad dose of electron beam irradiation (inferred from EDXA analysis), but the surface of blend B and EPDM rich blends (D and E) has smaller number of cracks. This is validated through the increase in oxygen and silicon concentrations from EDXA analysis. The surface smoothness of blend C is moderate. This may be due to the reduction in oxygen and silicon concentrations.

6. Conclusion

The blend C is found to have the improved BDV and DS values for all doses of electron beam irradiation. Also a significant improvement in DC has been noticed at 5 Mrad in blends C, D, and at 5 and 25 Mrad in blend E respectively. A considerable improvement in DF has been observed at 15 and 25 Mrad in blend C and at 25 Mrad in blend D respectively. Hence, it is concluded that blend C and EPDM rich blends are found to have improved dielectric performance after the electron beam exposure.

Acknowledgements

The authors gratefully acknowledge the financial support extended by SSN College of Engineering for carrying out this research work.

Author details

R. Deepalaxmi[1*], V. Rajini[1*] and C. Vaithilingam[2*]

*Address all correspondence to: deepalaxmir@ssn.edu.in

*Address all correspondence to: rajiniv@ssn.edu.in

*Address all correspondence to: cv_srm@yahoo.co.in

1 Department of EEE, SSN College of Engineering, Chennai, Tamilnadu, India

2 SELECT, VIT University, Chennai, Tamilnadu, India

References

[1] R. Deepalaxmi, M. Balaji and V. Rajini, Particle Swarm Optimization Based Selection of Optimal Polymeric Blend, IEEE Trans. Dielectr. Electr. Insul., 2013, Vol. 20, No. 3, pp. 922–931.

[2] R. Deepalaxmi and V. Rajini, Performance Evaluation of Electron Beam Irradiated SiR-EPDM Blends, IEEE Trans. Dielectr. Electr. Insul., 2015, Vol. 22, No. 6, pp. 3366–3375.

[3] R. Deepalaxmi and V. Rajini, Performance Evaluation of Gamma Irradiated SiR-EPDM Blends, Nucl. Eng. Des. (Elsevier), 2014, Vol. 273, pp. 602–614.

[4] R. Deepalaxmi and V. Rajini, Property Enhancement of SiR-EPDM Blend Using Electron Beam Irradiation, Int. J. Electr. Eng. Technol., 2014, Vol. 9, No. 3, pp. 984–990.

[5] V. Rajini and K. Udayakumar, Degradation in Silicone Rubber under AC or DC Voltages in Radiation Environment, IEEE Trans. Dielectr. Electr. Insul., 2009, Vol. 16, No. 3, pp. 834–841.

[6] V. Rajini and K. Udayakumar, Resistance to Tracking of EPDM Aged by Gamma Irradiation under AC and DC Voltages, Int. J. Emerg. Electr. Power Syst., 2007, Vol. 8, No. 3, pp. 1–207.

[7] R. Hackam, Outdoor HV Composite Polymeric Insulators, IEEE Trans. Dielectr. Electr. Insul., 1999, Vol. 6, No. 5, pp. 557–5851.

[8] M. Brown, Compounding of Ethylene Propylene Polymers for Electrical Applications, IEEE Electr. Insul. Mag., 1994, Vol. 13, No. 1, pp. 16–22.

[9] R. Raja Prabu, S. Usa and K. Udhyakumar, Electrical Insulation Characteristics of Silicone and EPDM Blends. IEEE Trans. Dielectr Electr. Insul., 2007, Vol. 14, No. 5, pp. 1207–1214.

[10] M. Ehsani, H. Borsi, E. Gockenbach, G.R. Bakhshande, I.J. Morshedian and I.N. Abedi, Study of Electrical, Dynamic Mechanical and Surface Properties of Silicone-EPDM Blends, IEEE Int. Conf. Solid Dielectr., Toilouse, France, 2004, pp. 1–4.

[11] S. Kole and K. Tripathy, Morphology and Ageing Behaviour of Silicon-EPDM Blends, J. Mater. Sci., 1994, pp. 2451–2455.

[12] P.D. Blackmore, D. Birtwhistle, G.A. Cash and G.A. George, Condition Assessment of EPDM Composite Insulators using FTIR Spectroscopy, IEEE Trans. Dielectr. Electr. Insul., 1998, Vol. 6, No. 5, pp. 132–141.

[13] M. Celina, K.T. Gillen, J. Wise and R.L. Clough, Anomalous Aging Phenomena in the Cross Linked Polyolefin Cable Insulation, IAEA-TECDOC-1012, Radiat. Phys. Chem., 1996, Vol. 48, pp. 613–626.

[14] W. Arayapranee and G.L. Rempel, Properties of NR/EPDM Blends with or without Methyl Methacrylate-Butadiene-Styrene (MBS) as a Compatibilizer, Int. J. Mater. Struct. Reliab., 2007, Vol. 5, No. 1, pp. 1–12.

[15] S. Simmons, M. Shah, J. Mackvich and R.J. Chang, Polymeric Outdoor Insulating Materials, Part-III-Silicone Elastomer Consideration, IEEE Electr. Insul. Mag., 1997, Vol. 13, No. 5, pp. 25–30.

[16] Z. Tvaaruzkova and V. Bosacek, Characterization of Hydroxyl Groups of Y Zeolites by Infrared Spectra, Chem. Zvesti, 1975, Vol. 29, No. 3, pp. 325–330.

[17] A.M. Shehap, Thermal and Spectroscopic Studies of Polyvinyl Alcohol/Sodium Carboxy Methyl Cellulose Blends, Egypt. J. Solids, 2008, Vol. 31, No. 1, pp. 75–91.

[18] B.Saikia and G. Parthasarathy, Fourier Transform Infrared Spectroscopic Characterization of Kaolinite from Assam and Mehalaya, Northern India, 2010, J. Mod. Phys., Vol. 1, pp. 206–210.

Section 2

Biomaterials

Chapter 5

Radiation and Environmental Biophysics: From Single Cells to Small Animals

Yanping Xu

Additional information is available at the end of the chapter

http://dx.doi.org/10.5772/62623

Abstract

In this chapter, two of very unique and novel radiation technologies for modern radiobiology studies are reviewed. First of all, it is concentrated on the developments of accelerator-based particle microbeam system, which has been effectively used for studying the puzzle of "radiation-induced bystander effect." In addition, a recent published single-cell microbeam study, which is aiming to directly measure a cell's radio-sensitivity combining microbeam system with self-referencing biosensor, is included. Then, toward the study of realistic irradiation scenarios in radiation biology in particular, such as a nuclear attack for homeland security concerns or a potential large-scale radiological event, there would be a major need to ascertain, within a few days, the radiation doses received by tens or hundreds of thousands of individuals. Specifically, biological tests would need to be established to estimate the likelihood of such radiation exposure to result in serious health consequences; tests that would then be applied to decide on the correct treatments that might mitigate the short- and long-term health effects of such radiation exposure. However, because of the complexity and difficulty of conducting tests in such circumstance, innovative irradiation systems and technology are required. So the new developments of small animal irradiation system for evaluating the radiation risk and carrying out animal model radiobiology experiments within the mimicked radiation scenarios are covered in the second half of this chapter.

Keywords: Microbeam, Micro-biosensor, Single-cell irradiation, Small animal irradiation, Accelerator

1. Introduction

Historically, cancer and other health risks stemming from exposure to low-level radiation have been difficult to evaluate due to the high spontaneous frequency of age-related cancer and degenerative diseases. With the dramatic increase in exposure of the human population to low-dose radiation either from diagnostic procedures, industrial applications, mining, cleanup of contaminated sites, and space travel, there has been a great scientific need for better estimates of the risks to such exposures. One of the driving forces behind the low-dose radiation research is developing specialized radiation technology and novel, versatile biophysics tools for such radiation biology and radiation physics studies.

2. Single-cell microbeam irradiation

2.1. Microbeam

For low-dose radiation circumstances which are extremely relevant to the environmental radiation exposure, individual cells hardly experience traversals by radiation particles. The biological effects of such low-dose radiation are unknown, and it is very difficult to study with conventional broad radiation beam exposure due to Poisson distribution of tracks. Also the cell-to-cell communication cannot be addressed directly because of the random radiation of broad beam particles. The microbeam system can overcome these difficulties. Microbeam is an irradiation system which delivers a certain number of particles with a micron-sized diameter spot to a chosen biological target, which allows damage to be precisely deposited within specific locations (e.g., nuclei or cytoplasm of single cells). Charged particle microbeams have been significant contributions to defining the biological targets of ionizing radiations. Also with microbeam, cell nuclear and cytoplasmic responses of targeted cells are established along with the responses of non-targeted bystander cells [1]. There is evidence that radiation-induced bystander signals between cells may originate with a diffusible mediator [2–4].

For more than 15 years, the Radiological Research Accelerator Facility (RARAF) of Columbia University has built and operated a charged particle microbeam facility capable of irradiating the nuclei of individual biological cells with as few as one helium ion or proton [5, 6]. The RARAF microbeam II is an updated system, and it is driven by a 5.5 MV HVEE Singleton Accelerator. The particles are ionized by a radio-frequency (RF) ion source inside the Singleton Accelerator and are accelerated from a DC high-voltage terminal to reach the desired energy. Then, the particle beam passes through a beam transport system to the beam end station where irradiation experiments take place. The beam transport system includes a few beam manipulation elements (**Figure 1**). The main beam slits and beam stop are used to eliminate unwanted ion beams, to limit the size of the beam entering a 90° bending magnet, and to stop the beam from entering the bending magnet when irradiations are not expected. The magnetic steering magnet is used to make fine adjustments, aiming the ion beam at the entrance of the bending magnet. The 90° bending magnet is used to bend the beam into the vertical direction. The beam deflector/shutter is an electrostatic system that steers the beam rapidly (~1 ms) to end the irradiation of a cell. The object aperture (~30 μm diameter) limits the initial beam size. A custom built electrostatic double-quadrupole triplet system [7] is used to focus the beam at the position of the cells to be irradiated. In front of the first triplet lens, an angular limiting aperture is used to eliminate particles entering at large angles to help reduce the diameter of the beam spot.

http://dx.doi.org/10.5772/62623

The beam exit window is a thin silicon nitride foil attached to a stainless steel disk, with thickness of either 500 nm or 100 nm (for a sub-micron beam spot) to minimize scattering the beam. The cell dish holder is designed to locate cells at the focusing plane of the microbeam and is in the view range of a customized microscope. An imaging/targeting system comprises a precision XYZ stage (MadCity Labs, Inc. WI), a Nikon Eclipse E600 microscope, and an attached PhotonMAX-512B EMCCD camera (Princeton Instruments, NJ). This equipment combination also allows us to image the wide range of fluorescent proteins that have been developed. A computer control program written in Visual Basic locates the cells, plated in a cell culture dish, and positions them for irradiation. While the RARAF microbeam is primarily used for charge particle irradiation, the accelerator can be used as source for neutral particle radiation, for example, neutron radiation [8–10].

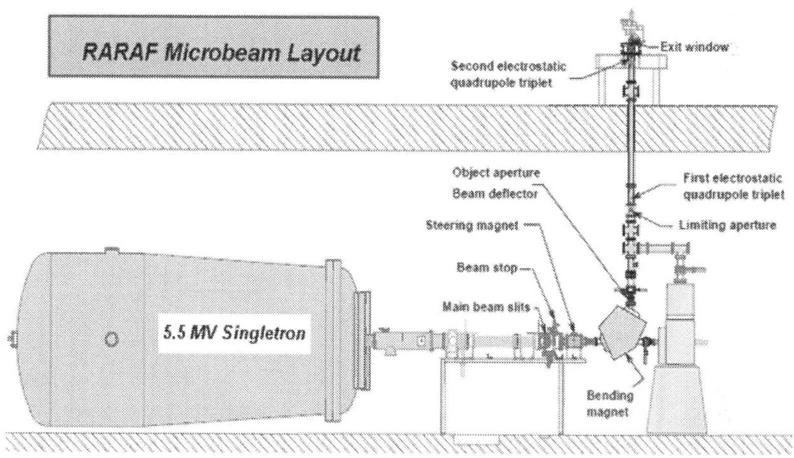

Figure 1. RARAF microbeam.

2.2. Single-cell microbeam irradiation with oxygen micro-biosensor

To physically demonstrate the radiation-induced oxygen flux changes outside the irradiated cell, a self-referencing amperometric electrochemical sensor was proposed for use with single-cell microbeam irradiation experiment. This type of micro-biosensor allows sensitive and non-invasive measurement of the flux of radical mediators, such as oxygen (or nitric oxide), in single cells. This is achieved by repeatedly moving the micro-biosensor/probe tip through the extracellular gradient at a known frequency and a known distance apart. This so-called self-referencing technique minimizes the measurement challenges caused by the random drift of the sensor output. In conjunction with a microbeam, it has the capacity to accurately detect selected ionic and molecular gradient changes surrounding a single cell with high spatial resolution. The first radiobiology experiment was to analyze metabolic oxygen consumption in individual living lung cell after sub-cellular irradiation and to explore the radiation response of such cells. The self-referencing oxygen electrochemical system was developed at the BioCurrents Shared Resource at the Marine Biological Laboratory (MBL), Woods Hole, MA. It was integrated with the RARAF single-particle, single-cell microbeam to form a single-cell irradiation response detection platform.

2.3. Oxygen self-referencing micro-biosensor

Self-referencing polarographic (SERP) micro-sensor technology was developed at the BioCurrents Shared Resource [11–13]. These sensors are made of borosilicate glass capillaries (1B150, WPI) which are pulled to outer tip diameters of ~3 μm using a Sutter P-97 (Sutter Instrument, CA). Then, a 25-μm diameter gold wire (Alfa Aesar, Ward Hill, MA) is electrochemically etched in an aqueous solution of 1N HCl to reduce the tip diameter to ~1 μm. After being rinsed with water and isopropyl alcohol, the etched Au wire is inserted into the capillary so that the wire protrudes slightly from the pipette tip. The electrode tip is dipped into UV curing epoxy (429, Dymax, Torrington, CT). The exposed Au electrode is then etched again (same conditions as above) to form a recessed electrode with a cavity 2–3 μm deep. Finally, the electrode is coated by dipping it in a solution of 10% cellulose acetate (30 kDa) for 60 s and drying for 5–10 min. The electrochemical sensor itself is attached to a BRC amperometric head stage via a modified BNC connecter. An L-shaped Ag/AgCl reference electrode, connected via a 3M KCl/5% agar bridge (reference probe) placed in the bulk solution, completes the circuit (**Figure 2**). Selectivity for amperometric electrodes is usually defined by the conditioning, operating voltage, and excluding membranes.

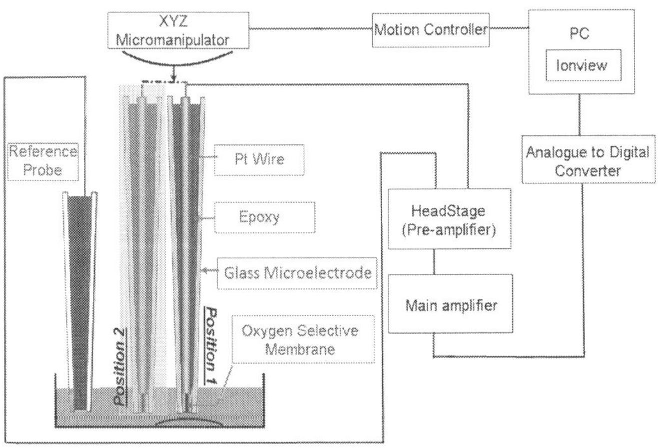

Figure 2. Self-referencing micro-biosensor detection system.

http://dx.doi.org/10.5772/62623

2.4. System integration

For single-cell irradiation response measurements, sensor access during and after irradiation for precision location of damage within single cells with the imaging system is crucial. Microprobe measurements without irradiation are usually performed on an inverted microscope giving open access from above for probe placement. With microbeam irradiation, there is a constraint from the exit window below the sample, so microscopy and probing must be done from above the sample. Access requires that the tip of the probe (microelectrode) approaches a single cell, at an angle of between 20° and 30°, to within microns of the plasma membrane. Access for the reference electrode is also required. Both the measuring probe and the reference electrode have body diameters of 1.5 mm, while the measuring tip is drawn to a point. A Nikon 10× long working distance (4 mm) dry microscope objective is placed with the probe in a sample dish. Because the probe approach angle is so severely constricted, the angle setting technique appropriate for an open-access system is completely inadequate. An offset hinge manipulator was designed (**Figure 3**) and built which allows rapid repeatable repositioning of the probe and simple angle adjustments. The hinge was constructed in a stacked configuration using high-precision flex pivots in such a way that angular settings between 10° and 60° can be set. In order to use the manipulator and the stacked hinge to satisfy our needs, a universal mounting car (Thorlabs, NJ) riding on an optical rail with an integrated robotic retraction mechanism is used. The probe manipulators are mounted accurately on the car for simple interchange. This robotic manipulator structure and the associated fully integrated control systems allow us to meet all the micromanipulation, and capillary probe placement needs in an efficient manner.

2.5. Measurement

2.5.1. Cell preparation

The human telomerase reverse transcriptase (hTERT) immortalized human small airway epithelial (SAE) cells were thawed from liquid nitrogen and cultured in fresh medium. Cells were diluted in fresh medium, and cultured cells were maintained at 37°C in a humidified 5%

Figure 3. Electrochemical micro-sensor mounted on microbeam end station by the offset hinge system.

CO_2 incubator. The microbeam cell culture dishes were custom-made for cell growth and cell irradiation. They are made of Falcon 60 mm Petri dish, and a 0.25-inch diameter hole is drilled into the center of the dish bottom. The polypropylene film covered on the bottom of microbeam dish wells was treated with Cell-Tak (BD Biosciences) to enhance cell attachment. Also it allows the chosen radiation to get through to the cells while allowing them to be placed upright on the microbeam end stations with minimal distance between the dish bottom and beam exit window. Dishes were incubated at 37°C for 30 min and were then rinsed. Then, the cells were trypsinized and diluted to 1.5 × 104/ml (about 30 cells in a total volume of 2 μl medium). A sterile 18–22 mm square coverslip covered the well after cells in a droplet were plated using a micropipetter as close as possible to the center of the dish. The dishes were placed in an incubator until cells attached to the polypropylene. After cell attachment, the coverslips were removed and 5 ml more of medium was replenished to the dishes. Cells will typically flatten out within 1–3 h. The cells were stained by exposure to a 50 nM solution of the vital DNA-binding stain Hoechst 33342 for 30 min prior to radiation. This low stain concentration necessitates the use of an EMCCD camera (Princeton Instrument) to obtain a high-contrast image and allows rapid location of the cell nuclei to be hit, or not hit, as the experiment calls for during irradiation.

2.5.2. Radiation beam setup

Beam size and beam location were measured with a nickel knife edge scan [7]. Beam location was registered using scanning of fluorescent beads. This was done by placing a microbeam dish containing fluorescent beads on the stage. Then, an isolated bead was moved to the approximate beam position. A spiral energy loss mapping scan was conducted with a solid-state charge particle detector. Once the center of the beam was identified, the bead was moved to that position (the center of beam mapping) and the beam location was registered with the imaging system. The low-magnification pictures of the bead were taken, and a center-of-

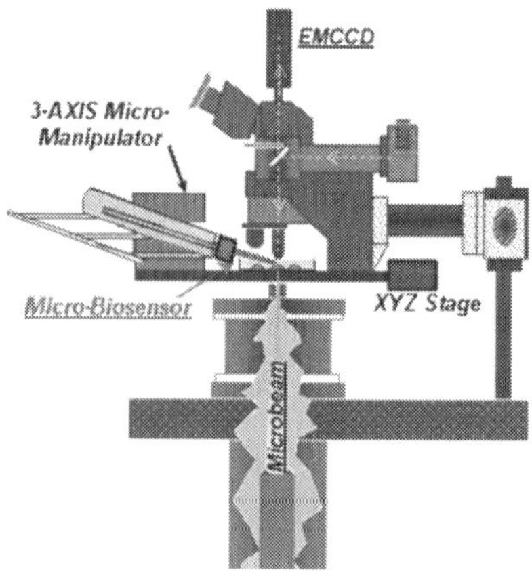

Figure 4. Measurement setup on microbeam end station.

http://dx.doi.org/10.5772/62623

gravity program routine was used to transfer the bead coordinates to the computer. Then, the cell dish was removed from incubator and was mounted on the microbeam stage. The chosen cell was moved to the registered beam position with an XYZ stage (MadCity Labs, Inc. WI), while monitored with a Nikon Eclipse E600 microscope and a reference picture was taken with an attached EMCCD camera (**Figure 4**). Then, the micro-electrochemical sensor and reference electrode were mounted on the XYZ stage, and the sensor/probe was carefully moved close to the cytoplasm (about 10–15 μm away) monitored with a Nikon 10× objective lens, and a background/control measurement was run for about 30 min with the sensor moved rapidly between two positions 15 μm apart at a frequency of 0.3 Hz. Data were collected at each pole position for approximately 1 s or 70% of the cycle time. The current signals were averaged at each position, and then, a differential current was obtained that can be converted into a directional measurement of flux using the Fick equation. Referencing the signals in this manner has the advantage that sources of interference caused by random drift and noise are effectively filtered from the signal and fluxes can be monitored in real time.

2.5.3. Radiation and real-time oxygen consumption measurement

During the single-cell microbeam irradiation, the current changes were monitored with biosensor in self-referencing mode. The number of helium ions was set at 20 or 30 in the microbeam irradiation protocol for this single-cell irradiation. The particle beam count rate was measured at about 200 per second. A pulser (Ortec Inc, TN) was used to simulate the real count rate and to control the beam shutter, because the 5.6 MeV helium ions cannot pass all the way through the cell and the medium (~1 mm thick) without being absorbed. In the cytoplasm irradiation experiments, an exclusion zone around each fluorescing nucleus is automatically generated to ensure that the cytoplasm target positions from one cell are not accidentally within the nucleus of an adjacent cell. Mutation induction caused by cytoplasmic irradiation has been reported using this technique. The image analysis system defines the long axis of each cell nucleus, after which the computer system delivers particles at two target positions along this axis, 6 μm away from each end of the cell nucleus. During the experiment, an in house code with Matrox Genesis imaging library has been used to handle the images (subtract background, correct for illumination variation, locate cells, and record location of nearest frame).

2.6. Results

To detect physiologically driven molecular movements around the cell membrane, which is normally very small (fA current), the drift and background have to be taken care of with self-referencing technique. This requires extracting small electrical signals, fA differences (AC) on top of large offset signals 10s of pA (DC). With cells exposed to 10 μM anti-mycin, the basal cellular O2 flux was tested and the changes in oxygen concentration dependent current (DC changes) were detected. Then, both cytoplasm and nucleus irradiations with 100% cell radiation were conducted. **Figure 5** shows the background and drifting (DC) during the radiation. A very obvious O2 current change (AC) was recorded within seconds after cytoplasm irradiation. **Figure 6** shows the AC current change from two measurements with two

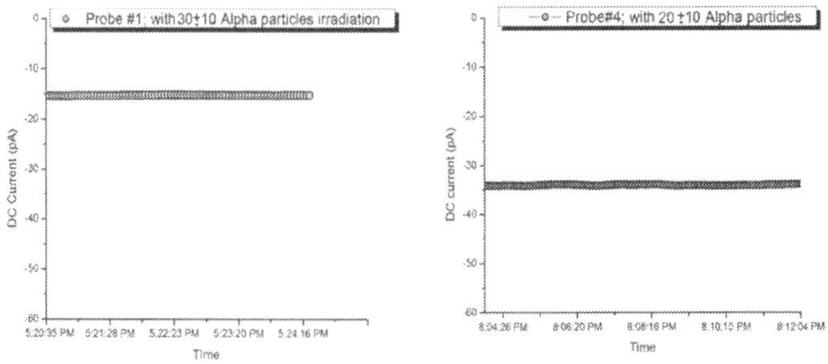

Figure 5. Self-referencing oxygen ion–selective probe result with alpha microbeam irradiation (DC).

different probes. The small changes in oxygen consumption could be extracted from the background oxygen concentration with the self-referencing method. A similar but relative small change was observed after nucleus irradiation. Without radiation, any of these large spikes cannot be seen on a long-time background measurement. Twelve measurements were conducted resulting in a success rate of ~30% as determined by the individual cell flux test because of the cell-to-cell variations and the uncertainty of the probe locations. The results indicate a role for mitochondrial damage following irradiation and enable further evaluations of the radiation-induced bystander effect. This effect is hypothetically due to the result of damage signals received by non-hit cells from hit cells. Establishing the mechanistic basis for such responses in the form of damage signaling from hit to non-hit cells and continued signaling has proven to be elusive. However, evidence for both oxygen- and nitrogen-based small molecules and mitochondrial dysfunction has been produced. The approach outlined in this study suggests biosensor mediators in the form of oxygen radicals, nitric oxide, and hydrogen peroxide can be directly measured at a single-cell level in both hit cells and bystander cells providing an incisive method of evaluating such evidence. This establishment of a non-invasive, self-referencing biosensor/probe system in conjunction with the RARAF microbeam provides an additional means for probing biological responsiveness at the level of individual cell, after precise sub-cellular targeting in hit cells and bystander cells.

3. An accelerator-based neutron small animal irradiation facility

3.1. Overview of IND-related radiation protection

Several scenarios of large-scale radiological events include the use of an improvised nuclear device (IND) that may produce a significant neutron component with the prompt radiation exposure [14]. Specifically, the prompt radiation from this type of detonation is expected to be qualitatively similar to that of the gun-type 15 kT device exploded over Hiroshima [15]. In order to assess the significance of the neutron exposure in dose reconstruction for this type of scenario and to allow characterization of novel neutron-specific biodosimetry assays, a new broad-energy neutron irradiator was designed [9] at the Columbia University RARAF.

http://dx.doi.org/10.5772/62623

This accelerator-driven neutron irradiator provides a broad-spectrum neutron field with energies from 0.2 to 9 MeV that mimics the evaluated energy spectrum produced in the detonation of the atomic bomb at Hiroshima at 1–1.5 km distance from ground zero [15]. At this distance, both survival and radiation exposure are expected to be sufficiently high to require triage for allocation of medical efforts; based on the Hiroshima data, the most survivors around this distance receive an appreciable neutron dose (up to 0.25 Gy [16]). However, the spectrum observed at this distance is significantly different from a standard reactor spectrum due to transport in the air, and has a larger component of low-energy neutrons. It is expected that this difference would have a significant impact on biodosimetric dose reconstruction.

The neutron field is produced by a mixed beam and composed of 5 MeV atomic and molecular ions of hydrogen and deuterium that is used to bombard a thick beryllium (Be) target. The latter is a well-known neutron-producing material not only because of its high neutron yield but also because of its stability and high specific heat. This mixed beam produces a neutron spectrum which is the sum of the spectra from the 9Be(d,n)10B and 9Be(p,n)9B reactions for all the incident ions (monatomic, diatomic and triatomic) and for energies from 5 MeV and down. In general, for monatomic 5 MeV projectiles, the 9Be(d,n)10B reaction provides a spectrum with higher-energy neutrons (above 1 MeV), while the 9Be(p,n)9B reaction primarily yields neutrons below 1 MeV. These nuclear reactions generate a combined neutron spectrum with a wide range of energies, which can then be used to irradiate biological samples and small animals (e.g., mice) for radiobiology studies. The beam composition in the present setup is approximately a 1:2 ratio of protons to deuterons. However, for other scenarios, the spectrum shape can be modified by adjusting the ratio of protons to deuterons and the incident beam energy.

As described elsewhere [17], the neutron spectra were evaluated by making combined measurements with a proton-recoil proportional counter [18] and liquid scintillator detector [19]. The measured recoil spectra were unfolded using maximum entropy deconvolution [20], based on Monte Carlo simulated detector response functions [21].

The dosimetry for the irradiations was performed using a custom tissue-equivalent (TE) gas ionization chamber, placed on the sample holder wheel. This chamber measures the total dose

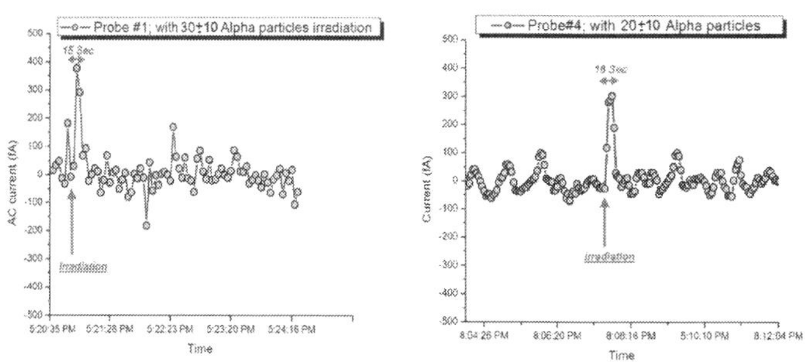

Figure 6. Self-referencing oxygen ion-selective probe result with alpha microbeam irradiation (AC).

in the mixed neutron and γ-ray field. To evaluate the ratio of neutron and γ doses, gamma-ray dosimetry was performed separately by replacing the ionization chamber with a compensated Geiger–Mueller dosimeter, which has a very low neutron response [22]. These measurements indicated that all neutron exposures, using this spectrum, are accompanied by a parasitic photon dose of about 21% of the total dose delivered.

In order to account for possible variations in the dose rate during irradiations, a second TE gas ionization chamber is placed in a fixed location on the beam axis, directly downstream of the neutron target and used as a monitor. All measurements are normalized to the signal from the monitor, which is used to determine the dose during irradiation.

In a realistic scenario, it is expected that the neutron dose will only be a small fraction of the total exposure (e.g., DS02 reported only 2% neutron dose at 1.5 km from the Hiroshima epicenter [16]). In order to mimic such a mixed field, a 250 kVp orthovoltage X-ray machine is located on site to allow irradiating samples with the necessary additional dose of photons.

To demonstrate the use of this facility, we present results from initial test using the in vitro cytokinesis-block micronucleus assay (CBMN, [23]) to measure the induction of micronuclei in peripheral human blood lymphocytes exposed to a range of neutron doses up to 1.5 Gy. The dose-response curves generated for micronuclei frequency indicated that the RBE of this neutron spectrum is between 3 and 5 compared to micronucleus yields induced by 250 kVp X-rays. As expected, these values fall between those for accelerator-generated energetic neutrons [24] and those for a reactor-based uranium fission spectrum [25].

3.2. Formation of a broad neutron spectrum

The key feature of our broad-energy neutron irradiator is its use of a mixed-gas ion source, with the spectrum depending on the ratio of gases in the mixture fed into the ion source. For this work, hydrogen and deuterium were combined at a ratio of 1:2 in one of the ion source gas supply cylinders and placed in the accelerator terminal. The ionization process of the mixed gas is complicated, as it generates many different ion combinations. To identify the actual ion beam ratios and to optimize the beam current, two different values of the gas valve control voltage were tested. The selected gas input parameter (percentage of maximum valve voltage) is used to control the pressure to provide sufficient beam current. To determine the ratio of the different ion species in our beam, we measured beam current at a 15° deflection, as a function of the field strength of the bending magnet. **Table 1** shows the fractions of the various atomic and molecular ions at two extremal pressures we can use (outside this pressure range, the beam current is too low or the ion source operation unstable). Several ion species (e.g., D^+ and H_2^+) cannot be separated magnetically because their magnetic rigidity is very close. As can be seen, increasing the valve voltage from 65.2 to 76.6% changed the ion ratios slightly for the molecular, but not atomic ions. As the major contributions to the spectrum come from the H^+ and D^+ ions, which vary by less than 2% over this range, we expect the spectrum will not vary significantly over this range of ion source parameters.

http://dx.doi.org/10.5772/62623

Gas valve control	**65.2%**	**76.6%**
Ion species		
D^+, H_2^+	35.75%	37.3%
D_2^+, H_2D	25.1%	16.7%
D_3^+	7.26%	12%
H^+	15%	16%
H_3^+	7.8%	2.2%
D_2H^+	8.9%	16%

Table 1. Ion species percentage for two different percentages of maximum gas control voltage.

Neutrons are generated as the particle beam impinges on a thick (500 μm) beryllium target. At this thickness, the 5 MeV deuterons and protons are completely stopped in the beryllium. The neutron energy spectrum obtained by this configuration was previously [9] modeled using MCNPX and more recently validated experimentally [17]. Briefly, an EJ-301 liquid-filled scintillation detector and a gas proportional counter filled with 3 atm of hydrogen were used for measuring neutron energies above and below 1.0 MeV, respectively. The combination of the two detection systems covers a wide-energy range, from 0.2 to >9 MeV. The recoil pulse height spectra acquired by the detector systems were carefully evaluated using different quasi-monoenergetic neutron beams (0.2–9 MeV) available at the RARAF accelerator, discriminating the γ-ray signals from the raw acquisition data with pulse rise time.

The two portions of the spectrum (obtained from the two detectors) were combined to form a kerma-weighted total spectrum (**Figure 7**). Overall, the obtained spectrum is similar to the one evaluated for Hiroshima [15], although it is slightly flatter. A more detailed discussion of the differences between our spectrum and Hiroshima appears elsewhere [17].

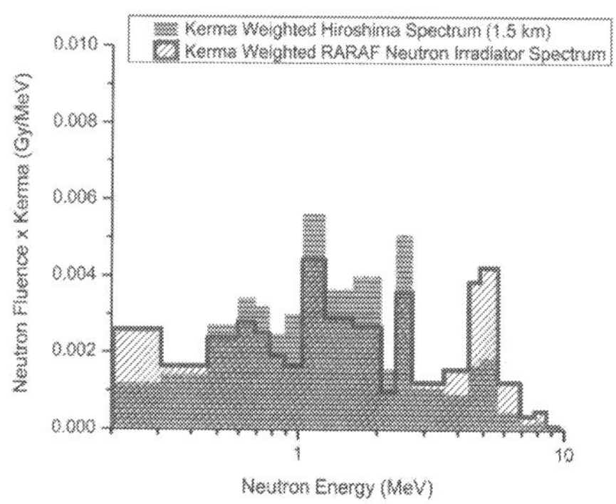

Figure 7. A comparison of the kerma-weighted neutron spectrum generated by us (hashed) with the one at 1.5 km from Hiroshima ground zero (gray).

3.3. Irradiation setup

Due to the requirement for a mixed H/D beam, the irradiation facility was set up using an undeflected (0° angle) beam line in the RARAF accelerator facility (**Figure 8a**). The target assembly, shown in Figure 8b consists of a 2-mm-thick Cu disk onto which a 500-μm-thick beryllium foil was diffusion bonded. The dimensions of the copper disk were chosen so that it can be used in lieu of a standard 2.75″ CF gasket. The target is clamped between two Conflat flanges and doubles as a vacuum window. The target is cooled by water impinging on the copper backing plate [9].

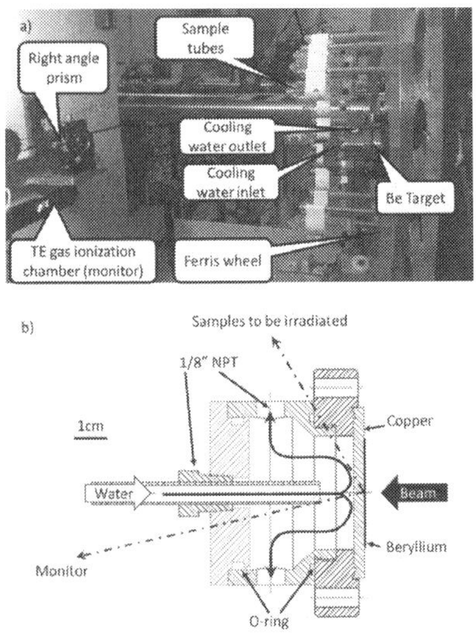

Figure 8. (a) Photo of the irradiation facility. Beam arrives from the right and impinges on the beryllium target generating neutrons. Samples to be irradiated (blood or mice) are placed in sample holders (shown here empty) and mounted on a Ferris wheel rotating around the target cooling chamber. At the left is the TE gas ionization chamber used as a beam monitor and the right angle prism used for aligning the beam line. (b) Cross-section of target and cooling chamber. Direction to the center of the sample holders (190 mm away) and the monitor chamber (610 mm away) are shown.

As the target cooling water lines, supporting structures, shielding, and other surrounding materials were seen to give about a 20% azimuthal variation in dose rate, a vertical Ferris wheel-like fixture (**Figure 8a**) is used to rotate the sample holder tubes (for either blood or mice) around the target. Customized tubes mounting from the rods on the wheel are used to hold the samples with a constant horizontal orientation at a distance of 190 mm from the target center. The fixture rotates up to 18 samples around the beam axis. The sample holders for both mice and ex-vivo irradiated blood are based on standard 50-ml conical centrifuge tubes (BD, Franklin Lakes, NJ) which were modified to allow hanging horizontally from rods on the wheel, so that the tubes maintain a constant orientation as the wheel rotates. This provides an isotropic irradiation, while maintaining the mice in an upright orientation, reducing stress.

http://dx.doi.org/10.5772/62623

During irradiations, the wheel is rotated at a speed of about 2 min per revolution and the dose rate adjusted so that the minimal dose is delivered in 10 rotations (20 min) with the sample tubes flipped end-to-end half-way through.

For pure photon exposure, a 250 kVp Westinghouse Coronado X-ray machine, placed within 15 m of the neutron irradiation facility, is used. This proximity allows for future mixed-field studies, where each sample may be immediately transported to the X-ray machine and irradiated after neutron exposure, with a time gap between the two irradiations of less than 5 min.

3.4. Neutron dosimetry

The total dose measurement for the IND-like neutron/gamma mixed-field irradiations was performed using a custom A-150 muscle TE gas ionization chamber (**Figure 9**), as described by Rossi et al. [26]. This chamber is intended for use in a mixed neutron and γ field measurement and features an interchangeable internal TE plastic sleeve. In the dosimetry measurements reported here, we used a 3.5-mm-thick sleeve to model the dose deposited at the center of the blood samples used. The chamber was filled with methane TE gas at ~700 mm Hg before the dosimetry measurement and sealed. The detector was then attached directly to an electrometer system and calibrated using a 50 mg 226Ra γ-ray source, which had been previously calibrated by the National Bureau of Standards. The dose rate was ~36 μGy/h at 1 m from the source. After calibration, the dosimeter was mounted on the sample wheel for the IND-like neutron irradiator at 60° with respect to the ion beam axis and 190 mm away from the target.

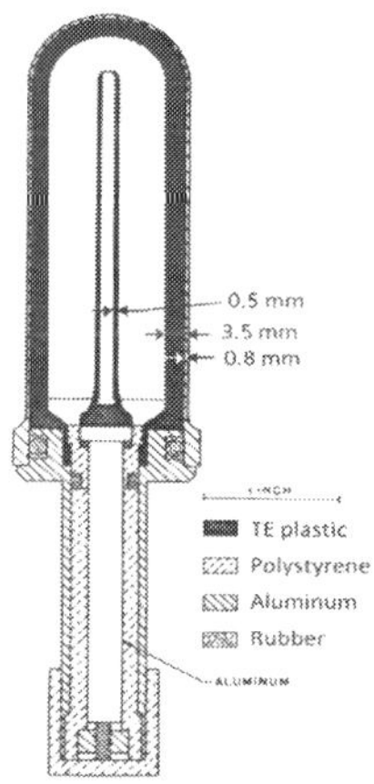

Figure 9. Custom TE gas ionization chamber for neutron dosimetry shown here with 3.5 mm TE plastic sleeve. This figure was originally published by Rossi et al. [26] and has been modified and re-published with the permission of *Radiation Research*.

To extract the dose due to neutrons, γ-ray dosimetry was performed separately with a compensated Geiger–Mueller dosimeter, which is 20 times more sensitive to photons than to monoenergetic neutrons in the range of 0.68–4.2 MeV [22]. The γ-ray dosimetry was conducted in the same manner as the total dose measurement and then subtracted from the latter. Since the γ-ray dose from the target rate is essentially isotropic, only inverse-square law corrections were performed.

The neutron dose rate at the sample position was ~8.6×10^{-2} Gy/h/μA, representing ~79% of the total dose rate, with the remaining 21% due to γ-rays. During the irradiation, the beam current was tuned to and kept at about 17.5 μA, which is equal to a neutron dose rate of ~1.5 Gy/h. Because of the possible variation of the dose rate relative to the beam current, a second TE gas ionization chamber was added as a monitor at a fixed location downstream of the neutron target at an angle of ~12° relative to the ion beam direction. The monitor ionization chamber was filled with flowing TE gas, which was regulated with a constant-density control system. The incident primary particle beam current was recorded with an electrometer coupled to the end of the beam line, which is a Faraday cup-like isolated beam pipe with the target at the end.

3.5. Micronucleus assay analysis

Micronucleus formation in peripheral blood lymphocytes is a well-established marker of ionizing-radiation-induced DNA damage. We have used a recently established cytokinesis-block micronucleus (CBMN) assay protocol by Fenech [23] for accelerated sample processing by performing a miniaturized version of the assay in a multi-tube plate system [27].

Peripheral blood samples were collected from three healthy donors after informed consent (IRB protocol #AAAF2671) and exposed (3 ml aliquots) to nominal neutron doses of approximately 0.25, 0.5, 1, and 1.5 Gy (plus the concomitant 0.06, 0.1, 0.2, and 0.3 Gy of γ-rays). Blood sample aliquots were also exposed to 1, 2, and 4 Gy of 250 kVp X-rays.

Two hours post-irradiation, triplicate blood sample aliquots (50 μl) from each dose point were placed into culture in 1.0 ml 2D-barcoded matrix storage tubes (Thermo Scientific, Waltham, MA) with 500 μl of PB-MAX Karyotyping medium (Life Technologies, Grand Island, NY). Following 44 h of incubation, cytochalasin-B (Sigma Aldrich, St Louis, MO) was added at a final concentration of 6 μg/ml to inhibit cell cytokinesis and the tubes returned to the incubator. After a total incubation period of 72 h, the cells were harvested. Following hypotonic treatment, the cells were fixed using ice cold 4:1 fixative (methanol–acetic acid). The fixed samples were stored at 4°C (at least overnight), dropped on slides and stained with Vectashield mounting medium containing DAPI (Vector Laboratories, Burlingame, CA). The slides were imaged using a Zeiss fluorescent microscope (Axioplan 2; Carl Zeiss MicroImaging Inc., Thornwood, NY) with a motorized stage and a 10× air objective. Quantification of lymphocyte micronuclei yields were determined by automatic scanning and analysis by the MetaferMN Score software (MetaSystems, Althaussen, Germany). Between 1800 and 6000, binucleate cells were analyzed for each data point.

Figure 10a shows the comparison of the micronucleus yields for lymphocytes exposed to different neutron or X-ray doses. Overall, a dose-dependent increase in micronuclei yields was observed with increasing dose with both radiation fields with the yields induced by the X-rays following a linear quadratic behavior with dose and the neutron data following a linear trend.

The neutron data are actually induced by a mixed neutron + photon field. The dashed line shows that what we would expect from a pure neutron irradiation. This estimate was obtained by calculating excess micronucleus yields over controls from the photon component (21% of the total dose) based on the linear quadratic fit to the X-ray data. This value was then subtracted from the mixed-field yields at the same total dose. The resultant difference corresponds to the micronuclei yields that would be seen in a pure neutron irradiation.

3.6. Relative biological effectiveness (RBE) calculations

The potential biological effects and damage caused by radiation depend not only on the radiation dose received but also on the type of radiation. RBE was introduced to normalize the radiobiological effects caused by different types of radiation. RBE is defined as the ratio of photon dose (in our case 250 kVp X-rays) to the dose of the radiation field of interest (in our case neutrons), providing the same biological effect [28]. The biological effects caused by neutrons vary with energy and produce greater damage than X- or γ-rays. In general, for the same dose, neutrons are much more effective in damaging cells because neutron-induced secondary particles, for example, low-energy protons, have high LET (linear energy transfer) and photon-induced particles are electrons having low-LET. In this paper, we describe experiments aimed at measuring the neutron RBE for micronucleus formation in peripheral blood lymphocytes as part of the irradiator testing. Ongoing experiments using a variety of cytogenetic, transcriptomic, and metabolomic endpoints, in human blood and in mice, will be published separately.

The RBE for the mixed-field irradiation (neutrons and γ-rays; **Figure 10b**) was calculated from the linear and linear-quadratic regression curves, fitted to the neutron and X-ray data,

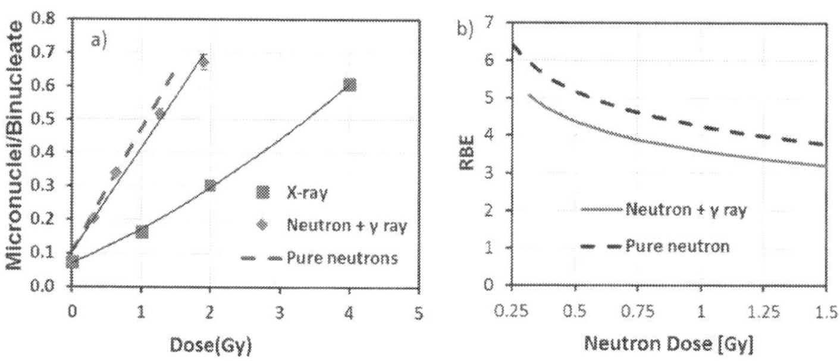

Figure 10. (a) Micronucleus frequency in human peripheral blood lymphocytes exposed *ex vivo* to neutrons or X-rays. The data show mean micronuclei per binucleate cell yields from three healthy volunteers. Error bars show ± SEM. The mixed-field yields plotted vs total dose (neutrons + γ-rays). The solid lines indicate a linear (neutrons) or linear-quadratic (X-rays) fit. The dashed line indicates the estimated pure neutron component. (b) RBE values for the mixed field and for the pure neutrons, calculated from panel a (see text for details).

respectively, by solving for the X-ray dose that would give the same micronucleus yield as a given dose of the mixed neutron/γ-ray field. The pure neutron RBE will be slightly higher than the RBE value of the mixed field. This was evaluated in the same way from the dashed line in **Figure 10a**.

The limiting RBE value of 4 is higher than the value of 2.5 reported for 6 MeV monoenergetic neutrons [24] but lower than the value of 6 calculated for a reactor fission spectrum [25], in accordance with what we would expect, based on the neutron energies.

4. Summary

A mixed beam of atomic and molecular deuterons and protons, accelerated to 5 MeV, impinges on a thick beryllium target, and the resultant neutron spectrum is the sum of the spectra from the 9Be(d,n)10B reaction (higher-energy neutrons) and the 9Be(p,n)9B reaction (lower-energy neutrons). The neutron energy spectrum is manipulated by adjusting the ratio of protons and deuterons to mimic the neutron spectra from an IND exposure for medical triage and bio-dosimetry studies. Specifically, it mimics the Hiroshima gun-type bomb spectrum at a relevant distance from the epicenter (1–1.5 km) and is significantly different from a standard reactor fission spectrum because the bomb spectrum changes as the neutrons are transported through air. The neutron spectrum of this irradiator was measured and is verified comparable with the Hiroshima bomb spectrum at 1.5 km. About 79% of the radiation dose is delivered by neutrons and He rest by γ-rays. A comparison of the biological effect of neutron and X-ray exposure on micronuclei yields in peripheral lymphocytes demonstrated that the IND-spectrum irradiator described above gives RBE values within the expected range.

Author details

Yanping Xu

Address all correspondence to: yx2132@cumc.columbia.edu

Radiological Research Accelerator Facility, Columbia University, Irvington, USA

References

[1] Hei TK, Ballas LK, Brenner DJ and Geard CR (2009) Advances in radiobiological studies using a microbeam. J Radiat Res, 50: A7–A12.

[2] Shao C, Furusawa Y, Aoki M, Matsumoto H and Ando K (2002) Nitric oxide mediated bystander effect induced by heavy-ions in human salivary gland tumor cells. Int J Radiat Biol, 78(9): 837–44

[3] Matsumoto H, Hayashi S, Hatashita M, Ohnishi K, Shioura H, Ohtsubo T, Kitai R,Ohnishi T and Kano E (2001) Induction of radioresistance by a nitric oxide-mediated bystander effect. Radiat Res, 155:387–396.

[4] Azzam EI, de Toledo SM, Spitz DR and Little JB (2002) Oxidative metabolism modulates signal transduction and micronucleus formation in bystander cells from alpha-particle irradiated normal human fibroblast cultures. Cancer Res, 62(19):5436–42.

[5] Gerard CR, Brenner DJ, Randers-Pehrson G and Marino SA (1991), Single-particle irradiation of mammalian cells at the radiological research accelerator facility: induction of chromosomal changes. Nucl Instrum Methods 11 B, 54, 411–416.

[6] Randers-Pehrson G, Geard CR, Johnson GW and Brenner DJ (2001), The Columbia University single-ion microbeam. Radiat Res, 156, 13 210–214

[7] Randers-Pehrson G, Johnson GW, Marino SA, Xu YP, Dymnikov AD and Brenner DJ (2009) The Columbia University sub-micron charged particle beam. Nucl Instrum Methods A, 609:294–299

[8] Xu YP, Randers-Pehrson G, Marino SA, Bigelow AW, Akselrod MS, Sykora JG and Brenner DJ (2011) An accelerator-based neutron microbeam system for studies of radiation effects. Radiat Prot Dosim, 145:373–376

[9] Xu Y, Garty G, Marino SA, Massey TN, Randers-Pehrson G, Johnson GW and Brenner DJ (2012) Novel neutron sources at the Radiological Research Accelerator Facility. J Instrum, 7, C03031, doi:10.1088/1748-0221/7/03/C03031.

[10] Bigelow AW, Randers-Pehrson G, Garty G, Geard CR, Xu Y, Harken AD, Johnson GW and Brenner DJ (2010) Ion, X-ray, UV and neutron microbeam systems for cell irradiation. AIP Conf Proc 1336: 351–355.

[11] Land SC, Porterfield DM, Sanger RH and Smith PJS (1999) The self-referencing oxygen-selective microelectrode: detection of transmembrane oxygen flux from single cells. J Exp Biol, 202: 211–218.

[12] Porterfield DM, Laskin JD, Jung SK, Malchow RP, Billack B, Smith PJS and Heck DE (2001) Proteins and lipids define the diffusional field of nitric oxide. Am J Physiol Lung C, 281: L904–L912.

[13] Kumar SM, Porterfield DM, Muller KJ, Smith PJS and Sahley CL (2001) Nerve injury induces a rapid efflux of nitric oxide (NO) detected with a novel NO microsensor. J Neurosci, 21: 215–220.

[14] Homeland Security Council, National Planning Scenarios (Final Version 21.3) (2006).

[15] Egbert SD, Kerr GD and Cullings HM (2007) DS02 fluence spectra for neutrons and gamma rays at Hiroshima and Nagasaki with fluence-to-kerma coefficients and transmission factors for sample measurements. Radiat Environ Biophys, 46: 311–25.

[16] Young RW, Egbert SD, Cullings HM, Kerr GD and Imanaka T (2005) Survivor dosimetry Part B. DS02 free-in-air neutron and gamma tissue kerma relative to DS86. Reassessment of the Atomic Bomb Radiation Dosimetry for Hiroshima and Nagasaki: Dosimetry System 2002.

[17] Xu Y, Randers-Pehrson G, Marino S, Garty G and Brenner D (2015) Broad energy range neutron spectroscopy with a liquid scintillator and a proportional counter: application to a neutron spectrum similar to that from an improvised nuclear device.. Nucl Instrum Methods A; 794, 234–239

[18] Tagziria H and Hansen W (2003) Neutron spectrometry in mixed fields: proportional counter spectrometers. Radiat Prot Dosim, 107, 73–93.

[19] Klein H (2003) Neutron spectrometry in mixed fields: NE213/BC501A liquid scintillation spectrometers. Radiat Prot Dosim, 107, 95–109.

[20] Reginatto M, Goldhagen P and Neumann S (2002) Spectrum unfolding, sensitivity analysis and propagation of uncertainties with the maximum entropy deconvolution code MAXED. Nucl Instrum Methods A, 476, 242–246.

[21] Pozzi SA, Flaska M, Enqvist A and Pázsit I (2007) Monte Carlo and analytical models of neutron detection with organic scintillation detectors. Nucl Instrum Methods A, 582, 629–637.

[22] Wagner EB and Hurst GS (1961) A Geiger–Mueller [gamma]-ray dosimeter with low neutron sensitivity. Health Phys, 5, 20–26.

[23] Fenech M (2007) Cytokinesis-block micronucleus cytome assay. Nat Protoc, 2, 1084–1104.

[24] Wuttke K, Müller W-U and Streffer C (1998) The sensitivity of the in vitro cytokinesis blocked micronucleus assay in lymphocytes for different and combined radiation qualities. Strahlentherapie und Onkologie (Radiotherapy and Oncology), 174, 262–268.

[25] Huber R, Schraube H, Nahrstedt U, Braselmann H and Bauchinger M (1994) Dose-response relationships of micronuclei in human lymphocytes induced by fission neutrons and by low LET radiations. Mutat Res Fundam Mol Mech Mutagen, 306, 135–141.

[26] Rossi HH, Bateman JL, Bond VP, Goodman LJ and Stickley EE (1960) The dependence of RBE on the energy of fast neutrons: 1. Physical design and measurement of absorbed dose. Radiat Res, 13, 503–520.

[27] Lue S, Repin M, Mahnke R and Brenner D (2015) Development of a high-throughput and miniaturized cytokinesis-block micronucleus assay for use as a biological dosimetry population triage tool. Radiat Res. 184(2): 134–142.

[28] Hall EJ and Giaccia AJ, Radiobiology for the radiologist. 7th ed. Philadelphia: Lippincott Williams & Wilkins; 2012.

Chapter 6

Radioactivity in Food: Experiences of the Food Control Authority of Basel-City since the Chernobyl Accident

Markus Zehringer

Additional information is available at the end of the chapter

http://dx.doi.org/ 10.5772/62460

Abstract

The contamination of our environment and of food with artificial radionuclides originates from several sources. First, nuclear powers spread contamination all over the Northern Hemisphere by carrying out more than 600 atmospheric bomb tests from 1945 to 1963. The peaceful use of nuclear fission brought several accidents in nuclear installations [nuclear power plant (NPP)]. This began in the late 1940s and ended recently with the NPP's core meltings at Fukushima-Daiji in 2011. The catastrophe at the Chernobyl NPP in 1986 spread enormous fallout over most parts of Europe. Besides the artificial contamination, one has to mention the exposure to naturally occurring radionuclides from the uranium and thorium decay series. From 1980 on, the State Laboratory Basel-City began a monitoring programme of food. Special equipment for the analysis of α-, β-, and γ-emitting radionuclides had to be built. In 1986/1987, the laboratory had to manage thousands of samples according to the accident at Chernobyl. The Government estimated the dose of the mean Swiss population from the ingestion of contaminated food to be 1 to 2 mSv. Today, the contamination of food has lowered significantly. The Office of Public Health estimated the total ingested dose of about 0.3 to 0.4 mSv/year. The main contribution comes from potassium-40 (^{40}K; 0.2 mSv/year) and from natural radionuclides of the uranium and thorium decay series. The remaining contamination from the bomb fallout is less than 0.1 mSv/year.

Keywords: Chernobyl, dose estimation, food, radioactive contamination, radiocaesium, radiostrontium

1. Introduction

In the 1950s, Otto Huber and his team from the University of Fribourg started the regular monitoring of radioactivity in Switzerland [1]. Some years later, the federal government initiated a countrywide monitoring programme. In 1969, the first nuclear power plant (NPP) of Beznau came on line to produce electric power. Today, five NPPs are producing about 38% of the electric power in Switzerland. In addition to the emission controls of the NPPs and other radiation-producing facilities, the monitoring of radioactive fallout from the atmospheric, nuclear bomb tests is of special concern. Over 600 bomb tests in the atmosphere led to a contamination of the Northern Hemisphere with long-lived radionuclides, such as radiocaesium, radiostrontium, and plutonium. The contamination situation was then aggravated by the reactor fire of the NPP of Chernobyl in late April of 1986. In Switzerland, the southern and eastern parts were more affected (in southern Switzerland, it rained on 3–5 May over 350 mm precipitation). On 5 May, the National Emergency Operations Centre (NAZ) started a monitoring programme for food with the help of the specialised laboratories in Dübendorf, Freiburg, Lausanne, Spiez, Würenlingen, and the State Laboratory Basel-City. In October 1986, a symposium on the measurements and their interpretation was held in Berne [2]. In 1987, other state laboratories took part in the monitoring programme. Five years later, when radiation levels were reduced considerably, most of the state laboratories reduced their monitoring programmes again; many state laboratories even cut off their survey activities. Until 2011, the radioactivity survey was mainly supported by the specialised laboratories. The core meltings of the Fukushima-Daiji NPPs gave a short increase in the survey activities for the years 2011/2012 in Switzerland.

2. Legislation of the radioactivity controls in Switzerland

In Switzerland, the radioactivity survey is a task of the Federal Office of Public Health (BAG), an office of the Federal Department of Home Affairs. One aspect is the monitoring of food, which is done in collaboration with the state laboratories of the cantons. In 1991, the Federal Assembly of the Swiss Confederation enforced the *Radiological Protection Act* [3] along with several Ordinances (e.g., the *Radiological Protection Ordinance*) [4]. The assessment of food is regulated in the *Ordinance on Contaminants and Constituents in Food* from 1994 (**Table 1**) [5]. This ordinance regulates the most important groups of radionuclides in a special way. For artificial radionuclides, two limits are set. The "tolerance values" should not be exceeded when food is grown or produced with good manufacturing practise. When values are over the tolerance limit, the food is considered as "reduced in its value". "Limit values" are of toxicological concern. When they are exceeded, the consumption of this food may lead to a dose of more than 1 mSv/year.

In emergency situations, when radioactive contamination may occur, special limit values are formulated. After the accident of Chernobyl, such special conditions were formulated for imports from East European countries [6]. Each importation of wild grown mushrooms has to be documented by an importation certificate, which documents a radiocaesium level below 600 Bq/kg. In 2011, the Federal Food Safety and Veterinary Office enforced precautions for the importation of Japanese food and feed origin from animals. Importations from specific prefectures have to be accompanied by a certificate of a radiocaesium analysis of the food [6, 7]. The Federal Office of Public Health will implement revised ordinances for Japan and

Chernobyl in the next few years. Even, the Government intends to annul most of the limit values for radionuclides in times of non-crisis but to implement special limit values for crises.

Radionuclides	Babyfood, infant formulas	Liquid food	Food of main importance	Food of minor importance
Radiostrontium $^{89+90}Sr$	1/75 75	1/125 125	1/750 750	1/7500
Radioiodine $^{131+132}I$	10/150 100	10/500 300	10/2000 2000	10/20,000
Radiocaesium $^{134+136+137}Cs$	10/400 200	10/1000 200	10/1250 500	10/20,000
Plutonium and trans-Pu-elements	0.1/1 1	0.1/20 1	0.1/80 10	0.1/800
Natural radionuclides 1 Σ ^{224}Ra, ^{228}Th, $^{234+235+238}U$	−/10	−/10	−/50	−/500
Natural radionuclides 2 Σ ^{210}Pb, ^{210}Po, $^{226+228}Ra$, $^{230+232}Th$, ^{231}Pa	−/1	−/1	−/5	−/50
Tritium	1000/3000	10/1000	1000/10,000	1000/100,000
Radiocarbon	200/1000	–	200/10,000	200/100,000
Others	10/400	10/1000	10/1250	10/12,500

Except for Brazil nuts (no limits).
Special tolerance and limit values for game: 600/1250 Bq/kg and wild berries: 100/1250 Bq/kg.
Special limit value for ^{210}Po in seafood: 150 Bq/kg.
All values in Bq/kg. First value of each row: tolerance value; second value: limit value. Second line: limits of the ordinance for imports of food from or origin of Japan (2011) [7].

Table 1. Ordinance on contaminants and constituents in food (1995). Appendix No. 6: "List of limit values for radionuclides" [5].

3. Radiation in the environment

Radioactive sources are in our body and everywhere outside in the environment. They can be divided in two groups of sources: naturally occurring radionuclides, which are part of naturally occurring radioactive materials (NORM), and technologically enhanced naturally occurring radioactive materials (TENORM). In addition, there are man-made (artificially produced) occurring radioactive materials.

Primordial radionuclides are radionuclides with long half-lives, even longer than the earth's age. They play an important role in the earth's radioactivity. The three natural decay series of

Uranium (^{235}U, ^{238}U) and thorium (^{232}Th) belong to them. Secondary radioactive elements are built by the decay of these primordial radionuclides and belong to the inventory of our soils. Some dose-relevant radionuclides of these decay chains are radium (^{224}Ra, ^{226}Ra, and ^{228}Ra), radon (^{220}Rn and ^{222}Rn), lead (^{210}Pb), polonium (^{210}Po), and thorium (^{228}Th, ^{230}Th, and ^{232}Th). Another ubiquitous radionuclide is potassium-40 (^{40}K), Which is a large part to the annual received dose. Cosmogenically produced radionuclides are, for example, beryllium-7 (^{7}Be), tritium (^{3}H), or radiocarbon (^{14}C) [8, 9].

Artificial radionuclides have been produced and released to the environment since the 1940s, when nuclear weapons development and tests began in the United States. These tests of nuclear fission in the United States, USSR, China, GB, and France led to fallout with a great number of radioactive fission products and activated radionuclides. Over 600 atmospheric bomb tests from 1945 to 1963 led to a contamination of the Northern Hemisphere with artificial radionuclides such as ^{14}C, ^{3}H, radiocaesium (^{134}Cs and ^{137}Cs), radiostrontium (^{90}Sr), and plutonium (^{239}Pu). [10]. It is because of scientists, such as Ernest Sternglass, who investigated and proved the negative effects of the bomb fallout on childhood mortality. Their warnings helped to enact the partial test ban treaty,[1] which most nuclear powers ratified [11]. After 1965, the fallout clearly decreased. One can clearly see this in sediment profiles of lakes (e.g., lake sediment investigations in Switzerland) [12].

NPPs are also emitters of artificial radionuclides. Several accidents caused the release of large quantities of radioactive fallout to the environment. Accidents were the burn of one pile at the Windscale reactor in Sellafield in 1957, the partial core melting of the Three Mile Island reactor in Harrisburg in 1979, the nuclear catastrophe of the Chernobyl NPP in 1986, and the core meltings of the NPPs of Fukushima-Daiji in Japan 2011. The catastrophe of Chernobyl affected many European countries, many thousands of miles away from the NPP ground.

The use of radionuclides in diagnoses and therapies against cancer leads to the release of short-lived radionuclides, such as iodine (^{131}I), technetium (^{99m}Tc), indium (^{111}In), lutetium (^{177}Lu), yttrium (^{90}Y), and others. They do not enter the food chain because of their short half-lives and are therefore of minor concern.

4. Radiation detection

The following overview of radioactivity laboratory equipment for food control is only a brief summary of the required equipment. For information about the theory of radiation detection and measurements, technical details, or special applications, we refer to the standard literature [13–16].

4.1 Dose rate monitors

Measurement devices to count the dose rate of radioactive sources are not sensitive enough for the precise contamination measurements of food. However, they can be adequate for

[1] Treaty banning of nuclear weapon tests in the atmosphere, in outer space, and under water. 5 August 1963.

screening analysis. In 2013 and 2014, the State Laboratory Ticino and the Federal Office of Public Health used dose rate monitors to screen wild boars shot in the southern parts of Switzerland. They only sampled animals over a certain dose rate for precise γ-spectrometric analysis for radiocaesium [17]. This kind of measurement equipment is not sensitive enough to detect contamination in the case of food imports from Ukraine, Japan, and other contaminated countries. Therefore, one cannot save the counting on a γ-spectrometer.

4.2. γ-Spectrometry

It is of primary importance to be equipped with a γ-ray detector. γ-Spectrometry can be operated with inorganic scintillators, such as pure crystals of sodium iodide (NaI) and caesium iodide (CsI) or doped crystals, such as NaI(Tl) or CsI(Tl). NaI(Tl) is the most used scintillator and has excellent light yield and a good linear response, but energy resolution is quite limited. Today, the best choice is semiconductor detectors. Crystals of silicon, germanium (Ge), cadmium-telluride, and others are the detector materials. Ge detectors have been widespread since 1980, when the production of high-purity Ge monocrystals became possible. They have to be kept under vacuum and cooled with liquid nitrogen. The detectors have an excellent energy resolution. Therefore, even complex spectra can be analysed without prior chemical separation steps. Two criteria are of importance: resolution and sensitivity. Energy resolution of Ge detectors is excellent: 1.5 to 2.5 keV at 1.33 MeV, below 100 keV, less than 1 keV FWHM (full width at half maximum). The efficiency of the detector (relative efficiency compared to a NaI detector at 1.33 MeV) and the relation of the peak to Compton background are the most important factors for the sensitivity. Today, Ge detectors with efficiencies of 25% and higher are available. The peak/Compton quotient is more than 46. In our laboratory, we use Ge detectors with 50% efficiency. Important factors for quantitative γ-spectrometry are the shielding of the detector and the efficient suppression of the electronic noise of the amplifier system. γ-Spectrometry has the advantage that γ-radiation can be measured without the elimination of the matrix. Therefore, sample preparation takes only a short time. γ-Spectrometers need an exact calibration over the whole energy range (e.g., 50–2000 keV). Calibrations with certified radioactive sources are necessary for every counting geometry used (volume and shape of the sample, and distance from the detector) Calibration solutions consist of a mix of short-lived radionuclide with emission lines over the whole energy range (e.g., ^{109}Cd, ^{57}Co, ^{113}Sn, ^{137}Cs, ^{88}Y, and ^{60}Co). After 1 year, the short-lived radionuclides are partly disintegrated; therefore, the calibration mix cannot be used anymore. This can be overcome using mixtures of ^{152}Eu (half-life of 13.5 years) in combination with a low-energy γ-nuclide, such as ^{210}Pb or ^{241}Am. When using ^{152}Eu, summation effects have to be corrected properly.

There are other important factors to consider besides the counting geometry. The sample matrix itself absorbs γ-rays before they arrive at the detector. These self-absorption effects are a function of the sample geometry and the elemental composition and the density of the sample matrix. Normally, calibrations and efficiency curves are produced with calibration standards in water or gels of density 1. Density corrections can be calculated for every material and geometry with means of a software based on Monte Carlo simulations. The background radiation has to be considered. Some background radiation remains, even with good shielding with lead and copper. This background consists of radionuclides of the natural decay series, such as ^{214}Pb and ^{214}Bi and others. For every counting geometry, the background has to be measured with water-filled containers of the needed counting geometries. The γ-spectra have to be subtracted by the specific background spectrum. Radionuclides with cascade emissions show coincidence summing effects (e.g., ^{134}Cs and ^{152}Eu). Spectra have to be corrected or the measurement must be repeated with a distance between sample and detector. For short-lived radionuclides, their partial decay has to be corrected to the reference date (e.g., the date of the sampling). Further advice and descriptions over quantitative γ-spectrometry are given in the literature [18].

4.3. β-Spectrometry

For the counting of β-rays, the sample matrix has to be eliminated. An exception is water samples (e.g., the measurement of ^{3}H needs not much sample preparation, only the mix of the water sample with a scintillator cocktail). Some important β-nuclides, such as ^{3}H, ^{14}C, ^{89}Sr, ^{90}Sr, and ^{90}Y, can be analysed with scintillation counting. Commercially available scintillation counters can detect α- and β-decays. This widens the spectrum of radionuclides (e.g., ^{222}Rn can be analysed in water samples or in charcoal air samples). When samples are analysed directly, the sensitivity is given by the small sample amount of typically some millilitres.

Radiostrontium, ^{89}Sr and ^{90}Sr, are important fission products. One possibility is to extract the ^{90}Sr with the use of specific crown ethers from the sample. In our laboratory, we have developed a fast analysis scheme for water samples [19]. Another possibility is to clean up extracts over a column filled with crown ethers. These methods are suitable for activity concentrations higher than 1 Bq/kg. For sensitive analyses, the β-spectrometers of choice are gas flow proportional counters. We use this technique for the analyses of ^{90}Sr traces in food, human, and environmental samples. The method is based on the counting of the daughter nuclide, ^{90}Y. Before the counting starts, a rigorous elimination of the matrix and disturbing β-nuclides, such as ^{40}K, is necessary. With an oxalate precipitation step, most of the ^{40}K is eliminated. Then, ^{90}Y is separated from ^{90}Sr by precipitation as hydroxide. The $Y(OH)_3$ is dissolved and precipitated as $Y_2(oxalate)_3$. These β-sources are pure enough for counting. Counting is performed in 10 consecutive runs, as ^{90}Y decays during the counting (half-life is 64 h). A good-quality criterion for the purity is the measured decay of the source. Decay should be near 64 h. When decay is slower, impurities are present. The conserved ^{90}Sr solution may be prepared and analysed again after 20 days (the built-up ^{90}Y will then arrive equilibrium with ^{90}Sr). Quite sensitive analyses may be performed down to 10 mBq/kg. Counting time is 3 days. Therefore, several detectors should be available. Our Canberra α/β-counter LB 4100 can take up to four drawers with four sample holders each [20].

4.4. α-Spectrometry

Like β-spectrometry, α-spectrometry requires an elimination of the sample matrix. Only water samples need a minimal preparation. Two counting techniques are common today: scintillation counting and passivated implanted planar silicon (PIPS) detectors. We use liquid scintillation counters in our laboratory for the analyses of uranium, thorium, radium, and polonium.

According to the methods published by W. Jack McDowell, the water sample is extracted once with some millilitres of a nuclide selective extractant, which contains the scintillator cocktail for the α-analysis. Very low activity concentrations (5–10 mBq/L) can be achieved [21]. The disadvantage is the poor energy resolution. Also, α/β-discrimination has to be set carefully. Analysis time is 24 hours for water samples containing 100 mBq.

Analyses with PIPS detectors show good energy resolution and efficiencies from 20% to 30%. Counting has to be performed under vacuum and very thin layer α-sources are needed. The sample matrix has to be destroyed (by ashing, etc.) following a clean-up (specific nuclide extraction, scavenging, etc.) and a preparation of a thin-layer source by means of electroplating or coprecipitation. Many actinides, such as uranium, thorium, and plutonium, may be analysed (well described in [22]). Some elements such as ^{210}Po and radium nuclides may be auto-deposited onto special surfaces. ^{210}Po is auto-deposited under reductive conditions onto silver or copper disks after a microwave digestion of the sample [23, 24]. Radium nuclides (^{224}Ra and ^{226}Ra) are auto-deposited onto MnO_2 surfaces at pH 8 as has been shown by Surbeck [25, 26]. These methods are suitable for drinking water and mineral water analyses. Some new developments were done for uranium and thorium analyses in honey and spices [27, 28] (**Figure 1**).

Figure 1. Equipment of a radioactivity laboratory: (A) γ-spectrometry; (B) α-liquid scintillation (PERALS); (C) α-Spectrometry (PIPS detectors); and (D) β-spectrometry (gas proportional counter), drawer with ^{90}Y oxalate sources.

4.5. Neutron activation analysis (NAA)

INAA (instrumental neutron activation analysis) is based on the production of radionuclides by nuclear reactions. Thermal neutrons can activate many elements. The efficiency of this irradiation process depends on the flux density of the neutrons and the cross-section of the nuclear reaction. The thermal neutrons are generated in a nuclear reactor, for example, at the University of Basel (AGN-211-P, a light water moderated swimming pool reactor).

The samples (1–5 g material) are inserted into the core over a cannula through the so-called glory hole and irradiated for 30 min with a power rate of 2 kW. After a cooling time of some hours, the samples are counted on an HPGe detector.

We used INAA for the analysis of total bromine content. Bromide is built by the decay of the fumigant methyl bromide. We used this technique for many years to determine the total bromine in spices, tea, and dried mushrooms. Another application is the determination of the total bromine content as a screening analysis for flame-retardants in plastic materials [29]. ^{238}U and ^{232}Th can be determined by INAA in suspended matter and sediments [30]. In addition, total iodine content in iodine rich food, such as algae, can be determined over the activation of ^{127}I to ^{128}I, which decays to ^{128}Xe [31].

We mentioned the use of INAA as a completion of the possibilities of γ-spectrometry. These applications will not be discussed further, because the analytes are not radionuclides.

5. Results from 35 years of food control

In Basel, the Government decided to buy the equipment for the monitoring of β- and γ-nuclides in 1980 because of the NPP accident at Harrisburg in 1979. Therefore, we were the only state laboratory that was prepared when the accident at Chernobyl happened. In 1986 and 1987, thousands of samples were analysed.

The fallout from bomb tests and the NPP accident of Chernobyl resulted in the ubiquitous contamination of the landscapes of the Northern Hemisphere. The core meltings at the NPP of Fukushima-Daiji also reached Europe, but the fallout was considerably lower than from Chernobyl. The situation after the Chernobyl fallout in Switzerland is well described in some papers [31–33, 98].

The contamination of farmland leads to contaminated food, such as milk or vegetables. These matrices are an important part of the Swiss survey programme. The fallout from Chernobyl affected the regions of southern Switzerland the most (total rain of 350 mm). This has to be compared to the washout in eastern Switzerland with 150 mm and the regions of Basel and Jura with 50 mm rain.

5.1. Milk and milk products

In 1981, the State Laboratory Basel-City started the first regular radioanalyses in Basel. Milk samples from local milk production centres in northwest Switzerland of the states (cantons) of Aargau, Basel-Campaign, Basel-City, Solothurn, and Jura were analysed with β- and γ-spectrometry. On 2 May 1986, after the accident in Chernobyl, the frequency of the survey was intensified. On 4 and 5 May, radioiodine activity concentrations between 220 and 650 Bq/L ^{131}I were measured. The milk from the mountainous region of the state of Jura showed lower values than in the states localised in the plain: 178 Bq/L ^{131}I (4 May). In the milk distribution centres of Basel-City, the milk from the different regions had to be mixed in such a way that the population received milk with activities below 350 Bq/L. Due to its short half-life, the activity of radioiodine fell under the detection limit (0.1 Bq/L) in July. The activities of caesium and strontium were quite lower but resisted longer. The activity line of total caesium in the milk of the state of Jura (**Figure 2**) shows a maximum value of 127 Bq/L. In the following 3 years, smaller peak values could be observed due to the fact that the cows were fed with contaminated hay from the year before. After 1990, the total caesium level fell below 1 Bq/L, except for some farms in southern Switzerland, where, even in 2013, the radiocaesium level of the milk of one farm was over the tolerance limit of 10 Bq/kg.

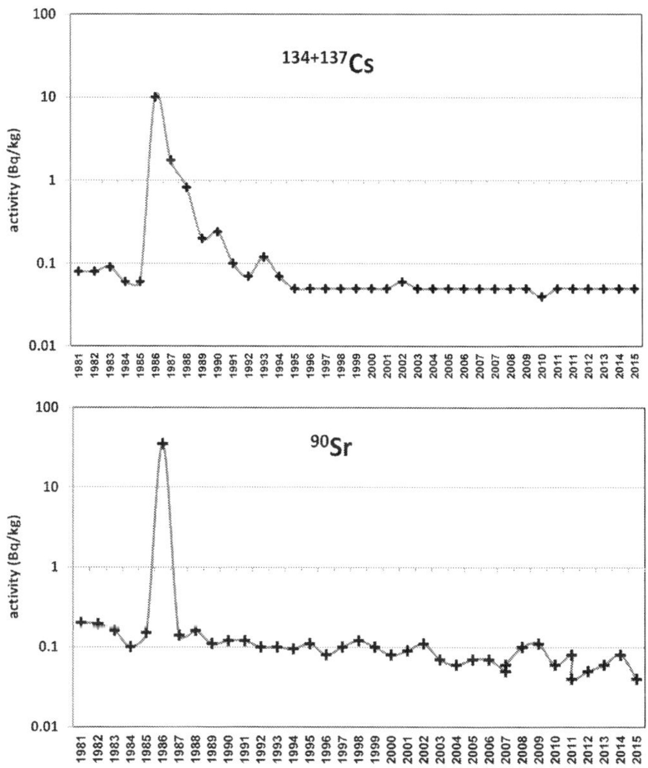

Figure 2. Results of the monitoring of milk from northwest Switzerland. Notice the sharp peaks due to the Chernobyl fallout. The ^{90}Sr activity concentration shows a steady decline from 1981 to 2015.

We found much lower activity of ^{90}Sr with a maximum value of 4.4 Bq/L on 11 March 1987 in Jura. In the Swiss Alps, the level reached 35 Bq/L. After 1988, the contamination of the milk from the canton of Jura reached a value of approximately 0.1 Bq/L. The present contamination of Swiss milk is highest in the Alpine regions of the states of Grison and Ticino, where the radiostrontium level is a factor 10 higher than in the rest of Switzerland (0.1–0.4 Bq/L), and radiocaesium reaches 9 Bq/L for the highest value (0.3–9 Bq/L) [34, 35].

Many milk products were analysed in 1986: Swiss cheese, milk powder, butter, yogurt, and cream [34, 36]. These investigations showed the same trends in somewhat lower activity concentrations. Two drinks also showed higher levels of radiocaesium (22–96 Bq/L in 1987), as they contained milk serum. The same contaminated milk from 1986 was used to produce milk powder. This milk powder was used for the production of chocolates. Milk chocolate contains 20 to 25 g milk powder (160–200 mL milk) per 100 g chocolate. We analysed the first samples in autumn 1986 for a local chocolate producer. Soon, we noticed higher levels when chocolates contained hazelnuts. These chocolate samples were in the range of 55±13 Bq/kg radiocaesium. In contrast, chocolates without nut ingredients showed lower activities (12±3 Bq/kg). These investigations showed that the contamination level of chocolates is even more from the use of hazelnuts (60%) than from the milk (~40%). In the following year, we found even higher radiocaesium values: 710 Bq/kg in chocolates without hazelnuts and 1.3 kBq/kg in chocolates with hazelnuts. Even chocolates without hazelnuts showed higher values. We explained this with higher values in milk. Hence, we started a special survey programme for hazelnuts.

5.2. Wild-grown vegetables

5.2.1. Hazelnuts and other nuts

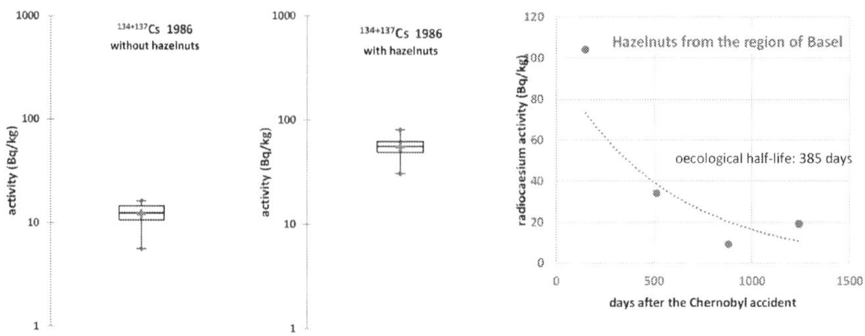

Figure 3. Activity concentrations of radiocaesium in hazelnuts. Plots to the left: Comparison of the radiocaesium level in chocolates with and without hazelnuts (1986). Plot to the right: Development of the activity in hazelnuts from Basel.

The investigation of chocolates containing hazelnuts showed that the latter were part of the contamination. In 1986 and 1987, more than 180 hazelnut samples from a chocolate producer of the region were analysed for radiocontaminants. In 1987, the contamination level reached a total radiocaesium activity of up to 17 kBq/kg. Most hazelnuts were imported from Turkey, a country that was seriously contaminated with fallout from Chernobyl. Since 1989, we also investigated different nuts from other countries. Hazelnuts from Turkey remained the only contaminated nut species. Even in 2007, 21 years after the catastrophe of Chernobyl, one hazelnut sample had to be rejected. It did not reach the limit value but was over the tolerance value of 10 Bq/kg. After 1990, the contamination of hazelnuts in Switzerland and abroad remains under 10 Bq/kg; meanwhile, in Turkey, the contamination level was reduced much more slowly due to the higher contaminated soils (**Figure 3**).

5.2.2. *Brazil nuts*

Brazil nuts (*Bertholletia excelsa*) are known to accumulate earth alkaline metals from soil. First, this was shown with barium. Then, Penna-Franca detected the radium nuclides ^{226}Ra and ^{228}Ra, both from the natural decay series of uranium and thorium, in amounts of 50 Bq/kg. In Switzerland, imported Brazil nuts showed 61 to 112 Bq/kg total radium. This means 1000-fold of the radium content of the total diet in Europe. The consummation of Brazil nuts is estimated to be 0.1 g/day and person. Therefore, the consummation of Brazil nuts is not relevant. In literature, it is advised to consume Brazil nuts to enhance the selenium level. Two Brazil nuts lead to a yearly dose of 160 μSv, which is relevant [8].

Another important alkaline earth metal is strontium. Its most important radioactive nuclide is ^{90}Sr. Froidevaux et al. [37] measured 11 to 15 Bq/kg in imported Brazil nuts. As Brazil nuts are often part of nut mixtures, the measured natural radioactivity is mainly produced by the Brazil nuts contained in the mixture.

5.2.3. *Mushrooms*

Mushrooms are known to be organisms that can accumulate heavy metals from soil. Therefore, mushrooms from abroad and from Switzerland were of concern in 1987. More than 200 samples were analysed in our laboratory for radiocaesium. 24 samples had to be rejected. They showed a contamination level of more than 600 Bq/kg. The range of all analysed mushrooms was from 107 to more than 8000 Bq/kg. Swiss mushrooms and mushrooms from abroad then did not show any significant difference in their contamination level [38]. From 1989 on, mushrooms were regularly analysed, but violations became rare.

A quite different situation could be observed in South Bavaria, Germany. At the end of April 1986, much Chernobyl fallout was washed out in southern Bavaria over several days. In Munich, the local dose rate increased up to 1.1 μSv/h (10-fold over the normal value). Therefore, the contamination level in wild-grown mushrooms and berries reached high values (five-fold the normal radiocaesium activities due to the bomb fallout). Mushrooms with mycorrhiza, such as boletus species (*Hydnum repandum* and *Boletus badius*), blueberries, mosses, and lichens, were most affected [39–41]. Other mushrooms, such as porcini, champignons, and chanterelles, did not show this effect. In 2002, bay boletes (*B. badius*) showed radiocaesium levels of 33 to 23,200 Bq/kg and porcini from 6 to 10,000 Bq/kg. In 2015, activities were reduced remarkably in boletes 13 to 900 Bq/kg and porcini 4 to 108 Bq/kg [39]. Similar values were reported in Austria [42] and Switzerland [43]. Even in 2015, bay boletes were found with maximum activities of more than 2600 Bq/kg, whereas porcini showed a reduced contamination level with a maximum of 416 Bq/kg. The reduction was slow compared to West European countries. For example, in Spain, values ranged from 1 to 122 Bq/kg for radiocaesium and from 0.2 to

3.5 Bq/kg for radiostrontium. Mushrooms of a special region in the canton of Aargau (Switzerland) show a continuous fall of the activity from 1986 to 2014 [44].

Wood soils build their own biosphere: dead leaves, pins, etc., build the humic fraction of the soil. The plants take up the radioactivity from the soil. The radioactivity is recycled to the soil through the fall of the leaves and pins.

5.2.4. Wild-grown berries

In 1986, the State Laboratory Basel-City analysed approximately 30 berry samples from Switzerland. Strawberries showed a quite low activity (2.7 Bq/kg radiocaesium), whereas currants, raspberries, elderberries, and gooseberries showed significantly higher activities (51±27 Bq/kg). The reason is that strawberries are cultivated in greenhouses, in the shelter. Only in 2009 was our focus set on blueberries and other wild-grown berries. These products are imported in big charges of tons from East European countries, such as Ukraine, Poland, Russia or Hungary. From 16 analysed samples, we registered eight violations because of too high activities of radiostrontium. In the following years, we analysed more than 100 samples with 10 further violations for products from Austria, Ukraine, or Poland. The products were blueberries and blueberry products, such as marmalades. Violations were mostly because of radiostrontium (>1 Bq/kg) and radiocaesium (>100 Bq/kg) [45]. Wild berries grow in woods. The soil is more acid and promotes the uptake of the contaminants. Because of the cycle soil-plant-soil, the residues in the plants are reduced only slowly compared to berries grown on farmland.

5.3. Vegetables and fruit

5.3.1. Vegetables

Leafy vegetables were, besides milk, the most contaminated food. Depositions from washouts on spinach and salad cultures happened just before harvest. Radioiodine level reached almost 4000 Bq/kg in spinach. Therefore, it was advised that small children, pregnant women, and nursing women should not consume this kind of food. The investigation of mother's milk showed considerable amounts of radioiodine (up to 35 Bq/L). We tried to reduce the contamination on salads with washings but with only poor success. In May 1986, leafy vegetables were most affected by radioiodine (^{131}I and ^{132}I): radioiodine was 65% of the measured total dose. Radiocaesium (^{134}Cs and ^{137}Cs) was 25% and other short-lived radionuclides, such as ^{103}Ru, ^{140}Ba, ^{140}La, and ^{99}Mo, gave approximately 10% to the total dose. We noticed many violations of the limit values for radioiodine and radiocaesium (24 and 10 violations for baby food from a total of 40 samples analysed). This survey focused on food with higher radiation levels as a screening before the γ-analyses were done. The investigation of deep-frozen vegetables did not show any elevated radiation. These products were produced before 1986 [46]. After 2010, we started up again with controls of leafy and root vegetables. The contamination level was low and caused by the fallout from the bombs. Radiocaesium is normally below the detection limit of 0.1 Bq/kg. Radiostrontium is detectable in small amounts of 0.05 to 0.5 Bq/kg [47].

5.3.2. Herbs

Herbs of 1986/1987 showed the highest contamination with radiocaesium of all investigated food. In 1986, 35 of 42 herbs contained up to 3.200 Bq/kg. In the following year, the contamination level was even higher: 12.300±32.000 Bq/kg [38].

5.3.3. Fruit

In 1986, fruit was less contaminated than vegetables. When harvesting time arrived, the radioiodine had already disintegrated. The radiocaesium level was below 100 Bq/kg. In 1990, no more contamination was found (<2 Bq/kg radiocaesium in dried fruit) [46].

5.3.4. Chestnuts

Chestnuts, the fruits from the chestnut tree, *Castanea sativa*, are cultivated in northern Italy and southern Switzerland. At harvest time, in autumn, they are roasted and eaten or processed to crèmes or purees. These were also the regions with high fallout and washout from Chernobyl. The soils and vegetation had elevated activities of radiocaesium. As a consequence, chestnuts contained elevated radiocaesium levels. In 1986, a level of 265 Bq/kg was reached. In the following years, the contamination was reduced but only slowly. Even in 1988, one sample violated the tolerance value of 10 Bq/kg. In 2005 and 2015, we again investigated chestnuts and chestnut products. The radiocaesium level was reduced by a factor of 10. The resulting level was approximately 1 to 4 Bq/kg. No more violations were noticed [48].

5.4. Flour, bread, and biscuits

One year after Chernobyl, corn, grains, bread, shortbreads, and pasta were seriously contaminated with radiocaesium. In 1987, cereals from 1986 were taken into production. Seven of 17 breads and shortbreads and one pasta were over the limit value of 1250 Bq/kg and had to be withdrawn. Twenty-one more samples were over the tolerance level of 100 Bq/kg. As a consequence, the control of flour and flour products was intensified over the next years. However, with the exception of one sample in 1988, no more violations of the tolerance value were noticed. In 2011, the radiation level fell down to approximately 0.5 Bq/kg radiocaesium in cereals.

Natural radionuclides are present in cereals (radium 1–5 Bq/kg and thorium 0.1–1 Bq/kg). The levels of radium are near the limit value of 5 Bq/kg.

5.5. Meat

Our own γ-analyses of meat in 1986 showed values in two categories: for cows and calves, the level was low (57±71 Bq/kg), whereas meat from sheep and lamb was quite higher contaminated (645±744 Bq/kg radiocaesium). In the following year, the radiocaesium levels were elevated to the same level: sheep and lamb 550±198 Bq/kg and cow and calf 417±384 Bq/kg. During 1986, the Federal Food Safety and Veterinary Office (former Federal Office for Veterinary Affairs) analysed more than 1700 meat samples of sheep, goat, cow, pig, and game mainly from eastern and southern Switzerland. The contamination levels in the meat were higher in regions with higher depositions of fallout. In southern Switzerland, the radiocaesium activity concentrations varied from 40 to 4400 Bq/kg. The highest value was found in a goat. In contrast to this, in eastern Switzerland, values were from 40 to 1300 Bq/kg. In the other parts of Switzerland, even lower values were found. Almost 2000 samples from imported meat were analysed. Here, 251 samples had to be rejected due to values more than 600 Bq/kg. During the year 1986, a reduction of the contamination level was observed, with the exception of southern Switzerland (cantons of Ticino and Grison). The biological half-life for radiocaesium was calculated to be approximately 50 days for sheep and goat and approximately 30 days for game. Pigs showed a half-life of 90 days [49].

5.6. Game

Game became of interest when high radioactive contaminations of reindeer were reported in Norway and Sweden. The northern European countries were more highly contaminated with radioactive fallout from Chernobyl than other European countries. Wild berries, mushrooms, and lichens are the main food of reindeers. These were seriously contaminated with this fallout. Thousands of animals had to be burnt because of a violation of radiocaesium values that were too high. The survey of game in Switzerland began in autumn 1986. We analysed meat from five roe deer and deer with γ-ray spectrometry. The activities were not high. Nevertheless, in 1987, the contamination of the game meat showed higher values up to 7 kBq/kg radiocaesium. From 1986 on, game was investigated yearly. In 1990, four objections had to be executed according to our measurements. It seemed that chamois were the most contaminated game in Switzerland. After 1995, the monitoring programme was reduced. The detected radioactivity was under the tolerance limit of 600 Bq/kg.

Wild boars are an exception. The southern parts of Switzerland, such as in Bavaria (Germany), are more contaminated landscapes. Here, the contamination of wild boar remains a problem up to today. Wild boars search for their food on the ground. Elaphomyces, a truffle species, grow underground and are able to enrich radiocaesium from soil. It is estimated that these fungi can be up to 20% to 30% of the food of wild boars. In 2012, in southern Bavaria and the Bavarian Wood, the contamination levels were up to 9.8 kBq/kg wild boar and up to 430 Bq/kg in roe deer [50]. In Switzerland, the State of Ticino, in collaboration with the Federal Office of Public Health, investigated wild boars in 2013 and 2014. In 2013, 28 wild boars of a total of 470 animals violated the limit value of 1250 Bq/kg. In 2014, they found 13 such contaminated animals. These animals had to be confiscated [17]. In contrast, in the State of Zurich, the State Laboratory Zurich found no violations when they analysed 80 wild boars. The mean activity was low with 28 Bq/kg radiocaesium [51]. In 2014, the Umweltinstitut München reported from their monitoring programmes for wild-grown vegetables and game. More than 2000 samples showed contamination levels over the limit value of 600 Bq/kg, and 141 samples showed radiocaesium activities of more than 10,000 Bq/kg. Ten samples contained more than 16,000 Bq/kg. The maximum value was 27,800 Bq/kg [52]. Therefore, the serious contamination rests a problem in Bavaria.

http://dx.doi.org/ 10.5772/62460

Wild boars have to be surveyed over the coming years. Fortunately, they are not a widely consumed game. More important are deer and roe deer, which show considerably lower contaminations [53, 54].

5.7. Seafood and fish

5.7.1. Fish

In Switzerland, we analysed fish from the Rhine River at Basel for radioactive contamination in 1986. Fifteen species caught by local fishermen contained 22 to 707 Bq/kg radiocaesium. These activities were not alarming, as fish are not an important part of the daily food consumption in Switzerland. Again, the most affected region was the southern part of Switzerland. The mean activity of 70 fish species caught in the lake of Lugano was 1.09±0.6 kBq/kg radiocaesium with a maximum value of more than 4.4 kBq/kg, approximately five times higher than in fish from other lakes of Switzerland. This lake has no major confluences and effluents such as the lake of Maggiore where the contamination level of fish was quite lower [49, 55]. After 1987, the contamination with radiocaesium from the Chernobyl fallout was reduced and reached a level of approximately 0.2 Bq/kg for ^{137}Cs [56].

The accidents at the Fukushima-Daiji NPPs gave us cause to investigate fish importation from the Pacific Ocean. Approximately 90% of the released fallout reached the sea (4–90 PBq ^{137}Cs). Radiocaesium levels in fish reached 200 kBq/kg. In 2011, the Japanese Government banned fishing in the coastal waters near Fukushima NPP and the fishing of fish species in some prefectures, which are severely contaminated. Local, private associations of fishermen voluntarily imposed a limit value for radiocaesium of 50 Bq/kg [57]. Also in territorial waters, such as lakes, ponds, and rivers, fish accumulated radioactivity from fallout. Here, radioactivity levels reached approximately 10 kBq/kg fish. In 2012, an intensified monitoring programme of Pacific blue tunas off the Californian coast showed a slight contamination of the fish (0.7±0.2 Bq/kg ^{134}Cs and 2.0±0.5 Bq/kg ^{137}Cs). The presence of the short-lived ^{134}Cs proves the contamination from the Fukushima fallout [58–61]. A received dose of 1 mSv/year was estimated from the consumption of 50 kg fish caught within a zone 3 km away from the Fukushima NPP. Our own investigations of imported fish from the Pacific show a relatively low contamination level below 1 Bq/kg radiocaesium, with a mean value of approximately 0.3 Bq/kg. Some fish samples also contained the short-lived caesium-nuclide ^{134}Cs [62]. Only 14% of the total dose of the consumption of fish and sea food originates from artificial radionuclides, and 86% is from natural radionuclides, such as polonium (^{210}Po).

5.7.2. Natural radionuclides

Natural radionuclides cause the main radiocontamination of fish and seafood. Mussels and molluscs are known to enrich ^{210}Po in the intestinal tract, whereas the mother nuclide lead-210 (^{210}Pb) is not enriched. Cherry and Shannon [63] published an excellent review. Both radionuclides are part of the natural decay series of uranium and are built at the end of the decay chain. ^{210}Po is a powerful α-emitter with an energy of 5,500 keV and a half-life of 183 days. ^{210}Pb is a β-emitter with a half-life of 23 years and acts as a reservoir for ^{210}Po. Activity concentrations range from 20 to 100 Bq/kg. In fish, the ^{210}Po level is much lower (1–20 Bq/kg) [64].

Our own investigations on imported seafood in 1998 resulted in 12 objections in mussels and 2 objections in sardines concerning too high levels of ^{210}Po. For food producers and food distribution agents, this was surprising. ^{210}Po was never seen as a problem. One consumes sardines as a whole fish, the intestinal tract included [65, 66]. This explains the higher contamination level of sardines and anchovies. A second survey in 2010 showed values equal to those in 1998. Since 1990, the limit value for ^{210}Po in fish was raised from 10 to 150 Bq/kg (the rate of fish and seafood consumed in Switzerland is of minor relevance). Therefore, since 1998, no more objections had to be raised [56]. A survey of the ^{210}Pb and ^{210}Po contamination of seafood in France over the last 15 years reports the same contamination levels [67]. Even higher ^{210}Po levels were found in anchovy from local fishers at the Turkish coast of the Aegean Sea. The annual dose by ingestion was calculated to be 15 μSv [68]. Low activity concentrations were found in fish caught in Swiss lakes. A mean value of 87 samples was 0.4±0.3 mBq/kg ^{210}Po. Such low values are not astonishing. Only the edible parts of the fish, without the intestinal tract and entrails, were analysed. Measurements of entrails of 34 fish samples showed a mean activity concentration of 25 Bq/kg [69].

5.8. Baby food

Baby food is infant follow-on formula that is industrially produced from cow's milk or soybeans. It is given to children up to 4 months after birth. For this kind of food, more restricted limit values are regulated concerning radionuclides. The given limit values are calculated to the final reconstructed constitution of the food (**table 1**). Radioactive contaminants are introduced through the milk into the products. Therefore, radiostrontium and radium are of special interest.

In 1987, an investigation of 56 samples of follow-on formulas showed a severe contamination with radiocaesium. Four samples exceeded the limit value of 400 Bq/kg; the highest value was more than 5.000 Bq/kg. Ten further samples contained radiocaesium in amounts greater than today's tolerance limit of 10 Bq/kg. However, no radiostrontium was analysed, so it is unknown how more violations were present concerning too high activities of ^{90}Sr. In 2007 and 2012, we analysed baby food for both radionuclides. Whereas radiocaesium levels were quite low (<0.05 up to 0.5 Bq/kg), the radiostrontium contamination reached almost the same values (0.3 Bq/kg). Radium belongs to the same element group of the earth alkaline metals as calcium and strontium. Therefore, it is not surprising to find contaminations with radium (^{226}Ra and ^{228}Ra) in infant formulas (0.1–0.8 Bq/kg) [70, 71].

5.9. Spices and salt

In 1986/1987, spices were of no special concern. They figure as a food of minor relevance, because the consumption rate of spices is relatively low in Switzerland. A second reason is the fact that spices are imported from the Middle and Far East, where they were not at all affected by the fallout from Chernobyl. Over the last 10 years, the radiocaesium content in spices was

http://dx.doi.org/ 10.5772/62460

stable: 1 to 5 Bq/kg for 200 samples analysed. Higher activity concentrations were found for natural radionuclides from the uranium and thorium decay series. For example, white and black pepper and cinnamon contain considerable amounts of thorium and radium (up to 40 Bq/kg). This is near the Swiss limit value of 50 Bq/kg [72].

Salts belong to the spices. Either they are produced from evaporation of seawater or are yielded from mines or salt fields. A survey of 23 products gave the following results: artificial radionuclides, such as radiocaesium, are not present. A major contaminant of salts is potassium chloride. Therefore, it is not surprising that ^{40}K activity concentrations are relatively high in salts: 330±30 Bq/kg. One salt from Persia contained 6 kBq/kg ^{40}K. Some radionuclides from the decay series, such as radium, are present in small amounts: 1.2±1.1 Bq/kg $^{226+228}Ra$ [73].

5.10. Honey

In 1986, 40 Swiss honey samples were investigated with γ-spectrometry. They contained ^{131}I with a mean of 40.3 Bq/kg (6 samples) and radiocaesium with 54±47 Bq/kg (35 samples). The highest value was 192 Bq/kg of ^{137}Cs. Six samples contained radioiodine over the tolerance limit of 10 Bq/kg, and six samples were over the tolerance level for radiocaesium of 10 Bq/kg. No violation of the limit values was observed [74].

Honey is considered as food of minor importance. Therefore, its survey was stopped in 2004. From 2005 on, we analysed more than 150 honey samples. They can be divided into two groups: honey from flowers and forest honeys (included chestnut honeys). Honey from flowers show only small amounts of contamination: 0.2 to 5 Bq/kg radiocaesium. In forest honey, one can find up to 25 Bq/kg ^{137}Cs. Sporadically, we found violation of the tolerance values for radiocaesium and radiostrontium. One honey from Austria contained 1.6 Bq/kg ^{90}Sr and 176 Bq/kg ^{137}Cs. These products, and especially products from East European countries, contain elevated contaminations even 30 years after the accident at Chernobyl [75, 76].

5.11. Tea

After the Chernobyl accident, we analysed tea with γ-ray spectrometry. From 21 samples, 12 teas exceeded the tolerance value of 500 Bq/kg. The mean activity found was 12.4±10 kBq/kg radiocaesium. One tea showed 429 Bq/kg, a clearly elevated contamination. In the following year, no sample exceeded 500 Bq/kg radiocaesium. After some years, the contamination levels were reduced to below 5 Bq Cs/kg with one exception. Contamination in black tea from Turkey was only slowly declining. Even in 2015, almost 30 years after the Chernobyl accident, the radiocaesium level reached 50 Bq/kg, and we found radiostrontium in amounts of 38 Bq/kg maximum. In 2011, our focus was set on imported tea from Japan. Until the end of 2015, we analysed more than 150 tea samples coming from different prefectures in Japan. The γ-analyses proved the contamination of green tea from the Fukushima-Daiji NPP's accident. At least, part of the measured radiocaesium originates from the fallout of the NPP's accident. This is proven by the presence of the short-lived radionuclide ^{134}Cs (2.1 years). In 30 of 157 investigated tea samples, ^{134}Cs was present in amounts of 13.4±24.8 Bq/kg. Besides tea, other food categories that are imported from Japan were analysed. Over the last 5 years, we analysed more than 350 food samples (**Table 2**). As can be noticed, the radiostrontium level of tea is approximately 5±7 Bq/kg and could be found in every tea sample analysed. This contamination mainly comes from the bomb's fallout. Teas from other countries of the Far East also contain radiostrontium [77, 78].

Food category	^{134}Cs	^{137}Cs	$^{134+137}Cs$	^{90}Sr
Tea (157)	13.4±24.8 (30)	9.9±28.6 (86)	8.0±33 (90)	5.2±7.4 (94)
	<0.5–87	<0.5–171	<0.5–258	<0.1–57
Soups, miso, (44)	<0.5	0.34±0.24 (9)	0.34±0.24 (9)	N/A
		<0.5–0.7	<0.5–0.7	
Algae (60)	<0.2	0.6±0.7 (9)	0.6±0.7 (9)	0.56±0.31 (12)
		<0.2–2.4	<0.2–2.4	<0.01–1.0
Rice and rice products (24)	<0.2	0.31±0.04 (2)	0.31±0.04 (2)	N/A
		<0.2–0.3	<0.2–0.3	
Soja and soja products (7)	<0.1	0.5±0.2 (1)	0.5±0.2 (1)	N/A
Cereals and cereal products (26)	<0.2	6.3±8.1 (2)	6.3±8.1 (2)	N/A
		<0.2–12	<0.2–12	
Vegetables, fruits (15)	3.9 (1)	5.0±6.9 (2)	6.9±9.6 (2)	0.5±0.4 (2)
	<0.5–3.9	<0.5–9.8	<0.5–14	
Fish and fish products (4)	<0.2	0.3±0.2 (1)	0.3±0.2 (1)	N/A
		<0.2–0.3	<0.2–0.3	
Divers (28)	<0.2	0.52±0.54 (2)	0.52±0.54 (2)	N/A
		<0.2–0.14	<0.2–0.14	

All values in Bq/kg. First line: mean ± standard deviation of the activity concentrations; the number of samples with values over the detection limit is bracketed. The total of analysed samples per food category is set in brackets after the food category name. Second line: activity concentration range of all samples. N/A, not analysed.

Table 2. Overview of investigated food imports from Japan from 2011 to 2015.

5.12. Mineral and tap waters

5.12.1. Artificial radionuclides

1981, when our laboratory started with the radioactivity survey of food, water, besides milk, was the first food category to be monitored. Before 1986, no bomb fallout was detectable in the drinking water of Basel (<0.01 Bq/L radiocaesium). Just after the accident at Chernobyl, radioiodine and radiocaesium were detectable in small amounts of 59±77 and 9±9 Bq/L in some drinking water reservoirs of the state of Jura. The production of drinking water of the city of Basel was never affected.

5.12.2. Natural radionuclides

After 1995, the focus was set on the natural radionuclides from the uranium and thorium decay series. First, uranium and radium were analysed in tap and mineral waters of Switzerland and of abroad. Uranium and radium were found in activity concentrations from <10 to 250 mBq/L and from <10 to 200 mBq/L, respectively. In 2005, the Federal Office of Health analysed more than 5,500 water samples for their uranium content [79]. The World Health Organisation (WHO) considers uranium as relevant in drinking water because of its toxicity as a heavy metal. The WHO recommends a limit value of 30 μg/L (376 mBq/L) for drinking water [80]. In Germany, the tap water from more than 500 drinking water plants were analysed for their natural radionuclides. According to this study, the drinking water of 10% of all plants was over the limit dose of 0.1 mSv/year [81]. One possible input of uranium is supposed to come from the use of phosphate fertilisers in agriculture. These fertilisers may contain uranium up to 50 mg U/kg P_2O_5, which was shown by a market survey in Basel [82]. Our investigation gave cause for a national investigation of the Federal Office of Agriculture. The uranium is relatively soluble and washed from the fertiliser into the soil. From there, it is transferred into groundwater. Surbeck [83] estimated that the use of fertilisers results in the increase of the uranium concentration in groundwater from <0.1 to 3 μg/kg. In 2014, we analysed the tap water of all villages of the states of Basel-Campaign and Basel-City. The spectrum of the relevant radionuclides was expanded with ^{222}Rn and ^{210}Po. For uranium and radium, we found 12±16 mBq/L (n=120) and 17±27 mBq/L (n=54). Radon was present in all samples in the Becquerel range (5±6 Bq/L). Also, ^{210}Po was present in 57 samples in the low mBq range of 26±30 mBq/L [84]. In the alpine regions of southern Switzerland, the activity concentrations were somewhat higher due to the geological underground [85].

5.13. Healing earths

Minerals and sediments consisting mainly of silicon dioxide (quartz) are known as siliceous earths. These fine, floury mineral mixtures are deposits of the silica shells of diatoms, the main constituent of marine phytoplankton. The dead cells sink to the ocean floor and form sediments. These layers of sediment are extracted in numerous mines all over the world. Siliceous earths have a wide variety of uses in, for example, the pharmaceutical and food industries (e.g., as a food supplement). Due to their special structural properties, foreign atoms and ions are incorporated during sediment formation, such as radionuclides of the natural decay series of uranium and thorium [86].

In 2008, we collected some siliceous earth products on the Swiss market and analysed them with γ-spectrometry. In two products, the limit value for natural radionuclides was exceeded (50 Bq/kg). Furthermore, the annual dose by regular consumption of one product reached half of the permitted yearly dose of 1 mSv. The Federal Office for Health Products Control, Swiss Medic, complained about these products. The company involved then withdrew the product from the market [86]. An inspection of the products in 2010 showed that two products from one producer in Germany slightly exceeded the limit value. Higher levels of ^{226}Ra and ^{228}Ra were the reason for this. The estimation of the received annual dose from the consumption of the product according to recommended amount per day as advised on the information leaflet enclosed would lead to 0.1 mSv/year. Healing earths, and also silica-based chemicals of chemical laboratories, remain a source for natural radionuclides [87].

5.14. Charcoal and briquettes

In 2009, more than 10,000 tons of wood pellets had to be withdrawn from the Italian market. The product was from Lithuania and was contaminated with radiocaesium (300 Bq/kg). The fact that such products and barbecue coals are imported from countries such as Ukraine or Poland motivated us to conduct this investigation. Charcoal is produced either by charcoal burning of wood (possibly contaminated by the fallout of Chernobyl) or from coal mining. The survey of barbecue coals over the last 6 years showed that there is some contamination with radiocaesium (13±20 Bq/kg). We could not verify the high values from Italy; instead, one also has to consider some radiation of the coals derived from natural radionuclides, such as radium, uranium, thorium, and lead (^{210}Pb). The latter is present in activities of around 65±68 Bq/kg, which is over the limit of 10 Bq/kg of the ordinance of radioprotection. The thorium activity concentrations also reach the permitted limit of 6 Bq/kg [88, 89]. Barbeque experiments with steaks grilled over charcoal showed only a slightly contamination of the meat. The main activity remains in the barbeque ashes [90].

5.15. Estimation of internal doses by the consumption of contaminated food in 1986/1987

Based on the results presented from our own investigations in 1986/1987, we estimated a received dose by ingestion of 4.6 mSv. A main contribution came from contaminated vegetables. It is not clear if the population followed an appeal and the recommendations by the government to avoid the consumption of such contaminated vegetables. If so, the dose would have been reduced to approximately 2.4 mSv. Our estimations seem to be too high. The basis of our calculations was dominantly on the first months after the Chernobyl accident. Thus, our mean values are not representative of the whole years 1986 and 1987. **Table 3** gives an overview of these investigations.

	Origin	Activity levels 1986/1987			Activity levels 1990/2015	
		$^{131+132}I$	$^{134+137}Cs$	^{90}Sr	$^{134+137}Cs$	^{90}Sr
Milk, CH	Cow	<1	<-1025	35	<0.05–0.2	0.04–0.2
	Sheep	<1–30	<1–624	<1–22	N/A	N/A
	Mother	<1–30	<1	N/A	N/A	N/A
Milk powder	CH	<1	<1–19,000	<1–22	N/A	N/A
Baby food	CH	<1	<1–5400	N/A	<0.05–0.5	0.03–0.3
Milk products	CH	0.4–2	1–2	0.05–1.1		
Chocolate	CH, D	<1	6–80		N/A	N/A
Nuts	TR	<1	100–17,100		0.1–30	N/A

	Origin	Activity levels 1986/1987			Activity levels 1990/2015	
		$^{131+132}I$	$^{134+137}Cs$	^{90}Sr	$^{134+137}Cs$	^{90}Sr
	CH	<1	0.6–120		N/A	
Mushrooms, wild	CH and imports	<1	107–8100	N/A	<1–1300	N/A
Fruit, dried	CH	<1	<1–1600		<1–1	
Vegetables	Leafy	<1–5500	<1–1100		<0.1–1	0.1–0.4
	Root					<0.05–0.5
Herbs	CH	<1–4	<5–123,000			
Chest nuts	CH, IT	<1	6–270		<0.17	N/A
Wild-grown berries	CH	N/A	<1–110	N/A	0.1–170	0.2–6
	East Europe		N/A			
Tea	Imp.	<2	<2–30,000	430	<0.5–100	2–40
Meat	CH, cow	<1–30	<1–1200			
	CH, sheep, lamb	<1–75	13–2200			
Game	CH, roe deer	<1	150		<1–400	
	CH, deer	<1	20		<1–600	
	CH, wild boar	<1	600		<1–>1250	
	Reindeer imp.	<1	3600		40–600	
Fish	Fish, Basel (CH)	<	22–710	N/A	0.32±0.16	0.17±0.10
	Fish, Ticino (CH)	<1	40–4400	N/A		
	Sea fish, imp.	<1	N/A	N/A		
Flour	CH	<1	20±37		<1–9	
Bread	CH	<1	30–1200		0.1–10	
Flakes, muesli	CH	<1	N/A		<0.2–5	
Honey	CH and abroad	<2–110	<2–192		<0.2–24	<0.05–2
Spices	Abroad	N/A	N/A	N/A	<0.5–5	N/A
Drinking water	CH, tap water, mineral water	<1–200	0.4–22	N/A	<0.05	N/A
	CH and abroad		N/A		<0.05	

Table 3. Summary of the artificial radioactivity levels found in different food of 1986/1987 compared to the data of 1990/2015. All data were generated at the State-Laboratory Basel-City.

No dose estimation was possible for ^{90}Sr. There were not sufficient ^{90}Sr data available for most of the food categories. Most Swiss agencies believe that ^{90}Sr activity was low in the fallout from Chernobyl.

The Federal Office of Public Health estimated the dose commitment for the Swiss population to be 0.2 mSv/year in 1986. The estimation was based on the results of the food categories milk, vegetables, and meat only. Other food categories were not taken into account (e.g., cereals or ^{90}Sr data of milk). According to the estimated dose from radioiodine in milk, they estimated that, for children, 75 new cases of thyroid cancer with seven or eight fatal cases would occur. For adults, there was no risk seen [91]. The genetically based radiation risk, which causes genetically anomalies, was estimated at 2 to 22 new cases between 1986 and 2086 [92]. The Swiss Federal Nuclear Safety Inspectorate (former HSK) published a more profound study. They came to the following conclusions: for children of age 1 year, children up to 10 years, and adults, a whole body dose of 0.6 to 1.6, 1.0, and 1.1 mSv was calculated, respectively [93]. The Association for Radioprotection of Germany and Switzerland calculated somewhat lower doses (0.4–1.0 mSv for children and 0.4–0.8 mSv for adults) [94]. The Swiss Federal Office of Public Health compared the calculated doses to the whole-body countings of Swiss people. They calculated 10-fold lower doses according to the whole-body countings [95].

In 2014, the Federal Office of Public Health estimated the mean dose through the consumption of food to be 0.35 mSv. The main contribution came from ^{40}K (0.2 mSv/year) and from natural radionuclides of the uranium and thorium series. The remaining contamination from the bomb fallout was less than 0.1 mSv/year [96].

The listed dose estimations fluctuate because of the use of different radioactivity concentrations in food and different assumptions of the consumption rates. All these estimations are based on the dose coefficients for inhalation and ingestion of the International Commission for Radiation Protection (ICRP). However, there are other commissions such as the European Committee on Radiation Risk (ECRR), which conducts more assessments on the effects of low doses. Their dose coefficients are different from ICRP data for some important radionuclides. The ingestion dose factors of the ECRR for ^{137}Cs and ^{90}Sr are 5- and 320-fold higher than the ICRP factors. The ECRR attaches more importance to aspects such as DNA damage by radionuclides [97]. The application of these dose factors would result in higher doses by ingestion of ^{137}Cs and ^{90}Sr. Furthermore, by respecting these dose factors, the limit values for radionuclides in food should be considerably lower for ^{90}Sr and ^{137}Cs.

Different countries have different limit values, even using the same dose coefficients from ICRP. This led to confusion in 1986 and was also a problem in 2011. Policymakers and the general public do not understand such differences. Why are there different threshold values for the same radionuclide in the same food? In Switzerland, the Government has the intention to let fall most limit values for radionuclides in food in "times of non-crisis". In times of crisis, the Government should enact ad hoc specific limit values, as was done in 1986 and 2011.

It is not clear how the state laboratories can sustain their activities and conserve their know-how in the field of radioactivity monitoring of food without limit values. However, we must always be prepared for emergency cases.

http://dx.doi.org/ 10.5772/62460

Acknowledgements

Special thanks go to the former Director of the State Laboratory Basel-City, Dr. Martin R. Schüpbach, who took the far-sighted decision to build up the equipment for radioactivity analyses of food in 1980. Dr. Werner Manz prepared and took into operation the first γ- and β-measurement equipment for food monitoring. Basel was the first Swiss food authority, which was able to analyse food for radiocontaminants on a regular basis. The first analysed food categories were milk and drinking water. When the accident of the NPP of Chernobyl occurred, our laboratory was able to analyse almost 2000 food and environmental samples, which were well managed by the team of Drs. André Herrmann and Claude Ramseier with their technicians Matthias Stöckli, Peter Schaltenbrand, Michael Wagmann, and others. After 1990, most Swiss food-control authorities again reduced their built-up analytical potential and manpower in this field. Fortunately, Verena Figueiredo continued the radioactivity work in Basel and even expanded the equipment and methods for the analysis of natural radionuclides in food. Over the last 15 years, we were able to upgrade our equipment, and the State Laboratory Basel-City now employs of several α-, β-, and γ-spectrometers. This allows us to analyse some thousand samples a year with only one team of three to four persons.

All references concerning the reports of the State-Laboratory of Basel-City are on the Internet. Available at: http://www.gesundheitsschutz.bs.ch/konsum-umwelt/berichte.html.

Author details

Markus Zehringer

Address all correspondence to: markus.zehringer@bs.ch

State Laboratory Basel-City, Basel, Switzerland

References

[1] Völkle H 40 years of radioactivity monitoring in Switzerland, history and stories. In: Freiburger naturforschende Gesellschaft, editor.

[2] Huber O, Michaud B, The campaign of the emergency organisation after the catastrophe of Chernobyl. In: André L, Born E, Fischer G, editors. Radioactivity survey in Switzerland after Chernobyl and its scientific interpretation.

[3] The Federal Assembly of the Swiss Confederation. Radiological Protection; 1991. Status: 1 January 2007.

[4] Swiss Federal Council. Radiological Protection Ordinance; 1994. Status: 1 January 2014.

[5] The Federal Department of Home Affairs. Ordinance on Contaminants and Constituents in Food; 1995. Status: 1 October 2015.

[6] Federal Office of Public Health: Information bulletin No128: prescriptions for imports of wild grown mushrooms from East Europe.

[7] Federal Food Safety and Veterinary Office. Import of food and feed of animal origin from Japan; 2011, Status: 14 April 2014.

[8] Bundesamt für Strahlenschutz. Federal Office for Radioprotection: Natural radionuclides in food [Internet]; 2015. Available at: www.bfs.de/DE/themen/ion/umwelt/lebensmittel/radioaktivitaet-nahrung/radioaktivitaet-nahrung.html [Accessed 18 December 2014].

[9] Pöschl M. What are radionuclides? In: Pöschl M, Nollet L, editors. Radionuclide concentrations in food and the environment. New York: CRC Taylor & Francis; 2007. pp. 9–10. ISBN: 0-8493-3594-9. Ch. 1.

[10] Feely H, Volchok H, Hardy E, Toonkel L Worldwide deposition of ^{90}Sr through 1976. Environ. Sci. Technol. 1978;12:808–809.

[11] Strahlentelex Obituary: Ernest J Sternglass is dead, Strahlentelex.2015;676/677:10-11.

[12] Thevenon F, Wirth S, Fujak M, Poté J, Girardcos S Human impact on the transport of terrigenous and anthropogenic elements to peri-alpine lakes (Switzerland) over the last decades. Aquat. Sci. 2013;75:413–424.

[13] Knoll G Radiation Detection and Measurement. 3rd ed. New York: John Wiley & Sons; 1999. ISBN 0-471-81504-7.

[14] L'Annunziata M, editor. Handbook of Radioactivity Analysis. 2nd ed. San Diego: Academic Press; 2003. ISBN 0-12-436603-1.

[15] Khater A Radiation detection methods. In: Pöschl M, Nollet L, editors. Radionuclide Concentrations in Food and the Environment. New York: CRC Taylor & Francis; 2007. pp. 269–332. ISBN: 0-8493-3594-9. Ch. 9.

[16] Valkovic V Determination of radionuclides in environmental samples. In: Barcelo D, editor. Environmental Analysis: Techniques, Applications and Quality. Amsterdam: Elsevier Science Publishers. Ch. 9.

[17] Palacios M, Estier S, Ferreri G On the trace of ^{137}Cs in wild boars in Ticino. In: Federal Office of Public Health, editor. Environmental radioactivity and radiation doses in Switzerland. Bern: BAG;2013. pp. 100-101.

[18] Wallbrink P, Walling D, He Q Radionuclide measurement using HPGE γ spectrometry. In: ZapataF, editor. Handbook for the Assessment of Soil Erosion and Sedimentation Using Environmental Radionuclides. Amsterdam: Springer; 2003. ISBN: 978-0-306-48054-6. pp.67-87.

[19] Zehringer M, Abraham J, Kammerer F, Syla V, Wagmann M A rapid extraction procedure for the determination of strontium-90 in water samples. Chimia 2016, in press.

[20] Wagmann M, Kammerer F, Zehringer M Trace analysis of radiostrontium in food samples by means of β-spectrometry at the sub Bequerel level. Proceedings of the Swiss Food Science Meeting (SFSM '13); 27–28 June 2013; Neuchatel.

[21] Zehringer M, Stöckli M. Alphaspectrometry of environmental and food samples with photon/electron rejecting α-liquid scintillation (PERALS). Proceedings of the 2nd European ICRP Congress on Radiation Protection; Paris; 2006. p. 113.

[22] Khater A Radiation detection methods. In: Pöschl M, Nollet L, editors. Radionuclide concentrations in food and the environment. New York: CRC Taylor & Francis; 2007. pp. 318–331. ISBN: 0-8493-3594-9. Ch. 9.

[23] Heyraud M, Cherry R. Polonium-210 and lead-210 in marine food chains. Marine Biol. 1979;52:227–236.

[24] Yamamoto M, Abe T, Kuwabara J, Komuro K, Ueno K, Takizawa Y Polonium-210 and lead-210 in marine organisms: intake levels for Japanese. J Radioanal. Nucl. Chem. 1994;178:81–90.

[25] Surbeck H Monitoring of radium, uranium and radon activity concentrations in water. In: Federal Office of Public Health, editor. Environmental radioactivity and radiation doses in Switzerland. Bern:BAG;1996.Ch.B.4.3.1.

[26] Surbeck H Determination of natural radionuclides in drinking water; a tentative protocol. Proceedings of the 7th International Symposium on Environmental Radiochemical Analysis (ERA '96); 1996; Bournemouth, UK.

[27] Baltzer C Determination of ^{238}U and ^{234}U in honey with means of α-spectrometry [Master's thesis]. University of Applied Sciences and Arts of Northwest Switzerland, FHBB; 2005.

[28] Thimmaiah D, Zehringer M Determination of actinides in spices by means of α spectrometry after enrichment by an electrodeposition procedure. Chimia. 2012;66:493.

[29] Zehringer M, Stöckli M Determination of total bromine residue in food and non-food samples. Chimia. 2004;59:112.

[30] Zehringer M, Mazacek J, Dolf R, Testa G, Jourdan J Neutron activation analysis—another approach to uranium and thorium analysis in environmental samples. Chimia. 2013;67:828.

[31] Zehringer M, Testa G, Jourdan J Determination of total Iodine content of food with means of Neutron Activation Analysis (NAA). Proceedings of the Swiss Food Science Meeting (SFSM '13); 27–28 June 2013; Neuchatel.

[32] Weiss W The accident of Chernobyl. Nucl.-Med.1991;30:226-232.

[33] Bundesamt für Gesundheit. Federal Office of Public Health Contamination situation in Switzerland after the accident of Chernobyl. September 1986;Bern.

[34] Herrmann A, Ramseier C, Schüpbach M Radioactivity in milk, cheese and other milk products, Mai-August 1096. . In: André L, Born E, Fischer G, editors. Radioactivity survey in Switzerland after Chernobyl and its scientific interpretation. Proceedings of the Federal Office of Public Health; 20-22 October 1986;Bern:BAG;1986.pp.403-410.

[35] Zehringer M Radioactivity survey of milk. In: Annual Report of the State-Laboratory Basel-City. Basel: KLBS; 2014. pp. 49–51

[36] Jörensen J, Byrde F, Meyer M, Rentsch F, Schmid E, Schweizer M, Steffen C, Steiger G Radioactive contamination in milk, cheese, and other milk products, Mai-August 1986. In: André L, Born E, Fischer G, editors. Radioactivity survey in Switzerland after Chernobyl and its scientific interpretation. Proceedings of the Federal Office of Public Health; 20-22 October 1986;Bern:BAG;1986.pp.308-403.

[37] Froidevaux P, Dell D, Tossell P Radionuclides in foodstuffs and food raw material. In: PöschlM, NolletL, editors. Radionuclide Concentrations in Food and the Environment. New York: CRC Taylor & Francis; 2007. pp. 259–261. ISBN: 0-8493-3594-9. Ch. 8.

[38] Herrmann A, Schüpbach M, Görlich W, Völkle H Herbs, herbal tea and mushrooms. In: André L, Born E, Fischer G, editors. Radioactivity survey in Switzerland after Chernobyl and its scientific interpretation. Proceedings of the Federal Office of Public Health; 20-22 October 1986;Bern:BAG;1986.pp.487-494.

[39] Umweltinstitut München Mushrooms and game. Chernobyl is not eaten yet [Internet]; 2005.Available at: http://www.umweltinstitut.org/fileadmin/Mediapool/Downloads/01_Themen/01_Radioaktivitaet/Messungen/umweltinstitut_pilze_und_wild.pdf [Accessed 9.2005].

[40] Kalac P A review of edible mushroom radioactivity. Food Chem. 2001;75:29–35.

[41] Baeza A, Guillen J, Mietelski J Uptake of α and β emitters by mushrooms collected and cultured in Spain. J. Radioanal. Nucl. Chem. 2004;61:375–380.

[42] Federal Ministery of Health: Radioactivity in Austrian mushrooms [Internet]; 2015 Available at: www.bmg.gv.at/home/Schwerpunkte/Gesundheitsfoerderung_Praevention/Strahlenschutz/Radioaktivität_in_heimischen_Wildpilzen [Accessed 2 July 2015].

[43] Steinmann P, Estier S, Zehringer M, Bajo C Food. In; Federal Office of Public Health, editor. Environmental Radioactivity and doses in Switzerland. Bern: BAG; 2014. pp. 86-87.

[44] Steinmann P, Estier S, Bajo C, Zehringer M Food. In; Federal Office of Public Health, editor. Environmental Radioactivity and doses in Switzerland. Bern: BAG; 2012. p. 86.

[45] Zehringer M Radionuclides in wild grown berries. In: Annual Report of the State Laboratory Basel-City. Basel: KLBS; 2014. pp. 41–45.

[46] Herrmann A, Schüpbach M The contamination of vegetables and fruit in the region of Basel. In: André L, Born E, Fischer G, editors. Radioactivity survey in Switzerland after Chernobyl and its scientific interpretation. Proceedings of the Federal Office of Public Health; 20-22 October 1986;Bern:BAG;1986.pp.477-481.

[47] Zehringer M Radiocaesium and -strontium in root vegetables. In: Annual Report of the State-Laboratory Basel-City. Basel: KLBS; 2014. pp. 46–48.

[48] Zehringer M Radioactivity in chestnuts and chestnut products. In: Annual Report of the State-Laboratory Basel-City. Basel: KLBS; 2015. pp.60-61.

[49] Häsler S, Heiz H, Schmid E, Zeller W Radioactivity survey of meat. Swiss Food. 1987;4:21-27.

[50] Federal Office for Radioprotection. The contamination of food after the catastrophe at Chernobyl [Internet]; 2013 (rev. January 2015). Available at: http://www.bfs.de/SharedDocs/Downloads/BfS/DE/broschueren/ion/stth-lebensmittel.pdf?

[51] State Laboratory Zurich. Wild boar meat from the state of Zurich only slightly contaminated [Internet]; 2014. Available at: http://www.kl.zh.ch/internet/gesundheitsdirektion/klz/de/aktuell/mitteilungen/2015/wildschweinefleisch_radioaktiv_nur_gering_belastet.html [Accessed 18 March 2015].

[52] Strahlentelex. Interactive map with radioactivity data of mushrooms and game. Strahlentelex 2014;670-671:7.

[53] Zehringer M Radioactivity in game. In: Annual Report of the State-Laboratory Basel-City. Basel: KLBS; 2014. pp. 40–41.

[54] Radioactivity in game. In: Annual report of the state laboratory Basel-City. Basel: KLBS; 2010, pp.54-56.

[55] Santschi P, et al. Radionuclides from the Chernobyl fallout in rivers: washout, dilution, elimination and enrichment processes. In: André L, Born E, Fischer G, editors. Radioactivity survey in Switzerland after Chernobyl and its scientific interpretation. Proceedings of the Federal Office of Public Health; 20-22 October 1986;Bern:BAG;1986.pp. 323-338.

[56] Zehringer M ^{210}Po, ◎-nuclides and heavy metals in fish and mussels. In: Annual report of the state laboratory Basel-City. Basel: KLBS;2010, pp.57-59.

[57] Strahlentelex. Radioactive contamination of fish and sea food from Japanese waters, Strahlentelex. 2013;626-627:5-6.

[58] Madigan D, Baumann Z, Snodgrass O, Ergül H, Dewar H, Fisher N. Radiocesium in pacific bluefin tuna *Thunnus orientalis* in 2012 validates new tracer technique. Environ. Sci. Technol. 2013;47:2287–2294.

[59] Neville D, Phillips A, Brodeur R, Higley K Trace levels of Fukushima disaster radionuclides in East Pacific Albacore. Environ. Sci. Technol. 2014;48:4739–4743.

[60] Johansen M, Ruedig E, Tagami K, Uchida S, Higley K, Beresford N Radiological dose rates to marine fish from the Fukushima Daiichi accident: the first three years across the North Pacific, Environ. Sci. Technol. 2015;49:1277–1285.

[61] Fujimoto K, Miki S, Kaeriyma H, Shigenobu Y, Takagi K, Abe D, Ono T, Watanabe T, Morinaga K, Nakata K, Morita T Use of otolith for detecting strontium-90 in fish from the harbor of Fukushima-Dai-ijchi Nuclear Power Plant, Environ. Sci. Technol. 2015;49:7294–7301.

[62] Zehringer M Fish from the Pacific Ocean. In: Annual report of the state laboratory Basel-City. Basel: KLBS;2015, pp. 61-63.

[63] Cherry R, Shannon L The α radioactivity of marine organisms. Atmos. Energy Rev. 1974; 12:3–45.

[64] Froidevaux P, Dell D, Tossell P Radionuclides in foodstuffs and food raw material. In: PöschlM, NolletL, editors. Radionuclide concentrations in food and the environment. New York: CRC Taylor & Francis; 2007. pp. 239–243. ISBN: 0-8493-3594-9. Ch. 8.

[65] Figueiredo V Pb-210 and Po-210 in molluscs, crustacean and fish. Annual report of the state laboratory Basel-City. Basel: KLBS;1998, pp.55-56.

[66] Zehringer M Radioactive contamination (^{210}Po) and phosphate content of shrimps and mussels. Annual report of the state laboratory Basel-City. Basel: KLBS;2004, pp.84-86.

[67] Renaud Ph, Roussel-Debet S, Pourcelot L, Thébault H, LoyenGurriaran R ^{210}Pb and ^{210}Po activities in French foodstuffs. Radioprot. 2015;50:123–128.

[68] Aközcan S Levels of ^{210}Po in some commercial fish species consumed in the Aegean Sea coast of Turkey and the related dose assessment to the coastal population. J. Environ. Rad. 2013;118:93–95.

[69] Sahli H, Holzer R, Balsiger B Determination of ^{210}Po in fish from Swiss lakes. In: Federal Office of Public Health, editor. Environmental Radioactivity and radiation doses in Switzerland. Bern: BAG; 2013. pp. 97-99.

[70] Zehringer M Radioactive contamination of baby food In: Annual report of the state laboratory Basel-City. Basel: KLBS;2007, pp.67-68.

[71] Zehringer M Radioactivity of baby food and infant formula In: Annual report of the state laboratory Basel-City. Basel: KLBS;2012, pp.43-46.

[72] Zehringer M Sudan dye content, radioactive contamination and heavy metals in spices In: Annual report of the state laboratory Basel-City. Basel: KLBS;2006, pp.89-91.

[73] Zehringer M Toxic metals and radionuclides in rock salts and table salts. In: Annual report of the state laboratory Basel-City. Basel: KLBS;2010.pp.67-70.

[74] Herrmann A, Schüpbach M, Görlich W, Völkle H The contamination of honey. In: André L, Born E, Fischer G, editors. Radioactivity survey in Switzerland after Chernobyl and its scientific interpretation. Proceedings of the Federal Office of Public Health; 20-22 October 1986;Bern:BAG;1986.pp.482-486.

[75] Zehringer M Radiocaesium in honey In: Annual report of the state laboratory Basel-City. Basel: KLBS;2004, pp.87-88.

[76] Frey T, Zehringer M Radioactive contamination, HMF and volatile pharmaceuticals in honey In: Annual report of the state laboratory Basel-City. Basel: KLBS;2008, pp.77-79.

[77] Frey T, Zehringer M Pesticide residues, fumigant residues and radioactive contamination in black and green tea In: Annual report of the state laboratory Basel-City. Basel: KLBS;2009, pp.34-38.

[78] Zehringer M Radioactivity in tea In: Annual report of the state laboratory Basel-City. Basel: KLBS;2015, pp.54-57.

[79] Haldimann M Distribution of uranium in Swiss drinking water, BAG-Bulletin. 2012;12:206-207.

[80] Giddings M Uranium. In: World Health Organization. Guidelines for Drinking-Water Quality. 2nd ed., Geneva: WHO; 1998. Vol. 2 abb., pp. 81–94.

[81] Bundesamt für Strahlenschutz. Federal Office for Radioprotection Radiation exposure caused by natural radionuclides in drinking water from Germany [Internet]; 2009. Available at: http://www.bfs.de/DE/themen/ion/umwelt/lebensmittel/trinkwasser/trinkwasser.html [Accessed 1 December 2015].

[82] Parrat Y Nutrient declaration, heavy metals, declaration, certification and security data sheets of fertilizers In: Annual report of the state laboratory Basel-City. Basel: KLBS; 2013, pp.208-210.

[83] Surbeck H Risk assessment concerning the immission of uranium to groundwater in Switzerland. In: Federal Office of Public Health, editor. Environmental Radioactivity and radiation doses in Switzerland. Bern: BAG; 2013. pp. 82-90.

[84] Zehringer M Artificial and natural radionuclides in drinking water. In: Annual Report of the State-Laboratory Basel-City. Basel: KLBS; 2014. pp. 31–37.

[85] Froidevaux P, Dell D, Tossell P Radionuclides in foodstuffs and food raw material. In: Pöschl M, Nollet L, editors. Radionuclide concentrations in food and the environment. New York: CRC Taylor & Francis; 2007. p. 242, ISBN: 0-8493-3594-9. Ch. 8.

[86] Zehringer M Radionuclides in siliceous and healing earths. In: Annual Report of the State-Laboratory Basel-City. Basel: KLBS; 2008. pp. 79–82.

[87] Zehringer M Radionuclides in siliceous and healing earths. In: Annual Report of the State-Laboratory Basel-City. Basel: KLBS; 2010. pp. 70–73.

[88] Zehringer M Radioactivity of barbeque charcoal and briquettes In: Annual report of the state laboratory Basel-City. Basel: KLBS;2009, pp.70-72.

[89] Zehringer M Radioactive contamination of barbeque charcoals In: Annual report of the state laboratory Basel-City. Basel: KLBS;2015, pp.96-98.

[90] Partl D Radioactive contamination of barbeque charcoals and briquettes [scholarly paper]. Zurich: State Grammar School of Zurich North; 2015.

[91] Stoll E Estimation of the radiation exposition of the Swiss population. In: Chernobyl: Estimation of the radiation dose and its problems, Federal Office of Public Health. Bern:BAG;1987. pp.2-11.

[92] Stoll E Possible health risks In: Chernobyl: Estimation of the radiation dose and its problems, Federal Office of Public Health. Bern:BAG;1987. pp.17-26.

[93] Bundesamt für Energiewirtschaft. Federal Office of Energy Economy The accident of Chernobyl. An overview of the reasons and impacts. 1986;HSK-AN-1816.pp.61-66.

[94] Winter M, Völkle H, Narrog J, Meyer P, Kirchhoff K Radioactivity in the Federal Republic of Germany and in Switzerland after the accident in Chernobyl. Fachausschuss für Strahlenschutz E.V. 1986;FS-86-39-AKU:60-61.

[95] Bundesamt für Gesundheit. Federal Office of Public Health Contamination situation in Switzerland after the accident in Chernobyl. Bern:BAG;1986.pp.30-38.

[96] Federal Office of Public Health Radioprotection and radioactivity monitoring in Switzerland. Results. Bern:BAG;2014. pp.32-33.

[97] Busby C Aspects of DNA damage from internal radionuclides. In: ChenC, editor. Biochemistry, Genetics and Molecular Biology. New Research Directions in DNA Repair. 2013. ISBN: 978-953-51-1114-6. Ch. 22.

[98] SantschiH Chernobyl - Radionuclides in the environment. Tracer for the linking of atmospheric, terrestial and aquatic systems. Mitt. EAWAG. 1987;23:2-9.

Chapter 7

Radiation Influence on Edible Materials

Nelida Lucia del Mastro

Additional information is available at the end of the chapter

http://dx.doi.org/10.5772/62430

Abstract

Innovations in the food industry are shaped both by new technologies available and by society's requirements. A good knowledge on the chemistry and biological role of the macro-nutrients (proteins, carbohydrates, lipids and also energy and water) and micro-nutrients (minerals and vitamins) is required. Food production foundations include not only the design of the food products but also the materials, mechanics, ingredients, conversion and transformation all must be taken into consideration. Edible polymers are polymeric materials that can be easily consumed by human beings or lower animals in whole or part via the oral cavity and given harmless effect to the health. An edible polymer is originated from natural products such as polysaccharides, proteins and lipids, with the addition of plasticizers and surfactants. Radiation-processing technologies are used currently for numerous applications of commercial and economic importance, but it is an emerging application with the use of ionizing radiation to enhance properties of edible polymers such as carbohydrates or proteins. This chapter aims at supplying the state of the art about the effects of ionizing radiation on edible polymers: starch and vegetal proteins and also on gelatin that comes from animal origin.

Keywords: ionizing radiation, polysaccharides, proteins, edible films, nanotechnology

1. Introduction

There is an increasing interest in green and bioactive materials that could be used in direct contact with aliments. Edible films can be produced from natural materials with film-forming ability that can be applied on food surfaces as a thin-layer edible film and can potentially extend the shelf life and improve the quality of food. Edible film as a solid sheet can be applied between food components or on the surface of the food system in order to inhibit migration moisture, oxygen, CO_2, aromas and lipids.

Edible polymer film is a thin layer of edible material formed on a food as a coating or placed (pre-formed) on or between food components; in this case, edible films can improve the quality of multicomponent foods. The sanitary condition of the edible packaging would need to be maintained during storage, transportation and marketing. The end result would be source reduction and or improved recyclability of the remaining elements of the packaging system.

One major advantage of using edible films and coatings is that several active ingredients (antimicrobials, antibrownings, texture enhancers and nutraceuticals) can be incorporated into the polymer matrix and consumed with the food, thus enhancing safety or even nutritional and sensory attributes [1].

Many potential uses of edible films can be described such as inhibit migration of moisture, oxygen, carbon dioxide, aromas; carry food ingredients (e.g., antioxidants, antimicrobials, flavor) and or improve mechanical integrity or handling characteristics of the food [2]. Practical uses of edible films include wrapping various products; individual protection of dried fruits, meat and fish; control of internal moisture transfer in pizzas, pies, which are based on the film's properties (e.g., sensory, mechanical, gas and solute barrier). Also, ingredients can be delivered to processors on water-soluble and edible packaging films in premeasured amounts of ingredients to food processors and foodservice operations. Future applications of the concept have been envisioned in consumer-sized pouches of dried products ready for reconstitution on water.

The functionality and performance of edible polymer mainly depend on their barrier, mechanical and color properties, which in turn depend on film composition and its formation process. Polysaccharide (cellulose, starch, dextrin, vegetable and other gums) and protein (gelatin, gluten, casein) based films can have suitable mechanical and sensory properties, while wax (beeswax, carnauba wax) and lipid or lipid derivative films have enhanced water vapor barrier properties. It was shown that the incorporation of oil between plasticized starch layers could reduce oxygen and water vapor permeability of the films [3]. The film-forming technology, solvent characteristics, plasticizing agents, temperature effects, solvent evaporation rate, coating operation and usage conditions of the film (relative humidity, temperature) can also substantially modify the ultimate properties of the film [4].

Important properties to be evaluated in an edible coating are its microbiological stability, adhesion, cohesion, wettability, solubility, transparency, mechanical properties, sensory and permeability to water vapor and gases. In order to choose the best edible film to be employed in each case, it is necessary to take in account the characteristics of the food intended to be protected. For coating a fresh fruit, for instance, it will be desired to induce low water vapor permeability to preserve texture and moderated O_2/CO_2 permeability to permit respiration. On the contrary, when it is intended to protect dried fruits and nuts, low water vapor permeability is required to maintain crispiness and low O_2 permeability to avoid oxidation.

Proteins as components of edible films (gelatin, zein, casein) has usually better performance than polysaccharides (chitosan, starch, pectin); lipids have excellent water vapor barrier but do not form stand-alone films, and can be used as coatings, such as waxes on fruit surfaces. Besides the barrier efficiency, edible films and coatings have to be also sensory acceptable.

In the food industry, there are commonly used starches, dextrins, alginates, protein and lipid materials edible films as encapsulating materials [5]. A wide variety of foods are encapsulated: flavoring agents, acids, bases, artificial sweeteners, colorants, preservatives, leavening agents, antioxidants, agents with undesirable flavors, odors and nutrients, among others. In order to protect the viability of the probiotic bacteria, for instance, several types of biopolymers such as alginate, chitosan, gelatin, whey protein isolate, cellulose derivatives are used for encapsulation and several methods of encapsulation such as spray drying, extrusion, emulsion have been reported [6].

Active edible films represent one of the current and future trends in the development of new polymers for selected applications, particularly food packaging [7]. The concept of active packaging is described as a type of packaging that alters conditions surrounding the food to maintain product quality and freshness, improve sensory properties or enhance product safety and shelf life.

Edible and biodegradable polymer-based films offer alternative packaging without the environmental cost. In that case, they must fulfil the food packaging requirements as help to contain, protect, preserve, distribute and describe food from fresh to highly processed ones. The potential of biopolymers for packaging application is often limited due to their poor processability, so edible polymers originated from natural products are added to plasticizers and surfactants. A careful selection of the plasticizer allows biopolymer films with improved processability and high or low permeabilities to be manufactured [8].

Shit and Shah in 2014 distinguished the edible polymers into any of the four categories: (1) hydrocolloids, (2) polypeptides, (3) lipids, (4) synthetic and composite edible polymers [9].

In the following sections, the interaction of ionizing radiation with polysaccharides and proteins, which are considered the most important edible materials, as well as the role that radiation plays on the development of new edible materials will be addressed.

2. High energy radiation and polymers

Gamma rays and electron beam are two commonly used ionizing radiation sources in industrial process. Gamma rays, 1.17 and 1.33 MeV, are emitted continuously from radioactive source such as cobalt-60, whereas electrons are generated from an accelerator to produce a stream of electrons called electron beam. The energy of electrons depends on the type of machines and can vary from 200 keV to 10 MeV. High-energy radiation using conventional gamma or e-beam sources is an adequate tool for the modification of polymer materials through degradation, grafting and cross-linking [10].

The absorbed dose of radiation is expressed in units of gray (Gy). The Gy was adopted by the International Commission of Radiation Units and measurements (ICRU) in 1975 as the special name for the standard international (SI) unit of absorbed dose, being 1Gy = 1 Joule/kg of matter. When processing food products very seldom the employed dose exceeded 20 kGy, and most of the cases remained below 10 kGy [11]. The adsorbed radiation dose is a consequence of the radiation exposure time and the distance between the product and the radiation source in the irradiation camera.

The long molecular chains of polymers can be broken by the absorption of a quantum of energy above the energy of the covalent bond of the main carbon chain, which typically is in the range of 5–10 eV. The energy of beta and gamma photons of 1 to 10 MeV surpasses by many orders of magnitude this minimum value, representing a high probability of affecting all kinds of polymers, naturals and synthetics alike [12]. Usually, there is a competition between cross-linking and chain scission reaction. If chain scission reaction predominates the material degrades.

Radiation, as a non-selective, highly efficient tool of ionization, may form excited sites, ions and free radicals in almost all kinds of materials. Radiation treatment of polymer mixtures, even if they are (partially) incompatible, gives a chance for bridge-forming bonds [13].

The key effect of radiation is the production of reactive oxygen species (ROS), which affects biomolecules (e.g., lipid, protein, polysaccharides). One of the important radiation-induced free-radical species is the hydroxyl radical, which indiscriminately attacks neighboring molecules often at near diffusion-controlled rates. Hydroxyl radicals are generated by ionizing radiation either directly by oxidation of water, or indirectly by the formation of secondary partially ROS. These may be subsequently converted to hydroxyl radicals by further reduction ("activation"). Secondary, radiation injury is therefore influenced by the surrounding antioxidant status and the amount and availability of activating mechanisms. The biological response to radiation may be modulated by alterations in factors affecting these secondary mechanisms [14].

3. Radiation and polysaccharides

Carbohydrates or saccharides are polyhydroxy aldehydes or ketones that have the empirical formula $(CH_2O)_n$. Most of the carbohydrates found in nature occur as polysaccharides of high molecular weight. Polysaccharides are polymers of monosaccharides linked with glycosidic bonds containing usually no more than two kinds of residues. The most abundant monosaccharide is the six-carbon sugar D-glucose; it is the primordial monosaccharide from which all others are derived. D-Glucose is the most important fuel molecule for most organisms, and also the basic building block of the most abundant polysaccharides. The monomeric units contain many hydroxyl groups, which can engage in intra- and inter-molecular formation of hydrogen bonds. This hydrogen bonding keeps the chains together and contributes to the high tensile strength of the polymeric material. Other forms of functionalization can also occur.

Polysaccharide films are made from starch, alginate, cellulose ethers, chitosan, carageenan, or pectins and impart hardness, crispness, compactness, thickening quality, viscosity, adhesiveness and gel-forming ability to a variety of films. These films usually exhibit good gas permeability properties, resulting in desirable modified atmospheres that enhance the shelf life of the product without creating anaerobic conditions. Additionally, polysaccharide films and coatings can be used to extend the shelf life of muscle foods by preventing dehydration, oxidative rancidity and surface browning, but their hydrophilic nature makes them poor barriers for water vapor.

Starch contains only D-glucose as monomeric units and occurs in two forms, α-amylose (a long unbranched chain) and amylopectin (highly branched). Films produced from pure starch are generally brittle and difficult to handle, although there are some exceptional reports that starches such as cassava can make transparent and colorless flexible films without any previous chemical treatment [15].

Other polysaccharides like pectin, pullulan and chitosan, coming from different origins, present different capability of acting as edible active coatings [16].

The poor mechanical properties and water stability of starch have restricted its industrial applications. However, they can often be functionalized with carboxyl groups, phosphate groups and/or sulfuric ester groups, and the combination of plasticizers and surfactants can bring the enhancement of properties of the starch films [17, 18].

The development of biodegradable materials based on starch has become an attractive option and the production of starch-based plastics is gradually obtained considerable importance in the world. Chemical modifications (e.g., cross-linking) or using a second biopolymer in the starch-based composite have been studied as strategies to produce low water-sensitive and relatively high-strength starch-based materials [19–22].

Particularly, citric acid showed to cross-link starch and improve the tensile strength, thermal stability and decrease the dissolution of starch films in water and formic acid [23].

Starch is considered easily depolymerizable by radiation treatment. Gamma and electron beam irradiation are used to induce radiation degradation of chitosan, alginate, carrageenan, cellulose, pectin, for recycling these bio-resources and reducing the environmental pollution. These carbohydrates, when degraded by radiation, present various kinds of biological activities such as promotion of plant growth, anti-microbial activity, phytoalexins induction, biocontrol elicitors and also can be added to aquaculture and animal feed to enhance immunity of animals [24, 25].

The radiation starch or dextrin degradation is a clean and safety tool. The process is very efficient and can be easily controlled by choosing a proper irradiation dose. A dose of 4.4 kGy, for instance, was enough to decrease the molar mass of wheat starches by one order of magnitude [26]. Other works describe the action of gamma radiation on potato, bean and maize starches [27–29].

Physical and structural characteristics of rice flour and starch obtained from gamma-irradiated white rice were determined. Pasting viscosities of the rice flour and starch decreased continuously with the increase in irradiation dosage. Gamma irradiation had no significant effect on the amylopectin branch chains, but produced more branch chains when the irradiation dosage was less than 9 kGy. It might be deduced that gamma irradiation caused the breakage of the amylopectin chains at the amorphous regions, but had little effects on the crystalline regions of starch granules, especially at low-dosage irradiation [30].

Several other reports were published about the beneficial use of gamma radiation on the improvement of properties of different biopolymeric materials. Ionizing radiation was employed for the synthesis of polysaccharide derivatives to be used as oral delivery system for a colon-specific drug carrier [31]; other works mention the increase of mechanical and barrier properties of polysaccharide materials and the induction of grafting of methylcellulose-based films [32].

Some authors reported the development of edible films from potato peel. High-pressure, gamma rays and ultrasound were applied to potato peel solutions to break down biopolymer particles in the solution small enough to allow for biopolymer film formation. Film properties, including moisture barrier and tensile properties, color and microstructures, were investigated from the films formed with different concentrations of plasticizer (glycerol) and emulsifier (soy lecithin). The authors concluded that the concentrations of both plasticizer (glycerol) and emulsifier (soy lecithin) were important variables in producing biopolymer films from potato peel [33].

4. Radiation and proteins

All proteins are polymers and their monomeric units are α-amino acids. Twenty chemically different amino acids are incorporated in proteins. Proteins in foods serve dual roles as nutrients and structural building blocks. The concept of protein functionality has historically been restricted to non-nutritive functions—such as creating emulsions, foams and gels—but this places sole emphasis on food quality considerations and potentially overlooks modifications that may also alter nutritional quality or allergenicity. Foegeding proposed recently a new model that addresses the function of proteins in foods based on the length scale(s) responsible for the function. Properties such as flavor binding, color, allergenicity and digestibility are explained based on the structure of individual molecules, placing this functionality at the nano/molecular scale. At the next higher scale, applications in foods involving gelation, emulsification and foam formation are based on how proteins form secondary structures that are seen at the nano- and microlength scales, collectively called the mesoscale. The macroscale structure represents the arrangements of molecules and mesoscale structures in a food. Macroscale properties determine the overall product appearance, stability and texture. For applications in food products, protein functionality should start with the identification of functional needs a0 scales. Those needs are then evaluated relative to how processing and other ingredients could alter desired molecular scale properties, or proper formation of mesoscale structures. This allows for a comprehensive approach to achieving the desired function of proteins in foods [34].

Proteins have long been empirically used to make biodegradable, renewable and edible packaging materials. Numerous cereal and vegetable proteins, such as corn zein, wheat gluten and soy proteins, and animal proteins (such as milk proteins, collagen, gelatin, keratin and myofibrillar proteins) are commonly used to form agricultural packaging materials.

Lacroix and Vu, among others, had described the methods to produce protein-based films and coatings, the properties of them and their applications in food systems. They focus on selected proteins originated from animal and plant sources consisting of caseins, whey proteins, collagen and gelatin, plasma proteins, myofibrillar proteins, egg white proteins, soy protein, wheat gluten and zein [35–37].

Soy protein isolate (SPI), a protein with good biocompatibility, biodegradability and processability, has a significant potential in the food industry, agriculture, bioscience and biotechnology. Up to now, several technologies have been applied to prepare SPI-based materials with equivalent or superior physical and mechanical properties with petroleum-based materials [38].

Among protein-based films, those made of gelatin are specially important and can be produced from a variety of origins, with or without the addition of other components as plasticizers, surfactants or mixture with others substances [39–42].

Fish gelatin is a potential alternative to current mammalian (beef and pork) gelatin. However, its physical and thermal properties limit its use in many applications. The treatment of microbial transglutaminase as a cross-linking agent could be a practical way to increase the use of fish gelatin films in various applications [43].

Edible protein film characteristics can be enhanced with chemical and enzymatic methods, combining with hydrophobic material or some polymers or using a physical method, and the resulting film properties will depend on modification methods and conditions [44].

The enzyme and chemical modifications are efficient in lowering water vapor permeability. Composite edible protein films in combination with lipids can result in better functionality than the films produced with only proteins, especially with respect to their barrier properties. Of the lipids, waxes produce the best water vapor barrier properties, but produce fragile and/or brittle films.

The preparation of gelatin and glycerol mixture with microbial transglutaminase as the cross-linking agent was described [45]. A composite casein-gelatin was prepared using also transglutaminase [46]. In some cases, however, the chemical/enzymatical cross-linking was not satisfactory [47].

Application of radiation has been extended to the modification of proteins. Depending on the adsorbed radiation dose or radiation exposure time, various effects can be achieved resulting in the polymerization (cross-linking) or depolymerization of protein molecules. Most food proteins, however, undergo irradiation-induced cross-linking and subsequent improvement on the film properties [48].

Radiation-induced lipid oxidation can be readily detected by our body's olfactory cells—rancidity, off-flavor and, sometimes on the positive aspect, aroma; similarly, it is easy to notice discoloration that results from the oxidation of pigments. In contrast, protein oxidation occurs undetected by sensory organs; instrumental analysis is therefore required. Yet, proteins are very susceptible to reactive oxygen species (ROS) and impart both desirable and undesirable consequences when oxidatively modified. For example, oxidant-initiated disulfide bond formation among gluten molecules is responsible for the desirable rheology and sponginess of bread. And functional myosin (or actomyosin) aggregates produced by the reaction of cysteine, lysine and tyrosine residues with low concentrations of free radicals promote protein gel networks in processed muscle foods, hence the products' firmness and mouthfeel.

Chemically, protein oxidation that can result from radiation treatment in the presence of O_2 involves the initial modification of amino acid side chain groups by ROS. The radiation sensitivity of the amino acids is the highest for cysteine and decrease following the sequence: cysteine, methionine, tyrosine, tryptophan, phenylalanine, valine, leucine, histidine, glutamyl, proline, threonine, arginine and lysine. The initial modification of amino acid side chain groups by ROS leads to the conversion to carbonyl and other derivatives. Electron-deficient carbonyl groups are highly reactive with amines and thiol moieties to produce cross-links between polypeptides or segments within the same protein molecule. Disulfide and dityrosine are other covalent linkages in oxidatively stressed proteins. Protein radicals, which usually have a long half-life, are precursors of polymers as well [49].

The field of health science has pioneered protein oxidation research. Similarly, research in food science over the past two decades has established ubiquitous occurrences of protein oxidation in both fresh and processed foods, especially meat products. Most of the studies point to the negative aspect because uncontrolled oxidation leads to deleterious consequences: tissue hardening (as in frozen fish), loss of water-binding potential and off-flavor due to thiol oxidation. For that reason, developments about novel antioxidants, such as phenolic derivatives, peptides/protein hydrolysates, phospholipids and polysaccharides, and their role in food quality preservation are going on [50].

Many peptides and protein hydrolysates, being preferred targets of ROS and excellent functional compounds (water-binding, foaming, rheology and so on), are qualified as "multifunctional" natural food additives. Aside from the negative impacts of oxidation, mildly oxidized lipids give us a complex, highly desirable aroma of foods as that coming from frying. Today, the beneficial effects of limited protein oxidation are also no longer ignored. That must be taking in account whenever any protein food is submitted to oxidative processes such as irradiation.

Gamma-irradiation affects proteins by causing conformational changes, oxidation of amino acids, rupture of covalent bonds and formation of protein free radicals that can be beneficial for specific further applications. Chemical changes in the proteins that are caused by gamma irradiation are fragmentation, cross-linking, aggregation and oxidation by oxygen radicals that are generated in the radiolysis of water. For example, the hydroxyl and super oxide anion radicals that are generated by radiation of film-forming solution could modify the molecular properties of the proteins, which results in the alteration of protein films by covalent cross-linkages formed in protein solution after irradiation.

Using gamma irradiation to induce cross-linking was found to be an effective method for the improvement of both barrier and mechanical properties of the edible films and coatings based on proteins. There are plenty of examples in the literature that corroborate the efficiency of radiation application on natural films, like those made of zein, the protein found in maize [51]. Different types of gelatin-based films with enhanced properties can be produced by means of application of ionizing radiation [52, 53].

The effect of electron beam accelerator doses on properties of plasticized fish gelatin film was studied. The electron spin resonance spectra indicated free radical formation during irradiation, which might induce intermolecular cross-linking. Tensile strength for gelatin film significantly increased after irradiation (improved by 30% for 60 kGy). The vapor permeability was weakly affected by irradiation. Surface tension and its polar component increased significantly in accordance with the increase of wettability. So, the authors suggest that irradiation may change the orientation of polar groups of gelatin at the film surface and cross-link the hydrophobic amino acids. They did not observed modification of the crystallinity of the film. So, they conclude that structure changes only occurs in the amorphous phase of the gelatin matrix. It is also observed that irradiation enhances the thermal stability of the gelatin film, by increasing the glass transition temperature and the degradation temperature [54].

Cross-linked copolymers of gelatin and poly(vinyl alcohol) (PVA) with excellent water absorption and water retention abilities were successfully synthesized using Co-60 gamma radiation [55]. Also, gamma irradiation was applied on milk protein to improve characteristics of milk protein films [56–59].

5. Radiation and nano technology on synthetic and composite edible polymers

Two classes of biodegradable polymers can be distinguished: synthetic or natural polymers. Recent developments in biopolymer-based food packaging materials include natural biopolymers (such as starches and proteins), synthetic biopolymers (such as poly lactic acid), biopolymer blends and nanocomposites based on natural and synthetic biopolymers [60]. The combination of synthetic and natural polymers to form films are of remarkable importance in producing composite polymers with permeability or mechanical properties according to the need of a specific application that can be applied either in the form of an emulsion, suspension or dispersion of the non-miscible constituents, or in successive layers, or in the form of a solution in a common solvent [61].

Biodegradable synthetic aliphatic polyester, like polylactide (PLA), has been studied extensively for a number of applications like drug delivery system. Frequently renewable resource-generated monomers possess better mechanical properties and easy processability by conventional methods like thermoforming, injection and blow molding with non-toxic degradation products, which have made it superior than the other conventional thermoplastics [62].

In addition to traditional plant materials for biodegradable polymer production, it is worthy to mention the advances in synthesizing novel polymers within transgenic plants, especially those in the polyhydroxyalkanoate class [63].

Application of nanoscience and nanotechnology to the agriculture and food sector is relatively recent compared with their use in drug delivery and pharmaceuticals [64, 65].

In the food industry, nanotechnology can be utilized in order to enhance the delivery of food ingredients to target sites, increase flavor, inhibit bacterial growth, extend product shelf life and improve food safety.

Applications of nanomaterials that do not involve direct addition of nanoparticles to consumed foods, and thus more likely to be marketed to the public in the short term, are related to food packaging and food safety. These applications include polymer/clay nanocomposites as high barrier packaging materials, silver nanoparticles as potent antimicrobial agents and nanosensors and nanomaterial-based assays for the detection of food-relevant analytes (gases, small organic molecules and food-borne pathogens) [66].

Other emerging topics of nanotechnology for food and agriculture are smart delivery of nutrients, bioseparation of proteins, rapid sampling of biological and chemical contaminants and nanoencapsulation of nutraceuticals, as well as advances in technologies, such as DNA microarrays, microelectromechanical systems and microfluidics [67, 68].

Nano-scale cellulose fiber materials (e.g., microfibrillated cellulose and bacterial cellulose) are promising candidates for bio-nanocomposite production due to their abundance, high strength and stiffness, low weight and biodegradability [69].

Several reinforcing nanoparticles such as clays, silica or silver have been used for industrial applications, but cellulose nanocrystals (CNCs) are a better choice for food industry due to their biodegradable and biocompatible nature as well as their outstanding potential in improving mechanical and barrier properties of nanocomposites [70–72]. Also, cellulose nanofibers (CNF) reinforcement improved mechanical properties, except elongation of mango puree edible films [73].

The development of nanoscale systems for the encapsulation, protection and delivery of lipophilic nutrients, vitamins and nutraceuticals was recently reported [74]. A promising route to the synthesis of protein-mimetic materials that are capable of complex functions, such as molecular recognition and catalysis, is provided by peptoid nanosheets polymers structurally related to biologically occurring polypeptides [75].

The conjunction of radiation and nanotechnology for the improvement of biomaterials are drawing special interest [76]. Gamma radiation and nanocrystalline cellulose was used for the reinforcement of poly(caprolactone) composites [77]. Also, the radiation synthesis of gelatin hydrogels containing Ag nanoparticles was reported [78]. Grasielli and his group synthesized nanoparticles of seroalbumin via intramolecular cross-linking using gamma rays technology [79]. The radiation synthesis of a composite prepared from the algae polysaccharide alginate and clay nanocomposite was also described [80].

6. Concluding remarks

To meet the growing demand of recyclable or natural packaging materials and consumer demands for safer and better quality foods, new and novel food-grade packaging materials or technologies have been, and continue to be, developed. Examples of these packaging materials include bio-based polymers, bioplastic or biopolymer packaging products made from raw materials originating from agricultural or marine sources. Under that view, the use of ionizing radiation to synthesize and or enhance properties of edible polymers based on polysaccharides or proteins appears as an emerging promising application. The use of edible coating previously cross-linked by irradiation has permitted a better control of the release of active compounds. So far, the industrial effectiveness of the processing accumulates enough evidences within the huge spectrum of possibilities that radiation technology can offer. In the meanwhile, in order to increase further applications of edible films prepared using radiation technology, an update of the sensory implications, regulatory status and future perspectives on innovative food preservation must still be examined carefully.

Author details

Nelida Lucia del Mastro

Address all correspondence to: nlmastro@ipen.br

Energy and Nuclear Research Institute IPEN-CNEN/SP, Sao Paulo, SP, Brazil

References

[1] Dhall RK. Advances in edible coatings for fresh fruits and vegetables: a review. Critical Reviews in Food Science and Nutrition. 2013;53(5):435-450. DOI: 10.1080/10408398.2010.541568.

[2] Zhu D, Jin L, Wang Y, Ren H. Water barrier property of gelatin-based films cross-linked with microbial transglutaminase. Advanced Materials Research. 2011;335-336:1391-1395. 10.4028/www.scientific.net/AMR.335-336.1391.

[3] Basiak E, Debeaufort F, Lenart A. Effect of oil lamination between plasticized starch layers on film properties. Food Chemistry. 2016;195:56-63. DOI:10.1016/j.foodchem.2015.04.098.

[4] Guilbert S, Gontard N, Cuq B. Technology and applications of edible protective films. Packaging Technology and Science. 1995;8(6):339-346. Article first published online: 28 APR 2006. DOI::10.1002/pts.2770080607.

[5] Gibbs BF, Kermasha S, Alli I, Mulligan CN. Encapsulation in the food industry: a review. International Journal of Food Sciences and Nutrition. 1999;50(3):213-224. DOI: 10.1080/096374899101256.

[6] Huq T, Khan A, Khan RA, Riedl B, Lacroix M. Encapsulation of probiotic bacteria in biopolymeric system. Critical Reviews of Food Science Nutrition. 2013;53(9):909-916. DOI:10.1080/10408398.2011.573152.

[7] Mellinas C, Valdés A, Ramos M, Burgos N, Garrigós MC, Jiménez A. Active edible films: current state and future trends. Journal of Applied Polymer Science. 2016; app. 42631, DOI:10.1002/app.42631.

[8] Jost V, Stramm C. Influence of plasticizers on the mechanical and barrier properties of cast biopolymer films. Journal of Applied Polymer Science. 2015;133(2). DOI:10.1002/app.42513.

[9] Shit SC, Shah PM. Edible polymers: challenges and opportunities. Journal of Polymers. 2014;2014(2014), Article ID 427259, 13 p. http://dx.doi.org/10.1155/2014/427259.

[10] Clough RL. High-energy radiation and polymers: a review of commercial processes and emerging applications. Nuclear Instruments & Methods in Physics Research. Section B-Beam Interactions with Materials and Atoms. 2001;185:8-33. DOI:10.1016/S0168-583X(01)00966-1.

[11] Mastro NL. Role of irradiation treatment on food industry. International Journal of Nuclear Governance, Economy and Ecology. 2011;3(3):266-273. DOI:http://dx.doi.org/10.1504/IJNGEE.2011.042213.

[12] Czvikovszky T. Degradation effects in polymers. In: IAEA. Advances in Radiation Chemistry of Polymers. TECDOC-1420, Vienna, IAEA; 2004. pp. 91-102. @ISBN 92–0–112504–6 @ISSN 1011–4289.

[13] Czvikovszky T. Radiation-assisted compatibilization of polymers. In: IAEA. Advances in Radiation Chemistry of Polymers. TECDOC-1420, IAEA, Vienna, IAEA; 2004. pp. 61-74. @ISBN 92–0–112504–6 @ISSN 1011–4289.

[14] Riley PA. Free radicals in biology: oxidative stress and the effects of ionizing radiation. International Journal of Radiation Biology. 1994;65 (1):27-33. DOI: 10.1080/09553009414550041.

[15] Vicentini NM, Dupuy N, Leitzelman M, Cereda MP, Sobral PJA. Prediction of cassava starch edible film properties by chemometric analysis of infrared spectra. Spectroscopy Letters. 2005:38(6):749-767. DOI:10.1080/00387010500316080.

[16] Treviño-Garza MZ, García S, del Socorro Flores-González M, Arévalo-Niño K. Edible active coatings based on pectin, pullulan, and chitosan increase quality and shelf life of strawberries (Fragaria ananassa). Journal of Food Science. 2015;80(8):M1823-M1830. DOI:10.1111/1750-3841.12938.

[17] Gutierrez MC, Nunez-Santiago MD, Romero-Bastida CA, Martinez-Bustos F. Effects of coconut oil concentration as a plasticizer and Yucca schidigera extract as a surfactant in the preparation of extruded corn starch films. Starch-Starke. 2014;66(11-12):1079–1088. DOI:10.1002/star.201400062.

[18] Nobrega MM, Olivato JB, Grossmann MEV, Bona E, Yamashita F. Effects of the incorporation of saturated fatty acids on the mechanical and barrier properties of biodegradable films. Journal of Applied Polymer Science. 2012;124(5):3695-3703. DOI: 10.1002/app.35250.

[19] Ghanbarzadeh B, Almasi H, Entezami AA. Improving the barrier and mechanical properties of corn starch-based edible films: effect of citric acid and carboxymethyl cellulose. Industrial Crops and Products. 2011;33:229-235. DOI:10.1016/j.indcrop. 2010.10.016.

[20] Ghanbarzadeh B, Almasi H, Entezami AA. Physical properties of edible modified starch/carboxymethyl cellulose films. Innovative Food Science & Emerging Technologies. 2010;11(4):697-702. DOI:10.1016/j.ifset.2010.06.001.

[21] Dick M, Pagno CH, Costa TMH, Gomaa A, Subirade M, Rios AO, Flores SH. Edible films based on chia flour: development and characterization. Journal of Applied Polymer Science. 2016;133(2) Article Number 42455. DOI:10.1002/app.42455.

[22] Fakhouri FM, Fontes LCB, Gonçalves PVM, Milanez CR, Steel CJ, Collares-Queiroz FP. Films and edible coatings based on native starches and gelatin in the conservation and sensory acceptance of Crimson grapes. Ciênc. Tecnol. Aliment. Campinas (Food Sc Technol. Campinas. 2007;27(2). DOI:10.1590/S0101-20612007000200027.

[23] Reddy N, Yang Y. Citric acid cross-linking of starch films. Food Chemistry. 2010;118(3): 702-711. DOI:10.1016/j.foodchem.2009.05.050.

[24] Makuuchi K, Cheng S. Radiation Processing of Polymer Materials and Its Industrial Applications. Hoboken, New Jersey, John Wiley and Sons, Inc. 2012. 325 p. ISBN978-0470-58769-0.

[25] Kume T, Nagasawa N, Yoshii F. Utilization of carbohydrates by radiation processing. Radiation Physics and Chemistry. 2002;63(3-6):625-627. DOI:10.1016/ S0969 806X(01)00558-8.

[26] Hu B, Huang M, Yin S, Zi M, Wen Q. Effects of electron-beam irradiation on physicochemical properties of starches separated from stored wheat. Starch/Staerke. 2011;63(3):121-127. DOI:10.1002/star.201000052.

[27] Chung HJ, Liu Q. Molecular structure and physicochemical properties of potato and bean starches as affected by gamma-irradiation. International Journal of Biological Macromolecules. 2010;47(2):214-222. DOI:10.1016/j.ijbiomac.2010.04.019.

[28] Gani A, Bashir M, Wani SM, Masoodi FA. Modification of bean starch by γ-irradiation: effects on functional and morphological properties. LWT - Food Science and Technology. 2012;49(1):162-169. DOI:10.1016/j.lwt.2012.04.028. ISSN: 0023-6438.

[29] Liu T, Ma Y, Xue S, Shi J. Modifications of structure and physicochemical properties of maize starch by γ-irradiation treatments. LWT - Food Science and Technology. 2012;46(1):156-163. DOI:10.1016/j.lwt.2011.10.012. ISSN: 0023-6438.

[30] Bao JS, Ao Z, Jane JL. Characterization of physical properties of flour and starch obtained from gamma-irradiated white rice. Starch/Stärke. 2005 57:480–487. DOI: 10.1002/star.200500422.

[31] Ali AEH, AlArifi A. Characterization and in vitro evaluation of starch-based hydrogels as carriers. Carbohydrate Polymers. 2009;78:725-730. DOI:10.1016/j.carbpol. 2009.06.009.

[32] Salmieri S, Khan RA, Safrany A, Lacroix M. Gamma rays-induced 2-hydroxyethyl methacrylate graft copolymerization on methylcellulose-based films: structure analysis and physicochemical properties. Industrial Crops and Products. 2015;70:64-71. DOI: 10.1016/j.indcrop.2015.02.056.

[33] Kang HJ, Min SC. Potato peel-based biopolymer film development using high-pressure homogenization, irradiation, and ultrasound. LWT—Food Science and Technology. 2010;43(6):903-909. DOI:10.1016/j.lwt.2010.01.025.

[34] Foegeding EA. Food protein functionality—a new model. Journal of Food Science. 2015;80(12):C2670-C2677. DOI:10.1111/1750-3841.13116.

[35] Lacroix M, Vu KD. Edible Coating and Film Materials: Proteins—Chapter 11. In: Han JH (Ed). Innovations in Food Packaging (2nd Edition) 2014, Elsevier, pp. 277-304. DOI: 10.1016/B978-0-12-394601-0.00011-4.

[36] Su JF, Huang Z, Yuan XY, Wuang XY, Li, M. Structure and properties of carboxymethyl cellulose/soy protein isolate blend edible films crosslinked by Maillard reactions. Carbohydrate Polymers. 2010;79(1):145-153. DOI:10.1016/j.carbpol.2009.07.035.

[37] Rakotonirainy AM, Padua GW. Effects of lamination and coating with drying oils on tensile and barrier properties of zein films. Journal of Agriculture and Food Chemistry. 2001;49(6):2860-2863. DOI:10.1021/jf000845u.

[38] Song F, Tang DL, Wang XL, Wang YZ. Biodegradable soy protein isolate-based materials: a review. Biomacromolecules. 2011;12:3369-3380. DOI:10.1021/bm200904x.

[39] Kim DY, Min SC. Trout skin gelatin-based edible film development. Journal of Food Science. 2012;77(9):E240-E246. DOI:10.1111/j.1750-3841.2012.02880.x.

[40] Ma H, Shen JJ, Yang Q, Zhou J, Xia SH, Cao J. Effect of the introduction of fish collagen on the thermal and mechanical properties of poly(lactic acid). Industrial Engineering and Chemistry Research. 2015;54(43):10945-10951. DOI:10.1021/acs.iecr.5b02969.

[41] Guerrero P, Beatty E, Kerry JP, de la Caba K. Extrusion of soy protein with gelatin and sugars at low moisture content. Journal of Food Engineering. 2012;110:53-59. DOI: 10.1016/j.jfoodeng.2011.12.009.

[42] Soradech S, Nunthanid J, Limmatvapirat S, Luangtanaanan M. An approach for the enhancement of the mechanical properties and film coating efficiency of shellac by the

formation of composite films based on shellac and gelatin. Journal of Food Engineering. 2012;108(1):94-102. DOI:10.1016/j.jfoodeng.2011.07.019.

[43] Yi JB, Kim YT, Bae HJ, Whiteside WS, Park HJ. Influence of transglutaminase-induced cross-linking on properties of fish gelatin films. Journal of Food Science. 2006;71(9):E376-E383. DOI:10.1111/j.1750-3841.2006.00191.x.

[44] Arabestani A, Kadivar M, Shahedi M, Goli, SAH, Porta R. The effect of oxidized ferulic acid on physicochemical properties of bitter vetch (Vicia ervilia) protein-based films. Journal of Applied Polymer Science. 2015;33(2). DOI:10.1002/app.42894.

[45] Zhu D, Jin L, Wang Y, Ren H. Water barrier property of gelatin-based films cross-linked with microbial transglutaminase. Advanced Materials Research. 2011;335-336:1391-1395. 10.4028/www.scientific.net/AMR.335-336.1391.

[46] Zhang YN, Liu N, Zhao XH. A study on the preparation and some functional properties of a cross-linked casein-gelatin composite by a microbial transglutaminase. International Journal of Food Science & Technology. 2011;46(12):2641-2647. DOI:10.1111/j.1365-2621.2011.02795.x.

[47] Alves PMA, Carvalho RA, Moraes ICF, Luciano CG, Bittante AMQB, Sobral PJA. Development of films based on blends of gelatin and poly(vinyl alcohol) cross linked with glutaraldehyde. Food Hydrocolloids. 2011;25(7):1751-1757. DOI:10.1016/j.foodhyd.2011.03.018.

[48] Kuan YH, Bhat R, Patras A, Karim AA. Radiation processing of food proteins-A review on the recent developments. Trends in Food Science & Technology. 2013;30:105-120. DOI:10.1016/j.tifs.2012.12.002.

[49] Xiong Y, Foegeding EA. Protein oxidation—the less appreciated sibling of lipid oxidation. Editorial. Journal of Food Science. 2015;80(7):iii. DOI: 10.1111/1750-3841.12625.

[50] Shahidi F, Zhong Y. Novel antioxidants in food quality preservation and health promotion. European Journal of Lipid Technology. 2010;112(9):930-940. DOI:10.1002/ejlt.201000044.

[51] Lacroix M, Khan R, Senna M, Sharmin N, Salmieri S, Safrany A. Radiation grafting on natural films. Radiation Physics and Chemistry. 2014;94:88-92. DOI:10.1016/j.radphyschem.2013.04.008.

[52] Khan MA, Rahman N, Rahman M. Preparation and characterization of gamma radiation cured gelatin-pva bio-blend. Multi-functional materials and structures III, Parts 1 and 2. In: Lee JH. Advanced Materials Research. 2010;123-125:347-350. The 3rd International Conference on Multi-Functional Materials and Structures, Chonbuk Natl Univ, Jeollabuk do, Jeonju, South Korea, September 14–18, 2010. DOI:10.4028/www.scientific.net/AMR.123-125.347.

[53] Vu KD, Hollingsworth RG, Salmieri S, Takala PN, Lacroix M. Development of bioactive coatings based on γ-irradiated proteins to preserve strawberries. Radiation Physics and Chemistry. 2012;81(8):1211-1214. DOI:10.1016/j.radphyschem.2011.11.071.

[54] Benbettaieb N, Karbowiak T, Brachais CH, Debeaufort F. Impact of electron beam irradiation on fish gelatin film properties. Food Chemistry. 2016;195:11-18. DOI: 10.1016/j.foodchem.2015.03.034.

[55] Kaur I, Bhati P, Bala K, Kanwar SS. Effective immobilization of lipase onto a porous gelatin-co-poly (vinyl alcohol) copolymer and evaluation of its hydrolytic properties. Journal of Applied Polymer Science. 2014;131(10), article # 39622. DOI:10.1002/app. 39622.

[56] Vachon C, Yu HL, Yefsah R, Alain R, St-Gelais D, Lacroix M. Mechanical and structural properties of milk protein edible films cross-linked by heating and γ-irradiation. Journal of Agriculture and Food Chemistry. 2000;48(8):3202-3209. DOI:10.1021/jf991055r.

[57] Ciesla K, Salmieri S, Lacroix M. Gamma-irradiation influence on the structure and properties of calcium caseinate–whey protein isolate based films. Part 1. Radiation effect on proteins gels and films. Journal of Agriculture and Food Chemistry. 2006;54(17):6374-6384. DOI:10.1021/jf060486o.

[58] Ciesla K, Salmieri S, Lacroix M. Gamma-irradiation influence on the structure and properties of calcium caseinate–whey protein isolate based films. Part 2. Influence of polysaccharide addition and radiation treatment on the structure and functional properties of the films. Journal of Agriculture and Food Chemistry. 2006;54(23): 8899-8908. DOI:10.1021/jf060981k.

[59] Ramos OL, Fernandes JC, Silva SI, Pintado ME, Malcata FX. Edible films and coatings from whey proteins: a review on formulation, and on mechanical and bioactive properties. Critical Reviews in Food Science and Nutrition. 2012;52(6):533-552. DOI: 10.1080/10408398.2010.500528.

[60] Tang XZ, Kumar P, Alavi S, Sandeep KP. Recent advances in biopolymers and biopolymer-based nanocomposites for food packaging materials. Critical Reviews in Food Science and Nutrition. 2012;52(5):426-444. DOI:10.1080/10408398.2010.500508.

[61] Arvanitoyannis IS. Totally and partially biodegradable polymer blends based on natural and synthetic macromolecules: preparation, physical properties, and potential as food packaging materials. Journal of Macromolecular Science. 1999;39(2):205-271. DOI:10.1081/MC-100101420.

[62] Gupta AP, Kumar V. New emerging trends in synthetic biodegradable polymers – polylactide: a critique. European Polymer Journal. 2007;43(10):4053-4074. DOI:10.1016/j.eurpolymj.2007.06.045.

[63] Mooney BP. The second green revolution? Production of plant-based biodegradable plastics. Biochemistry Journal. 2009;418(2):219-232. DOI:10.1042/BJ20081769.

[64] Sanguansri P, Augustin MA. Nanoscale materials development: a food industry perspective. Trends in Food Science and Technology. 2006;17:547-556. DOI:10.1016/j.tifs.2006.04.010.

[65] Karim S, Dufresne A, Tahir PM, Karimi A, Abdulkhani A. Biodegradable starch-based composites: effect of micro and nanoreinforcements on composite properties. Journal of Materials Science. 2014;49(13):4513-4521. DOI:10.1007/s10853.014.8151.1.

[66] Duncan TV. Applications of nanotechnology in food packaging and food safety: barrier materials, antimicrobials and sensors. Journal of Colloid and Interface Science. 2011;363(1):1-24. DOI:10.1016/j.jcis.2011.07.017.

[67] Sozer N, Kokini JL. Nanotechnology and its applications in the food sector. Trends in Biotechnology. 2009;27(2):82-89. DOI:10.1016/j.tibtech.2008.10.010.

[68] Hannon JC, Kerry J, Cruz-Romero M, Morris M, Cummins E. Advances and challenges for the use of engineered nanoparticles in food contact materials. Trends in Food Science & Technology. 2015;43(1):43-62. DOI:10.1016/jifs.2015.01.008.

[69] Siró I, Plackett D. Microfibrillated cellulose and new nanocomposite materials: a review. Cellulose. 2010;17(3):459-494. DOI:10.1007/s10570-010-9405-y.

[70] Criado P, Fraschini C, Salmieri S, Lacroix M. Modification of nanocrystalline cellulose for bioactive loaded films. Journal of Research Updates in Polymer Science. 2014;3:122-135. DOI:http://dx.doi.org/10.6000/1929-5995.2014.03.02.7.

[71] Khan A, Huq T, Khan RA, Riedl B, Lacroix M. Nanocellulose-based composites and bioactive agents for food packaging. Critical Reviews of Food Science Nutrition. 2014;54(2):163-174. DOI:10.1080/10408398.2011.578765].

[72] Silva JBA, Pereira FV, Druzian JI. Cassava starch-based films plasticized with sucrose and inverted sugar and reinforced with cellulose nanocrystals. Journal of Food Science. 2012; 77(6):N14–N19. DOI:10.1111/j.1750-3841.2012.02710.x.

[73] Azeredo HMC, Mattoso LHC, Wood D, Williams TG, Avena-Bustillos RJ, Mchugh TH. Nanocomposite edible films from mango puree reinforced with cellulose nanofiber. Journal of Food Science. 2009;74(5):N31-N35. DOI:10.1111/j.1750-3841.2009.01186.x.

[74] McClements DJ. Nanoscale nutrient delivery systems for food applications: improving bioactive dispersibility, stability and bioavailability. Journal of Food Science. 2015:80(7):N1602-N1611. DOI:10.1111/1750-3841.129.

[75] Mannige RV, Haxton TK, Proulx C, Robertson EJ, Batigelli A, Butterfoss GL, Zuckerman RN, Whitelam S. Peptoid nanosheets exhibit a new secondary-structure motif. Nature. 2015;526:415-420. DOI:10.1038/nature15363.

[76] Kharisov BI, Kharissova OV, Mendez UO, editors. Radiation Synthesis of Materials and Compounds. Boca Raton, 2013, CRC Press, Taylor & Francis Group; 560 p. ISBN 13.978-1-4665-0523-0.

[77] Khan RA, Beck S, Dussault D, Salmieri S, Lacroix M. Mechanical and barrier properties of nanocrystalline cellulose reinforced poly(caprolactone) composites: effect of gamma radiation. Journal of Applied Polymer Science. 2013;129(5):3038-3046. DOI:10.1002/app.38896.

[78] Leawhiran N, Pavasant P, Soontornvipart K, Supaphol P. Gamma irradiation synthesis and characterization of AgNP/gelatin/PVA hydrogels for antibacterial wound dressings. Journal of Applied Polymer Science. 2014;131(23), article #41138. DOI:10.1002/app.41138.

[79] Espinoza SLS, Sanchez ML, Risso VA, Grasielli M. Radiation synthesis of seroalbumin nanoparticles. Radiation Physics and Chemistry. 2012;81(9):1417-1421. DOI:10.1016/j.radphyschem.2011.11.040.

[80] Taleb MFA, Hegazy DE, Ismail SA. Radiation synthesis, characterization and dye adsorption of alginate–organophilic montmorillonite nanocomposite. Carbohydrate Polymers. 2012;87(3):2263-2269. DOI:10.1016/j.carbpol.2011.10.058.

Chapter 8

Transient Anions in Radiobiology and Radiotherapy: From Gaseous Biomolecules to Condensed Organic and Biomolecular Solids

Elahe Alizadeh, Sylwia Ptasińska and Léon Sanche

Additional information is available at the end of the chapter

http://dx.doi.org/10.5772/63293

Abstract

This chapter focuses on the fundamental processes that govern interactions of low-energy (1–30 eV) electrons with biological systems. These interactions have been investigated in the gas phase and within complex arrangements in the condensed phase. They often lead to the formation of transient molecular anions (TMAs), and their decay by autoionization or dissociation accompanied by bond dissociation. The damage caused to biomolecules via TMAs is emphasized in all sections. Such damage, which depends on a large number of factors, including electron energy, molecular environment, and type of biomolecule, and its physical and chemical interactions with radiosensitizing agents are extensively discussed. A majority of recent findings resulting from experimental and theoretical endeavors are presented. They encompass broad research areas to elucidate important roles of TMAs in irradiated biological systems, from the molecular level to nanoscale cellular dimensions. Fundamental aspects of TMA formation are stressed in this chapter, but many practical applications in a variety of radiation-related fields such as radiobiology and radiotherapy are addressed.

Keywords: ionizing radiation, low-energy electrons, dissociative electron attachment, DNA, strand breaks

1. Introduction

High-energy ionizing radiation (e.g., γ- and X-rays, electrons, and ions) affects biological materials, via a chain of physical, chemical, and biological processes. A complete understanding of these processes in living cells and tissues is a challenging task because of the multiple sequences of events, which lead to cell mutation or death. Nonetheless, such knowledge enhances our ability to cause death or inhibit growth of cancer cells in radiation therapy and to save healthy cells by radiation protection. As shown by many studies [1], cellular deoxyribose nucleic acid (DNA), containing genomic information, is the primary target for cell damage from ionizing radiation. The fundamental mechanisms involved in the induction of damage to DNA by radiation have therefore been subjects of intense investigations during the past decades [2, 3]. When exposed to ionizing radiation, large biomolecules such as DNA and proteins in the cell can be ionized and/or excited. This may effectively cause changes in their molecular structures by inducing bond ruptures and successive fragmentations, which then affect the function and metabolism of the cell. In DNA, the resulting damages may lead to incomplete repair, misrepair, or unrepair of the molecule. The displaced, mismatched, or damaged DNA bases may be misinterpreted during the replication cycle, deterring cell replications and causing accumulation of cancer predispositions for mutations [4–6].

Ionizing radiation damage to DNA can be induced directly by the interactions of primary quanta of radiation via ionization or excitation of individual components of the DNA itself and by secondary particles, including radicals, electrons, and ions generated along the track, after the interaction of primary radiation with molecules surrounding DNA, that is, water and other cellular components [7, 8]. It is by now well established that the consequences of radiation exposure of biological matter at the molecular level are largely due to secondary electrons (SEs), which are formed with a yield of about 5×10^4 per MeV of deposited radiation energy. SEs are the most abundant secondary species generated by the transfer of energy from ionizing radiation into the medium and essentially comprise slow electrons with kinetic energies below 30 eV. The energy distribution of SEs has a most probable energy around 9–10 eV [9], and those electrons of higher initial energy undergo successive energy losses via inelastic collisions, for example, electronic excitation and ionization. These later create further generations of electrons of significantly lower energies. As all electrons necessarily reach the low-energy range ($E < 30$ eV), a detailed knowledge of reactions involving such low-energy electrons (LEEs) with DNA is thus crucial to understand and accurately describe radiobiological damage. LEEs have been shown to induce genotoxic damage, for example, single- and double-strand breaks (SSBs and DSBs) and other multiple damage sites by bond cleavage, chiefly through formation of a transient molecular anion (TMA) of DNA subunit, followed by dissociative electron attachment (DEA) or autoionization of TMA [10].

The main purpose of this chapter is to describe the phenomena related to reactions of LEEs, which may produce biological effects in the cell, such as apoptosis and cell cycle arrest. Since utmost of the harmful mutagenic and lethal damages of ionizing radiation result from chemical modifications in the nucleus of living cells, sustained studies have been focused on the ultrafast mechanisms involved in the direct interaction of LEEs with DNA and its different subunits, as well as indirect processes which are associated to the interactions of electrons with the principal cellular components nearby DNA. An ultimate understanding of LEE damage mechanisms and their role in DNA damage due to radiation can be obtained from experiments with molecular targets of increasing complexity, that is, from simple gaseous and condensed phase biomolecules to plasmid and cellular DNA. This wide range of target structures is essential to systematically understand how the fundamental principles of the LEE interaction with simple biomolecules and DNA components intervene in more complex ones up to and including cellular DNA [11].

In the first two sections of this chapter, the formation of TMAs and their decay into DEA and autoionization processes are extensively reviewed for simple molecules in the gas and condensed phases. The next section exclusively concerns the interactions of LEEs with basic DNA subunits, that is, the bases, the sugar–phosphate unit, and its two basic constituents in the gas phase. Such studies are necessary to understand how SSBs and DSBs and base release in the much more complex DNA molecule can occur by LEE impact. Additionally, gas-phase DEA to radiosenisitizers (halogenated nucleobase, Pt- and nitrogen-based compounds) is discussed in Section 4.

While most of the simple DNA building blocks can be readily vaporized for experimental study to the gas phase, most of the larger units, that is, nucleosides (containing a DNA base + sugar) and entire nucleotides (sugar + base + phosphate group) undergo decomposition during evaporation [12]. Electron attachment to the short oligonucleotides and single-stranded oligomers containing different bases is reviewed in Section 5. Such molecules with a strong tendency to capture electrons and formation of electronically stable anions simplified the analysis of degradation products relative to longer single- and double-stranded configurations [13]. Since histones and the other chromosomal proteins present in the nucleus are in close contact with DNA, reactive species resulting from the interactions of LEEs with nearby amino acids may also interact with DNA, causing indirect damage. There is thus considerable interest in studying the fragmentation of chromosomal proteins induced by LEEs [14], and Section 6 is devoted to the investigations of the action of LEEs on building blocks of proteins, more particularly on amino acids and peptides.

Despite the significance of the gas-phase and condensed-phase experiments in revealing the major interactions of LEE with DNA, the results of these experiments do not essentially correspond to those obtained in the dynamic existent situation of the cell, where cellular DNA lies in a medium containing essentially water with proteins, ions, and vitamins dissolved in the aqueous environment. Section 7 thus reviews recent studies in more complex systems, where a DNA molecule is embedded into more realistic environments containing water, oxygen, histones, and DNA-binding proteins that mimic cellular conditions.

The role of secondary LEEs in radiosensitization and radiation therapy is discussed in the final section of this chapter. LEEs have subcellular ranges (on the order of 10 nm) in biological materials and interact strongly and destructively with chemical bonds; so, they are ideal for promoting local (i.e., nanoscopic) increases of radiation damage in cells, particularly for targeted cancer therapies. We review a wealth of experimental data on LEE-induced lesions in DNA bound to radiosensitizing gold nanoparticles and the platinum-chemotherapeutic agents. This final section links the effects of radiation and chemotherapy, showing that by modulating the radiation chemistry, chemotherapeutic agents can become radiosensitizers. It also explains how our fundamental understanding of LEE-induced DNA damage can be applied to optimize concomitant chemoradiation therapy (CRT) by modifying the action of LEEs or by increasing their numbers in cancer cells.

2. Decay of Gaseous Transient Anions into Dissociative Electron Attachment (DEA) and Autoionization

In electron–molecule collisions, a TMA is formed, when an incoming electron occupies a previously unfilled orbital of a molecule for duration greater than the usual scattering time [15–17]. Since such an orbital exists at a precise energy [15], TMAs are formed at specific energies usually below 15 eV and rarely above 30 eV [15, 16]. Because of the uncertainty principle, the transient state has a width in energy, which characterizes and identifies the process in the dependence on incident electron energy of cross sections for particular energy-loss processes or the formation of the products or damage yields (i.e., the yield functions). Thus, at the resonance energy, corresponding to the formation of the TMA, yield functions exhibit pronounced maxima that can be superimposed on monotonically increasing background, which results from nonresonant or direct scattering.

The formation of TMAs is well described and reviewed in the literature [15–21]. There are two major types of TMAs or "resonances" [15–17]. The first, known as a single-particle resonance, occurs when the additional electron occupies a previously unfilled orbital of a molecule (or subunit of a large biomolecule) in its ground state. Here, the electron is temporarily trapped within an angular momentum barrier by the shape of the electron–molecule potential. Such TMAs are thus also termed shape resonances. Core-excited resonances or "two-particle, one-hole" states form when electron capture is accompanied by electronic excitation, such that two electrons occupy previously unfilled orbitals. The incident electron is in effect captured by the positive electron affinity of an excited state of the molecule or basic subunit in the case of a large biomolecule, which for DNA might include a base, sugar, or phosphate group. If a momentum barrier in the electron–molecule or electron-subunit potential contributes to the retention of the electron in the electronically excited molecule or subunit, the transitory anion is referred to as core-excited shape resonance. If the TMA state is dissociative and the resonance lifetime is greater than about half of the vibration period of the anion, the latter dissociates. This process is called DEA.

The decay of a TMA into dissociative channels can be understood by considering the hypothetical internuclear potential-energy curve of a diatomic molecule AB and one of its TMA state AB^- shown in **Figure 1**. While the following description is rigorously applicable only to diatomic molecules, it is still qualitatively valid along a specific bond of a polyatomic molecule. Assuming that only Franck–Condon (F–C) transitions are possible and that the AB^- state is dissociative, we see from the consideration of the ground-state nuclear wave function that electrons with energies of between E_1 and E_2 are required to fragment AB^-. However, its fragmentation into $A^- + B$ is only possible if its life time is long enough to survive autodetachment, which can occur for internuclear separations, $R < R_C$. For $R > R_C$, AB^- is stable against autodetachment, as electron emission is endothermic. If the TMA does not dissociate, the

http://dx.doi.org/10.5772/63293

electron is re-emitted into the continuum, leaving the target in vibrational, rotational, or even electronically excited states in the case of a core-excited TMA.

When the TMA state lies above the electronically excited states of the molecule, this later can acquire electronic energy, after autoionization of the anion, in addition to vibrational and rotational motion. If the electronic excited state is dissociative, then fragments A and B (**Figure 1**) are produced. Thus, both decay by autoionization into dissociative electronically excited states and DEA cause the molecule to fragment.

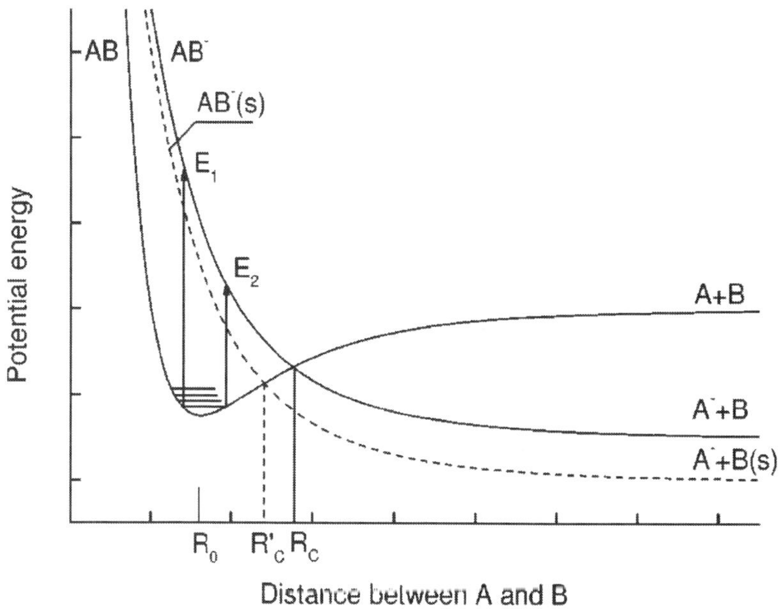

Figure 1. Born–Oppenheimer potential-energy curves associated with dissociative electron attachment. AB represents the potential-energy curve of the ground state of a diatomic molecule and AB^- represents a dissociative state of a corresponding transient anion. The dashed line, $AB^-(s)$, represents the potential-energy curve of AB^- within a molecular solid. R_0 is the equilibrium distance between A and B in the ground-state AB. AB^- is stable against autoionization for $R > R_C$.

Within a local complex potential–curve–crossing model, the DEA cross section may be expressed as

$$\sigma_{DEA}(E) = \sigma_{CAP} . P_s \quad (1)$$

where P_s represents the survival probability of the anion against autodetachment of the electron. The capture cross section σ_{CAP} is given by:

$$\sigma_{CAP}(E) = \lambda_e \, g \, |\chi_v|^2 \left[\frac{\tilde{A}_a}{\tilde{A}_b} \right] \tag{2}$$

where λ_e is the de Broglie wavelength of the incident electron, g is a statistical factor, and χ_v is the normalized vibrational nuclear wave function. Γ_a and Γ_b are the local energy widths of the AB^- state in the F–C region and the extent of the AB^- curve in the F–C region, respectively. The width of the transient anion state in the autodetaching region defines the lifetime τ_a toward autodetachment, $\tau_a(R)=\hbar / \Gamma_a(R)$, such that the survival probability of the TMA, after electron capture, is given by

$$P_s = exp\left[-\int_{R_0}^{R_c} \frac{dt}{\tilde{A}_a(R)} \right] \tag{3}$$

where R_0 is the equilibrium bond length of the anion at energy E and R_c is the internuclear separation beyond which autodetachment is no longer possible. Hence, the DEA cross section depends exponentially on the lifetime of the TMA and the velocities of the fragments.

For further information on the mechanism of TMA formation and its effects on isolated electron–molecule systems, the reader is referred to previous works [15, 16, 22–26]. Information on resonance scattering from single layer and submonolayers of molecules physisorbed or chemisorbed on conductive surfaces can be found in the review by Palmer and Rous [20]. The following section provides information essentially on TMA formation in the condensed phase (i.e., in molecules in solids, condensed onto a dielectric surface or forming a molecular or biomolecular thin film).

3. Modification of electron capture and decay of transient anions in the condensed phase

In principle, the formation and decay of TMAs of condensed molecules can be described using a modified gas-phase picture. For molecular solids or sufficiently thick molecular films condensed onto a metallic substrate or a dielectric surface, the target molecules are unaffected by the substrate, and they exist in the physisorbed state [27]. This weak form of adsorption is characterized by a lack of a chemical bond between molecules, so that the electronic structure and vibrational frequencies of the condensed molecule are essentially unchanged from those in the gas phase [17, 27]. Conversely, electron–molecule scattering is modified in the condensed phase as well as the properties of TMAs [17, 20].

Low-energy (0 – 30 eV) electrons have wavelengths comparable to the distance between molecules in condensed media. Hence, they interact within molecular solids via delocalized processes, predominantly including static and correlation interactions with neighboring molecules, excitation transfer, and coherent scattering [28–31]. Such conditions make it difficult to transfer electron scattering and attachment data from the gas phase to the condensed phase. Even though theoretical models have tried to approximate these processes, the resulting calculations differ substantially from the available experimental data [31–34]. For example, in the gas phase, the incoming electron wave function is a plane wave, whereas scattering events in the condensed phase are those of a diffracted electron wave function that depends on the ordering of the solid. It can be readily seen from Eq. (2) that this change in the partial wave content of the scattered electron wave modifies the capture cross section. Furthermore, Γ_a in Eq. (2) changes in the condensed phase, since new decay channels (e.g., phonon modes) appear and the TMA is formed at lower energy due to the polarization potential induced by the temporarily localized electron [17, 35] and possible lowering of the symmetry of the anion state [20]. The dash curve in **Figure 1** shows the lower energy of the potential-energy curve of the condensed-phase transient anion AB^-. The lower energy causes the curves AB and AB^- to cross at a shorter internuclear distance R_C' than that in the gas phase (R_C). This leaves less time for autoionization of the TMA. In other words, the value of the integral in Eq. (3) becomes smaller, and P_s becomes larger. Moreover, lowering the potential curve of the TMA changes the number of decay channels. The intramolecular channels are decreased because of the lower TMA energy, but new intermolecular channels must be added to take into account decay into collective vibrations (phonon modes). Hence, the resonance lifetime may increase or decrease, and so the DEA cross section (i.e., the DEA intensity depends on the details of AB and AB^-potential-energy curves and the number of decay channels). In addition, electron transfer from one molecule to another may occur, and hence provide additional decay pathways for TMAs [36]. For very large biomolecules, such as DNA, electron transfer between elemental subunits also impedes electron localization [35]. Hence, due to intramolecular electron transfer, the probability of TMAs forming on specific subunits can also be reduced.

The increase in DEA cross section resulting from the shift of the curve crossing point in **Figure 1** from R_C to R_C' can be illustrated experimentally by covering a metal surface with a multilayer film of a condensed rare gas and depositing a molecule on the film surface. As an example, **Figure 2** shows the result of such an experiment in which a 0.1 monolayer (ML) of CH_3Cl was condensed onto a 20 ML thick Kr film [37]. The variation of a surface charging coefficient A_s, which is directly proportional to the absolute cross section (μ) for the reaction (1) recorded between incident electron energies 0 and 2.5 eV, is shown in the inset of **Figure 2**.

$$e^- + CH_3Cl \rightarrow CH_3Cl^- \rightarrow CH_3 + Cl^- \tag{4}$$

Within this energy range there exists a single structure in the A_S energy dependence, the maximum of which lies at approximately 0.5 eV for large Kr coverage. The peak denotes the energy of the TMA CH_3Cl^-. As the Kr film thickness is reduced, the transitory CH_3Cl^- anion moves closer to the metal substrate, and the energy of the maximum in the inset lowers owing to the larger polarizability of the metal compared to Kr. The lower curve in **Figure 2** shows this shift in energy of $CH_3Cl^-$with decreasing thickness. However, as the energy of transitory CH_3Cl^- on the Kr film lowers, according to **Figure 1**, R_C' becomes smaller and P_S increases. Thus, as seen in the experimental curve with the full squares in **Figure 2**, the magnitude of the absolute cross section for Cl^- production at the peak values increases with decreasing thickness of the Kr film. When CH_3Cl^- is formed too close to the metal substrate, the additional electron transfers to the metal, and μ sharply decreases.

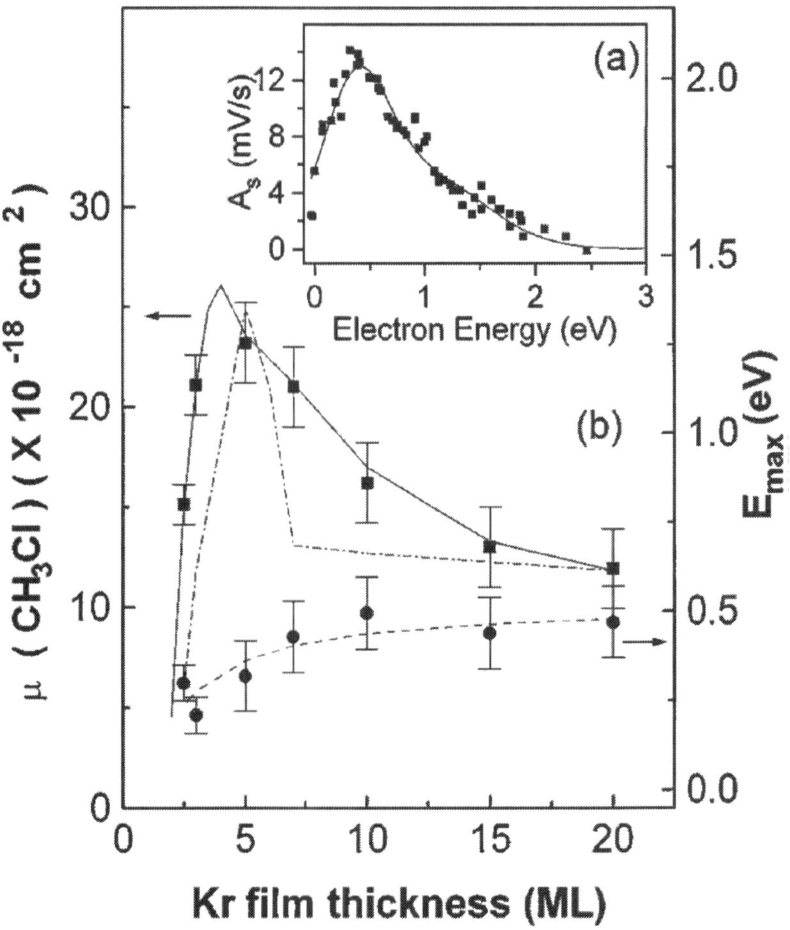

Figure 2. CH_3Cl^- formation and dissociation by electrons of 0–2.5 eV incident on submonolayer amounts of CH_3Cl physisorbed on a multilayer film of Kr. (a) Variation of the charging coefficient of the film A_s due to CH_3Cl^- dissociation. (b) Variation with film thickness of: (■) the amplitude of the maximum in the charging cross section (μ); (-··-) the amplitude of the maximum in μ calculated with the *R*-matrix method [37]; (•) variation of the energy of maximum in μ and A_s; and (---) a parametric fit of this maximum using the image charge model [38].

In the condensed phase, TMAs differ from their gas-phase counterparts, in the following ways: (1) the electron energies required for their formation are usually lower by 0.5–1.5 eV, dependent

http://dx.doi.org/10.5772/63293

on the local polarization of the solid and/or changes to the anion's symmetry; (2) due to their lower energies, they usually have fewer intramolecular decay channels, although, new intermolecular channels via electron emission into the dielectric may appear; (3) the lifetimes will be longer or shorter due to the changes in the number of decay channels, energy, and symmetry; (4) the initial electron capture probability, and the cross sections for decay into particular intermolecular and intramolecular excitations or for DEA may vary by orders of magnitude, as these are dependent on energy, intramolecular and intermolecular electron transfer, and symmetry. In summary, when a TMA is formed on a molecule located inside or at the surface of a molecular or biomolecular solid, its gas-phase characteristics are usually considerably affected by the local environment.

4. DEA to gaseous DNA subunits and radiosensitizers

4.1. DNA bases

A large number of DEA studies have been performed on gas-phase DNA bases and their derivatives over the last two decades [2, 39]. Briefly, DEA is the resonant process that involves the LEE capture by a molecule (AB) to produce gaseous TMAs ($(AB)^-$), described in Section 2, which then dissociate into an anion (A^-) and a neutral radical or radicals ($B^\bullet$), according to the following reaction:

$$e^- + AB \leftrightarrow (AB)^- \rightarrow A^- + B^\bullet \qquad (5)$$

In general, the low-energy resonances in nucleobases are present either at subexcitation (< 3 eV) energies or in the energy range 5–12 eV [39]. The yield function for the DEA processes for thymine resulting in multiple fragment formation is shown in **Figure 3**. To analyze the formation of the negative ions, yield functions were usually recorded by scanning the incident electron energy, while potential voltages applied to the quadrupole mass spectrometer were set for a given ion mass. The ion yields were detected by a channeltorn and plotted as a function of the incident electron energy.

The high-energy resonances lead to transient anion fragmentation via opening of the ring structure, while the low-energy resonances are primarily due to loss of one or two neutral hydrogen, which maintains the ring structure.

The DEA yield functions for nucleobases and their related compounds show a remarkable feature that can be recognized as a common phenomenon, that is, site selectivity [40–42]. By tuning the energy of the incoming electron, it is possible to control the location of the bond cleavage. That is, a specific chemical bond in a molecule can be targeted by electrons followed by fragmentation. As an illustration of this site selectivity in nucleobases, DEA to thymine with deuterated and methylated substitutions is described. This phenomenon was observed for other nucleobases and their derivatives, for example, adenine [43] and hypoxanthine [44].

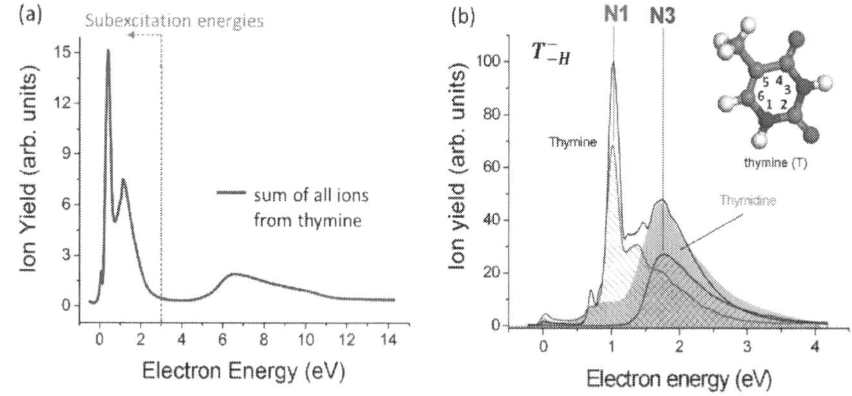

Figure 3. (a) Total ion yield as a function of electron energy for DEA process of thymine. (b) Formation of dehydrogenated anions from thymine (black curve), methylated thymine at N3 site (red curve), methylated thymine at N1 site (blue curve), and thymidine (gray area) [45]. Molecular structure of thymine with numbering and atom labeling.

At subexcitation energies, DEA leads to thymine dehydrogenation by loss of a neutral hydrogen atom [40, 42]. This reaction can be expressed as follows:

$$e^- + T \rightarrow T^- \rightarrow T^-_{-H} + H^{\bullet} \tag{6}$$

where T^- is the TMA of thymine (T) and T^-_{-H} is the closed-shell anion formed by the ejection of a neutral hydrogen radical <i>$H^{\bullet}$</i>. This dehydrogenation process depends on the site from which the H atom is removed. Experimental studies with partially deuterated thymine, in which the deuterium is at either nitrogen or carbon sites, showed that hydrogen loss occurs exclusively from the N sites. H loss from the C sites is thermodynamically accessible within this energy range, but has not been observed experimentally. Moreover, in employing methylated thymine and uracil, it has been shown that by adjusting the electron energy, the loss of H can be made even site-selective with respect to the N1 and N3 positions. Although 1 eV electrons induce H loss at the N1 position (N1-H), the process can be switched at 1.8 eV to N3-H (**Figure 3b**). These results have significant consequences for the molecular mechanism of DNA strand breaks induced by LEEs. Within DNA, the N1 position of thymine is coupled with the sugar moiety and thus forms thymidine, which is one of the nucleosides. Because the shapes of the signals from thymine and the more complex thymidine resemble each other, it can be concluded that H abstraction in thymidine predominantly occurs at the thymine moiety and, more precisely, at the N3 position (**Figure 3b**) [45].

In addition to the detection of anions and the energies at which they are formed, much effort has been expended to matching particular types of DEA process to specific resonant peaks observed in DEA ion yields. In the case of the most abundant anion formed for all nucleobases,

http://dx.doi.org/10.5772/63293

it has been proposed that these resonant peaks can be assigned to vibrational Feshbach resonances (VFRs) [46, 47]. VFRs usually occur at low energies, when vibrational levels of the transient anion lie below the corresponding vibrational states of the neutral, and are more expected in greatly polarized molecules with very large dipole moment, which leads to a long-range attractive interaction. They may serve as a gateway for dissociation at low energies if they are coupled with a dissociative valence state. This can be the case for the formation of dehydrogenated ions from nucleobases, where resonances arise from coupling between the dipole bound state and the transient anion state associated with the occupation of the lowest σ^* orbital. Recently, the nucleobase fragmentation of N–H bonds induced by LEEs was studied by employing the CASPT2//CASSCF computational approach [48]. These calculations showed that the two lowest lying π^* states can be determined at energies below 1.0 eV and above 2.0 eV for pyrimidines, whereas for purines, this energy gap between the two anionic states was less pronounced. These calculations also suggested the possibility of coexistence of dipole-bound and valence-bound processes in the low-electron energy range.

Further to the observations of site selectivity in DEA processes leading to single-bond cleavage within a nucleobase, site selectivity also occurred in multiple-bond cleavage. As in the cases of both dehydrogenation of nucleobases and its complementary channels, which resulted in the H^- formation, site selectivity was demonstrated when multiple-bond cleavage was involved, for example, for the formation of NCO^- from thymine and its derivatives [49]. This anionic fragment was formed in a sequential decay reaction, in which the dehydrogenated anionic nucleobase acts as an intermediate product. In this case, the remarkable resonances, which were observed for dehydrogenation and for H reaction channels in nucleobases, were preserved for the subsequent decay reaction, leading to the formation of NCO^- as the final product.

In general, the total cross sections for DEA to nucleobases exhibited comparable magnitudes in the energy range for TMA formation [50]. However, these cross sections were up to 10 times smaller than those for the formation of single-strand breaks, while the cross sections for sugar and phosphate group analogs (see Section 4.2) were even smaller in magnitude.

4.2. Sugar and phosphate group

The high fragility of the DNA backbone with respect to the impact of LEEs with low kinetic energy was observed for 2-deoxy-D-ribose and its RNA equivalent (i.e., ribose), along with their analogs [2]. In principle, the dissociation of any of P–O–C bonds in the sugar–phosphate backbone or C–C bond within the sugar could result in a DNA strand break. If such breakages were to occur via the DEA process in DNA, then DEA would represent an important pathway through which the direct interaction of LEEs could affect biologically significant damage.

The DEA to 2-deoxy-D-ribose results in a strong decomposition of the sugar at electron energies near 0 eV, indicating the loss of one or more molecules of water [51]. Similar findings were observed for D-ribose and other sugars [2], indicating that DEA at 0 eV is a common property of all monosaccharides. However, the mechanisms for DEA reactions leading to loss of neutral water are more complex in comparison to the dehydrogenation of the nucleobases, because they involve the dissociation of multiple bonds and/or atom rearrangement with simultaneous formation of new bonds. Therefore, the mechanism of DEA to sugars near 0 eV is not fully understood. It is however proposed to occur via the formation of a "shape" resonance. In a sugar molecule, the extra orbital can be a σ^* orbital of the O–H bond. As was observed for alcohols [52], the σ^* orbital of the hydroxyl group for the dehydrogenation channels appears at higher energy for simple alcohols than for cyclic alcohols. Moreover, it was found that larger numbers of hydroxyl groups present in a molecule could enhance the dissociation of an O–H bond, by decreasing the energy of the thermodynamic threshold. This mechanism has been suggested for 2-deoxy-D-ribose and D-ribose, which contain three and four hydroxyl groups, respectively [2]. In addition, experiments with the ribose analogs tetrahydrofuran and 3-hydroxytetrahydrofuran showed that DEA cross sections were greatly enhanced by the presence of OH groups [50]. However, for alcohols, their molecular dissociation involved simple bond cleavage, while in sugars, fragmentation of several different bonds occurs. One of the proposed models for sugar dissociation was provided from ab initio calculations of VFRs formed initially by a dipole-bound state of sugar due to a large dipole moment [53]. Other quantum chemical calculations confirmed this model, showing that the sugar ring can convert into an open chain by intramolecular charge transfer. This mechanism can lead to dissociation by loss of a water molecule, assuming that the barrier for such a transfer is sufficiently low [54]. It was also calculated through quantum dynamics scattering theory that the formation of shape resonances for D-ribose is excluded at low energies, but they can be formed at higher energies [55].

In the case of thymidine, in which sugar is covalently bound to thymine, the DEA study showed two resonant structures (**Figure 3b**) [45]. The one at lower energy was associated with a reaction in which the excess electron is initially localized in the sugar moiety, leading to the glycosidic bond cleavage. The second resonance was attributed to a reaction in which the excess electron was localized on the thymine moiety, resulting in the loss of a neutral H atom from the N3 site, as was mentioned for thymine. Since nucleosides can be easily decomposed due to the elevated temperatures necessary for evaporating samples, no experimental data for other gas-phase nucleosides or nucleotides are reported, besides those for thymidine [45], cytidine [56], and 2-deoxycytidine 5-monophosphate [56].

Similarly, due to experimental difficulties, the phosphate group in the gas phase could not easily be investigated as an isolated compound. Its simplest analog, H_3PO_4 (phosphoric acid), is not easily vaporized for gas-phase experiments or molecular deposition for thin film experiments [11]. Therefore, to understand the DEA process within the phosphate group, several compounds involving phosphoric acid derivatives, for example, dibutylphosphate and triethylphosphate [57], were examined. DEA to these compounds lead to P–O and C–O bond cleavages, which correspond to a direct single-strand break. As for sugars, many fragmentation channels occurred close to 0 eV; however, these low-energy channels are most likely driven by the large electron affinity of PO_3 (4.95 eV). The cross sections for DEA to the sugar and phosphate group analogs were relatively small, that is, about one magnitude lower than those for nucleobases [50]. These gas-phase results on sugars and phosphate units revealed that LEE attachment can induce single-strand breaks by electron localization either on the sugar moiety followed by the electron transfer to the backbone or directly on the phosphate group.

4.3. Radiosensitizers

An important characteristic of many current and potential radiosensitizers used in radiotherapy (or potential ones) is a high cross section for DEA. Since halogenated pyrimidines, mainly substituted uracil derivatives, exhibit high sensitivity to electron attachment and a rich fragmentation pattern from DEA, they have attracted considerable interest as radiosensitizers. From a medical point of view, the substitution of pyrimidines in the genetic sequence of cellular DNA does not affect the gene expression, and additionally enhances the sensitivity of living cells to radiation. A large number of gas-phase experimental and theoretical studies of several halogenated pyrimidines (e.g., 5-bromouracil [58–63], 5-chlorouracil [58, 59, 61, 64, 65], 5-fluorouracil [58, 59, 61, 65], 5-iodouracil [59, 62], 6-chlorouracil [58,66]) were performed in recent years and report orders of magnitude of higher cross sections for DEA relative to their nonsubstituted precursors. Further to the DEA studies, other electron spectroscopic techniques and theoretical calculations at the ab initio and density functional theory levels were utilized to characterize electronic structure and reveal the fragmentation mechanisms of halogenated pyrimidines [67]. These studies elucidated the energies of vertical transitions to π^* and σ^* orbitals, showing that the ground TMA state of pyrimidine with the additional π^* electron is a few tens of eV more unstable than the neutral ground state, whereas the vertical electron affnities of the halogenated derivatives were found to lie close to 0 eV. Moreover, DEA studies revealed that the lowest π^* anion states of the halogenated pyrimidines follow similar fragmentation channels, resulting in the formation of the halide fragment anion. These studies also revealed that the total anion yields for bromopyrimidine were much larger than those measured for the chloro-derivatives. These results indicate that bromopyrimidines carry the greatest potential as radiosensitizers for damage by SEs, which, via DEA to bromo-substituted DNA, will enhance radiation-induced damage to the cell. Recently, gas-phase DEA studies on halogenated purines (e.g., chloroadenine [68]) and fluorinated nucleosides (2-deoxy-5-fluorocytidine and 2,2-difluorocytidine (gemcitabine) [69]) have been initiated to determine in what ways their radiosensitizing properties are derived from LEE-driven chemistry.

In addition to the halogenated nucleobases, several aromatic compounds containing nitro groups have been recently investigated in the gas phase. For instance, DEA studies performed for 5-nitrouracil showed the formation of a long-lived parent anion, as well as a rich fragmentation pattern via formation of either "shape" or "core-excited" resonances at low electron energies [70, 71]. The properties of 5-nitrouracil showed a radiosensitizing nature similar to that of the halogenated pyrimidines. Interestingly, while in the case of halogenated pyrimidines, the most dominant fragment formed was a halide anion, that for 5-nitrouracil is an anion of the pyrimidine without a nitro group. Therefore, the counterpart fragment of this dissociation channel is the formation of the NO_2 radical, which is formed in close vicinity to DNA and can lead to the activation of lethal cluster damage in living cells.

There is also a great potential for other nitro-containing compounds such as nitroimidazolic compounds to be used in radiotherapy, since LEEs effectively induce their dissociation [72, 73]. Similarly, their decomposition via DEA involves a range of unimolecular fragmentation channels from simgle-bond cleavages to complex reactions, possibly leading to a complete degradation of the target molecule. However, these studies revealed that the entire rich chemistry induced by DEA was completely suppressed by methylation in the electron energy range below 2 eV.

In recent years, platinum-based drugs were also investigated regarding their decomposition by LEEs. It was suggested that in concomitant treatment in which chemotherapeutic drugs and radiotherapy are combined, one possible mechanism responsible for the observed synergy between treatments is the enhancement in the number of secondary species induced by primary radiation in the vicinity of the binding site of the platinum compounds in DNA (see Section 8). The gas-phase DEA studies of $PtBr_2$ in the electron energy range between 0 and 10 eV showed the formation of the Br anion via two possible channels. The most dominant channels were assigned to the Br^- + PtBr dissociation limit reached at ~1 eV and the higher energy channel to Br^- + Pt + Br [74].

The observation that all these radiosenisitizers exhibit DEA with high effciency, even close to 0 eV, may have significant implications for the development and use of these drugs in tumor radiation therapy. Considering their use as radiosensitizers, their fragmentation and the resulting generation of radicals at very low electron energies may be a key in understanding their action and the molecular mechanisms necessary to improve radiotherapy.

5. Electron attachment to short single-stranded and plasmid DNA

Cellular DNA consists of a double-stranded helical structure, composed of two long polynucleotide chains [75]. Thus, as already mentioned in the Introduction section, in order to systematically understand LEE damage mechanisms and their role in radiation DNA damage, molecular targets of increasing complexity were studied, from simple molecules containing just two of the basic subunits (e.g., a phosphate group coupled with a sugar or a nucleoside having a DNA base + sugar), via synthetic, single- and double-stranded oligonucleotides, containing multiple nucleotides to plasmid and other cellular DNA with many thousands of base pairs.

Even though most simple DNA components may be easily vaporized for experimental investigation in the gas phase, the larger units such as nucleosides and nucleotides usually decompose during evaporation [12]. In any case, the condensed phase is certainly the more appropriate environment to study problems relevant to radiation damage in biomolecular systems. The experimental methods and techniques, used in the condensed phase, differ from those in the gas phase. Most condensed phase experiments are achieved by bombarding thin films (2–10 nm) of oligonucleotides or plasmid DNA with an energy-selected beam of LEEs from an electron gun or an electron monochromator. To prevent excessive charging, these thin-film biological samples are deposited onto a metal substrate by spin-coating, lyophilisation (freeze-drying), or molecular self-assembly, as in the case of thiolated DNA on gold substrates [10] and 1,3-diaminopropane layer plasmid on graphite [76]. The LEE-induced damage to plasmid and linear DNA films has then been investigated by (1) measuring electron-stimulated desorption (ESD) of anions, (2) imaging the breaks by atomic force and scanning tunneling microscopies, and (3) analyzing, after bombardment, the change of DNA topology by gel electrophoresis or the molecular content by high- performance liquid chromatography (HPLC) and mass spectroscopy [35, 77].

Oligomers of single-stranded DNA containing the four bases (e.g., G, C, A, and T), which are among the simplest forms of DNA, have made the analysis of degradation products much simpler than would be the case for longer single- and double-stranded configurations. Short oligomers deposited onto metal surfaces (e.g., tantalum, platinum, and gold) as films of different thicknesses (1–5 ML) were bombarded with LEEs and produced fragments analyzed by HPLC [77]. The results for the GCAT oligonucleotide indicated that strand breaks occur preferentially by cleavage of the C–O bond rather than the P–O bond, with two maxima at electron energies of 6 and 10 eV [78, 79].

Recently, Bald and co-workers demonstrated the visualization of LEE-induced bond cleavage in DNA origami-based DNA nanoarrays on the single-molecule level using atomic force microscopy (AFM) [80–82]. This novel method has a number of advantages: (1) only miniscule amounts of material are required to create submonolayer surface coverage, because of the facility to detect the DNA strand breaks at a single-molecule level; (2) within a single experiment, more than one oligonucleotide sequence with various arrangements can be irradiated to efficiently compare a number of different DNA structures; (3) the method represents a simple way to obtain absolute strand break cross sections, thus providing benchmark values for further experimental and theoretical studies, and finally (4) this technique is not limited to single strands, but can be extended to quantify DSBs and to investigate higher order DNA structures.

Applying this technique, Bald and coworkers compared the absolute strand break cross sections of different 13-mer oligonucleotide sequences (i.e., 5'-TT(*XTX*)3TT, with *X* = A, C, or G) to evaluate the role of the different DNA nucleobases in DNA strand breakage. They also studied the sensitizing effect of incorporation of 5-bromouracil (BrU) by comparing the absolute strand break cross sections for the sequences 5'-TT(*X*BrU*X*)3TT, with *X* = A, C, or G. The observed trend in the absolute strand break cross sections agrees qualitatively with the previous HPLC studies investigating the fragmentation of oligonucleotide trimers of the sequence T*X*T, with *X* = A, C, G, irradiated with 10 eV electrons [83]. Additionally, the cross sections measured with this method are comparable in magnitude with the cross sections for strand breaks in different plasmid DNA molecules induced by 1–10 eV electrons, as determined by agarose gel electrophoresis [84, 85]. The DNA nanoarray technique thus bridges the gap between very large genomic double-stranded DNA and very short oligonucleotides, and enables the detailed investigation of sequence-dependent processes in DNA radiation damage. Further experimental and theoretical studies are carried out covering a broad range of electron energies and DNA sequences to elucidate the most relevant damage mechanisms [86].

In order to increase the complexity of targeted biomolecules, several studies have investigated the damage induced by LEEs in double-stranded plasmid DNA. Due to the supercoiled arrangement of plasmid DNA, a single-bond rupture in a DNA with a few thousand base pairs can produce a conformational change in the topology of the entire molecule. These changes include base alterations, abasic sites, intra- and inter-strand base cross-links, DNA adducts, and SSBs or DSBs; hence, these can be detected efficiently by techniques such as gel electrophoresis. This technique can identify supercoiled (SC), nicked circular (C), full-length linear (L), cross-linked (CL), and short linear forms of DNA, which can be assigned to undamaged DNA, SSBs, DSBs, several types of cross-linked DNA, and multiple double-strand breaks (MDSBs), respectively [87].

Though it has been established that most of the strand breaks induced by ionizing radiation have been repaired by a DNA ligation step, a DSB represents a particularly detrimental lesion that poses a serious threat to the cell, since it usually cannot be easily repaired [88]. Indeed, even a single DSB can lead to cell death if left unrepaired or, more worryingly, it can cause mutagenesis and cancer if repaired improperly [89].

The results obtained for LEE-irradiated supercoiled plasmid DNA in several investigations are well described in the literature and summarized in authoritative review articles [10, 11, 35]. These studies have shown that SSBs can occur as a result of DEA at electron energies well below electronic excitation and ionization thresholds (0.8–10 eV) [83, 90]. The results of Martin et al. [90] reveal two resonant peaks at 0.8 and 2.2 eV in the SSB yield function (i.e., the number of strand breaks versus the incident electron energy) via the formation of TMAs. These findings are consistent with theoretical calculations indicating that SSBs induced by near-zero energy electrons are thermodynamically feasible [91–93]. Theoretical simulations of electron scattering and electron capture via "shape" resonances support the role of LEEs in DNA strand breaks [94]. Theoretical calculations on scattering and attachment of LEEs to DNA components up to supercoiled plasmid DNA have been intensively reviewed in recent years [95, 96].

Another spatially resolved technique that exploits the use of graphene-coated Au thin films and surface-enhanced Raman spectroscopy (SERS) has recently emerged. Utilizing this technique, the sequence dependence of DNA damage at excitation energies < 5 eV can be studied [97]. Currently, Ptasińska and coworkers are performing a quantitative and qualitative investigation of the various types of damages to dry and hydrated DNA induced by exposure to helium and nitrogen atmospheric pressure plasma jets (APPJs). Since an APPJ contains multiple reactive species, including LEEs, also found in radiation chemistry, exposure to these plasma jets provides information on both the direct and indirect pathways to damaging DNA. Ptasińska and coworkers have employed nitrogen APPJ to induce DNA damage in SCC-25 oral cancer cells, and have thus provided new insight into radiation damage to a cellular system [98].

6. LEEs interaction with protein building blocks

It is well known that within the cells, DNA is in close contact with, and packed by, chromosomal proteins (histones). The attachment of proteins protects DNA from damage by compaction (e.g., which restricts easy access by free radicals to DNA) and repairs some of the damage of electron/hydrogen donation [99]. LEE damage to proteins within cells should not, by itself, cause significant long-term biological damages, because proteins can be replaced. However, due to the presence of histones and other chromosomal proteins in the vicinity of DNA, reactive species produced from LEE interactions with protein constituents (e.g., nearby amino acids) may in turn interact with DNA, causing indirect damage. Thus, from a radiobiological point of view, there is considerable interest in studying the action of LEEs on this important class of biomolecules [100]. Recent work has focused on the building blocks of proteins, that is, amino acids and small peptides, since the size and complexity of chromosomal proteins prevent direct detailed analysis of the fragmentation processes induced by LEEs [11, 39]. Indeed, measuring the fragmentation of amino acids and their analogs is no more complex than it is for DNA constituents (see Sections 4 and 5) [101–103], and can help elucidate the effects of electron irradiation in larger more complex proteins [103].

In the recent years, several investigations have employed soft ionization techniques, such as matrix-assisted laser desorption ionization (MALDI) [104–107], electrospray ionization (ESI) [108, 109], and collision-induced dissociation (CID) [110–114], to study the ionization and fragmentation of different amino acids and small peptides in the gas phase. Gas-phase investigations of LEE-induced damage to protein subunits have been reported for the amino acids alanine [115], tyrosine [116], glycine [117, 118], proline [119, 120], cysteine [121], and serine [122,123], as well as small peptides, such as dialanine [124] and amino acid esters [125]. For all cases, the anion yield functions (i.e., ion yields measured as a function of electron energy) exhibited localized maxima at energies below 15 eV, indicating the formation of TMAs. It has been established that no intact parent anion is observable on mass spectrometric timescales after capture of a free electron, and that the most probable reaction corresponds to the loss of a hydrogen atom from a carboxyl group to form for a molecule "M," the dehydrogenated anion $(M–H)^-$ at energies of around 1.5 eV [120, 123, 126, 127]. Early DEA studies ascribed this process to initial electron attachment into a π^* orbital of the (C=O) bond in the COOH group, which couples to the repulsive σ^* (O–H) orbital [118]. However, recent calculations questioned this DEA mechanism [126]; instead, it was suggested that direct electron capture into the purely repulsive short-lived σ^* (O–H) orbital, which is a very broad resonance of more than 5 eV width, could be responsible for the loss of the hydrogen [126].

In the condensed phase, analyzing LEE-stimulated desorption of anions from physisorbed thin films of glycine, alanine, cysteine, tryptophan, histidine, and proline [128, 129] indicated that H^- was the major desorption fragment, as CH_3^-, O^-, and OH^- were the fragments produced with lower signals in all named amino acids. Similar results were observed in ESD experiments from LEE-bombarded chemisorbed films, prepared by self-assembled monolayers (SAMs) of two different chains of Lys amide molecules [129]. For this model of a segment of a peptide backbone, the desorbed signals were dependent on the length of the amino acid sequence.

Amino acids are also suitable model molecules for investigating the interactions of biomolecules with metallic surfaces, particularly silver and gold. Of the 20 naturally occurring amino acids, only cysteine contains a thiol (-SH) group, which allows it to bind to the metal by forming a S-Metal bond [130, 131]. This characteristic makes cysteine an ideal model to investigate protein interactions with gold surfaces including those of gold nanoparticles [132, 133]. A detailed study on electron attachment to L-cysteine/Au(111) was recently reported by Alizadeh et al. [134, 135] who measured anion yields desorbed from chemisorbed (SAMs) and physisorbed thin films bombarded with sub-20 eV energy electrons. These ESD measurements showed that LEEs are able to effciently decompose this amino acid via DEA and dipolar dissociation (DD), when the molecule is chemisorbed via the SH group to a gold surface.

Regarding the protective effect of amino acids on DNA against LEEs, Solomun et al. [136] reported that the single-strand DNA-binding *E. coli* protein can effectively inhibit the formation of SSBs by 3-eV electrons in oligonucleotides. Ptasińska et al. [137] subsequently investigated by post-irradiation analysis with HPLC-UV, the molecular fragmentation induced by 1-eV electrons in films comprising the GCAT tetramer and one of the two amino acids, glycine and arginine. At low ratios (R) of amino acid to GCAT (i.e., $R < 1$), particularly for glycine, the total oligonucleotide fragmentation yield unexpectedly increased. At higher ratios ($1 \leq R \leq 4$), protection of DNA from damage by electrons was observed for both glycine and arginine. Therefore, the amino acid probably reduced electron capture by GCAT and/or the lifetime of the TMA that initiates DEA process. A similar conclusion regarding the stability of the amino acid side chain–nucleobase complexes can be deduced from the theoretical studies of Wang et al. [138]. Wang and coworkers performed calculations at the B3LYP/6-311G(d,p)-level anionic hydrogen-bonded complexes formed between the amino acid side chains and the nucleobase guanine.

Furthermore, by studying via first-principles molecular dynamics simulations a model system composed of thymine and glycine, Kohanoff et al. [139] recently investigated the protection of DNA by amino acids against the effects of LEEs. They considered thymine–glycine dimers and a condensed-phase model consisting of one thymine molecule solvated in amorphous glycine. These results indicated that at room temperature, the amino acid chemically and physically performs the role of a protective agent for the nucleobase. In a chemical mechanism, the excess electron is first captured by the thymine; then, a proton is transferred in a barrierless way from a neighboring hydrogen-bonded glycine. Reducing the net partial charge on the thymine molecule stabilizes the excess electron. In the physical mechanism, glycine molecule acts as an electron scavenger to capture the excess electron directly, which prevents the electron to be localized in DNA. Protecting the nucleobase via the latter mechanism requires a predisposition for proton transfer to the oxygen in the carboxylic acid group of one of the involved amino acids. Consequently, raising the free-energy barrier associated with strand breaks, prompted by these mechanisms, can halt further reactions of the excess electron within the strand of DNA, for instance, transferring the electron to the backbone which leads to induce a strand break in DNA. Increasing the ratio of amino acid to nucleic acid will enhance the protecting role of amino acids, and accordingly will decrease the induction of DNA strand breaks by LEEs, as shown experimentally [137, 139].

7. LEEs interaction and induced damage under cellular conditions

The gas- and condensed-phase experiments with DNA and its constituents discussed previously were performed under ultrahigh vacuum (UHV) conditions to permit use of electron beams and mass spectrometry, and to better control the molecular environment. While such experiments provide information on the direct effects of LEEs, they do not reveal how LEEs can indirectly damage DNA. Comparatively, due to the experimental difficulties related to the production and observation of LEEs in aqueous media, studies on the indirect damage of LEEs to DNA have not been greatly developed.

Ideally, to understand how the fundamental mechanisms in LEE–DNA interactions are adapted in living cells, the experimental studies should be extended to the more complex dynamic molecular environment of the cell, or more realistic ones, for the DNA molecule that contains essentially water, oxygen, histones, and DNA-binding proteins [99]. For instance, in the work of Ptasińska and Sanche [140], the ESD yields of different anions desorbed by 3–20 eV electron impact on GCAT films were measured under an aqueous condition, corresponding to 5.25 molecules of water per nucleotide. Their experiments demonstrated that adding water to dry DNA results in the binding of the molecule to the phosphate group at the negatively charged oxygen [141], and then formation of a complex of tetramer and a water molecule (DNA•H_2O). This complex permits the formation of a new type of dissociative core-excited TMA located on the phosphate group, which decays by O^- desorption under electron impact via a resonance at 11–12 eV and by OH^- desorption from breaking the P–O bond. H^- also desorbs by dissociation of a TMA of the complex which causes bond cleavage on the H_2O portion. Moreover, LEE-induced damage to DNA via DEA enhances by a factor of about 1.6 when an amount of water corresponding to 60% of the first hydration layer is added to vacuum-dried DNA. Although the magnitude of this enhancement is considerable, it is still much smaller than the modification in yields of products produced by the first hydration layer surrounding the DNA during the radiochemical events that follow the deposition of the energy of LEE in irradiated cells. Theoretical and experimental studies were concurrently carried out on the diffraction of 5–30 eV electrons in hydrated B-DNA 5'-CCGGCGCCGG-3′ and A-DNA 5'-CGCGAATTCGCG-3′ sequences by Orlando et al. [142]. They postulated that compound H_2O•DNA states may contribute to the modification of strand breaks yield functions [142, 143]. Furthermore, Orlando et al. noted that lowering of the threshold energy for DSBs below 5 eV may be correlated with the presence of these compound states. In this case, an initial "core-excited" resonance would autoionize, yielding electronically excited water-derived states and a low-energy electron. The electronically excited state dissociates forming reactive O, OH, and H, which can lead to sugar–phosphate bond breakage. The slow electron could moreover scatter inelastically within a limited mean free path and excite a "shape" resonance of a base on the opposite strand. The combination of these two energy-loss channels could lead to a DSB. This type of DSB requires the presence of water and is difficult to be repaired due to the close proximity of damage sites.

Recent work using graphene-coated gold thin films also signaled the significance of the existence of water molecules in DNA damage mediated by "shape" resonances [144]. This is likely due to the influence of water on lowering the barrier for charge transfer from the base to the sugar–phosphate bond. In addition, the binding interaction of DNA with graphene allows direct coupling to the phosphates as well as more direct scattering with the guanine and adenine bases. Electrons that have not been captured by DNA bases can be captured by graphene and immediately transferred over 200 nm within < 0.36 ps. The environmental or graphene substrate interactions are critical, and at least two mechanisms occur simultaneously during DNA damage on monolayer graphene: direct base capture and ballistic transfer from the graphene.

An alternative approach to simulate cellular conditions has been recently developed by Alizadeh et al. [145] to investigate LEE-induced DNA damage under atmospheric conditions and at various levels of humidity and oxygen. Thin films of plasmid DNA deposited on tantalum and glass substrates were exposed to Al K_α X-rays of 1.5 keV. The general features of the photo-ejected SE from the metallic surfaces exposed by primary X-ray photons are well understood; in particular, more than 96% of SEs emitted from tantalum lie below 30 eV and the energy distribution peaks around 1.4 eV, with an average energy of 5.85 eV [145]. Whereas the damages induced in DNA deposited on glass are due to soft X-rays, those arising from DNA deposited on tantalum result from the interaction of X-rays + LEEs. The difference in the damage yields measured in the samples deposited on two different substrates is ascribed to the interaction of LEEs with the DNA and its nearby atmosphere.

Alizadeh and Sanche [146] employed this technique to examine how the presence of several cellular components (such as, O_2, H_2O and O_2/H_2O) modulates the LEE-induced damage to DNA molecules. They observed that for hydrated DNA films in an oxygenated environment, the additional LEE-induced damage that results from the combination of water and oxygen exhibits a super-additive effect, which produces a yield of DSB almost seven times higher than that obtained by X-ray photons. More recently, they reported the formation of four radiation-induced products from thymidine by soft X-rays and LEEs, specifically base release, and base modification including 5-HMdUrd, 5-FordUrd, and 5,6-DHT [147]. Of the products analyzed, thymine release was the dominant channel arising from N-glycosidic bond cleavage involving π^* low-lying TMA. A LEE-mediated mechanism was proposed to explain observation of 5-HMdUrd and 5-FordUrd products, which involve loss of hydride (-H^-) from the methyl group site via DEA. *G*-values derived from the yield functions indicate that formation of free thymine, 5-HMdUrd, and 5-FordUrd are promoted by an oxygen environment rather than a nitrogenous atmosphere, since the numbers and reactivity of radicals and ions are formed via interactions of radiation with O_2, and are considerably larger than under N_2. Moreover, O_2 can additionally react with C-centered radicals, thereby "fixing" or rendering the damage permanent. In contrast, no 5,6-DHT was detected when samples were irradiated under an O_2 atmosphere, indicating that O_2 molecules react with an intermediate radical compound, thereby inhibiting the pathway for 5,6-DHT formation [147].

Recently, novel decay mechanisms for electronic excitations and correlated electron interactions have become subjects of intense study. Just over a decade ago, Cederbaum et al. [148–150] proposed an ultrafast relaxation process in inner valence levels, which occurs in molecular systems with weakly bound forces, such as van der Waals forces or hydrogen bonding. This mechanism referred to as intermolecular Coulomb decay (ICD) is possible mainly due to the couplings and interactions induced by the local environment. Unlike most ionization processes, ICD results in the ejection of an electron from the neighbor of an initially ionized atom, molecule, or cluster [151]. The energy of the ICD electron is low, typically less than 10 eV. ICD is expected to be a universal phenomenon in weakly bound aggregates that contain light atoms and may represent a hitherto unappreciated source of LEEs. Though most ICD measurements

http://dx.doi.org/10.5772/63293

have concentrated on rare gas clusters, new sophisticated experimental approaches have detected ICD in large water clusters [152] or at condensed-phase interfaces containing water dimers and clusters [151].

Random damage to cellular biomolecules such as DNA is associated with the onset of cancer, whereas the controlled targeted local release and interactions of LEEs can be used as effective therapeutic cancer treatment agents. Since ICD is a source for the ejection of slow electrons, it has been proposed that ICD could play a role in the induction of SSB and DSB in DNA [153]. Estimation by Grieves and Orlando [152] indicated that ICD may represent up to 50% of the SSB probability for energy depositions >20 eV and ionization events directly at the DNA–water interface. Since the formation of DSBs requires excitation energies >5 eV, the impact on DSBs is expected to be much lower. If ICD contributes significantly to DNA damage, this could be exploited during X-ray treatment of cancer. **Figure 4** schematically shows that how utilizing of X-ray interactions with gold nanoclusters within living cells, which subsequently results in releasing both Auger and ICD electrons, has been suggested as a potential strategy for targeted cancer treatment [148].

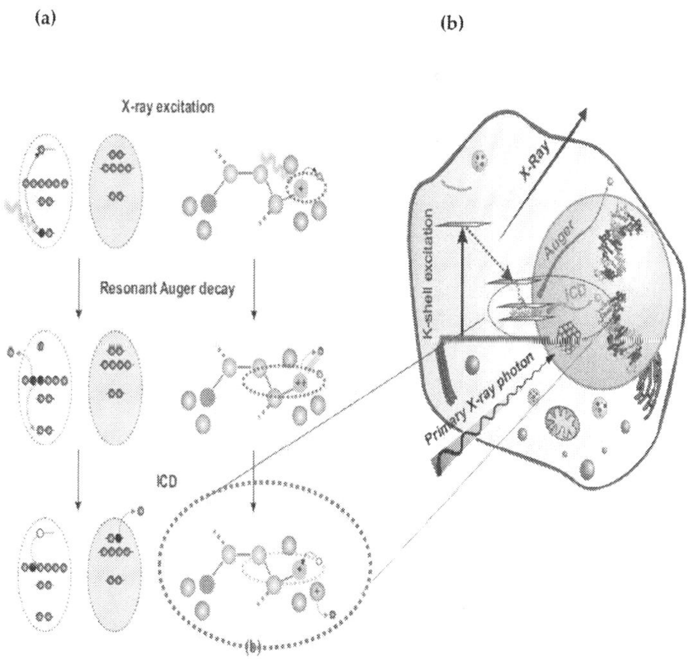

Figure 4. (a) Resonant Auger decay process following X-ray excitation. A second process known as interatomic or intermolecular Coulomb decay (ICD) can also occur, leading to the ejection of slow electrons and adjacent holes. (b) Possible exploitation of Au nanoparticles and ICD in the controlled radiation damage of cells [148].

After such extensive studies on LEE-induced damage under "near"-cellular conditions, it was only very recently that the lethal effects of LEEs in cells have been demonstrated by Sahbani et al. [154], who investigated the biological functionality of DNA, via a simple model system comprising *E. coli* bacteria and plasmid DNA bombarded by LEEs. In these experiments, highly ordered DNA films were arranged on pyrolytic graphite surface by molecular self-assembly technique using 1,3-diaminopropane ions to bind together the plasmid DNAs [155]. This assembly technique mimics somewhat the action of amino groups of the lysine and arginine amino acids within the histone proteins. These authors measured the transformation efficiency

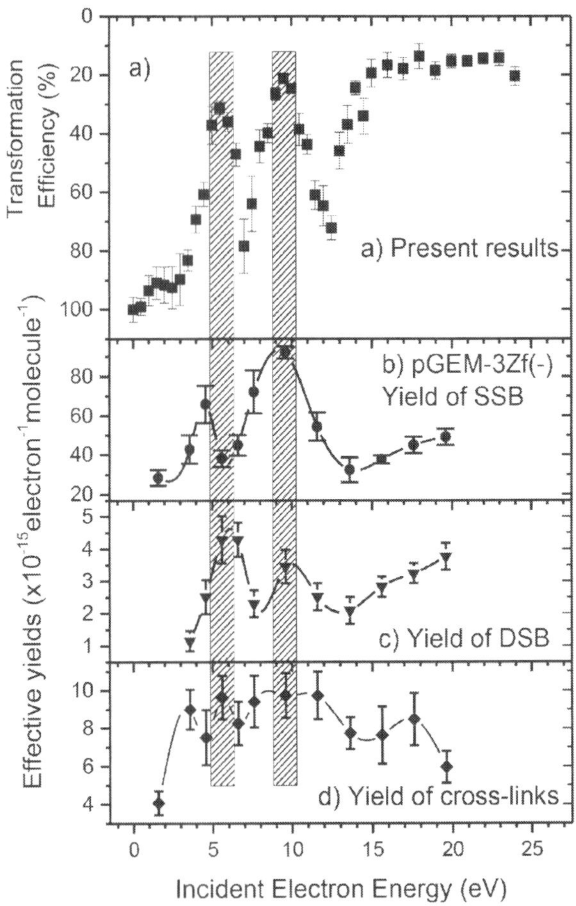

Figure 5. (a) Variation of transformation efficiency of *E coli* by pGEM 3Zf(-) plasmids irradiated by 0.5–18 eV electrons at a fluence of 27 × 10^{13} electrons/cm^2. The vertical axis is inverted. Effective yield functions for (b) single-strand breaks (SSBs), (c) double-strand breaks (DSBs), and (d) DNA cross-links [183].

of *E. coli* JM109 bacteria (essentially the number of bacterial colonies grown in an antibiotic environment) after insertion into the cells of [pGEM-3Zf (-)] plasmid, which when undamaged, can confer resistance to the antibiotic ampicillin. Before transformation, the plasmids were irradiated with electrons of specific energies in the energy range 0.5–18 eV [156]. Cells receiving severely damaged plasmids will not grow, and the transformation efficiency will be reduced. The loss of transformation efficiency plotted as a function of electron energy is shown in **Figure 5**. It reveals maxima at 5.5 and 9.5 eV, coincident with the maxima observed in the yields of DNA DSBs, which were attributed to the formation of core-excited TMAs. These results indicated that the effects of TMAs are observable in the electron-energy dependence of biological processes with negative consequences for cell viability. The result provides further evidence that LEEs play important roles in cell mutagenesis and death during radiotherapeutic cancer treatment [156].

8. Role of LEEs in radiosensitization and radiation therapy

A major impetus for achieving a better understanding of the action of ionizing radiation in biological systems relates to applications in radioprotection and radiotherapy. Since LEEs play a major role in transferring the energy of the initial high-energy particle or photon to initiate all subsequent chemistry in irradiated media, understanding their interaction with biomolecules is now being recognized as a crucial and essential step toward such applications. As seen from Section 4, in the last decades, our expanding knowledge of LEE–DNA interactions has been applied to experiments involving known radiosensitizers. Both at the theoretical and experimental levels, this work served to suggest new compounds having radiation-damage enhancing properties and to explain the details of their response to high-energy radiation either alone or when bound to DNA (i.e., the main target for cell killing in radiotherapy).

8.1. Transient anions in halogen compounds

Bromouracil, which can replace thymine in DNA during cell replication, and bromouridine were the first radiosensitizing candidates to be investigated theoretically and experimentally with LEEs. The studies [157–167] confirmed the prediction of Zimbrick and coworkers [168] that the radiosensitizing properties of these compounds arose from DEA of solvated electrons, and further showed that DEA of higher energy (0–7eV) electrons was also involved in radiosensitization. Platinum bromide, aromatic compounds containing nitro group and other halogenated thymidine derivatives were found to play similar roles [58, 70, 71, 163–165, 169, 170]. Following early investigations with solvated electrons [168], a relatively large number of experiments have been performed both in the gas (see Section 4.4) and condensed phases [160, 161, 165] to study electron scattering from—and attachment to—halogenated pyrimidines. Several experiments were performed using SAMs of BrdU-containing oligonucleotides [157, 158, 171]. These included the detection of the electron-stimulated desorption of ion and neutral species and HPLC analysis of damaged films, as well as electronic and vibrational electron-energy loss spectra for gaseous bromouracil [159]. These studies revealed that the radiosensitization properties of halogen compounds are more complicated than previously anticipated [168]. Within the 0–7 eV energy range, resonant electron scattering mechanisms with halouracils lead to more complex molecular fragmentation than that occurs with thymine, which produces a different range of anionic and neutral radical fragments. When formed within DNA, such fragments could react with local subunits, and thus lead to lethal clustered damage, further to that already occurring in unsensitized DNA. The most striking evidence of a huge enhancement of LEE damage obtained upon Br substitution in thymine is seen in the early results of Klyachko et al. [160], who found that, in the presence of water, DEA to bromouracil could be enhanced by orders of magnitude compared to the dry compound. Differences between wet and dry TMA states of halogenated pyrimidines have recently been investigated by Cheng et al. [172]. They applied Koopman's theorem in the framework of long-range corrected density functional theory for calculation of the TMA states and self-consistent reaction field methods in a polarized continuum to account for the solvent. Their results indicate that the TMAs of these molecules are more stable in water, but to differing degrees.

The radiosensitization properties of halouracils depend not only on hydrated electrons, but also on LEEs and on DEA. However, the high propensity of LEEs of very low energies (i.e., <1 eV) to fragment bromouracil and deoxybromouridine (BrUdR) may, according to the theory, exist only in single-stranded DNA [165]. This important prediction was confirmed by Cecchini et al. [173] for the case of solvated electrons and was commented upon by Sevilla [174]. Solutions of single- and double-stranded oligonucleotides, and of double-stranded oligos containing mismatched bubble regions, were irradiated with γ-rays, and the concentrations of various reactive species produced, including solvated electrons, were controlled with scavengers. When in the absence of oxygen, OH radicals were scavenged, BrUdR was shown to sensitize single-stranded DNA, but could not sensitize complementary double-stranded DNA. However, when BrUdR was incorporated in one strand within a mismatch bubble, the nonbase-paired nucleotides adjacent to the BrUdR, as well as several unpaired sites on the opposite unsubstituted strand, were highly sensitive to γ-irradiation. Since LEEs and solvated electrons fragment BrUdR by the same DEA mechanism [162–165, 168], these results imply that the strong sensitizing action of BrUdR to electron-induced damage is limited to single-stranded DNA, which can be found in transcription bubbles, replication forks, DNA bulges, and the loop region of telomeres. These results are clinically relevant since they suggest that BrUdR sensitization should be greatest for rapidly proliferating cells [173, 174]. When injected into a patient being treated for cancer, BrUdR quickly replaces a portion of the thymidine in the DNA of the fast-growing malignant cells, but radiosensitization occurs only when DNA is in a single-stranded configuration (e.g., at the replication forks during irradiation). From this conclusion, it appears advantageous to administer to patients receiving BrUdR, another approved drug, such as hydroxyurea, to increase the duration of the S-phase of cancer cells (i.e., the replication cycle). This addition would increase the probability that SEs would interact with bromouracil while bound to DNA in its single-strand form. Such a modality provides an example of how our understanding the mechanisms of LEE-induced damage can help to improve radiotherapy [174].

8.2. Transient anions in DNA bound to platinum chemotherapeutic agents

Considering that it can often take years, if not decades, before potential new radiosensitizers arrive in the clinic, Zheng et al. [175] hypothesized that present clinical protocols involving high-energy radiation and platinum (Pt) chemotherapeutic agents could be improved by considering the fundamental principles of energy disposition, including the results of LEE experiments. Their initial goal was to explain the superadditive effect occurring in tumor treatments, when cisplatin and radiation were administered in concomitance [176, 177]. Zheng et al. [175] found that, with cisplatin bound to DNA as in the cancer cells, damage to the molecule increases by factors varying from 1.3 for high-energy electrons to 4.4 at 10 eV. Considering the much higher enhancement factor (EFs) at 10 eV, the increase in bond dissociation was interpreted as being triggered by an increase in DNA damage induced by LEEs.

In their experiments, Zheng et al. [175] deposited lyophilized films of pure plasmid and plasmid–cisplatin complexes on a clean tantalum foil. The films were bombarded under UHV with electrons of 1–60 keV. Under these conditions, 90% of the plasmid–cisplatin complexes consisted of a cisplatin molecule chemically bound to DNA, preferentially at the N7 atom of two guanines producing an interstrand adduct. The films had the necessary thickness to absorb most of the energy of the electrons. The different forms of DNA corresponding to SSBs and DSBs were separated by gel electrophoresis, and the percentage of each form quantified by fluorescence. Exposure response curves were obtained for several incident electron energies for cisplatin bound or not to plasmid DNA. **Table 1** gives the results for exposure to 1, 10, 100, and 60,000 eV electrons of films of pure DNA and cisplatin/plasmid complexes with a ratio (R) of 2:1 and 8:1. For both R values, cisplatin binding to DNA increases the production of SSBs and DSBs, but in quite different proportions depending on electron energy. Considering that it takes about 5 eV to produce a DSB with electrons [90], the most striking result of **Table 1** is clearly the production of DSBs by 1 eV electrons. Later, Rezaee et al. [178] demonstrated that even 0.5 eV electrons could induce DSBs in DNA containing Pt adducts in similar proportions and more efficiently than other types of radiation, including X-rays and high-energy electrons. The formation of DSBs by 0.5 eV electrons resulted from a single-hit process. Gamma radiolysis experiments with plasmid DNA dissolved in water, further demonstrated that even solvated electrons could react with cisplatin–DNA complexes to induce DSBs [179]. The results of Zheng et al. [175] at higher energy were later confirmed by those of Rezaee et al. [180], who showed that increased damage via the formation of TMA could explain, at least partially, the concomitance effect in chemoradiation therapy for cisplatin, as well as for the other platinated chemotherapeutic drugs such as oxaliplatin and carboplatin.

This type of radiosensitization was investigated in more detail by irradiating with a γ source the oligonucleotide TTTTTGTTGTTT with or without cisplatin bound to the guanines [181]. Using scavengers and by eliminating oxygen, the oligonucleotide was shown to react with hydrated electrons. Prior to irradiation, the structure of the initial cisplatin adduct was identified by mass spectrometry as G-cisplatin-G. Radiation damage to DNA bases was quantified by HPLC, after enzymatic digestion of the TTTTTGTGTTT–cisplatin complex to deoxyribonucleosides. Platinum adducts were following digestion and separation by HPLC, quantified by mass spectrometry. The results demonstrated that hydrated electrons induce damage to thymines as well as detachment of the cisplatin moiety from both guanines in the oligonucleotide. The amount of free cisplatin (i.e., the cleavage of two Pt–G bonds) was found to be much larger than that of the products resulting from the cleavage of a single bond [181,182].

Form of damage	SSB				DSB			
Energy (eV)	1	10	100	60,000	1	10	100	60,000
Thickness	5 ML			2900 nm	5 ML			2900 nm
DNA	27± 3	33 ± 3	57 ± 5.5	1.2 ±0.1	ND	10 ± 1	13 ± 2	0.4 ± 0.2
Cisplatin:DNA = 2:1	38 ± 3	120 ± 11	150 ± 15	2.4 ±0.3	5 ±1	17 ± 1	36 ± 4	0.5 ± 0.2
Cisplatin:DNA = 8:1	52 ± 5	143 ± 14	199 ± 18	3.0 ±0.4	5 ±2	29 ± 2	44 ± 4	0.7 ± 0.1

ND, Not detected.

The errors represent the deviation of three identical measurements.

Table 1. Yields (in 10^{-15} electron^{-1} molecule^{-1}) for the formation of SSB and DSB induced by 1, 10, and 100 eV electron impact on 5 ML DNA films and 60 keV electron impact on 2900 nm DNA films deposited on a tantalum substrate.

These results suggest two major pathways by which hydrated electrons interact destructively with TTTTTGTGTTT–cisplatin [181, 182]. First, the hydrated electron is captured initially on a thymine base and is transferred to the guanine site by base to base electron hopping, where DEA detaches the cisplatin moiety from the oligonucleotide. Alternatively, the hydrated electron interacts directly with the platinum–guanine adduct, and cisplatin is detached via DEA. These hypotheses were consistent with those proposed by Rezaee et al. [178] for LEE-induced damage to plasmid DNA. Additionally, Rezaee et al. suggested that in the double-stranded configuration, the cisplatin molecule weakens many of the DNA chemical bonds and changes the topology of the molecule; these modifications render DNA much more sensitive to damage over large distances [180]. Of course, under high-energy irradiation conditions, the increase in ionization cross section, due to the presence of the Pt atom, also increases the quantity of LEEs near cisplatin and therefore may indirectly contribute to the increase in damage.

More recently, the energy dependence of conformational damage induced to pure plasmid DNA [183] and cisplatin–plasmid DNA complexes [184] was investigated in the range 2–20 eV. In addition to the strong resonances (i.e., TMAs) in pure DNA around 5 and 10 eV, further TMA specific to cisplatin-modified DNA were observed in the yield function of SSBs at 13.6 and 17.6 eV. Moreover, the presence of cisplatin lowered the threshold energy for the formation of DSBs to 1.6 eV, considerably below that observed with electrons in pure DNA films. In all cases, the measured yields were larger than those measured with nonmodified DNA. To reconcile all existing results starting from those obtained with hydrated electrons to those generated up to 20 eV, Bao et al. [184] suggested a single mechanism that could apply to shape and core-excited resonances, depending or not if electronic excitation of the Pt or guanines was involved in TMA formation. This mechanism, previously proposed for shape resonances by

Rezaee et al. [178], can be explained with reference to **Figure 6**. When the TMA is formed on the Pt adduct, the extra electron is delocalized and occupies simultaneously, with identical wave functions, the two bonds linking the Pt atom to guanine bases on opposite strands. Occupancy of the dissociative σ^* orbitals induces equal repulsive impulses on the two bonds between platinum and guanines (Pt–G), due to the symmetrical delocalization of the excess electron. If the extra electron autodetaches when the gained kinetic energy is larger than the energy barrier to dissociate the Pt–G bonds, both bonds can be simultaneously broken. The extra energy for dissociation is supplied to the complex by autodetachment from the σ^* bond, leaving the additional electron stabilized at the bottom of the potential well of the Pt. The simultaneous cleavage of two Pt–G bonds and formation of two guanine radicals are followed

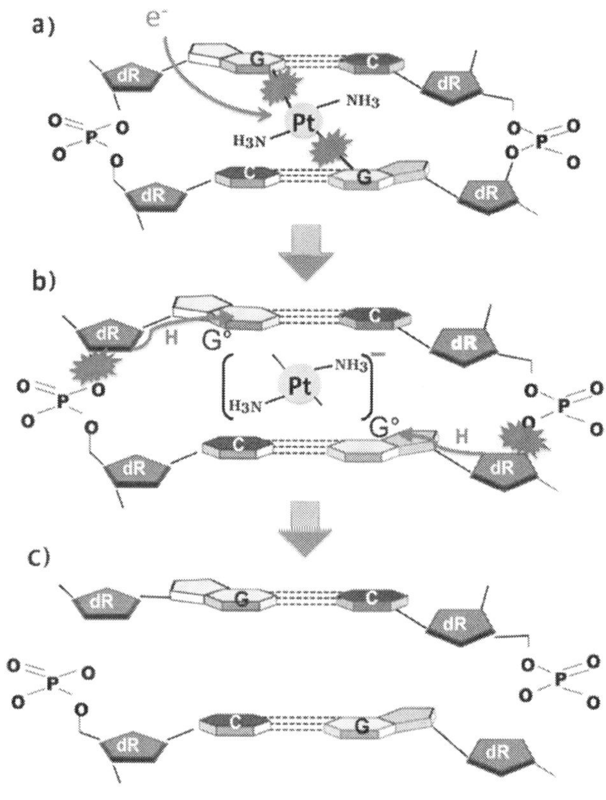

Figure 6. Possible mechanism for the formation of a DSB by a single electron, when cisplatin links two guanine (G) bases on opposite strands. (a) Electron capture into two identical dissociative orbitals between Pt and two Gs. (b) The transient anion thus formed dissociates, leaving the electron on the $(NH_3)_2Pt$ moiety and causing simultaneous cleavage of the two symmetrical Pt–G bonds. The resulting two guanine radicals ($G^\bullet$) abstract hydrogen from the backbones, causing cleavage of phosphodiester bonds on opposite strands. (c) Resulting DSB [178].

by hydrogen abstraction from the backbone. This abstraction cleaves the phosphodiester bonds in opposite strands, forming a DSB. Considering the results obtained with carboplatin and oxaliplatin [180], which are similar to those obtained with cisplatin, the mechanism depicted in the diagram of **Figure 6** is likely to apply also to these chemotherapeutic drugs. Since these latter behave as cisplatin and bind similarly to DNA, we can replace cisplatin by carboplatin in **Figure 6**; to represent oxaliplatin in the figure, NH_3 has to be replaced by $C_6H_{10}(NH_2)_2$.

The LEE enhancement mechanism of damage in DNA–Pt drug complexes acts on a femtosecond timescale, which quite unlike other biological mechanisms of radiosensitization, act over macroscopic times that can range from hours to days. These considerations imply that the mechanism (e.g., physicochemical vs biological) of radiosensitization by Pt agents in concomitant chemoradiation therapy may be sensitive to the timing between the injection of the drug to the patient and the irradiation. Thus, if TMA formation in DNA plays a major role in radiosentization by Pt drugs, maximal cancer cell killing should be achieved, if these cells are irradiated when the maximum amount of Pt is bound to their nuclear DNA.

Led by this hypothesis, Tippayamontri et al. [185, 186] determined the optimal conditions for concomitant chemoradiation treatment of colorectal cancer with cisplatin, oxaliplatin, and their liposomal formulations Lipoplatin and Lipoxal [187, 188]. Using an animal model of human colorectal cancer, they determined the time window for maximum radiosensitization and synergy with irradiation, by studying the pharmacokinetics and time-dependent intracellular distribution of the Pt drugs. This, in turn, is determined by the reaction kinetics of the drug with DNA and the DNA repair kinetics.

In nude mice bearing HCT116 colorectal carcinoma, treated with the Pt drugs, they measured by inductively coupled plasma mass spectrometry, the platinum accumulation in blood, serum, different normal tissues, tumor, and different tumor cell compartments, including the amount of Pt bound to nuclear DNA [185, 186] **Figure 7a** indicates the positions of binding of cispelatin to DNA. Examples of the amount of cisplatin and Lipoplatin binding to the DNA of HCT116 colorectal cancer cells in mice are shown in **Figure 7b** as a function of time after injection of the drug. Radiation treatment (15 Gy) was given 4, 24, and 48 h after drug administration. The resulting tumor growth delay was reported and correlated with apoptosis analyses. Optimal survival of the mice and highest apoptosis were observed when radiation was given at 4 or 48 h after drug injection. These times corresponded to the times of maximal platinum binding to tumor DNA, as shown in **Figure 7b** for cisplatin and Lipoplatin. When tumor irradiation was performed at 48 h, the ratio of tumor growth delay for the group having the combined treatment compared to delay for the group treated with chemotherapy alone varied from 4.09 to 13.00, depending on the drug. The most efficient combination treatment was observed when the amount of Pt drug binding to DNA was highest, as predicted from fundamental considerations [178–182]. Such results testify our fundamental understanding of the mechanisms of platinum-induced radiosensitization and should have significant impact on the design of more efficient treatment protocols.

http://dx.doi.org/10.5772/63293

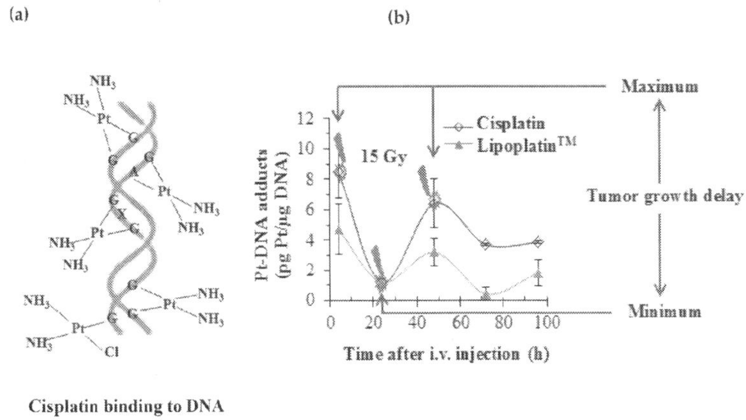

Figure 7. (a) Diverse sites of intrastrand and interstrand binding of cisplatin to cellular DNA. (b) Concentration of Pt–DNA adducts in the nucleus of human colorectal cancer cells of mice bearing HCT116 xenografts, as a function of time after administration of cisplatin and Lipoplatin™. The mice were irradiated at 4, 24, and 48 h after injection of the chemotherapeutic agents.

8.3. Interaction of LEEs with DNA bound to gold nanoparticles

So far in this section, we have shown that cancer cells can be made more sensitive to high-energy radiation by chemically modifying their nuclear DNA with small molecules. These latter provide at least some of their radiosensitizing action, by increasing the interaction of LEEs with DNA, the products of DEA, and the resulting induced damage. Another approach consists of simply increasing the numbers of LEEs near the DNA of cancer cells. The best examples of this type of radiosensitization have been provided by the numerous fundamental, in vitro, and in vivo investigations of enhanced radiation absorption by gold nanoparticles (GNPs).

Both in vitro and in vivo experiments [189–204] have shown radiation enhancement effects due to the presence of GNPs. Several models have been developed to account for dose enhancement in cells by considering the increase in radiation energy deposition [205–211], due to additional energy absorption by the GNPs, as a function of their size. As expected, the energy of electrons emanating from the GNPs is inversely proportional to their diameter. Many models [206–211] take into account localized effects of Auger-electron cascades. They consider the huge enhancement of energy deposited in the vicinity of GNPs, as arising from the considerable increase in photoelectric absorption cross section of gold in comparison to that of tissue [200, 208, 210, 211]. The increase in this cross section produces an additional local generation of photoelectrons, Auger electrons, and characteristic X-rays [208, 212]. The major portion of the energy absorbed by the GNPs is converted into electrons, most of which escapes the GNPs with low energy (0–30 eV) [213–215].

The indirect effect of emitted electrons was investigated in water solutions containing GNPs, where the nanoparticle-induced OH concentration from radiolysis was measured. Relevant literature and details can be found in the paper of Sicard-Roselli et al. [189], who also proposed a new mechanism for hydroxyl radical production in irradiated GNP solutions.

The direct effect of high-energy radiation on DNA, resulting from the presence of GNPs, was first investigated by Zheng and coworkers [35, 214–218]. Relatively thick (~0.3 and 2.9 μm) films of plasmid DNA with or without electrostatically bound GNPs were bombarded with 60 keV electrons. The probabilities of formation of SSBs and DSBs from the exposure of 1:1 and 2:1 GNP–plasmid mixtures to fast electrons increased by a factor of about 2.5, compared to DNA alone. It was suggested that the additional damage in the presence of GNPs was generated by LEEs escaping the nanoparticles. This hypothesis was later verified experimentally by the work of Xiao et al. [214]. These authors investigated the radiosensitization efficiency in terms of DNA damage as a function of the length of a ligand bound at one end to the surface of the GNP and at the other to DNA. They used the same DNA film preparation as in the experiments of Zheng et al. [215] and measured the ratio of induced damage with GNPs to that without GNPs (i.e., the enhancement factor, EF) for different lengths of the ligand. As indicated in **Figure 8** from their work, the corresponding EFs induced by 60 keV electrons on plasmid DNA bound to GNPs of various coatings range from 2.3 to 1.6 and 1.2, depending on the length of ligand separating the gold surface from the plasmid. This length ranged from 0 to 2.5 and 4 nm, respectively. The attenuation by the coating of short-range LEEs emitted from the GNPs could explain the decrease in radiosensitization with increasing length of the ligand [214]. Since the attenuation range of LEEs is shorter than about 10 nm, it is obvious that the emission of LEEs from the GNPs and LEE-interaction with DNA plays a major role in the mechanism of GNP radiosensitization.

Later, similar DNA–GNP films were bombarded with electrons of energies below the ionization potential of DNA. In this case, essentially no secondary LEEs were emitted from the DNA and the gold surface, so that Yao et al. [218] could investigate the purely chemical radiosensitization induced by GNPs. They showed that even without the emission of photoelectrons, direct electrostatic binding of an average of 0.2–2 GNPs to DNA could increase sensitization to LEEs by factors varying from 1.5 to 4.

Since GNPs increase the local density of LEEs and cisplatin enhances LEE interactions with DNA and damage to the molecule, it seemed likely that binding GNPs to a cisplatin–DNA complex would further boost radiosensitization and DNA damage induced by cisplatin [216]. This hypothesis was verified by irradiating with 60 keV electrons, GNPs electrostatically bound to a cisplatin–DNA complex [216]. Dry films of bare plasmid DNA and DNA–cisplatin, DNA–GNP, and DNA–cisplatin–GNP complexes were irradiated [216]. The yields of SSBs and DSBs were measured as described in the protocol established by Zheng et al. [215]. When the ratio of GNP to DNA was 1:1 and that for cisplatin to DNA was 2:1, the EFs for SSBs were between 2 and 2.5. With a cisplatin to GNP to plasmid ratio of 2:1:1, the EF increased to 3. This small increase could only be additive and unrelated to the interaction of additional LEEs with cisplatin. For DSB formation, however, the binding of both GNPs and cisplatin to a DNA molecule produced an impressive increase in the EF, that is, DSBs were increased by a factor of 7.5 with respect to pure DNA. It appeared quite obvious that the additional DSBs in the cisplatin–DNA–GNP complex arose from the generation of additional secondary electrons from the GNPs. The synergy between GNPs and cisplatin could arise from a number of basic phenomena, including the possibility of two or multiple event processes triggered by the interaction of a single 60 keV electron with a GNP. Within 10 nm of its location, a single gold atom increases the density of LEEs by a large factor [207, 212], and hence, a GNP that contains thousands of gold atoms is expected to generate a dramatic increase in this density [213]. Combined with the fact that cisplatin considerably lowers the energy threshold for DSB formation, a single or multiple LEE interactions on opposite strands within a distance of 10 base pairs could increase considerably the number of DSBs formed in GNP–cisplatin–DNA complexes.

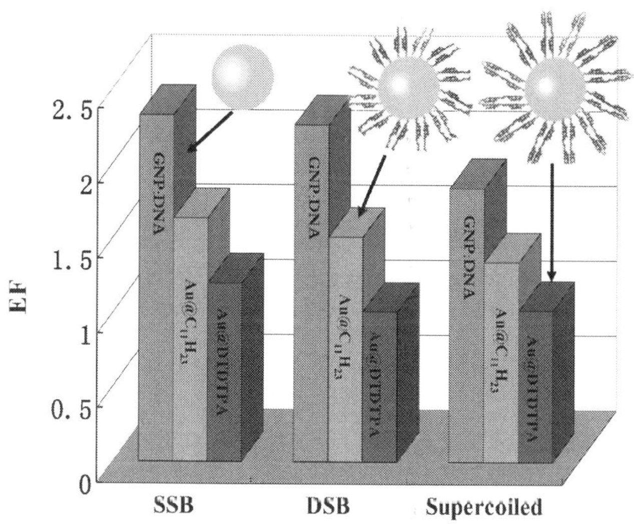

Figure 8. Enhancement factors (EFs) for the formation of SSB, DSB, and loss of supercoiled DNA induced by 60 keV electrons, obtained with GNP–DNA complexes of ratio 1:1. The groups of three histograms represent the respective damages. In each group, the EF corresponds to the damage when the GNP alone is bound to DNA or when the GNP has been coated with ligands 2.5 and 4 nm in lengths corresponding to GNP@$C_{11}H_{23}$ or GNP@DTDTPA (i.e., dithiolated diethylenetriaminepentaacetic acid), respectively in the figure.

As shown by Zheng et al., only one GNP per DNA molecule is on average necessary to increase DNA damage considerably [216]. Thus, as long as the nanoparticles reach the DNA of cancer cells, the amount to be administered to patients to obtain significant radiosensitization should be at most the same as that of the Pt-drugs routinely administered in chemotherapy [176, 177]. In recent in vitro experiments, GNPs were targeted to the DNA in the cell nucleus by linking peptides to the gold surface [197, 202]. Such vectored GNPs, targeting the DNA of cancer cells, should be applicable in the clinic and may accordingly offer a new approach to radiotherapy treatments. However, this type of radiotherapy is expected to be limited to superficial tumors, owing to the requirement for that low-energy (<100 keV) X-rays be used to optimize LEE production and hence radiosensitization by the photoelectric effect. To treat deep tumors, a radioactive source may have to be encapsulated inside a gold nanoparticle (i.e., in a gold nanocage) [219]. Furthermore, if DNA specificity cannot be achieved in patients, successful treatment may still be possible by intratumor injection of GNPs, as recently shown by Shi et al. [220] and Bobyk et al. [204].

9. Summary and Conclusions

The experimental and theoretical results of LEE impact on single- and double-stranded DNA, its basic constituents, protein subunits, as well as radiosensitizers and chemotherapeutic agents alone or bound to DNA were reviewed. Experimental details of LEE interactions with these biomolecules were obtained in both the condensed and gas phases. The condensed-phase experiments were conducted in UHV and at atmospheric pressure under environments closer to those of the cell. From these studies, which provide a fundamental comprehension of the role of TMAs in irradiated biological systems, we can arrive with considerable certainty at the following conclusions on LEE-induced damage to biomolecules. In the low-energy range (i.e., below the threshold for dipolar dissociation (~15 eV)), bond rupture in biomolecules occurs essentially via the formation of TMAs that decay either via autoionization with the accompanied production of dissociative electronically excited states, or into the DEA channel. The induced damage depends on a large number of factors, including electron energy, the environment and topology of the molecule, and the electrostatic or chemical binding of small radiosensitizing molecules. Such factors inevitably modify the lifetime and decay channels of transient anions, which often increase the damage cross sections.

Since secondary electrons of low energy possess a large portion of the energy deposited by high-energy radiation, any modification of how their energy deposits at crucial cellular sites is expected to have a strong radioprotective or radiosensitizing action. With DNA being the main target in radiotherapy, parameters that affect LEE-induced DNA damage are necessarily of relevance to radiosensitivity, and the mechanisms involved must be well understood to control and modulate the biological effects of ionizing radiation.

Many of these mechanisms are now well established as seen from the experiments and theoretical treatments reviewed in this chapter. Moreover, it has been shown that applying fundamental principles of action of LEEs to radiosensitizers or chemotherapeutic agents can lead to new strategies on how to improve radiotherapy outcomes. In particular, the role of LEEs in radiation damage was related to enhancement of the destruction of cancer cells by Pt-drugs and gold nanoparticles. LEEs were found to play an important role in providing guidelines in chemoradiation cancer treatment, as well as in the development of more efficient clinical protocols. Such applications point out the need for multidisciplinary studies in this field, where LEE–biomolecule interactions have become an area of intensive investigation that encompasses many aspects of cancer therapy.

http://dx.doi.org/10.5772/63293

Acknowledgements

The authors gratefully express thanks for helpful comments of Dr. Andrew D. Bass and Dr. Darel J. Hunting. LS acknowledges the financial support from Canadian Institute of Health Research (CIHR, via MOP 81356 and 86676) and from Natural Sciences and Engineering Research Council of Canada (NSERC). S.P. acknowledges the financial support from the U.S. Department of Energy Office of Science, Office of Basic Energy Sciences under award number DE-FC02-04ER15533.

Abbreviations

A	adenine
AFM	atomic force microscopy
APPJ	atmospheric pressure plasma jet
C	cytosine
C	circular
CL	cross-link
CRT	chemoradiation therapy
DD	dipolar dissociation
DEA	dissociative electron attachment
DNA	deoxyribonucleic acid
DSB	double-strand break
EF	enhancement factor
ESD	electron-stimulated desorption
ESI	electrospray ionization
eV	electron volt
F–C	Franck–Condon
G	guanine
GNP	gold nanoparticle
HPLC	high-performance liquid chromatography
ICD	intermolecular Coulombic decay
keV	kilo electron volt
Kr	krypton
L	linear
LEE	low-energy electron
MALDI	matrix-assisted laser desorption ionization

MDSB	multiple double-strand break
MeV	mega electron volt
ML	monolayer
nm	nanometer
ps	picosecond
Pt	platinum
SAM	self-assembled monolayer
SC	supercoiled
SE	secondary electron
SERS	surface-enhanced Raman spectroscopy
SSB	single-strand break
T	thymine
TMA	transient molecular anion
U	uracil
UHV	ultrahigh vacuum
UV	ultraviolet
VFR	vibrational Feshbach resonance

Author details

Elahe Alizadeh[1*], Sylwia Ptasińska[2] and Léon Sanche[3]

*Address all correspondence to: ealizade@uoguelph.ca

1 Department of Chemistry and Biochemistry, University of Guelph, Guelph, ON, Canada

2 Radiation Laboratory and Department of Physics, University of Notre Dame, Notre Dame, IN, USA

3 Groupe en Sciences des Radiations, Département de Médecine Nucléaire et Radiobiologie, Faculté de médecine et des sciences de la santé, Université de Sherbrooke, Sherbrooke, QC, Canada

References

[1] Lehnert S. Biomolecular Action of Ionizing Radiation. 1st ed. New York: Taylor and Francis; 2007. DOI: 10.1201/9781420011920

[2] Baccarelli I, Bald I, Gianturco FA, Illenberger E, Kopyra J. Electron-induced damage of DNA and its components: experiments and theoretical models. Phys. Rep. 2011; 508: 1–44. DOI: 10.1016/j.physrep.2011.06.004

[3] Denifl SP, Märk TD, Scheier P. The role of secondary electrons in radiation damage. In: Gomez-Tejedor GG, Fuss MC (eds). Radiation Damage in Biomolecular Systems. 1st ed. New York: Springer Science & Business Media; 2012. p. 45–58. DOI: 10.1007/978-94-007-2564-5. Ch 2.

[4] Hall, EJ, Giaccia AJ. Radiobiology for the Radiologist. 6th ed. Philadelphia: Lippincott Williams and Wilkins; 2006. DOI: 10.1667/RR0771.1

[5] Kufe DW, Pollock RE, Weichselbaum RR, Bast RC, Gansler TS, Holland JF, Frei E. Holland-Frei Cancer Medicine. 6th ed. Hamilton: BC Decker; 2003. ISBN-10: 1-55009-213-8

[6] O'Driscoll M, Jeggo PA. The role of double-strand break repair: insights from human genetics. Nat. Rev. Genet. 2006; 7: 45–54. DOI: doi :10.1038/nrg1746

[7] Alizadeh E, Orlando TM, Sanche L. Biomolecular damage induced by ionizing radiation: the direct and indirect effects of low-energy electrons on DNA. Ann. Rev. Phys. Chem. 2015; 66: 379–398. DOI: 10.1146/annurev-physchem-040513-103605

[8] Sevilla MD, Bernhard WA. Mechanisms of direct radiation damage to DNA. In: Spotheim-Maurizot M, Mostafavi M, Douki T, Belloni J (eds). Radiation Chemistry: From Basics to Applications in Material and Life Sciences. 1st ed. Les Ulis: EDP Sciences; 2008. p. 191–201. DOI:10.1002/cphc.201300915. ch13

[9] Pimblott SM, La Verne JA. Production of low-energy electrons by ionizing radiation. Radiat. Phys. Chem. 2007; 76: 1244–1247. DOI: 10.1016/j.radphyschem.2007.02.012

[10] Sanche L. Low-energy electron interaction with DNA: bond dissociation and formation of transient anions, radicals, and radical anions. In: Greenberg MM (ed). Radical and Radical Ion Reactivity in Nucleic Acid Chemistry. 1st ed. New Jersey: John Wiley and Sons; 2009. p. 239–293. DOI: 10.1002/9780470526279. ch9

[11] Alizadeh E, Sanche L. Precursors of solvated electrons in radiobiological physics and chemistry. Chem. Rev. 2012; 112: 5578–5602. DOI: 10.1021/cr300063r

[12] Sulzer P, Mauracher A, Denifl S, Zappa F, Ptasińska S, Beikircher M, Bacher A, Wendt N, Aleem A, Rondino F, Matejcik S, Probst M, Märk TD, Scheier P. Identification of isomers of nitrotoluene via free electron attachment. Anal. Chem. 2007; 79: 6585–6591. DOI: 10.1021/ac070656b

[13] Gu J, Xie Y, Schaefer HF. Electron attachment to hydrated oligonucleotide dimers: Guanylyl-3',5'-cytidine and cytidylyl-3',5'-guanosine. Chem. Eur. J. 2010; 16: 5089–5096. DOI: 10.1002/chem.200902977

[14] Arumainayagam, CR, Lee HL, Nelson RB, Haines DR, Gunawardane RP. Low-energy electron-induced reactions in condensed matter. Surf. Sci. Rep. 2010; 65: 1–44. DOI: 10.1016/j.surfrep.2009.09.001

[15] Schulz GJ. Resonances in electron impact on diatomic molecules. Rev. Mod. Phys. 1973; 45: 423–486. DOI: 10.1103/RevModPhys.45.423

[16] Allan M. Study of the triplet-states and short-lived negative-ions by means of electron impact sceptroscopy. J. Electron. Spectrosc. Relat. Phenom. 1989; 48: 219–351. DOI: 10.1016/0368-2048(89)80018-0

[17] Sanche L. Primary interactions of low-energy electrons in condensed matter. In: Jay-Gerin J-P, Ferradini C (eds). Excess Electrons in Dielectric Media. 1st ed. Boca Raton: CRC Press; 1991. p. 1–43. ISBN: 9780849369629 - CAT# 6962. chp1

[18] Christophorou LG. Electron-Molecule Interactions and their Applications. 1st ed. Orlando: Academic Press; 1984. DOI: 10.1109/TDEI.2010.5492245

[19] Massey HSW. Negative Ions. 5th ed. London: Cambridge University Press; 1976. ISBN: 978-0-521-20775-1

[20] Palmer RE, Rous P. Resonances in electron-scattering by molecules on surface. Rev. Mod. Phys. 1992; 64: 383–479. DOI: 10.1103/RevModPhys.64.383

[21] Sanche L, Electron resonances in desorption induced by electronic transitions. Surf. Sci. 2000; 451: 82–90. DOI: 10.1016/S0039-6028(00)00011-X

[22] O'Malley TE. Theory of dissociative attachment. Phys. Rev. 1966; 150: 1429. DOI: 10.1103/PhysRev.150.14

[23] Shimamura I, Takayanagi K. Electron-Molecule Collisions. 1st ed. New York: Plenum Press; 1984. DOI: 10.1103/RevModPhys.52.29

[24] Hinze J. Electron-Atom and Electron-Molecule Collisions. 1st ed. New York: Plenum Press; 1983. DOI: 10.1007/978-1-4899–2148–2

[25] Christophorou LG. The lifetime of metastable negative ions. Adv. Electron. Electron Phys. 1978; 46: 55–129. DOI: 10.1016/S0065-2539(08)60411-4

[26] Illenberger E, Momigny J. Gaseous Molecular Ions: An Introduction to Elementary Processes Induced by Ionization. 1st ed. New York: Springer Science & Business Media; 1992. DOI: 10.1007/978-3-662-07383-4

[27] Zangwill A. Physics at Surfaces. 1st ed. Cambridge: Cambridge University Press; 1988. DOI: 10.1017/CBQ9780511622564

[28] Sanche L, Michaud M. Interaction of low-energy electrons (1—30 eV) with condensed molecules: II. Vibrational-librational excitation and shape resonances in thin N_2 and CO films. Phys. Rev. B 1984; 30: 6078–6092. DOI: 10.1103/PhysRevB.30.6078

[29] Fano U, Stephens JA. Slow electrons in condensed matter. Phys. Rev. B 1986; 34: 438. DOI: 10.1103/PhysRevB.34.438

[30] Liljequist D. A model calculation of coherence effects in the elastic backscattering of very low energy electrons (1–20 eV) from amorphous ice. Int. J. Radiat. Biol. 2012; 88: 50–53. DOI: 10.3109/09553002.2011.577506

[31] Michaud M, Bazin M, Sanche L. Measurement of inelastic cross sections for low-energy electron scattering from DNA bases. Int. J. Radiat. Biol. 2012; 88: 15–21. DOI: 10.3109/09553002.2011.577505

[32] Michaud M,Wen A, Sanche L. Cross sections for low-energy (1–100 eV) electron elastic and inelastic scattering in amorphous ice. Radiat. Res. 2003; 159: 3–22. DOI: 0033-7587/03

[33] Liljequist D. A study of errors in trajectory simulation with relevance for 0.2–50 eV electrons in liquid water. Radiat. Phys. Chem. 2008; 77: 835–853. DOI: 10.1016/j.radphyschem.2008.03.004

[34] Toburen LH. Challenges in Monte Carlo track structure modelling. Int. J. Radiat. Biol. 2012; 88: 2–9. DOI:10.3109/09553002.2011.574781

[35] Sanche L. Nanoscale dynamics of radiosensitivity: role of low energy electrons. In: Gomez-Tejedor GG, Fuss MC (eds). Radiation Damage in Biomolecular Systems. 1st ed. New York: Springer Science & Business Media; 2012. p. 3–43. DOI: 10.1007/978-94-007-2564-5. ch1

[36] Keszei E, Jay-Gerin J-P, Perluzzo G, Sanche L. Quasielastic hot electron transport in solid N_2 films. J. Chem. Phys. 1986; 85: 7396–7402. DOI: 10.1063/1.451328

[37] Sanche L, Bass AD, Ayotte P, Fabrikant II. Effect of the condensed phase on dissociative electron attachment: CH_3Cl condensed on a Kr surface. Phys. Rev. Lett. 1995; 75: 3568. DOI: 10.1103/PhysRevLett.75.3568

[38] Michaud M, Sanche L. The $2\Pi_g$ shape resonance of N_2 near a metal surface and in rare gas solids. J. Electron Spectrosc. Relat. Phenom. 1990; 51: 237–248. DOI: 10.1016/0368-2048(90)80155-4

[39] Sanche L. Low energy electron-driven damage in biomolecules. Eur. Phys. J. D. 2005; 35: 367–390. DOI: 10.1140/epjd/e2005-00206-6

[40] Abdoul-Carime H, Gholke S, Illenberger E. Site-specific dissociation of DNA bases by slow electrons at early stage of irradiation. Phys. Rev. Lett. 2004; 92: 168103. DOI: 10.1103/PhysRevLett.92.168103

[41] Ptasińska S, Denifl S, Grill V, Märk TD, Illenberger E, Scheier P. Bond- and site-selective loss of H^- from pyrimidine bases. Phys. Rev. Lett. 2005; 95: 093201. DOI: 10.1103/PhysRevLett.95.093201

[42] Ptasińska S, Denifl S, Scheier P, Märk TD, Illenberger E. Bond and site selective loss of H atom from nucleobases by very low-energy electrons (<3 eV). Angew. Chem. Int. Ed. 2005; 44: 6941–6943. DOI: 10.1002/anie.200502040

[43] Denifl S, Sulzer P, Huber D, Zappa F, Probst M, Märk TD, Scheier P, Injan N, Limtrakul J, Abouaf R, Dunet H. Influence of functional groups on the site-selective dissociation of adenine upon low-energy electron attachment. Angew. Chem. Int. Ed. 2007; 46: 5238–5241. DOI: 10.1002/anie.200700032

[44] Dawley MM, Tanzer M, Carmichael I, Denifl S, Ptasińska S. Dissociative electron attachment to the gas-phase nucleobase hypoxanthine. J. Chem. Phys. 2015; 142: 215101. DOI: 10.1016/j.ijms.2013.12.005

[45] Ptasińska S, Denifl S, Scheier P, Märk TD, Gohlke S, Illenberger E. Decomposition of thymidine by low energy electrons: Implications for the molecular mechanisms of single strand breaks in DNA. Angew. Chem. Int. Ed. 2006; 45: 1893–1896. DOI: 10.1002/anie.200503930

[46] Burrow PD, Gallup GA, Scheer AM, Denifl S, Ptasińska S, Märk TD, Scheier P. Vibrational Feshbach resonances in uracil and thymine. J. Chem. Phys. 2006; 124: 124310. DOI: 10.1063/1.2181570

[47] Gallup GA, Fabrikant II. Vibrational Feshbach resonances in dissociative electron attachment to uracil. Phys. Rev. A 2011; 83: 012706. DOI: 10.1103/PhysRevA.84.012701

[48] Francés-Monerris A, Segarra-Martí J, Merchán M, Roca-Sanjuán D. Complete-active-space second-order perturbation theory (CASPT2//CASSCF) study of the dissociative electron attachment in canonical DNA nucleobases caused by low-energy electrons (0–3 eV). J. Chem. Phys. 2015; 143: 215101. DOI: 10.1063/1.4936574

[49] Ferreira da Silva F, Matias C, Almeida D, García G, Ingólfsson O, Dögg Flosadóttir O, Ómarsson B, Ptasińska S, Puschnigg B, Scheier P, Limão-Vieira P, Denifl S. NCO^-, a key fagment upon dissociative electron attachment and electron transfer to pyrimidine bases: site selectivity for a slow decay process. J. Am. Soc. Mass Spectrom. 2013; 24: 1787–1797. DOI: 10.1007/s13361-013-0715-9

[50] Aflatooni K, Scheer AM, Burrow PD. Total dissociative electron attachment cross sections for molecular constituents of DNA. J. Chem. Phys. 2006; 125: 054301. DOI: 10.1063/1.2229209

[51] Ptasińska S, Denifl S, Scheier P, Märk TD. Inelastic electron interaction (attachment/ionization) with deoxyribose. J. Chem. Phys. 2004; 120: 8505–8511. DOI: 10.1063/1.1690231

[52] Ibanescu BC, May O, Monney A, Allan M. Electron induced chemistry of alcohols. Phys. Chem. Chem. Phys. 2007; 9: 3163–3173. DOI: 10.1039/B704656A

[53] Fujita T, Kondo M, Takayanagi T. Quantum chemical study of dissociative electron attachment to D-ribose and D-fructose. Comp. Theor. Chem.. 2016; 1075: 70–76. DOI: 10.1140/epjd/e2005-00206-6

[54] Sommerfeld T, Doorway mechanism for dissociative electron attachment to fructose. J. Chem. Phys. 2007; 126: 124301. DOI: 10.1063/1.2710275

[55] Baccarelli I, Gianturco FA, Grandi A, Sanna N, Lucchese RR, Bald I, Kopyra J, Illenberger E. Selective bond breaking in β-D-ribose by gas-phase electron attachment around 8 eV. J. Am. Chem. Soc. 2007; 129: 6269–6277. DOI: 10.1021/ja070542h

[56] Kopyra J. Low energy electron attachment to the nucleotide deoxycytidine monophosphate: direct evidence for the molecular mechanisms of electron-induced DNA strand breaks. Phys. Chem. Chem. Phys. 2012; 14: 8287–8289. DOI: 10.1039/c2cp40847c

[57] König C, Kopyra J, Bald I, Illenberger E. Dissociative electron attachment to phosphoric acid esters: the direct mechanism for single strand breaks in DNA. Phys. Rev. Lett. 2006; 97: 018105. DOI: 10.1103/PhysRevLett.97.018105

[58] Abouaf R, Dunet H. Structures in dissociative electron attachment cross-sections in thymine, uracil and halouracils. Eur. Phys. J. D. 2005; 35: 405–410. DOI: 10.1140/epjd/e2005-00239-9

[59] Abdoul-Carime H, Huels MA, Illenberger E, Sanche L. Formation of negative ions from gas phase halo-uracils by low-energy (0–18 eV) electron impact. Int. J. Mass Spectrom. 2003; 228: 703–716. DOI: 10.1016/S1387-3806(03)00139-8

[60] Abouaf R, Pommier J, Dunet H. Negative ions in thymine and 5-bromouracil produced by low energy electrons. Int. J. Mass Spectrom. 2003; 226: 397–403. DOI: 10.1016/S1387-3806(03)00085-X

[61] Scheer AM, Aflatooni K, Gallup GA, Burrow PD. Bond breaking and temporary anion states in uracil and halouracils: implications for the DNA bases. Phys. Rev. Lett. 2004; 92: 068102. DOI: 10.1103/PhysRevLett.92.068102

[62] Kossoski F, do N. Varella MT. Negative ion states of 5-bromouracil and 5-iodouracil. Phys. Chem. Chem. Phys. 2015; 17: 17271. DOI: 10.1039/C5CP01475A

[63] Denifl S, Candori P, Ptasińska S, Limão-Vieira P, Grill V, Märk TD, Scheier P. Positive and negative ion formation via slow electron collisions with 5-bromouridine. Eur. Phys. J. D. 2005; 35: 391–398. DOI: 10.1140/epjd/e2005-00205-7

[64] Denifl S, Matejcik S, Gstir B, Hanel G, Probst M, Scheier P, Märk TD. Electron attachment to 5-chloro uracil. J. Chem. Phys. 2003; 118: 4107. DOI: 10.1063/1.1540108

[65] Kossoski F, Bettega MHF, do N. Varella MT. Shape resonance spectra of uracil, 5-fluorouracil, and 5-chlorouracil. J. Chem. Phys. 2014; 140: 024317. DOI: 10.1063/1.4861589

[66] Denifl S, Matejcik S, Ptasińska S, Gstir B, Probst M, Scheier P, Illenberger P, Märk TD. Electron attachment to chlorouracil: a comparison between 6-ClU and 5-ClU. J. Chem. Phys. 2004; 120: 704–709. DOI: 10.1063/1.1630959

[67] Modelli A, Bolognesi P, Avaldi L. Temporary anion states of pyrimidine and halopyrimidines. J. Phys. Chem. A. 2011; 115: 10775–10782. DOI: 10.1021/jp206559d

[68] Kossoski F, Kopyra J, do N. Varella MT. Anion states and fragmentation of 2-chloroadenine upon low-energy electron collisions. Phys. Chem. Chem. Phys. 2015; 17: 28958–28965. DOI: 10.1039/C5CP04967A

[69] Kopyra J, Keller A, Bald I. On the role of fluoro-substituted nucleotides in DNA radiosensitization for tumor radiation therapy. RSC Adv. 2014; 4: 6825–6829. DOI: 10.1039/C3RA46735J

[70] Abouaf R, Ptasińska S, Teillet-Billy D. Low energy electron impact on gas phase 5-nitrouracil. Chem. Phys. Lett. 2008; 455: 169–173. DOI: 10.1016/j.cplett.2008.02.104

[71] Ptasińska S, Alizadeh E, Sulzer P, Abouaf R, Mason NJ, Märk TD, Scheier P. Negative ion formation by low energy electron attachment to gas phase 5-nitrouracil. Int. J. Mass Spectrom. 2008; 277: 291–295. DOI: 10.1016/j.ijms.2008.06.008

[72] Tanzer K, Feketeova L, Puschnigg B, Scheier P, Illenberger E, Denifl S. Reactions in nitroimidazole and methylnitroimidazole triggered by low-energy (0–8 eV) electrons. J. Phys. Chem. A. 2014; 119: 6668–6675. DOI: 10.1021/acs.jpca.5b02721

[73] Tanzer K, Feketeova L, Puschnigg B, Scheier P, Illenberger E, Denifl S. Reactions in nitroimidazole triggered by low-energy (0-2 eV) electrons: methylation at N1-H completely blocks reactivity. Angew. Chem. Int. Ed. 2015; 53: 12240–12243. DOI: 10.1002/anie.201407452

[74] Tanzer K, Pelc A, Huber SE, Smialek MA, Scheier P, Probst M, Denifl S. Low energy electron attachment to platinum(II) bromide (PtBr2). Int. J. Mass Spectrom. 2014; 365: 152–156. DOI: 10.1016/j.ijms.2013.11.016

[75] Glick BR, Pasternak JJ, Patten CL. Molecular Biotechnology: Principles and Applications of Recombinant DNA. 4th ed. Washington: ASM Science; 2010. DOI: 10.1128/9781555816124

[76] Boulanouar O, Fromm M, Mavon C, Cloutier P, Sanche L. Dissociative electron attachement to DNA-diamine thin films: impact of the DNA close environment on the OH and O decay channels. J. Chem. Phys. 2013; 139: 055101. DOI: 10.1063/1.4815967

[77] Zheng Y, Cloutier P, Hunting DJ, Sanche L, Wagner JR. Chemical basis of DNA sugar-phosphate cleavage by low-energy electrons. J. Am. Chem. Soc. 2005; 127: 16592–16598. DOI: 10.1021/ja054129q

[78] Zheng Y, Wagner JR, Sanche L. DNA damage induced by low-energy electrons: electron transfer and diffraction. Phys. Rev. Lett. 2006; 96: 208101. DOI: 10.1103/PhysRevLett.96.208101

[79] Zheng Y, Cloutier P, Hunting DJ, Wagner JR, Sanche L. Phosphodiester and N-glycosidic bond cleavage in DNA induced by 4–15 eV electrons. J. Chem. Phys. 2006; 124: 064710. DOI: 10.1063/1.2166364

[80] Keller A, Bald I, Rotaru A, Cauët E, Gothelf CV, Besenbacher F. Probing Electron-induced bond cleavage at the single-molecule level using DNA origami templates. ACS Nano 2012; 6: 4392-4399. DOI: 10.1021/nn3010747

[81] Keller A, Kopyra J, Gothelf KV, Bald I. Electron-induced damage of biotin studied in the gas phase and in the condensed phase at a single-molecule level. New J. Phys. 2013; 15: 083045. DOI: 10.1088/1367-2630/15/8/083045

[82] Bald I, Keller A. Molecular processes studied at a single-molecule level using DNA origami nanostructures and atomic force microscopy. Molecules 2014; 19: 13803–13823. DOI: 10.3390/molecules190913803

[83] Park Y, Polska K, Rak J, Wagner JR, Sanche L. Fundamental mechanisms of DNA radiosensitization: damage induced by low-energy electrons in brominated oligonucleotide trimers. J. Phys. Chem. B 2012; 116: 9676–9682. DOI: 10.1021/jp304964r

[84] Panajotovic R, Martin F, Cloutier P, Hunting DJ, Sanche L. Effective cross sections for production of single-strand breaks in plasmid DNA by 0.1 to 4.7 eV electrons. Radiat. Res. 2006; 165: 452–459. DOI: 10.1667/RR3521.1

[85] Boulanouar O, Fromm M, Bass AD, Cloutier P, Sanche L. Absolute cross section for loss of supercoiled topology induced by 10 eV electrons in highly uniform/DNA/1,3-diaminopropane films deposited on highly ordered pyrolitic graphite. J. Chem. Phys. 2013; 139: 5. DOI: 10.1063/1.4817323

[86] Keller A, Rackwitz J, Cauët E, Liévin J, Körzdörfer T, Rotaru A, Gothelf KV, Besenbacher F, Bald I. Sequence dependence of electron-induced DNA strand breakage revealed by DNA nanoarrays. Sci. Rep. 2014; 4: 7391. DOI: 10.1038/srep07391

[87] Huels M, Boudaïffa B, Cloutier P, Hunting DJ, Sanche L. Single, double, and multiple double strand breaks induced in DNA by 3-100 eV electrons. J. Am. Chem. Soc. 2003; 125: 4467–4477. DOI: 10.1021/ja029527x

[88] Von Sonntag C. The Chemical Basis of Radiation Biology. 1st ed. London: Taylor and Francis; 1987. DOI: 10.1016/1011-1344(89)80053-3

[89] Podhorecka M, Skladanowski A, Bozko P. H2AX phosphorylation: its role in DNA damage response and cancer therapy. J. Nucl. Acids. 2010; 2010: 920161. DOI: 10.4061/2011/920161

[90] Martin F, Burrow PD, Cai Z, Cloutier P, Hunting DJ, Sanche L. DNA strand breaks induced by 0-4 eV electrons: The role of shape resonances. Phys. Rev. Lett. 2004; 93: 068101. DOI: 10.1103/PhysRevLett.93.068101

[91] Li X, Sevilla MD, Sanche L. Density functional theory studies of electron interaction with DNA: can zero eV electrons induce strand breaks? J. Am. Chem. Soc. 2003; 125: 13668–13669. DOI: 10.1021/ja036509m

[92] Simons J. How do low-energy (0.1-2 eV) electrons cause DNA-strand breaks? Acc. Chem. Res. 2006; 39: 772–779. DOI: 10.1021/ar0680769

[93] Bao X, Wang J, Gu J, Leszczynski J. DNA strand breaks induced by near-zero-electronvolt electron attachment to pyrimidine nucleotides. PNAS 2006; 103: 5658–5663. DOI: 10.1073/pnas.0510406103

[94] Caron LG, Sanche L. Diffraction in resonant electron scattering from helical macromolecules: A- and B-type DNA. Phys. Rev. A: At. Mol. Opt. Phys. 2004; 70: 032719. DOI: 10.1103/PhysRevA.70.032719

[95] Caron LG, Sanche L. Theoretical studies of electron interactions with DNA and its subunits: from tetrahydrofuran to plasmid DNA. In: Čársky P, Čurík R (eds). Low-energy Electron Scattering from Molecules, Biomolecules and Surfaces. 1st ed. Boca Raton: CRC Press (Taylor and Francis Group); 2012. p. 161–230. ISBN: 978-1-4398-3910-2. Ch. 6.

[96] Gu J, Leszczynski J, Schaefer III F. Interactions of electrons with bare and hydrated biomolecules: from nucleic acid bases to DNA segments. Chem Rev. 2012; 112: 5603–5640. DOI: 10.1021/cr3000219

[97] Sidorov AN, Orlando TM. Monolayer graphene platform for the study of DNA damage by low-energy electron irradiation. Phys. Chem. Lett. 2013; 4: 2328–2333. DOI: 10.1021/jz4010416

[98] Han X, Klas M, Liu Y, Stack MS, Ptasinska S. DNA damage in oral cancer cells induced by nitrogen atmospheric pressure plasma jets. Appl. Phys. Lett. 2013; 102: 233703. DOI: 10.1063/1.4809830

[99] Dame RT. The role of nucleoid-associated proteins in the organization and compaction of bacterial chromatin. Mol. Microbiol. 2005; 56: 858–870. DOI: 10.1111/j.1365-2958.2005.04598.x

[100] Mehmood S, Allison TM, Robinson CV. Mass spectrometry of protein complexes: from origins to applications. Ann. Rev. Phys. Chem. 2015; 66: 453–474. DOI: 10.1146/annurev-physchem-040214-121732

[101] Lin SD. Electron radiation damage of thin films of glycine, diglycine, and aromatic amino acids. Radiat. Res. 1974; 59: 521–536. DOI: 10.2307/3574071

[102] Abdoul-Carime H, Cecchini S, Sanche L. Alteration of protein structure induced by low-energy (<18 eV) electrons: I. The peptide and disulfide bridges. Radiat. Res. 2002; 158: 23–31. DOI: 10.1016/j.mrfmmm.2009.06.016

[103] Abdoul-Carime H, Sanche L. Alteration of protein constituents induced by low energy (<35 eV) electrons: II. Dissociative electron attachment to amino acids containing cyclic groups. Radiat. Res. 2003; 160: 86–94. DOI: 10.1039/b814219j

[104] Karas M, Hillenkamp F. Laser desorption ionization of proteins with molecular masses exceeding 10,000 Daltons. Anal. Chem. 1988; 60: 2299–2301. DOI: 10.1021/ac00171a028

[105] Hillenkamp F, Karas M. Matrix-assisted laser desorption/ionisation, an experience. Int. J. Mass Spectrom. 2000; 200: 71–77. DOI: 10.1111/j.1469-0691.2010.03274.x

[106] Karas M, Bahr U, Fournier I, Glückmann M, Pfenninger A. The initial ion velocity as a marker for different desorption-ionization mechanisms in MALDI. Int. J. Mass Spectrom. 2003; 226: 239–248. DOI: 10.1016/S1387-3806(02)01062-X

[107] Wind M, Lehmann W. Element and molecular mass spectrometry—An emerging analytical dream team in the life sciences. J. Anal. At. Spectrom. 2004; 19: 20–25. DOI: 10.1039/B309482K

[108] Brøndsted Nielsen S, Andersen JU, Hvelplund P, Liu B, Tomita S. Biomolecular ions in accelerators and storage rings. J. Phys. B: At. Mol. Opt. Phys. 2004; 37: R25–R56. DOI: 10.1088/0953-4075/37/8/R01

[109] Fenn JB, Mann M, Meng CK, Wong SF, Whitehouse CM. Electrospray ionization for mass spectrometry of large biomolecules. Science 1989; 246: 64–71. DOI: 10.1126/science.2675315

[110] Lucas B, Grégoire G, Lemaire J, Maître P, Ortega JM, Rupenyan A, Reimann B, Schermann JP, Desfrançois C. Investigation of the protonation site in the dialanine peptide by infrared multiphoton dissociation spectroscopy. Phys. Chem. Chem. Phys. 2004; 6: 2659–2663. DOI: 10.1039/B316053J

[111] Laskin J, Denisov E, Futrell JH. A comparative study of collision-induced and surface-induced dissociation. 1. Fragmentation of protonated dialanine. J. Am. Chem. Soc. 2000; 122: 9703–9714. DOI: 10.1021/ja001384w

[112] Polce MJ, Ren D, Wesdemiotis C. Dissociation of the peptide bond in protonated peptides. Int. J. Mass Spectrom. 2000; 35: 1391–1398. DOI: 10.1002/1096-9888(200012)35:12

[113] Laskin J, Futrell JH. On the efficiency of energy transfer in collisional activation of small peptides. J. Chem. Phys. 2002: 116: 4302–4310. DOI: 10.1016/S1387-3806(02)01017-5

[114] Wang J, Meroueh SO, Wang Y, Hase WL. Efficiency of energy transfer in protonated diglycine and dialanine SID: Effects of collision angle, peptide ion size, and intramolecular potential. Int. J. Mass Spectrom. 2003; 230: 57–63. DOI: 10.1016/j.ijms.2003.08.005

[115] Ptasińska S, Denifl S, Candor P, Matejcik S, Scheier P, Märk TD. Dissociative electron attachment to gas phase alanine. Chem. Phys. Lett. 2005; 403: 107–112. DOI: 10.1016/j.cplett.2004.12.115

[116] Abdoul-Carime H, Gohlke S, Illenberger E. Fragmentation of tryptophan by low-energy electrons. Chem. Phys. Lett. 2005; 402: 497–502. DOI: 10.1016/j.cplett.2004.12.073

[117] Gohlke S, Rosa A, Illenberger E, Brüning F, Huels MA. Formation of anion fragments from gas-phase glycine by low energy (0–15 eV) electron impact. J. Chem. Phys. 2002; 116: 10164–10169. DOI: 10.1063/1.1479348

[118] Ptasińska S, Denifl S, Abedi A, Scheier P, Märk TD. Dissociative electron attachment to gas-phase glycine. Anal. Bioanal. Chem. 2003; 377: 1115–1119. DOI: 10.1007/s00216-003-2254-x

[119] Abdoul-Carime H, Illenberger E. Fragmentation of proline induced by slow electrons. Chem. Phys. Lett. 2004; 397: 309–313. DOI: 10.1016/j.cplett.2004.08.119

[120] Sulzer P, Alizadeh E, Mauracher A, Scheier P, Märk TD. Detailed dissociative electron attachment studies on the amino acid proline. Int. J. Mass. Spectrom. 2008; 277: 274–278. DOI:10.1016/j.ijms.2009.10.003

[121] Abdoul-Carime, H.; Gohlke, S.; Illenberger, E. Conversion of amino-acids by electrons at subexcitation energies. Phys. Chem. Chem. Phys. 2004; 6: 161–164. DOI: 10.1039/B311675A

[122] Alizadeh, E. Dissociative Electron Attachment to Biomolecules [thesis], Innsbruck: University of Innsbruck; 2009.

[123] Kocisek J, Papp P, Mach P, Vasil'ev YV, Deinzer ML, Matejcik S. Resonance electron capture by serine. J. Phys. Chem. A 2010; 114: 1677–1683. DOI: 10.1021/jp906636b

[124] Alizadeh E, Gschliesser D, Bartl P, Edtbauer A, Vizcaino V, Mauracher A, Probst M, Märk TD, Ptasińska S, Mason NJ, Denifl S, Scheier P. Bond dissociation of the dipeptide dialanine and its derivative alanine anhydride induced by low energy electrons. J. Chem. Phys. 2011; 134: 054305. DOI: 10.1063/1.3544217

[125] Vasil'ev YV, Figard BJ, Barofsky DF, Deinzer ML. Resonant electron capture by some amino acids esters. Int. J. Mass Spectrom. 2007; 268: 106–121. DOI: 10.1016/j.ijms.2007.07.006

[126] Gallup GA, Burrow PD, Fabrikant II. Electron-induced bond breakage at low energies in HCOOH and glycine: the role of very short-lived short-lived σ^* anion states. Phys. Rev. A 2009; 79: 042701. DOI: 10.1103/PhysRevA.79.042701

[127] Abdoul-Carime H, König-Lehmann C, Kopyra J, Farizon B, Farizon M, Illenberger E. Dissociative electron attachment to amino-acids: the case of leucine. Chem. Phys. Lett. 2009; 477: 245–248. DOI:10.1016/j.cplett.2009.07.021

[128] Abdoul-Carime H, Sanche, L. Alteration of protein constituents induced by low-energy (<40 eV) electrons. III. The aliphatic amino acids. J. Phys. Chem. B 2004; 108: 457–464. DOI: 10.1021/jp030413x

[129] Cloutier P, Sicard-Roselli C, Escher E, Sanche L. Low-energy (3–24 eV) electron damage to the peptide backbone. J. Phys. Chem. B 2007; 111: 1620–1624. DOI: 10.1021/jp066947q

[130] Uvdal K, Bodo P, Liedberg B. l-cysteine adsorbed on gold and copper: An X-ray photoelectron spectroscopy study. J. Colloid Interface Sci. 1992; 149: 162–173. DOI: 10.1016/0021-9797(92)90401-7

[131] Kühnle A, Linderoth TR, Hammer B, Besenbacher F. Chiral recognition in dimerization of adsorbed cysteine observed by scanning tunnelling microscopy. Nature 2002; 415: 891–893. DOI: 10.1021/nn9012803

[132] L. Buimaga-Iarinca L, Calborean A. Electronic structure of the ll-cysteine dimers adsorbed on Au(111): a density functional theory study. Phys. Scr. 2012; 86: 035707. DOI:10.1088/0031-8949/86/03/035707

[133] Jain PK, Qian W, El-Sayed MA. Ultrafast cooling of photoexcited electrons in gold nanoparticle-thiolated DNA conjugates involves the dissociation of the gold-thiol bond. J. Am. Chem. Soc. 2006; 128: 2426–2433. DOI: 10.1021/ja056769z

[134] Alizadeh E, Massey S, Rowntree PA, Sanche L. Low-energy electron-induced dissociation in condensed-phase L-cysteine I: desorption of anions from chemisorbed films. J. Phys.: Conf. Ser. 2015; 635: 012001. DOI: 10.1088/1742-6596/635/1/012001

[135] Alizadeh E, Massey S, Sanche L, Rowntree PA. Low-energy electron-induced dissociation in condensed-phase L-cysteine II: a comparative study on desorption from chemisorbed and physisorbed films. Eur. Phys. J. D 2016; 70: 1–8 DOI: 10.1140/epjd/e2016-60739-y

[136] Solomun T, Skalicky T. The interaction of a protein-DNA surface complex with low-energy electrons. Chem. Phys. Lett. 2008; 453: 101–104. DOI: 10.1016/j.cplett.2007.12.078

[137] Ptasińska S, Li Z, Mason NJ, Sanche L. Damage to amino acid–nucleotide pairs induced by 1 eV electrons. Phys. Chem. Chem. Phys. 2010; 12: 9367–9372. DOI: 10.1039/B926267A

[138] Wang J. Gu J, Leszczynski J. Electron detachment of the hydrogen-bonded amino acid side-chain–guanine complexes. Chem. Phys. Lett. 2007; 442: 124–127. DOI: 10.1016/j.cplett.2007.05.071

[139] Gu B, Smyth M, Kohanoff J. Protection of DNA against low-energy electrons by amino acids: a first-principles molecular dynamics study. Phys. Chem. Chem. Phys. 2014; 16: 24350–24358. DOI: 10.1039/C4CP03906H

[140] Ptasińska S, Sanche L. Dissociative electron attachment to hydrated single DNA strands. Phys. Rev. E 2007; 75: 031915. DOI: 10.1103/PhysRevE.75.031915

[141] Falk M, Hartman KA, Lord RC. Hydration of deoxyribonucleic acid. II. An infrared study. J. Am. Chem. Soc. 1963; 85: 387–391. DOI: 10.1021/ja00887a004

[142] Orlando TM, Oh D, Chen Y. Alexandrov A. Low-energy electron diffraction and induced damage in hydrated DNA. J. Chem. Phys. 2008; 128: 195102. DOI: 10.1063/1.2907722

[143] Grieves GA, McLain JL, Orlando TM. Low-energy electron-stimulated reactions in nanoscale water films and water-DNA interfaces. In: Hatano Y, Katsumura Y, Mozumder A (eds). Charged Particle and Photon Interactions with Matter, Recent Advances, Applications, and Interfaces. 1st ed. Boca Raton: CRC Press; 2011. p 473–501. ISBN 9781439811771. Ch. 18.

[144] Sidorov AN, Orlando TM. Monolayer graphene platform for the study of DNA damage by low-energy electron irradiation. Phys. Chem. Lett. 2013; 4: 2328–2333. DOI: 10.1021/jz4010416

[145] Alizadeh E, Cloutier P, Hunting DJ, Sanche L. Soft X-ray and low energy electron-induced damage to DNA under N_2 and O_2 atmospheres. J. Phys. Chem. B 2011; 115: 4523–4531. DOI: 10.1021/jp200947g

[146] Alizadeh E, Sanche L. The role of humidity and oxygen level on damage to DNA induced by soft X-rays and low-energy electrons. J. Phys. Chem. C 2013; 117: 22445–22453. DOI: 10.1021/jp403350j

[147] Alizadeh E, Sanz AG, Madugundu GS, Garcia G, Wagner JR, Sanche L. Thymidine decomposition induced by low-energy electrons and soft X-rays under N_2 and O_2 atmospheres. Radiat. Res. 2014; 181: 629–640. DOI: 10.1667/RR13584.1

[148] Gokhberg K, Kolorenč P, Kuleff AI, Cederbaum LS. Site- and energy-selective slow-electron production through intermolecular Coulombic decay. Nature 2014; 505: 661–663. DOI: 10.1038/nature12927

[149] Cederbaum LS, Zobeley J, Tarantelli F. Giant intermolecular decay and fragmentation of clusters. Phys. Rev. Lett. 1997; 79: 4778–4781. DOI: 10.1103/PhysRevLett.79.4778.

[150] Santra R, Zobeley J, Cederbaum LS, Moiseyev N. Interatomic coulombic decay in van der Waals clusters and impact of nuclear motion. Phys. Rev. Lett. 2000; 85: 4490–4493. DOI: 10.1103/PhysRevLett.85.4490

[151] Grieves GA, Orlando TM. Intermolecular coulomb decay at weakly coupled heterogeneous interfaces. Phys. Rev. Lett. 2011; 107: 016104. DOI: 10.1103/PhysRevLett.107.016104

[152] Mucke M, Braune M, Barth S, Förstel M, Lischke T, Ulrich V, Arion T, Becker U, Bradshaw A, Hergenhahn U. A hitherto unrecognized source of low-energy electrons in water. Nat. Phys. 2010; 6: 143–146. DOI: 10.1038/nphys1500

[153] Hergenhan U. Production of low kinetic energy electrons and energetic ion pairs by intermolecular Coulombic decay. Int. J. Radiat. Biol. 2012; 88: 871–883. DOI: 10.3109/09553002.2012.698031

[154] Kouass Sahbani S, Sanche L, Cloutier P, Bass AD, Hunting DJ. Loss of cellular transformation efficiency induced by DNA irradiation with low-energy (10 eV) electrons. J. Phys. Chem. B 2014; 118:13123–13131. DOI: 10.1021/jp508170c

[155] Boulanouar O, Khatyr A, Herlem G, Palmino F, Sanche L, Fromm M. Soft adsorption of densely packed layers of DNA-plasmid • 1,3-diaminopropane complexes onto highly oriented pyrolitic graphite designed to erode in water. J. Phys. Chem. C 2011; 115: 21291–21298. DOI: 10.1021/jp207083r

[156] Kouass Sahbani S, Sanche L, Cloutier P, Bass AD, Hunting DJ. Electron resonance decay into a biological function: decrease in viability of *E. coli* transformed by plasmid DNA irradiated with 0.5–18 eV electrons. J. Phys. Chem. Lett. 2015; 6: 3911–3914. DOI: 10.1021/acs.jpclett.5b01585

[157] Dugal P-C, Abdoul-Carime H, Sanche L. Mechanisms of low energy (0.5–30 eV) electron-induced pyrimidine ring fragmentation within thymine and halogen-substituted single strands of DNA. J. Phys. Chem. B. 2000; 104: 5610–5617. DOI: 10.1021/jp9938112

[158] Abdoul-Carime H, Dugal P-C, Sanche L. Damage induced by 1–30 eV electrons on thymine and bromouracil substituted oligonucleotides. Radiat. Res. 2000; 153: 23–28. DOI: 10.1667/0033-7587(2000)153

[159] Abouaf R, Pommier J, Dunet H. Electronic and vibrational excitation in gas phase thymine and 5-bromouracil by electron impact. Chem. Phys. Lett. 2003; 381: 486–494. DOI: 10.1016/j.cplett.2003.09.121

[160] Klyachko DV, Huels MA, Sanche L. Halogen anion formation in 5-halouracil films: X-rays vs subionization electrons. Radiat. Res. 1999; 151: 177–187. DOI: 10.2307/3579945

[161] Li Z, Cloutier P, Sanche L, Wagner JR. Low energy electron induced DNA damage in a trinucleotide containing 5-bromouracil. J. Phys. Chem. B. 2011; 115: 13668–13673. DOI: 10.1021/jp205428j

[162] Abdoul-Carime H, Limao-Vieira P, Gohlke S, Petrushko I, Mason NJ, Illenberger E. Sensitization of 5-bromouridine by slow electrons. Chem. Phys. Lett. 2004; 393: 442–447. DOI: 10.1016/j.cplett.2004.06.081

[163] du Penhoat H, Huels MA, Cloutier P, Jay-Gerin J-P, Sanche L. Electron stimulated desorption of H- from thin films of 5-halouracils. Phys. Chem. 2003; 5: 3270–3277. DOI: 10.1039/B212552H

[164] du Penhoat H, Huels MA, Cloutier P, Jay-Gerin J-P, Sanche L. Anion fragment formation in 5-halouracil films induced by 1–20 eV electron impact. J. Phys. Chem. B 2004; 108: 17251–17260. DOI: 10.1021/jp0478817

[165] Li X, Sanche L, Sevilla MD. Dehalogenation of 5-halouracils after low energy electron attachments: a density functional theory investigation. J. Phys. Chem. A. 2002; 106: 11248–11253. DOI: 10.1021/ja034286u

[166] Park Y, Polska K, Rak J, Wagner JR, Sanche L. Fundamental mechanisms of DNA radiosensitization: damage induced by low energy electrons in brominated oligonucleotide trimers. J. Phys. Chem. B. 2012; 116: 9676–9682. DOI: 10.1021/jp304964r

[167] Polska K, Rak J, Bass AD, Cloutier P, Sanche L. Electron simulated desorption of anions from native and brominated single stranded oligonucleotide trimers. J. Chem. Phys. 2012; 136: 075101. DOI: 10.1063/1.3685587

[168] Zimbrick JD, Ward JF, Myers LS. Studies on the chemical basis of cellular radiosensitization by 5-bromouracil substitution in DNA. Int. J. Radiat. Biol. 1969; 16: 505–523. DOI: 10.1080/09553006914551571

[169] Tanzer K, Pelc A, Huber SE, Śmiałek MA, Scheier P, Probst M, Denifl S. Low energy electron attachment to platinum(II) bromide ($PtBr_2$). Int. J. Mass Spectrom. 2013; 365: 152–156. DOI: 10.1016/j.ijms.2013.11.016

[170] Scheer AM, Aflatooni K, Gallup GA, Burrow PD. Bond breaking and temporary anion states in uracil and halouracils: implications for the DNA bases. Phys. Rev. Lett. 2004; 92: 068102. DOI: 10.1103/PhysRevLett.92.068102

[171] Abdoul-Carime H, Dugal P. C, Sanche L. DIET of neutral fragments from chemisorbed biological molecular systems. Surf. Sci. 2000; 451: 102–107. DOI: 10.1016/S0039-028(00)00014-5

[172] Cheng HY, Hsieh SH, Chen YC, Lin CJ, Liu WC. Temporary anion states of radiosensitive halopyrimidines: shape and core-excited resonances. Comp. Theor. Chem. 2016; 1075: 18–29. DOI: 10.1016/j.comptc.2015.10.031

[173] Cecchini S, Girouard S, Huels MA, Sanche L, Hunting DJ. Single-strand-specific radiosensitization of DNA by bromodeoxyuridine. Radiat. Res. 2004; 162: 604–615. DOI: 10.1667/RR3267

[174] Sevilla M. Research breakthrough: DNA strandedness controls halouracil radiosensitization. Radiat. Res. 2004; 162: 603–603. DOI:10.1140/epjd/e2005-00206-6

[175] Zheng Y, Hunting DJ, Ayotte P, Sanche L. Role of secondary low-energy electrons in the concomitant chemoradiation therapy of cancer. Phys. Rev. Lett. 2008; 100: 198101. DOI: 10.1103/PhysRevLett.100.198101

[176] Tannock IF. Treatment of cancer with radiation and drugs. J. Clin. Oncol. 1996; 14: 3156–3174. DOI: 10.1200/JCO.2015.635540

[177] Prestwich RJ, Shaakespeare D, Waters S. The rationale for and the current role of chemoradiotherapy. J. Radiother. Pract. 2007; 6: 11–19. DOI: 10.1097/CND.0b013e3181629396

[178] Rezaee M, Alizadeh E, Cloutier P, Hunting DJ, Sanche L. A single subexcitation-energy electron can induce a double strand break in DNA modified by platinum chemotherapeutic drugs. Chem. Med. Chem. 2014; 9: 1145–1149. DOI: 10.1002/cmdc.201300462

[179] Rezaee M, Sanche L, Hunting DJ. Cisplatin enhances the formation of DNA single and double strand breaks by hydrated electrons and hydroxyl radicals. Radiat. Res. 2013; 179: 323–331. DOI: 10.1667/RR3185.1

[180] Rezaee M, Hunting DJ, Sanche L. New insights into the mechanism underlying the synergistic action of ionizing radiation with platinum chemotherapeutic drugs: the role of low-energy electrons. Int. J. Radiat. Oncol. Biol. Phys. 2013; 87: 847–853. DOI: 10.1016/j.ijrobp.2013.06.2037

[181] Behmand B, Wagner JR, Sanche L, Hunting DJ. Cisplatin intrastrand adducts sensitive DNA to base by hydrated electrons. J. Phys. Chem. B. 2014; 118: 4803–4808. DOI: 10.1021/jp5014913

[182] Behmand B, Wagner J. R, Hunting D, Marignier JL, Mostafavi M, Sanche L. Rate constant of reaction between the GTG-cisplatin complex and hydrated electrons. J. Phys. Chem. B. 2015; 119: 9496–9500. DOI: 10.1021/acs.jpcb.5b01752

[183] Luo X, Zheng Y, Sanche L. DNA strand breaks and crosslinks induced by transient anions in the electron-energy range 2-20 eV. J. Chem. Phys. 2014; 140: 155101. DOI: 10.1063/1.4870519

[184] Bao Q, Chen Y, Zheng Y, Sanche L. Cisplatin radiosensitization of DNA irradiated with 2-20 eV electrons: role of transient anions. J. Phys. Chem. C. 2014; 118: 15516–15524. DOI: 10.1021/jp503706h

[185] Tippayamontri T, Kotb R, Paquette B, Sanche L. Optimal timing in concomitant chemoradiation therapy of colorectal tumors in nude mouse treated with Cisplatin and Lipoplatin™. Anticancer Res. 2013; 33: 3005–3014. DOI: 10.1155/2013/409179

[186] Tippayamontri T, Kotb R, Sanche L, Paquette B. New therapeutic possibilities of combined treatment of radiotherapy with oxaliplatin and its liposomal formulations Lipoxal™ in rectal cancer using nude mouse xenograft. Anticancer Res. 2014; 34: 5303–5312.

[187] Boulikas T, Vougiouka M. Cisplatin and platinum drugs at the molecular level. Oncol. Rep. 2003; 10: 1663–1682. DOI: 10.3892/or.10.6.1663

[188] Andre N, Schmiegel W. Chemoradiotherapy for colorectal cancer. Gut. 2005; 54: 1194–1202. DOI: 10.1136/gut.2004.062745

[189] Sicard-Roselli C, Brun E, Gilles M, Baldacchino G, Kelsey C, McQuaid H, Polin C, Wardlow N, Currell F. A new mechanism for hydroxyl radical production in irradiated nanoparticle solutions. Small 2014; 10: 3338-3346. DOI: 10.1002/smll.201400110.

[190] Coulter JA, Jain S, Forker J, McMahon SJ, Schettino G, Prise KM, Currell FJ, Hirst DG. Evaluation of cytotoxicity and radiation enhancement using 1.9 nm gold particles:

potential application for cancer therapy. Nanotechnology. 2010; 29: 295101. DOI: 10.1088/0957-4484/21/29/295101

[191] Butterworth KT, Wyer JA, Brennan-Fournet M, Latimer CJ, Shah MB, Currell FJ, Hirst DG. Variation of strand break yield for plasmid DNA irradiated with high-Z metal nanoparticles. Radiat. Res. 2008; 170: 381–387. DOI: 10.1667/RR1320.1

[192] Chang MY, Shiau AL, Chen YH, Chang CJ, Chen HH, Wu CL. Increased apoptotic potential and dose-enhancing effect of gold nanoparticles in combination with single-dose clinical electron beams on tumor-bearing mice. Cancer Sci. 2008; 99: 1479–1484. DOI: 10.1111/j.1349-7006.2008.00827.x

[193] Coulter JA, Hyland WB, Nicol J, Currell FJ. Radiosensitising nanoparticles as novel cancer therapeutics—Pipe dream or realistic prospect? Clin. Oncol. 2013; 25: 593–603. DOI: 10.1016/j.clon.2013.06.011

[194] Hainfeld JF, Slatkin DN, Smilowitz HM. The use of gold nanoparticles to enhance radiotherapy in mice. Phys. Med. Biol. 2004; 49: N309–N315. DOI:10.1088/0031-9155/49/18/N03

[195] Jain S, Coulter JA, Hounsell AR, Butterworth KT, McMahon SJ, Hyland WB, Muir MF, Dickson GR, Prise KM, Currell FJ, O'Sullivan JM, Hirst DG. Cell-specific radiosensitization by gold nanoparticles at megavoltage radiation energies. Int. J. Radiat. Oncol. Biol. Phys. 2011; 79: 531–539. DOI: 10.1016/j.ijrobp.2010.08.044

[196] Chithrani BD, Stewart J, Allen C, Jaffray DA. Intracellular uptake, transport, and processing of nanostructures in cancer cells. Nanomed.: Nanotechnol. Biol. Med. 2009; 5: 118–127. DOI: 10.1016/j.nano.2009.01.008

[197] Chithrani DB. Nanoparticles for improved therapeutics and imaging in cancer therapy. Recent Patents Nanotechnol. 2010; 4: 171–180. DOI: 10.2174/187221010792483726

[198] Chithrani DB. Gold nanoparticles as radiation sensitizers in cancer therapy. Radiat. Res. 2010; 173: 719–728. DOI: 10.1667/RR1984.1

[199] Liu CJ. Enhancement of cell radiation sensitivity by pegylated gold nanoparticles. Phys. Med. Biol. 2010; 55: 931–945. DOI: 10.1088/0031-9155/55/4/002

[200] Hainfeld JF, Slatkin DN, Focella TM, Smilowitz HM. Gold nanoparticles: a new X-ray contrast agent. Br. J. Radiol. 2006; 79: 248–253. DOI: 10.1259/bjr/13169882

[201] Brun E, Duchambon P, Blouquit Y, Keller G, Sanche L, Sicard-Roselli C. Gold nanoparticles enhance the X-ray-induced degradation of human centrin 2 protein. Radiation Physics and Chemistry. 2009; 78: 177–183. DOI: 10.1016/j.radphyschem.2008.11.003

[202] Kang B, Mackey MA, El-Sayed MA. Nuclear targeting of gold nanoparticles in cancer cells induces DNA damage, causing cytokinesis arrest and apoptosis. J. Am. Chem. Soc. 2010; 132: 1517–1519. DOI: 10.1021/ja9102698

[203] Hébert E, Debouttière P-J, Lepage M, Sanche L, Hunting DJ. Preferential tumor accumulation of gold nanoparticles, visualized by magnetic resonance imaging:

radiosensitization studies in vivo and in vitro. Int. J. Radiat. Biol. 2010; 86: 692–700. DOI: 10.3109/09553001003746067

[204] Bobyk L, Edouard M, Deman P, Vautrin M, Pernet-Gallay K, Delaroche J, Adam JF, Estève F, Ravanat JL, Elleaume H. Photoactivation of gold nanoparticles for glioma treatment. Nanomedicine. 2013; 9: 1089–1097. DOI: 10.1016/j.nano.2013.04.007

[205] Hyun Cho S, Jones BL, Krishnan S. The dosimetric feasibility of gold nanoparticle-aided radiation therapy (GNRT) via brachytherapy using low-energy gamma-/x-ray sources. Phys. Med. Biol. 2009; 54: 4889–4905. DOI: 10.1088/0031-9155/54/16/004

[206] Lechtman E, Mashouf S, Chattopadhyay N, Keller BM, Lai P, Cai Z, Reilly RM, Pignol JP. A Monte Carlo-based model of gold nanoparticle radiosensitization accounting for increased radiobiological effectiveness. Phys. Med. Biol. 2013; 58: 3075–3085. DOI: 10.1088/0031-9155/58/10/3075

[207] McMahon SJ, Mendenhall, MH, Jain S, Currell F. Radiotherapy in the presence of contrast agents: a general figure of merit and its application to gold nanoparticles. Phys. Med. Biol. 2008; 53: 5635–5651. DOI: 10.1088/0031-9155/53/20/005

[208] McQuaid HN, Muir MF, Taggart LE, McMahon SJ, Coulter JA, Hyland WB, Jain S, Butterworth KT, Schettino G, Prise KM, Hirst DJ, Botchway SW, Currell FJ. Imaging and radiation effects of gold nanoparticles in tumour cells. Sci. Rep. 2016; 6: 19442. DOI: 10.1038/srep19442

[209] Cho S. Estimation of tumour dose enhancement due to gold nanoparticles during typical radiation treatments: a preliminary Monte Carlo study. Phys. Med. Biol. 2005; 50: N163–173. DOI: 10.1088/0031-9155/50/15/N01

[210] Rahman WN. Enhancement of radiation effects by gold nanoparticles for superficial radiation therapy. Nanomedicine. 2009; 5: 136–142. DOI: 10.1016/j.nano.2009.01.014

[211] Carter JD, Cheng NN, Qu Y, Suarez GD, Guo T. Nanoscale energy deposition by X-ray absorbing nanostructures. J. Phys. Chem. B. 2007; 111: 11622–11625. DOI: 10.1021/jp075253u

[212] Casta R, Champeaux J-P, Moretto-Capelle P, Sence M, Cafarelli P. Electron and photon emissions from gold nanoparticles irradiated by X-ray photons. J. Nanoparticle Res. 2015; 17: 1–17. DOI: 10.1007/s11051-014-2807-2

[213] Casta R, Champeaux JP, Sence M, Moretto-Capelle P, Cafarelli P. Comparison between gold nanoparticle and gold plane electron emissions: a way to identify secondary electron emission. Phys. Med. Biol. 2015; 60: 9095–9105. DOI: 10.1088/0031-9155/60/23/9095

[214] Xiao F, Zheng Y, Cloutier P, He Y, Hunting DJ, Sanche L. On the role of low-energy electrons in the radiosensitization of DNA by gold nanoparticles. Nanotechnology. 2011; 22: 465101. DOI: 10.1088/0957-4484/22/46/465101

[215] Zheng Y, Hunting DJ, Ayotte P, Sanche L. Radiosensitization of DNA by gold nanoparticles irradiated with high-energy electrons. Radiat. Res. 2008; 169: 19. DOI: 10.1667/RR1080.1

[216] Zheng Y, Sanche L. Gold nanoparticles enhance DNA damage induced by anti-cancer drugs and radiation. Radiat. Res. 2009; 172: 114–119. DOI: 10.1667/RR1689.1.

[217] Zheng Y, Cloutier P, Hunting D. J, Sanche L. Radiosensitization by gold nanoparticles: comparison of DNA damage induced by low and high-energy electrons. J. Biomed. Nanotechnol. 2008; 4: 469–473. DOI: 10.1166/jbn.2008.012

[218] Yao X, Huang C, Chen X, Zheng Y, Sanche L. Chemical radiosensitivity of DNA induced by gold nanoparticles. J. Biomed. Nanotechnol. 2015; 11: 478–485. DOI: 10.1166/jbn.2015.1922

[219] Sanche L. Cancer treatment: low-energy electron therapy. Nat. Mater. 2015; 14: 861–863. DOI: 10.1038/nmat4333

[220] Shi M, Thippayamontri T, Gendron L, Guérin B, Paquette B, Sanche L. Increased radiosensitivity of colorectal tumor with intra-tumoral injection of low dose of gold nanoparticles. (in press).

Chapter 9

Elimination of Potential Pathogenic Microorganisms in Sewage Sludge Using Electron Beam Irradiation

Jean Engohang-Ndong and Roberto M. Uribe

Additional information is available at the end of the chapter

http://dx.doi.org/10.5772/62705

Abstract

Microbiological analyses on municipal sewage sludge sample treated in a pilot plant process utilizing an electron accelerator with a beam energy of 3 MeV were conducted as a way to show the potential of this technology to decontaminate sludge containing 15% solids. Bacterial counts including total heterotrophic bacterial, total coliform, and fecal coliform counts were performed on sewage sludge samples pre- and postirradiation with the electron beam at doses ranging between 2.7 and 30.7 kGy. At each irradiation dose, bacterial and *Ascaris* ova counts and survival were measured in triplicate as colony forming units (CFUs) per milliliter (ml) of sewage sludge. Experimental results obtained revealed that a dose of 6.7 kGy is enough to reduce bacterial load to consider the treated sewage sludge safe for both the environment and human according to the Environmental Protection Agency standards. However, a dose of 25.7 kGy was needed to reduce the concentration of *Ascaris* ova at levels deemed safe for land applications. This study also showed that electron beam treatment is less energy consuming with shorter processing times than conventional techniques used to decontaminate sludge. Taken altogether, these observations open new avenues for large urban agglomerations to save money on sewage sludge treatment.

Keywords: Electron beam, Irradiation, Microorganisms, Sewage sludge, Decontamination

1. Introduction

Radiation processing has been used in biotechnological applications for more than 50 years. The effect of radiation on pathogenic microorganisms was first initiated in 1956 by Ethicon Inc. (a subsidiary of Johnson and Johnson) together with High Voltage Engineering Corp. (a manufacturer of accelerators) in order to sterilize single-use medical devices such as gloves, hypodermic needles, sutures, surgical drapes. Nowadays, it is worldwide used, not only for medical devices, but also for cosmetics. The exact mechanism by which radiation kills microorganisms is not very well understood, but it is certainly related to the damage caused by the radiation to the DNA molecule of the microorganisms. Also it is generally accepted that the smaller the microorganism the larger the dose of radiation needed to kill it. So, the radiation dose needed to kill bacteria will be larger than the dose needed to kill human cells, and it will be smaller than the dose needed to kill a virus.

Sterilization is not the only area in which radiation can be used in biotechnological applications. Radiation is being used to develop new implant materials which are biocompatible. An example of this is the irradiation of water-soluble polymers in aqueous solutions, with or without the addition of another monomer which gives rise to a variety of cross-linked gels which can be used in the biomedical field. Some of these hydrogels can be used to hydrate the skin of patients with severe burns.

Radiation is also used in the area of food preservation. Depending on the dose used on a food commodity, the radiation can either sterilize (e.g., in meat products), kill bacteria including *Salmonella* and *Streptococcus* species, disinfest (e.g., in fruits and grains), kill insects in adult, larvae, or pupae stages, or delay maturation (e.g., in some fruits and vegetables) by decomposing the enzymes responsible for ripening.

2. Application of electron beam technology

The commercial use of irradiation to disinfect sludge started in 1973 when an industrial gamma ray facility from Geiselbullach near Munich (Germany) used Co-60 and Cs-137 sources [1]. The facility used 90,000 Ci of Co-60 and 570,000 Ci of Cs-137 and treated up to 180 m^3/day of sludge. More recently, a new technology using electron beam accelerators was developed in Miami, Florida where a 1.5 MeV 50 mA accelerator with a throughput of 645 m^3/day, and in Brazil where a 1.5 MeV 37.5 kW accelerator, with a maximum throughput of 45 l/min were described [2]. However, these two studies only addressed the engineering aspects of the facilities and the possibilities to use electron beam irradiation for environmental applications. The radiation effects on the bacterial load and removal of noxious chemical compounds have been performed mostly in small samples of sludge irradiated under laboratory conditions and mainly address either only the microbiological or the chemical effect of radiation in a sample of sludge [3–5]. Processing and disposal of wastewater sludge are a critical problem worldwide [5]; therefore, new technologies to solve this problem are constantly being sought.

Sludge is commonly used as a soil amendment and fertilizer but must be treated in order to remove various bacteria, toxic compounds, parasites, and viruses. Many researchers have shown that exposing sludge to high-energy radiation successfully removes all the bacteria and other organisms from the sludge. Thus, the right dose of radiation will ensure proper sludge disinfection. It has been shown that even a small dose of radiation will remove 99.9% of all bacteria in sludge [6]. In addition to disinfection, irradiation of sludge often accelerates sedimentation and filtration, which helps facilitate removal of water from the sludge. Even while changing the physical makeup of sludge, this does not affect the ability of using sludge as a good fertilizer.

The generation of offensive odors from sewage sludge is also a concern in the subsequent disposal and/or use of sludge. Volatile sulfur-containing compounds (carbon disulfide CS_2, dimethylsulfide $[(CH_3)_2S]$, dimethyldisulfide $[(CH_3S)_2]$ and volatile carboxylic acids (acetic acid, propanoic acid, butanoic acid) have been identified as odor causing compounds in sewage sludge [7].

3. Sludge irradiation dosimetry

3.1. Dose mapping: experimental procedure

Prior to the determination of the dose using different accelerator parameters, the scanned beam on top of the sludge delivery system needed to be mapped in order to verify that all the water coming through the weir length of the sludge delivery system would be exposed to the electron beam. The delivery system consists of a stainless steel box 152 cm (59.8 in) long with two compartments, one for the incoming sludge and the other one to drain the irradiated sludge. The sludge is transferred from one compartment to the other through a weir located in the center of the box. Irradiation takes place at the top of the weir (**Figure 1**).

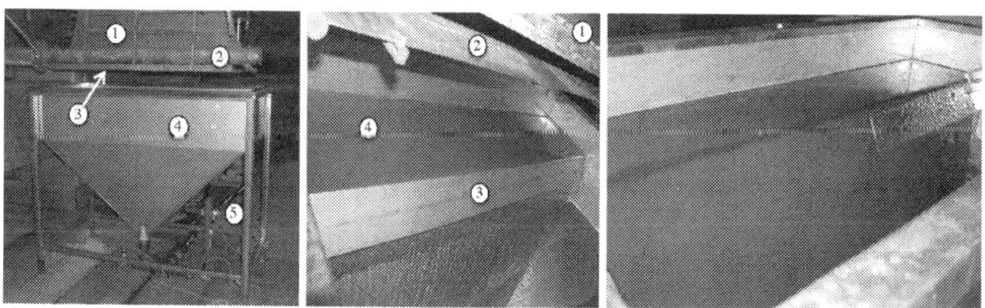

a) b) c)

Figure 1. Irradiation dispositive used to irradiate sludge. (a) Overall view of the system, showing the accelerator scanner (1), the window blower (2), the electron beam shutter (3), the sludge delivery system (4), and the pipes delivering the sludge to the system (5). (b) Simulation of the process using tap water. (1) Lower portion of the accelerator scanner, (2) beam shutter, (3) weir, and (4) incoming water. The depth of the sludge layer at the top of the weir was about 0.3–0.4 cm. (c) sludge sample (adapted from [7]).

To map the extent of the irradiation zone along the weir, a CTA film was taped to the top of the delivery system just underneath the scanner system of the accelerator and irradiated for 5 s using the following accelerator conditions: E = 3 MeV, I = 15 mA, and S = 100%. After irradiation, the optical absorption at 280 nm was measured along the film using a Genesys 5 spectrophotometer fitted with a driving mechanism to measure film strips. Aerial™ software determined the dose from the absorbance measurements.

3.2. Dose mapping: results and discussion

Figure 2 shows a graph of the dose along the top surface of the weir to treat the sludge. The graph shows two features, the extent of the irradiation zone on top of the weir system and the dose uniformity along the weir. The graph shows an effective irradiation length of 127.5 cm which is shorter than the length of the weir itself (152 cm). In order to ensure that all sludge falling over the weir was irradiated, two pieces of aluminum tabs 10 cm long were fastened to each edge of the weir using C-clamps. These tabs shortened the effective weir by a total of 20 cm and allowed for all the sludge falling over the weir to be irradiated by the electron beam given the fact that the scanning angle of the electron beam is 18.5°. The graph also gives information about the uniformity of the electron beam on top of the weir and shows that the dose at the two ends of the weir is about 25% lower than the dose in the idle section of the weir. However, this measurement was taken under static conditions, and in the case of water or sludge, the liquid will not move on a laminar flow fashion and might have been receiving an average dose with variations of up to ± 12.5% assuming that part of the liquid moved on the middle section and another part on the extreme end of the weir. So, then it is reasonable to assume that with the movement of the sludge as well as with its mixing, this might be the maximum difference in dose achieved by the sludge.

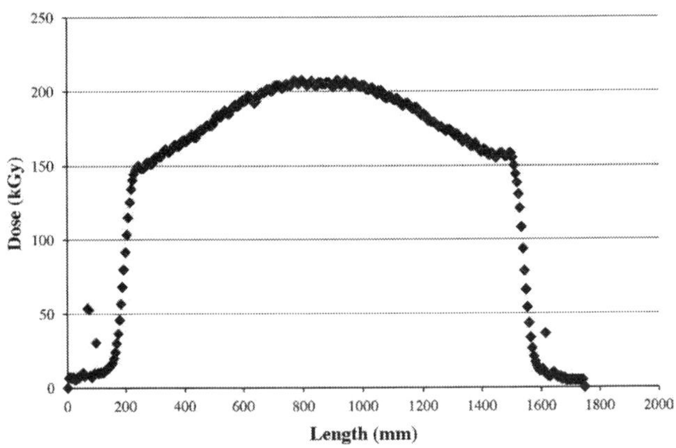

Figure 2. Dose along the sludge delivery system. The measurements were taken using cellulose triacetate film irradiated with 3 MeV electrons, 15 mA of current, 5 s of exposure time and 100% scanning aperture [7].

3.3. Dose measurements: experimental procedure

The dose delivered to the sludge was determined from temperature measurements made on the sludge before and after irradiation, after calibration in terms of dose with alanine pellets

http://dx.doi.org/10.5772/62705

and films. The irradiation of alanine dosimeters produces free radicals that become trapped inside the solid matrix of the dosimeter and can be measured by electron spin resonance (ESR) spectrometry. The trapped radicals are stable over long periods of time, and their concentration can be directly related to the absorbed dose as determined from a calibration curve.

For the experiments described in this chapter, a Bruker eScan ESR spectrometer using an insert FL0041 to measure the alanine films and an insert PH0027 to measure the alanine pellets were used to measure the free radicals. In a first experiment, 40 alanine pellets and 40 alanine film strips were randomly selected and irradiated in order to make a calibration curve of the pellets. The pellets would be used to measure the dose in sludge once they were calibrated with the alanine films.

The 40 alanine pellets were divided into ten groups of four and were placed in small plastic bags 1.5 cm long and 0.5 cm wide, and sealed with heat. The bags with the pellets were placed on a piece of cardboard, one at a time, on top of one of the carts that would be conveyed through the electron beam. On the side of the individual bags, four alanine film strips were placed, to measure the dose. The cardboard was irradiated in the cart conveyor system of the NEO Beam facility Dynamitron electron accelerator and irradiated using the following beam parameters: 3 MeV electron energy and 100% scanning angle; the dosimeters were moving under the beam at a constant speed of 20.32 cm/s, and the current changed to give different dose values ranging from 2 to 40 kGy, according to **Table 1**. Once all the dosimeters were irradiated, the alanine films were measured to determine the dose in each run, and with this information and the measurement of the intensity of the ESR signal of the irradiated pellets, a calibration curve was constructed.

Dose (kGy)	Beam current (mA)
2	3.4
5	8.4
8	13.4
10	16.8
13	21.8
16	26.9
20	33.6
24	40.3
27	45.3
30	50.4

Table 1. Electron beam current values needed to produce the selected doses for alanine pellets running under the beam at 20.32 cm/s.

A second experiment consisted in irradiating a set of pellets in a pyrex baking dish containing cold tap water at a depth of 1.1 cm (7/16 in). The purpose of this experiment was to simulate the accelerator conditions needed to irradiate the sludge. Four vials per run were used, each containing three alanine pellets. These vials were placed into the baking dish and floated on top of the water. The electron beam parameters were set up such that the electron energy was 3 MeV and five runs were conducted underneath the electron beam. Each run had a constant speed of the samples equal to 23.3 cm/s and the following beam currents: 3.8, 9.6, 19.1, 38.3, and 45.9 mA. After irradiations, the dose from the ESR intensity of each alanine pellet using the Bruker eScan instrument was determined.

The next experiment measured the dose for a sample of water running through the delivery system to irradiate sludge and to relate those measurements to the temperature of the water coming in and going out of the system as measured by a set of thermocouples installed near the sludge delivery system in the influent and effluent pipes. Small sealed plastic bags containing two alanine pellets were introduced into the system through a "Tee" connection into the pipe where the water flowed and sent them through the irradiation zone. At this point, 300 gallons of water was being recirculated through the system at 50 GPM. Two sets of five runs were conducted. Beam conditions for each set were as follows for each run: E = 3 MeV, S = 100%, I = 3.8, 9.6, 19.1, 38.5, and 46.2 mA, respectively. After irradiation, the plastic bags with the alanine pellets were collected in a catch basket that would separate the sealed bags from the water. Some of the bags leaked water when they passed through the water pump that removed the irradiated water from the system. The bags that did not show water leaks were used to measure the dose. Dose measurements were then related to the temperature measurements from the thermocouples.

Finally, the dose in the sludge was determined from the temperature measurements with the thermocouples, after correcting for the dose measured by the alanine pellets.

3.4. Dose measurements: results and discussion

As mentioned earlier, the dose absorbed by the sludge was determined from temperature measurements in the sludge after a calibration with alanine pellets was performed. **Figure 3** shows the result of the dose calibration of the pellets when irradiated with alanine films in the cart conveyor system of the NEO Beam facility.

As stated in the experimental section, the pellets were calibrated using alanine films calibrated at Risø National Laboratory and then used to calibrate the in-house Bruker eScan spectrometer that measured the doses. After this, the calibrated pellets were used to determine the dose in the experimental setup to irradiate the sludge with the electron accelerator using different beam currents. Thus, the graph in **Figure 4** shows the dose recorded by the pellets run through the irradiation dispositive using water at different beam currents of the electron accelerator.

http://dx.doi.org/10.5772/62705

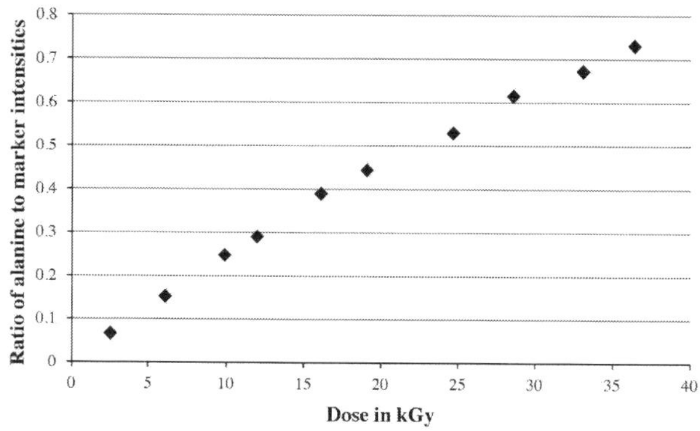

Figure 3. Calibration curve for the alanine pellets used to measure dose in this experiment. The dose was measured by alanine films and the response of the pellets as the ratio of the ESR intensity of the alanine to the internal marker of the pellet holder. The eScan instrument performed a trendline analysis on the experimental data obtaining a 4° polynomial as the best fit to the experimental data with a standard error of 0.0076 and an $R^2 = 0.9989$.

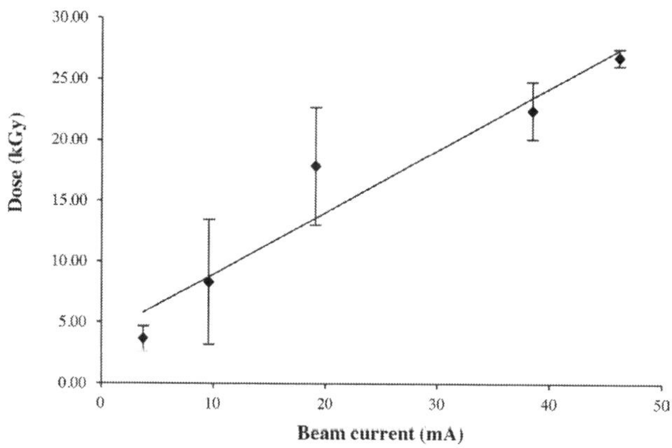

Figure 4. Doses of electron beam irradiation in water. Doses were measured by alanine pellets as a function of the electron beam current of the accelerator. Water was running in the system at a rate of 50 gpm, and the electron energy was 3.0 MeV [7].

At the same time, the increase in temperature of the water running through the sludge delivery system at constant flow rate of 50 gpm and different beam currents was recorded and compared with the dose given by the alanine pellets. This relationship was later used to determine the dose absorbed by the sludge when irradiated with the electron beam.

The flow rate during the irradiation of the sludge sample was 30 gpm instead of the 50 gpm originally selected for this experiment. In order to keep the doses within the interval selected for this experiment, it was decided to run the experiment at a reduced level of electron beam currents to compensate for this effect. The dose was determined then from temperature increase of the sludge by measuring the temperatures at the input and exit ports of the irradiation setup. **Table 2** presents results of the temperature increments and dose measurements as a function of the beam currents for the sludge sample running through the delivery system.

Beam current (mA)	Temperature increase (°c)	Dose (kGy)
2.3	0.6	2.7
5.8	1.6	6.7
11.5	3.1	13.2
23.0	6.1	25.7
27.6	7.3	30.7

Table 2. Irradiation conditions used to achieve targeted doses. Sludge samples were flowing at a rate of 30 gpm [7].

4. Elimination of potential pathogenic bacteria in municipal sludge

4.1. Sampling: experimental procedure

Sample collection, transport, and storage are crucial when studying the effect of electron beam irradiation on microbial population found in municipal sewage sludge. In these experiments, sewage sludge samples were collected in two separate batches. First batch contains pretreated municipal sewage sludge or influent samples, and a second batch is made of municipal sewage sludge treated with electron beam irradiation or effluent samples. In the case reported here, since the sludge was treated with different doses of electron beam irradiation, samples were collected prior to (influent) and after (effluent) irradiation of sludge at each dose. Several 100 ml influent and effluent samples of sewage sludge were harvested in sterile-caped plastic vials for bacterial count and survival. Each sample was then placed on ice immediately after collection and transported in an isotherm ice container (a cooler) from the electron beam irradiation facility to the microbiology laboratory for microbial analysis. For accurate observation of the direct effect of electron beam irradiation on bacterial population, samples should be analyzed as soon as possible after treatment.

4.2. Sample analysis using membrane filtration method: experimental procedure

Each sample was thoroughly mixed, and serial dilutions were performed in 1× phosphate-buffered saline. Influent samples were diluted up to 10^{-8}, while effluent samples were diluted up to 10^{-6}. Diluted samples were filtered using disposable filter funnels. For filtration, 10 ml of the diluted sample was transferred with a sterile pipette into the middle of a sterile 45 mm (diameter) and 0.45 μm (pore size) gridded membrane filter. After filtration, the filter was washed with three volumes of 1× phosphate-buffered saline. The filter was then removed and transferred on a 50-mm (diameter)-padded Petri dish plate containing 2 ml of culture medium for total heterotrophic bacterial (THB), total coliform (TC), and fecal coliform (FC) counts. In order to perform THB counts, mHPC Heterotrophic medium was used and for TC counts, mEndo medium was used, while m-FC medium supplemented with Rosolic acid was used for FC counts. Plates were placed in plastic bags containing moistened paper towels and transferred in an incubator. Heterotrophic plates were placed in an incubator for 48 h at 35 ± 0.5°C, and TC plates were incubated for 22–14 h at 35 ± 0.5°C, while FC were incubated at 44.5 ± 0.2°C. Known positive and negative controls were used in order to verify accuracy of analytical procedures for identification and counts of heterotrophic, TCs, and FCs. Thus, for TC media, *Escherichia coli* and *Enterobacter aerogenes* were used as positive controls, while *Staphylococcus aureus* and *Pseudomonas aeruginosa* were used as negative controls. The positive control for the FC media was *E. coli*, and the negative control was *E. aerogenes*. Prior to testing, test organisms were grown in tryptone soy broth and incubated overnight at 37°C. After growth, these cultures were treated according to the procedure used for sewage sludge samples. Only dilution plates with a density of 20–100 colonies were counted. FC colonies appeared as different shades of blue, while other non-FC colonies appeared as gray or cream-colored. Bacterial counts were performed in triplicate to verify the reproducibility of results for total heterotrophic counts, TC counts, and FC counts.

4.3. Results and discussion

Bacterial counts before and after irradiation were performed with the electron beam at doses 2.7, 6.7, 13.2, 25.7, and 30.7 kGy. The counts were done specifically for THB, TC, and FC. **Figure 5** shows the effect of electron beam irradiation on bacterial survival in municipal sewage sludge after treatment. It appears that THB, TC, and FC counts decreased in a dose-dependent manner. This decrease in bacterial population is directly associated with the ionizing effect of electron beam irradiation that damages bacterial DNA and biomembranes, and the production of reactive oxygen species which also damage cell components. A similar observation was recently made by Cao and Wang [8] when they treated municipal sludge with electron beam irradiation. However, these authors did not count specific types of bacteria.

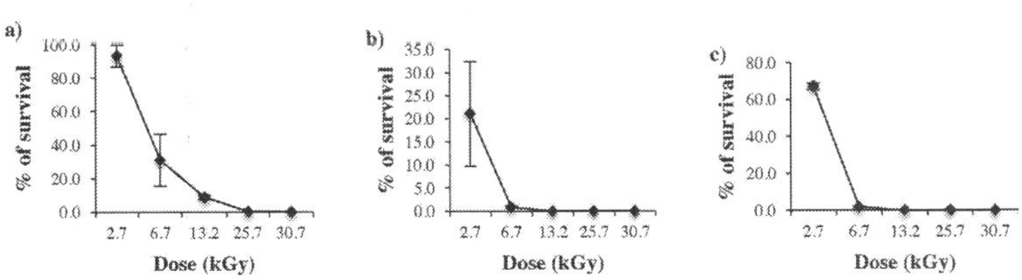

Figure 5. Effect of electron beam irradiation on bacterial survival in municipal sewage sludge samples. (a) Survival of total heterotrophic bacteria, (b) survival of total coliforms, and (c) survival of fecal coliforms [7].

Looking more into details, it was shown that when irradiating sludge with electron beam, a dose of 2.7 kGy, 93.3 ± 8.5% THB survived the treatment, while only 21.1 ± 11.4% of TC and 67.2 ± 1.8% of FC survived at the same irradiation dose. At a dose of 6.7 kGy, while 31 ± 15%

of THB survived the treatment, only 0.85 ± 0.23% and 1.85 ± 0.65% of the initial populations of TC and FC survived, respectively. At doses of 13.2 kGy and above, neither TC bacteria nor FC were detected. Nevertheless, at a 13.2 kGy irradiation dose, 8.9 ± 1.3% of THB from the initial population survived the treatment. At a dose of 25.7 kGy and above, no significant THB from the initial population were left in treated sewage sludge samples [7]. **Table 3** summarizes bacterial counts per gram of sludge dry weight at different electron beam doses. From these results, D_{10}-values were determined as 8.94, 3.16, and 3.17 kGy for THB, TC, and FC respectively. D_{10}-values are defined as doses necessary to kill 90% of the bacterial populations in the sample for irradiation conditions applied, or the dose needed to reduce the bacterial population by a factor of 10. A close look at **Table 2** shows that dose 6.7 kGy reduces the FC counts to 180 colony forming unit (CFU) per gram of sludge dry weight, a count that is within the Environmental Protection Agency (EPA) norm to classify such treated municipal sewage sludge as class A sludge utilizable for land application in agriculture [9]. However, from the D_{10}-value determined for FC, based on initial population of FC in influent samples, the dose required to convert this sludge to class A was estimated to be 4.5 kGy. Although no previous work similar to this one is known to perform a comparison with our estimated D_{10}-value, nevertheless, water-based and surface membrane *Bacillus* spore killing D_{10}-values were reported to be 1.3 and 1.53 kGy, respectively [10, 11]. These values are about twice lower than the 3.17 kGy determined in our case. This difference could be attributed to the presence of a large amount of organic and inorganic materials that make our sample relatively thick and slightly viscous compared to water and a surface membrane.

Dose (kGy)	Sluge dry weight (gram) percent	Total heterotrophic bacteria	Total coliforms	Fecal coliforms
			Counts (CFU) per gram of dry weight	
0.0	15.00%	1.4×10^6	1.7×10^5	2.0×10^4
2.7	25.75%	8.6×10^5	8.2×10^4	1.5×10^4
6.7	20.46%	3.2×10^5	9.3×10^2	1.8×10^2
13.2	12.29%	4.5×10^4	0.0	0.0
25.7	3.67%	1.3×10^4	0.0	0.0
30.7	3.25%	6.1×10^2	0.0	0.0

Table 3. Bacterial counts in sludge samples at different irradiation doses [7].

5. Elimination of *Ascaris* ova in municipal sludge

5.1. Sampling: experimental procedure

Compared to bacterial analyses of influent and effluent sewage sludge samples, *Ascaris* ova analyses required much larger volumes of biosolids (sewage sludge). Thus, several 1 l samples

http://dx.doi.org/10.5772/62705

of sludge were collected in sterile glass bottles. Some samples were obtained as influent samples, while others were obtained as effluent samples, transferred on ice immediately after collection, and transported to the microbiology laboratory for prompt *Ascaris* ova counts.

5.2. Sample analysis: experimental procedure

From each 1 l sample, 500 ml of well-mixed sludge was transferred in a blender, then 200 ml of sterile water was added, and the mixture was blended for 1 min at high speed. The blended mixture was transferred to a 1-l tall beaker to which 1% 7× detergent was added in order to reach 900 ml final volume. The same procedure was repeated for the second half the sludge sample, and the homogenized mixtures were combined and allowed to settle overnight in a cold (4°C) room or in a refrigerator. At this stage, some floating materials may be observed; therefore, stirring occasionally the mixture with a wooden applicator has shown to help settle the material. The supernatant was discarded by vacuum aspirating it to right above the layer of biosolids. The settled sediments were then transferred into a blender to which 500 ml of sterile water, blended again for 1 min at high speed, and transferred to a beaker. The blender was rinsed, and 1% 7× detergent was added to reach 900 ml final volume. Samples were allowed to settle for 2 h at 4°C after which the supernatant was discarded by vacuum aspirating it to right above the layer of biosolids. The biosolids were resuspended into 300 ml of 1% 7× detergent and stirred for 5 min using a magnetic stirrer. Homogenized sample was then strained through a 50 mesh (300 μm) sieve placed in a funnel over a beaker. Samples were washed through the sieve with a spray of 1% 7× detergent from a spray bottle. The sample volume in the beaker was adjusted to 900 ml by adding the necessary amount of 1% 7× detergent and allowed to settle for 2 h at 4°C. The supernatant was discarded using a vacuum, while the sediments were mixed and equally distributed in 50-ml centrifuge sterile tubes. In each tube, the sample volumes were adjusted to 50 ml with sterile water and centrifuged for 10 min at 1000×g. The supernatant was then discarded, and the pellet (biosolids) that should not exceed 5 ml was resuspend in 10 ml of $MgSO_4$ (specific gravity 1.2). Each tube was vortexed for 2 min, and more $MgSO_4$ was added to each tube to reach a volume of 50 ml. The tubes were then centrifuged for 10 min at 1000×g. The top 25–35 ml of supernatant of each tube was poured through a 400 mesh (38 μm) sieve supported in a funnel over a beaker. Biosolids retained on the sieve were washed, rinsed, and collected into a 100 ml beaker. The suspension of biosolids was then transferred into 15 ml centrifuge tubes. Tubes were centrifuged for 3 min at 800×g, and supernatants were discarded. If the previous step generated more than one tube for one initial sample, the sediments should be transferred into one single 15 ml tube and the centrifugation step repeated. Finally, after discarding the supernatant, the biosolids were resuspended in 4 ml 0.1 N H_2SO_4. The vials were incubated at 26°C for 3 weeks. After 24 days of incubation, when the majority of the controls were fully embryonated, samples were ready to be examined microscopically (10×) using a Sedgwick Rafter cell to enumerate the detected ova. Ova were classified as either nonviable (unembryonated) or viable (embryonated to the first, second, or third larval stage, those with the potential to become adult *Ascaris*). The percent moisture of the sample was determined by analyzing a separate portion of the sample, so that the final calculation of ova per gram dry weight could be determined. This was done by measuring the weight of the sludge samples before and after incubating at 45°C for 4 days, until dry by observation. Categories of ova per 4 grams per weight were calculated in the following manner:

Ova/g dry wt = (NO) × (CV) × (FV) / (SP) × (TS)

where NO = no ova, CV = chamber volume (=1 ml), FV = final volume in ml, SP = sample processed in ml or g, TS = % total solids.

5.3. Results and discussion

In order to determine *Ascaris* ova viability per four grams of sludge dry weight, the average dry weight of untreated sewage sludge samples was first determined. Hence, the dry weight of untreated sewage sludge samples was 15 ± 3% total solids. On average, untreated sewage sludge contained 312 ± 24 *Ascaris* ova per four grams of dry weight. **Figure 6** shows percentages of *Ascaris* ova viability after treatment of sewage sludge samples with electron-beam doses of 2.7, 6.7, 13.2, and 25.7 kGy.

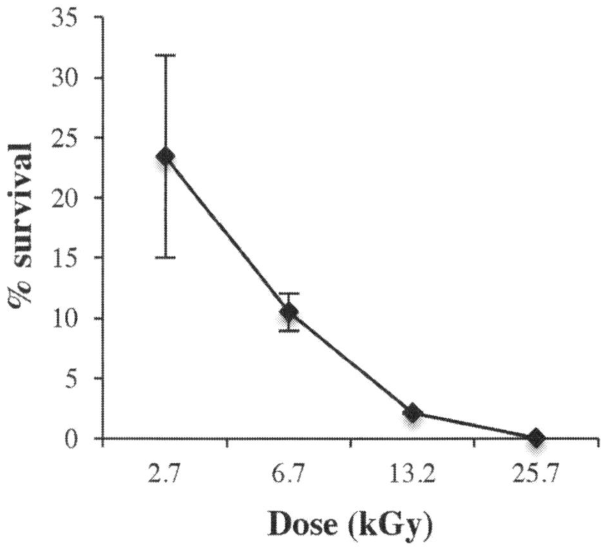

Figure 6. Effect of electron beam irradiation on *Ascaris* ova survival in municipal sewage sludge. Percentages of *Ascaris* ova survival after irradiation of samples [7].

Similar to bacterial counts, the results indicate that the viability of *Ascaris* ova decreases in a dose-dependent manner [7] with a D_{10}-value of 7.93 kGy. At dose 2.7 kGy, 23 ± 8% (72 ± 24 *Ascaris* ova per four grams of dry weight) of viable ova survived the treatment, while only 11 ± 1.6% (34 ± 4 *Ascaris* ova per four grams of dry weight) ova survived at dose 6.7 kGy. At dose 13.2 kGy, the survival rate dropped to 2 ± 0.03% (6 ± 1 *Ascaris* ova per four grams of dry weight). No *Ascaris* ova were detected in sewage sludge samples irradiated at 25.7 kGy. However from our counts, we estimated the electron-beam dose 14.5 kGy to be necessary to obtain a sewage sludge containing less than one *Ascaris* ova per four grams of sludge dry weight, meaning that the dose of 25.7 kGy applied during our experiments was high enough to achieve a class A sludge. Indeed, according to EPA standards, to be considered class A, sewage sludge must contain less than one *Ascaris* ovum per four grams of sludge dry weight [9].

6. Economic benefits of using electron beam irradiation for the treatment of municipal sewage sludge

An important aspect in the implementation of a new technology such as an electron accelerator in a wastewater treatment plant is to anticipate its impact on the operation costs of the facility and on the environment. In a recent investigation, the number of kWh used during the irradiation process of the sewage sludge was considered. Data were obtained from the electrical supply company delivering power to the NEO Beam accelerator facility (Toledo Edison, Toledo, Ohio) on the day of the experiment. Energy consumed by the electron accelerator recorded in **Figure 7** shows a relatively stable plateau in the power consumed at the facility prior to sample irradiation. The dosimetry calibration of the irradiation setup started at 9:30 am with an increase of the beam current from 0 to 46.2 mA in equal time intervals. From 9:30 am until the end of the irradiation procedure, we observed a constant increase in power consumption. However, the graph only shows electricity consumption from 9:30 am until 10:00 am.

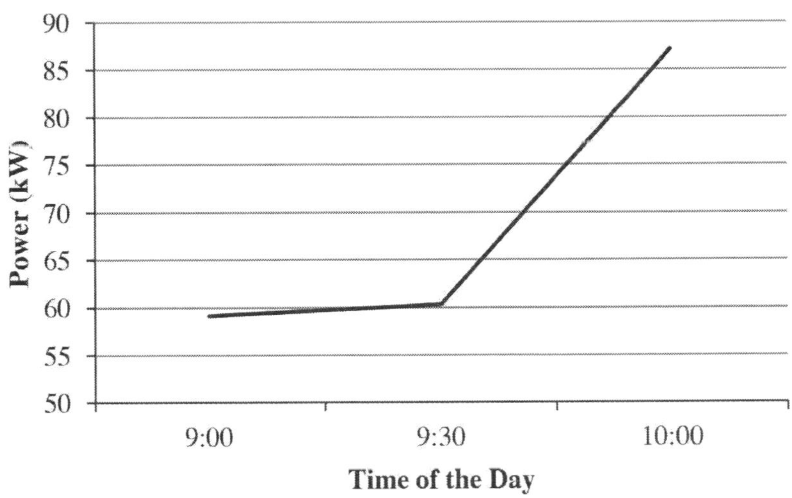

Figure 7. Average power consumed at the NEO Beam electron facility on the day of the experiment (information provided by Toledo Edison, Ohio, USA) [7].

The power consumed at a specific irradiation dose was obtained in terms of the beam current used, and the cost was determined to be $0.115/kWh. Therefore, at irradiation doses 6.7 and 25.7 kGy, the costs were $1.10 and $1.26 per m^3 of sludge, respectively. This represents only 15% of increase of the cost when quadrupling the dose of irradiation to achieve the required *Ascaris* ova reduction to a class A biosolids, suggesting that when selecting a higher dose of irradiation of sludge, the increase in cost due to the use of higher beam currents should not be a concern [7]. Similar results were obtained by other authors who estimated the cost of using electron beam and gamma radiation for the disinfection of sludge. However, experiments developed were performed under different considerations. Indeed, a team in Florida reported a cost of $2.50 per 1000 gallons of sludge for a 1.5 MeV electron irradiation facility running at 160 gallons per minute, while another group compared gamma and electron beam irradiations for a sample of activated sludge and obtained treatment costs of $4.20/$m^3$ for gamma irradiation and $2.10/$m^3$ for electron beam irradiation, which are lower compared with $4.85–$5.19 when using conventional technology at the Central District Wastewater Treatment Facility in Miami Dade County [12]. Furthermore, a comparison was made between irradiation at dose 6 kGy and incineration of sludge samples and showed a cost of $60.87/$m^3$ for this latter compared with $3.12/$m^3$ when using gamma radiation. In both instances, gamma and electron beam irradiations prove to be more economic than incineration [13]. Taken altogether, these observations show that electron beam irradiation of sludge is less energy consuming, with shorter processing times, and a more environmental friendly technology compared to methods such as incineration.

7. Conclusion

Electron beam irradiation technology is able to decrease microbial populations in a dose-dependent manner. In the experiments described in this chapter, it has been estimated that 4.5 kGy of irradiation is sufficient to reduce bacterial populations to safe levels for agricultural use. However, a dose of 14.5 kGy is required to eliminate risks of infection by helminths. Altogether, these observations suggest that irradiation of municipal sludge with electron beam requires at least a dose of 14.5 kGy to eliminate risks of microbial infection. Furthermore, electron beam technology is more cost-effective and less time-consuming than incineration in order to achieve a class A sludge according to EPA standards.

Author details

Jean Engohang-Ndong[1*] and Roberto M. Uribe[2]

*Address all correspondence to: jengohan@kent.edu

1 Kent State University at Tuscarawas,, New Philadelphia, Ohio, USA

2 Kent State University, Kent, Ohio, USA

http://dx.doi.org/10.5772/62705

References

[1] Lessel T. and Suess A., *Ten year experience in operation of a sewage sludge treatment plant using gamma irradiation*. Radiation Physics and Chemistry (1977), 1984. 24(1): pp. 3–16. doi:10.1155/2012/579593

[2] Borrely S.I., et al., *Radiation processing of sewage and sludge. A review*. Progress in Nuclear Energy, 1998. 33(1–2): pp. 3–21. doi:10.1016/S0149-1970(97)87287-3

[3] Reinthaler F.F., et al., *Resistance patterns of Escherichia coli isolated from sewage sludge in comparison with those isolated from human patients in 2000 and 2009*. Journal of Water and Health, 2013. 11(1): pp. 13–20. doi:10.2166/wh.2012.207

[4] Young S., Juhl A., and O'Mullan G.D., *Antibiotic-resistant bacteria in the Hudson River Estuary linked to wet weather sewage contamination*. Journal of Water and Health, 2013. 11(2): pp. 297–310. doi:10.2166/wh.2013.131

[5] Zhou B.W., et al., *Effect of microwave irradiation on cellular disintegration of Gram positive and negative cells*. Applied Microbiology and Biotechnology, 2010. 87(2): pp. 765–70. doi: 10.1007/s00253-010-2574-7

[6] Pikaev A.K. and WoodsR.J., ed. *Applied Radiation Chemistry: Radiation Processing*. New York, United States of America, 1993, Wiley. pp. 552.

[7] Engohang-Ndong J., et al., *Effect of electron beam irradiation on bacterial and Ascaris ova loads and volatile organic compounds in municipal sewage sludge*. Radiation Physics and Chemistry, 2015. 112: pp. 6–12. doi:10.1016/j.radphyschem.2015.02.013

[8] Cao C. and Wang M., *Treatment of municipal sludge by electron beam irradiation*. Nuclear Science and Techniques, 2012. 23(1): pp. 29–33. doi:10.13538/j.1001-8042/nst.23.29-33

[9] US EPA. *40 CFR Part 503—Standards for the Use or Disposal of Sewage Sludge*. 2007. Environmental Protection Agency. Washington, D.C.

[10] Fiester S.E., et al., *Electron beam irradiation dose dependently damages the bacillus spore coat and spore membrane*. International Journal of Microbiology, 2012. 2012: p. 579593. doi: 10.1155/2012/579593

[11] Helfinstine S.L., et al., *Inactivation of Bacillus endospores in envelopes by electron beam irradiation*. Applied and Environmental Microbiology, 2005. 71(11): pp. 7029–32. doi: 10.1128/AEM.71.11.7029-7032.2005

[12] Meeroff D., et al., *Radiation-assisted process enhancement in wastewater treatment*. Journal of Environmental Engineering, 2004. 130(2): pp. 155–66. doi:10.1061/(ASCE)0733-9372(2004)130:2(155)

[13] Swinwood J.F. and Kotler J., *Sewage sludge pasteurization by gamma radiation: Financial viability case studies*. Radiation Physics and Chemistry, 1990. 35(1–3): pp. 445–50. doi: 10.1016/1359-0197(90)90136-6

Section 3

Polymeric Materials

Chapter 10

Radiation Effects in Polyamides

Mária Porubská

Additional information is available at the end of the chapter

http://dx.doi.org/10.5772/62464

Abstract

Polyamides (PAs) are largely used either as engineering materials in virgin form or as composites and a component of polymer blends. Various processes have been used to modify some properties of polymers to improve their utility. For this purpose, radiation technologies present clear, one-step procedures and offer improvement to the performance of PA materials. Irradiation by accelerated electron beams, γ-rays, and accelerated protons is applied on PAs, particularly PA-6, as well as PA composites. Variations of important characteristics, such as chemical structure, supermolecular structure, mechanical properties, thermal resistance, water absorption, and other parameters, are analyzed involving results obtained by other authors. The application of irradiation on incompatible polymer blends involving PA is presented as well. The selection of radiation treatment of PAs has to be considered to obtain optimal results.

Keywords: Polyamide, Properties, Electron beam, γ-rays, Proton beam

1. Introduction

Polyamides (PAs) are linear semicrystalline polymers containing amide groups -CO-NH- in the chains. The nomenclature applied for PA uses numbers to describe the number of carbons between acid and amine function groups including the carbon of the carboxylic acid. Two types of polyreactions are used for their preparation. The first, polyaddition, is based on a lactam as the starting reactant. After opening the lactam ring breaking peptide bond, the created end groups -C(O)- and -N(H)- are coupled with -N(H)- and -C(O)- groups originating from other cleaved molecules of the lactam. This is the case of PA-6 when ε-caprolactam is used.

$$n[(CH_2)_5 - CO - NH] \xrightarrow{H_2O} [-HN - (CH_2)_5 - CO -]_n \quad (1)$$

The same result can be reached using the corresponding amino acid. However, the cleavage of water molecules is required, which is the characteristic for a polycondensation mechanism. Therefore, polycondensation is the second polyreaction used to prepare PAs. The most commonly applied reaction is the polycondensation of dicarboxylic acids with diamines (e.g., adipic acid with hexamethylene diamine giving PA-66).

$$H_2N - (CH_2)_6 - NH_2 + HOOC - (CH_2)_4 - COOH \rightarrow [-HN - (CH_2)_6 - NH - OC - (CH_2)_4 - CO-]_n \quad (2)$$

In such a case, the structure of the PA corresponds to the repetition of structure unit consisting of one of each monomer, so that they alternate in the chain unlike for a PA chain synthesized from a monomer as single starting reactant. Many demanding applications require a careful control of the synthesis as well as processing conditions considering the resulting molecular mass.

Variation of the starting reactants enables a large scale of PAs differing in properties from hard and tough PA to soft and flexible to be acquired. Depending on the type, PAs absorb different amounts of moisture, which affect the mechanical as well as dimensional characteristics. In general, PAs are characterized by high rigidity, hardness, abrasion resistance, thermal stability, good sliding properties, stress cracking resistance, barrier properties against oxygen, smells, and oils. The disadvantages of PA types involve a weak stability in the presence of UV radiation, oxidizing agents, strong acids, and bases, high shrinkage in molten section, degradation in electrical and mechanical properties due to high moisture absorptivity, and high notch sensitivity. The various particular properties of PAs enable various PAs to be processed into various items. They are most frequently used in the textile industry, electrotechnics, and automotive industry as engineering plastics either virgin or composites with modifying fillers. Depending on the type of the product application, the manufacturers can adjust the properties of initial materials to some extent. Besides physical modification, involving admixture of proper additives, the chemical modification of PA can also be applied. According to PA type, a high melting temperature within 170°C up to 290°C occurs. Therefore, the homogenous admixture of intended agents into the melt is possible only for sufficiently thermally stable substances. When the chemical modification of PA requires the generation of radicals (e.g., grafting various functional groups or crosslinking initiated by organic peroxides), then this modification cannot be always carried out in melt due to a weak thermal stability of some substances participating in the process. In particular, peroxides will be thermally destroyed well before the PA is melted. This handicap can be overcome using a proper radiation technology enabling such a modifying process in the solid state. Presenting a technology of one step, the radiation technologies have been largely applied during the last decades and have brought a wider spectrum of properties or design variations.

Because PA-6 and PA-66 comprise most of the world's market (~80%) [1], the effects of radiation modification involving them have been frequently investigated. Concerning some conflicting reports on electron beam effect on polymers found in scientific sources, in general, it could be said that variations in gel content in PAs irradiated with identical doses may not be the same. The reason is that the competing reactions, chain scission and recombination, can be affected by several factors, among others polymer type with its molecular characteristics. However, under irradiation, the most decisive factors are time and dose rate. The presence of oxygen provokes some side reactions and the consequence is a complex process. The oxygen effect can be hardly eliminated completely even if the irradiation is carried out in an inert atmosphere. Within the simultaneous process of recombination and scission macroradicals, the most important factor is the stationary concentration of the macroradicals.

Therefore, although the doses are the same, the results may differ if different electron beam sources and PA types are used. From this aspect, every comparison can be of framing character only. The same can be said regarding γ or proton exposure.

2. Electron beam irradiation of PAs

The effect of electron beam on PAs depends on several factors, such as installed parameters of the accelerator, absorbed dose, rate of the dose, environment of irradiation, geometry of irradiated object, temperature, and postradiation treatment. Concerning the environment, air is used mostly and an inert atmosphere is less common but more thrifty towards the polymer. Under irradiation of polymers, free radicals are created. Subsequently, two main actions take place, that is, recombination of the radicals resulting in crosslinking and oxidation resulting in polymer degradation. Both processes run in parallel [2], competing mutually, and the results depend on concrete conditions.

What exactly occurs when a PA is irradiated with an electron beam? An interaction occurs primarily in hydrocarbon sequences (R). The complete action involves three basic steps: initiation (generation of free radicals), propagation, and termination. In general, the following scheme can describe such processes.

Initiation:

$$R \rightsquigarrow R^{*} \rightarrow R^{\circ} + H^{\circ} \tag{3}$$

Propagation:

$$R^{\circ} + O_2 \rightarrow ROO^{\circ} \tag{4}$$

$$ROO^{\circ} + RH \rightarrow R^{\circ} + ROOH \tag{5}$$

$$ROOH \xrightarrow{\Delta} RO^{\circ} + HO^{\circ} \tag{6}$$

$$ROOR \xrightarrow{\Delta} 2RO^{\circ} \tag{7}$$

The generated radicals can take part in the destruction of the next hydrocarbon sequences, and hydroperoxides start the formation of new radicals and intermediates. In the presence of oxygen, the creation of carbonyl and aldehyde groups is significant as well.

Termination:

$$ROO^{\circ} + H^{\circ} \rightarrow ROOH \tag{8}$$

$$ROO^{\circ} + R^{\circ} \rightarrow ROOR \tag{9}$$

In the case of PA-6, the splitting of H^0 from ethylene group occurs most probably in the vicinity of the –NH–CO– sequence:

$$-NH-\left[CO-(CH_2)_4-CH_2-NH\right]_n - \rightsquigarrow - NH-\left[CO-(CH_2)_4-C^{\circ}H-NH\right]_n - + H^{\circ} \tag{10}$$

However, abstraction of hydrogen can occur in any other place within the hydrocarbon sequence as well.

Propagation takes place according to the above-mentioned steps. Termination of the radicals can occur as follows:

a) Crosslinking:

$$\begin{aligned} 2-NH-[CO-(CH_2)_4-C^{\circ}H-NH]_n- \rightarrow -NH-[CO-(CH_2)_4-&CH-NH]_n- \\ &| \\ -NH-[CO-(CH_2)_4-&CH-NH]_n- \end{aligned} \tag{11}$$

b) Disproportionation:

$$\begin{aligned} &2-NH-[CO-(CH_2)_4-C^{\circ}H-NH]_n- \rightarrow \\ &\rightarrow NH-[CO-(CH_2)_5-NH]_x-+-NH-[CO-(CH_2)_4-CH=N]_{n-x}- \end{aligned} \tag{12}$$

c) Oxidation and following degradation:

$$-NH-[CO-(CH_2)_4-C^{\circ}H-NH]_n- + \frac{1}{2}O_2 \rightarrow -NH-[CO-(CH_2)_4-CH(O^{\circ})-NH]_n- \quad (13)$$

$$\begin{aligned} &-NH-[CO-(CH_2)_4-CH(O^{\circ})-NH]_n- + H^{\circ} \rightarrow \\ &\rightarrow -NH-[CO-(CH_2)_4-CH{=}O]_y + NH_2-[CO-(CH_2)_5-NH-]_{(n-y)} \end{aligned} \quad (14)$$

Besides the above-mentioned irradiation conditions, the formation of intermediates and final products can be affected by the polymer structure, so that the products can be variable because of some side reactions.

2.1. Variations in chemical structure

In general, the modification of semicrystalline polymers by energetic radiation in the solid phase below the melting temperature of the crystallites is characterized by changes proceeding preferentially or, in many cases, almost exclusively in the amorphous phase [3,4]. Within the pre-crosslinking phase, there is the branching in the polymer [5] as revealed by the solution viscosity measurement for PA-6 [6,7] and glass fiber-reinforced PA-6 (PA-6/GF) with 30% GF [7], indicating an increase in the molecular weight due to the recombination of macroradicals at the polymer chain centers (**Figure 1**). Consequently also, viscosity increases in normal cases.

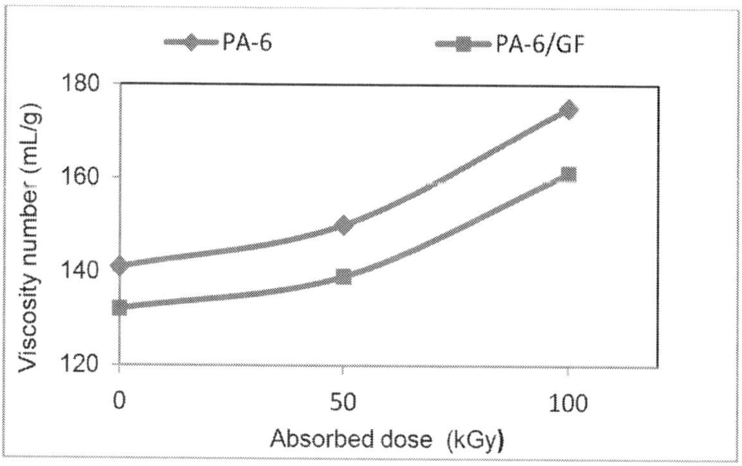

Figure 1. Dependence of viscosity number for virgin PA-6 and PA-6 in PA-6/GF on electron beam absorbed dose. Adapted from Porubská et al. [7].

A measurable portion of insoluble gel occurs at approximately 200 kGy (gel point) when the irradiation is conducted in air (**Figure 2**).

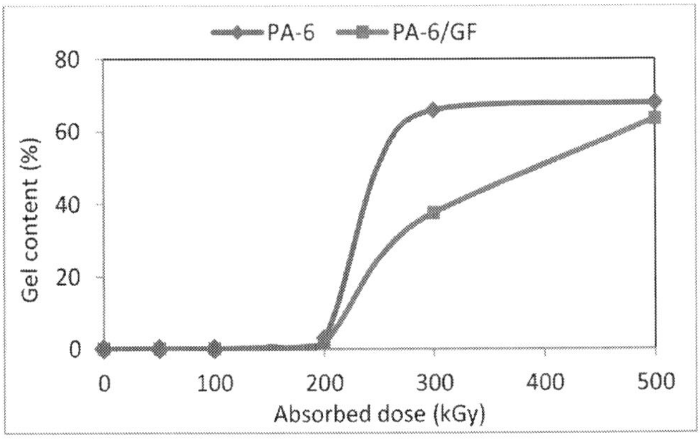

Figure 2. Dependence of gel content on absorbed electron beam dose for virgin PA-6 and PA-6/GF (30%) composite [7]. With permission of Elsevier.

This gel point is confirmed in several studies [7–10], although a low gel content for PA-6 was measured under the 150 kGy dose as well [11,12]. Some specific results in experiments with electron beam irradiation of PA-6 and PA-6,6 tyre cords and calendered fabrics are published by Aytaç et al. [13] when, contrariwise, some decrease in limiting viscosity number with increasing dose was observed within doses of 0 to 75 kGy. The measured data are explained by sufficient time for oxygen to diffuse into the samples during the foregoing calendering as well as by a low dose rate (25 Gy/pass). Under those conditions, radiolytic and oxidative degradation occur. The following strength testing confirmed the objectivity of the results.

Irradiation temperature	p_0/q_0	G_s	G_c	G_s/G_c
RT	0.63	0.49	0.39	1.3
50°C	0.60	0.61	0.50	1.2
70°C	0.49	0.46	0.50	0.91
80°C	0.29	0.23	0.40	0.58

Table 1. Radiochemical G_s scission yield and G_c crosslinking yield of PA-6 irradiated at different temperatures [10]. With permission of Elsevier.

Besides dose, temperature at the irradiation site also plays a role. As study [10] demonstrated that the crosslinking effect grows with increasing dose and temperature. Corresponding experiments with PA-6 film irradiated over a range of 15 to 1200 kGy (and at a dose rate of 4.48 kGy/min) were carried out at different temperatures from room temperature (RT) to 80°C involving the glass transition temperature (T_g) of approximately 50°C. Although the PA-6 gel point for all samples was observed more or less at 200 kGy, the gel content increased with temperature and the highest value (75%) was observed for a temperature of 80°C and dose of

http://dx.doi.org/10.5772/62464

800 kGy. The crosslinking rates of PA-6 irradiated above T_g are higher than those samples irradiated at temperatures below T_g. The increasing tendency in crosslinking with increasing temperature is attributed to the enhanced mobility of the PA-6 molecules above T_g; therefore, the probability of radical recombination is higher. The radiochemical yields of crosslinking and degradation determined according to the Charlesby-Pinner equation [14] are given in **Table 1**.

The ratio p_0/q_0 [(ration of main chain fractures to chain units)/(units crosslinked to chain units)] for PA-6 decreases with the increase in radiation temperature of the samples. This is a consequence of higher mobility of the chains and the recombination of radicals generated by the main chain scission as well as by the release of hydrogen radicals due to electron beam energy [Eq. (3)], indicating that crosslinking is the predominant process at temperatures above T_g. In contrast, the scission prevails at RT, below T_g [10,11]. Compared to the G_s/G_c values reported for PA-66, the scission dominates more in PA-66 than in PA-6 [8,9,15,16]. This demonstrates that a small difference in PA structure can play a role.

The final amount of the gel in polymer is limited. Generally, in virgin PA, the gel portion grows dramatically beyond the gel point and, in a certain phase, becomes stabilized. The reason is that a balance between the scission and recombination occurs, as the increasing viscosity in the matrix due to crosslinks breaks the mobility of the radicals to recombine. Although the gel content remains the same, the decrease of the swelling [17] or molecular weight between crossbonds [10] with increasing doses indicates that the network becomes denser [7] as seen in **Table 2**.

Dose (kGy)	RT	50°C	70°C	80°C
300	41,000	31,500	31,900	40,300
400	30,800	23,600	23,900	30,250
600	20,500	15,700	15,900	20,100
800	15,400	11,800	11,900	15,100
1000	12,300	9500	9500	12,100

Table 2. Average molecular weight between crosslinks (Mc) in PA-6 samples irradiated at different temperatures and doses [10]. With permission of Elsevier.

Irradiation experiments were also conducted in an inert (nitrogen) atmosphere and also with the annealing what led to lower chain scission and increased crosslinking reaction [18].

The crosslinking agents for polymers, including PAs, are molecules that contain two or more double bonds per molecule, such as triallyl cyanurate (TAC), triallyl isocyanurate (TAIC), and trimethylolpropane trimethacrylate (TMPTMA). The energy of electron beam easily cleaves them into radicals formatting cross-bonds. Then, the crosslinking agent can increase the crosslinking rate and shift the gel point to a much lower absorbed dose [6] because of the higher efficiency of free radical production. The presence of a filler (e.g., GF reinforcement) in the PA matrix can also modify the crosslinking rate, even if contrarily [7] in comparison to the crosslinking agents. The filler increases the viscosity of the matrix, which leads to a retardation of the cross-bond formation when compared to virgin PA due to slower macroradical recombination, leaving more time for disproportionation.

Concerning gel formation, the application of electron beam irradiation on incompatible polymeric blends or layers in laminates to generate free radicals and the following interaction, the grafting, between the components to improve their compatibility, can vary from case to case. Gel point shifts depending on the character of the polymeric components as well as possible crosslinking agent addition are shown in several works involving PA-6 [19–21]. Rare data on electron beam effect on gel creation in PA-12 can be found for ethylene propylene diene monomer (EPDM)/PA-12/maleated EPDM blend and these do not show any PA-in origin gel up to 200 kGy [22].

It is worth mentioning that crosslinking becomes easier as the number of methylene groups between the amide groups increases [23,24]. This finding is based on the gel measurement on PA-6, PA-610, and PA-12 irradiated under the same conditions. This conclusion is understandable because, as mentioned above, the initiation starts through a hydrocarbon sequence. The longer the hydrocarbon sequence, the more probable the hydrogen abstraction occurs.

2.2. Thermal properties related to supermolecular structure

Differential scanning calorimetry (DSC) is a useful tool to examine the structural variations in irradiated polymeric materials. Two basic types of data can be obtained from DSC curves, melting or crystallization temperature (T_m or T_c) and melting or crystallization enthalpy (ΔH_m or ΔH_c). The former is connected with the size of the crystalline units, whereas the last corresponds to the crystalline portion. The larger the crystalline unit, the higher T_m observed. Similarly, the more the crystalline phase in the polymer, the higher the melting enthalpy measured. In general, for most polymers fully or partially crosslinkable by electron beam irradiation, the dependences of both melting temperature and melting enthalpy on the dose measured from the first DSC heating reflect immediate changes in the original crystalline structure occurring during irradiation. The data from the second melting demonstrates the overall changes particularly due to irradiation-induced crosslinking and partial degradation, both occurring during irradiation mainly in the amorphous phase. The original crystalline structure melts within first heating. The following cooling leads to the crystallization of the melt of the irradiated polymer, so that the effect is related to the overall degree of crosslinking, representing defects hindering the crystalline structure formation after melting during the first heating [2]. The kinetic parameters of processes linked to the heat exchange can be determined as well.

Because PAs are semicrystalline polymers, DSC is used often to characterize them. As mentioned previously, the crosslinking mainly occurs in the amorphous phase. The effects of dose on the T_m determined for virgin PA-6 in the first and second runs (melting T_{m1} and remelting T_{m2}) are displayed in **Figure 3**, whereas **Figure 4** shows the dependence of melting and remelting enthalpies. A monotonous decrease in the T_m is observed over the 0 to 500 kGy dose range and the second run (T_{m2}) is lower than the first run (T_{m1}) [7,16].

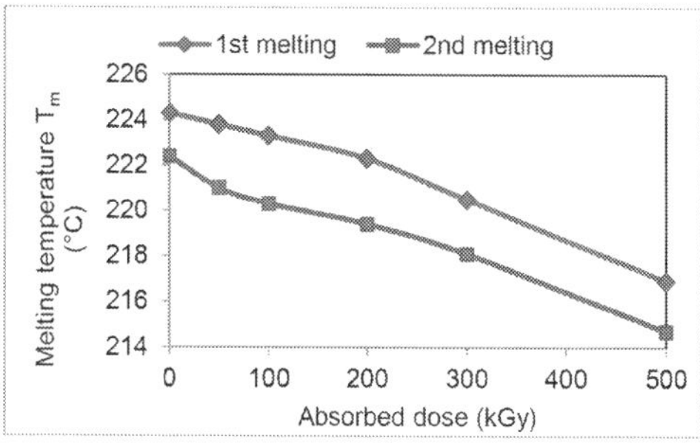

Figure 3. Variation of the first and second melting temperatures with absorbed electron beam dose for virgin PA-6 [7]. With permission of Elsevier.

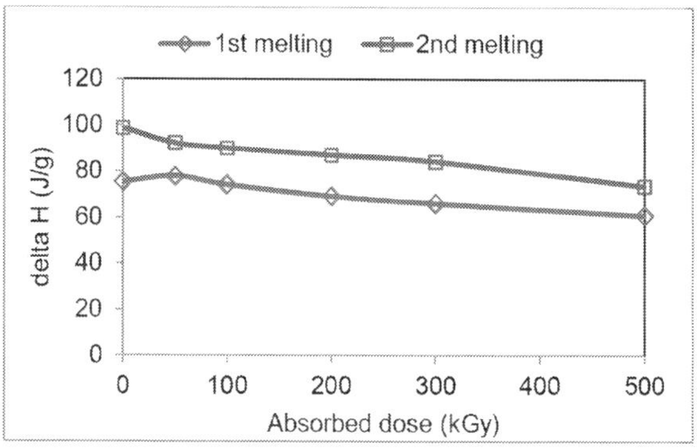

Figure 4. Variation of the first and second melting enthalpy with absorbed electron beam dose for virgin PA-6 [7]. With permission of Elsevier.

Such behavior is typical for the second run of irradiated crosslinkable polymers after melting and following the crystallization of a previously crosslinked material. The decrease of T_{m1} points to the thinning of the lamellae as a consequence of crystalline phase disruption within irradiation. The decrease in ΔH_1 indicates this fact, too. Consequently, it can be implicated that, during irradiation also, a considerable degradation occurs along with crosslinking. Similar facts are observed, irradiating PA-6 at temperatures above T_g with doses of 0 to 1200 kGy. T_m decreased from 224°C for 0 kGy up to 213°C for 1000 kGy. However, irradiation at RT leads to a decrease of T_m up to 600 kGy only with a subsequent mild increase [10]. One possible explanation may be the formation of larger crystallites due to a higher generation of smaller fragments from PA chains at higher doses and those being more mobile incorporate into existing crystallites more easy. This suggestion is in compliance with the finding that the samples irradiated above T_g showed a predominance of the crosslinking over scission (Section 2.1).

The effect of irradiation on crystallinity is linked with the values of the melting enthalpy directly proportionally, and some decrease is observed for virgin PA-6 with rising dose (**Figure 4**). The higher value of ΔH_{m2} than ΔH_{m1} for PA-6 (**Figure 4**) is a consequence of different cooling rates between cooling the testing specimens after preparation by injection molding (using a higher initial cooling rate from the processing temperature of ~290°C) and cooling of the sample after first heating in the DSC cell (applied cooling rate of 10°C/min). Therefore, a lower crystalline portion is the result for rapidly cooled materials. The irradiation effect on PA-6 is also observable on the shape of the DSC melting curves (**Figure 5**).

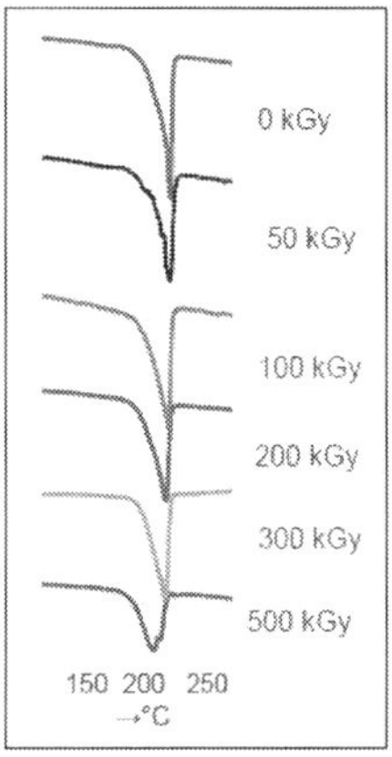

Figure 5. DSC melting endotherms of first melting (150–250°C) for PA-6 irradiated with electron beam.

The endotherms of irradiated samples become wider with increasing dose as a consequence of the progressive lamella thinning and amorphization that took place under electron beam irradiation. Additionally, the enlargement in the peaks indicates that the distribution of crystallite size becomes broader while the unit endotherm surface (ΔH_m) decreases.

PA-6 included in a multilayer film and irradiated in nitrogen applying a dose in the range of 0 to 150 kGy shows only a small total decrease in melting temperature at approximately 1°C and the crystallinity is diminished from 25.3% to 21.5% [18].

The influence of the crosslinking agent TAC on PA-6 irradiated with relatively low doses up to 100 kGy demonstrates a more pronounced effect on T_m and crystallinity than present in virgin PA-6. The decrease in both T_m and crystallinity for the doped PA is clearly observable already for the 40 kGy dose, whereas, for the virgin PA-6, the decrease is moderate, if any. The decrease in T_m seems to be dependent on TAC content (the more TAC, the lower the T_m), but the measure of the decrease in crystallinity is not affected by TAC content in the range of 1% to 3% [6].

The development of thermal characteristics with dose can somewhat vary when PA is part of a blend or composite. PAs are often used as reinforced composites or filled with various fillers, with GF being the most common reinforcing additive. There is not much information on the crosslinking of reinforced PAs (PA/GF) in the scientific literature. Concerning the changes in the melting temperature (**Figure 6**) and enthalpy of PA-6/GF (**Figure 7**) with rising dose, they are much less pronounced for PA-6/GF composite compared to virgin PA-6 (Figures **3** and **4**), indicating that the presence of GF partially eliminates the irradiation effects on PA. This conclusion arises from the dependences in **Figure 2** as well.

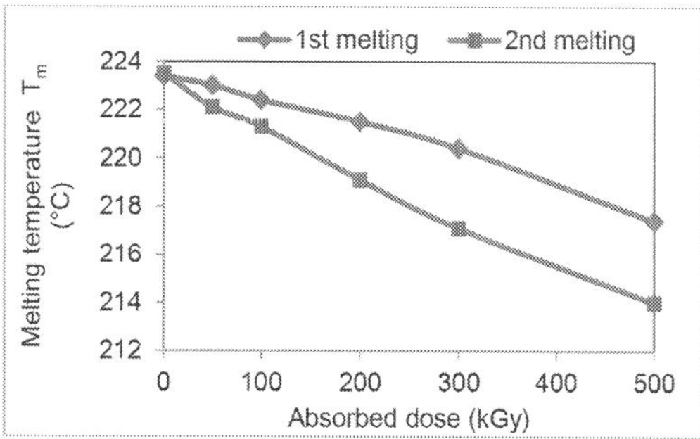

Figure 6. Variation of the first and second melting temperatures with absorbed electron beam dose for PA-6/GF(30%) composite [7]. With permission of Elsevier.

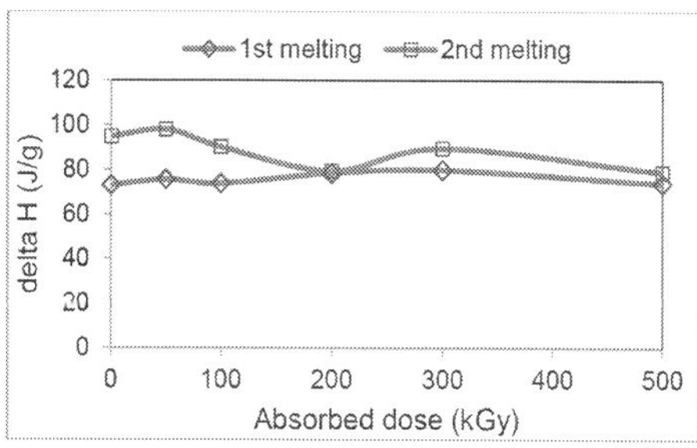

Figure 7. Variation of the first and second melting enthalpy with absorbed electron beam dose for PA-6/GF (30%) composite [7]. With permission of Elsevier.

The examination of (PP/PA-6+talc) composite with or without a compatibilizer or a crosslinking agent TAIC, applying a dose in range of 0 to 200 kGy [25], reveals a decrease in T_m for PA-6 component only when doses 50 and 100 kGy are used. A dose of 200 kGy provokes already a minor T_m growth and that is evident for the formulation containing TAIC. However, all T_ms are lower than the T_ms for the corresponding unexposed composite. Concerning the degree of PA-6 crystallinity, the percentage for (PP/PA-6+talc) formulation varies mildly being lower for the initial crystallinity; for the formulation involving the compatibilizer and TAIC, the decrease is definite.

The behavior of PA-66 is similar to PA-6 in principle. Injection-molded PA-66 samples irradiated with 200 and 500 kGy doses show the decrease in initial T_m regardless of the addition of TAC or the exposure at RT or 120°C. Nor does the water annealing affect this tendency, but the decrease for the samples with no TAC irradiated at RT is a little less when compared to the others. The same can be said about the crystallinity [15]. Thus, the dependences of the crystallinity on dose for PA-66 films [8] and for the injection-molded pieces are the same essentially regardless of the mode of sample preparation.

Data on PA-12 melting characteristics are found for EPDM/PA-12/maleated EPDM blend after being irradiated with 25 to 100 kGy doses [22]. Melting temperature does not show any change and the crystallinity is more or less also the same. However, such a narrow dose range does not allow the estimation of further development under higher doses.

2.3. Mechanical properties

The molecular and supermolecular structure of a polymer determines its properties. When any variation occurs in the structure, it should manifest itself in some changes of the properties. Electron beam activates significant structural changes, namely, scission and crosslinking as well as oxidative degradation if irradiated in air. Each particular property of a polymer is unequally sensitive towards the structural modification; therefore, it is important for examination to be focused exactly. Tensile properties, such as Young's modulus, strength at yield or break, elongation at yield or break, and flexural and impact parameters, reflect these variations well.

In general, when virgin PA is irradiated, the development of the tensile parameters follows the general framework that modulus and stress at yield are progressive, whereas stress at break and elongation are regressive. The scale of measured values is a question of the quality of the irradiated PA. The shape of stress-strain curves for the same PA remains similar irrespective of the absorbed dose [16].

The dimensional stability on load, the stiffness, is one important parameter for design engineers. In thermoplastics, the shape stability is affected by the crystallinity content to a great extent. Within the tensile properties, Young's modulus is the corresponding testing parameter for stiffness. As mentioned in Section 2.1, electron beam irradiation in the solid state results in crosslink formation primarily in the amorphous phase or at the crystal boundaries [3,4]. In addition, for thermoplastics, an observable increase in the strength parameters occurs when the gel content is at or more than 50 wt% [4]. Thus, unless the crystalline phase is substantially degraded, notable changes in the modulus will not occur as evident from the dependences of Young's modulus for virgin PA-6 and PA-6/GF composite (**Figure 8**).

In reality, regarding a large fluctuation in the standard deviation (brought on by GF dispersion within the PA matrix) for the composite, any variations of Young's modulus are statistically nonsignificant to be presented. Concerning virgin PA-6, the increase in the modulus is evident at 500 kGy dose when the gel content is more than 50%. In another study [11], a steady increase in the modulus of virgin PA-6 is measured within the dose range of 100 to 500 kGy. However, in this case, the gel point is measureable already at approximately 100 kGy and 50% growth

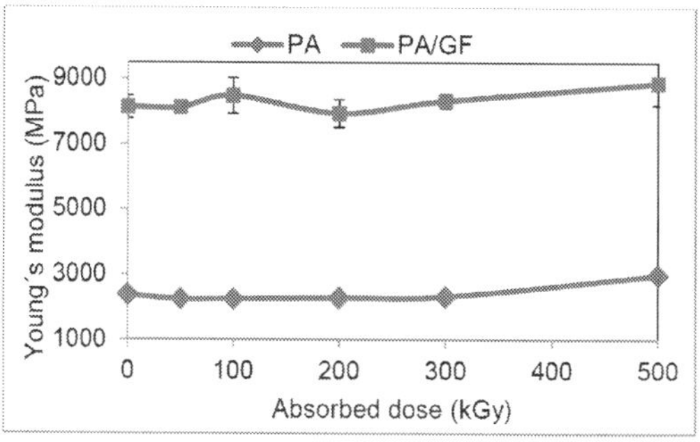

Figure 8. Variation of Young's modulus with absorbed electron beam dose for virgin PA-6 and PA-6/GF (30%) composite [7]. With permission of Elsevier.

in the gel content can be estimated at approximately 150 kGy. The relative increase in the modulus at a 500 kGy dose is approximately the same (28% over the initial rate) as the abovementioned result. When the crosslinking agent is added to the polymer to increase the gel formation, the limit of 50% gel is achieved earlier. Therefore, the higher modulus is observed when compared to the virgin polymer [11], as the crystalline phase, being not yet impaired, is supported by the sufficient gel portion.

The modulus and yield tensile strength of injection-molded PA-66 irradiated in the range of 0 to 500 kGy are found to increase over the unexposed samples with dose but displays a maximum at 200 kGy. Similar results are obtained in the elongation development; however, beyond 200 kGy, the elongation is reduced below the initial value [15] due to crosslinking.

The influence of the crosslinking agent can also be demonstrated on tensile strength and elongation of PA-6 exposed to irradiation within the range of 0 to 100 kGy. These characteristics are kept the same [6], because this dose range does not involve the gel point yet. However, the addition of the crosslinking agent increases the tensile strength and a decrease in the elongation immediately from 40 kGy, indicating that the corresponding gel point is lower than 40 kGy in this case. The higher the crosslinking agent concentration, the greater the effect is found.

The importance of the selection of the crosslinking agent is reflected by the experiment where PA-66 is doped with 1% TAC, TAIC, or TMPTMA when irradiated with doses up to 600 kGy [12] in air. The tensile strength for virgin PA-66 and PA-66 with TAC exhibits a similar pattern of initial rise up to 200 kGy followed by a gradual reduction up to 600 kGy. In the case of PA with TMPTMA, the tensile strength reduces continually. In the case of TAIC, in contrast, the strength rises throughout the whole range of irradiation. The elongation of PA-66 virgin as well as PA-66 with TAC or TMPTMA reduces displaying a sharp decrease at 100 kGy for the PA containing the agents and a more moderate decrease for virgin PA-66 up to 300 kGy. Beyond these doses, the corresponding values change negligibly. The elongation of PA-66 with TAIC follows a completely different course. It gradually drops over the entire range of dosage and the resulting elongation is the highest when compared to others (**Figure 9**). This possibly indicates that, despite having the highest gel content, the network is less dense, with TAIC saving more elasticity compared to others.

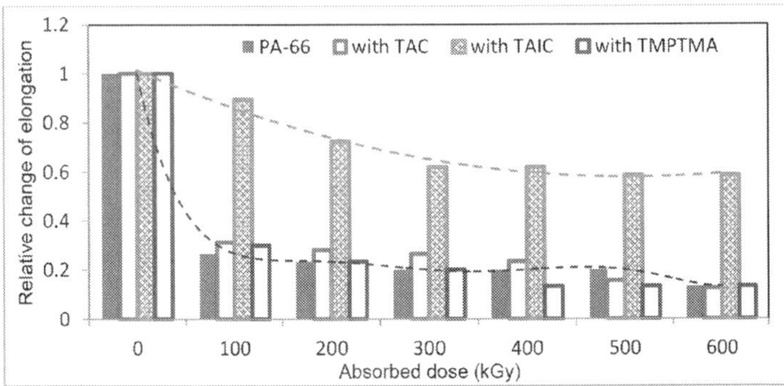

Figure 9. Relative variation of elongation of PA-66 with and without crosslinking agents with dose of e-beam. Adapted from Pramanik et al. [12].

Concerning the modulus (**Figure 10**), this increases beyond 200 kGy sharply for virgin PA-66 in compliance with the above-mentioned finding that at least 50% gel is needed to contribute towards the modulus increase [4].

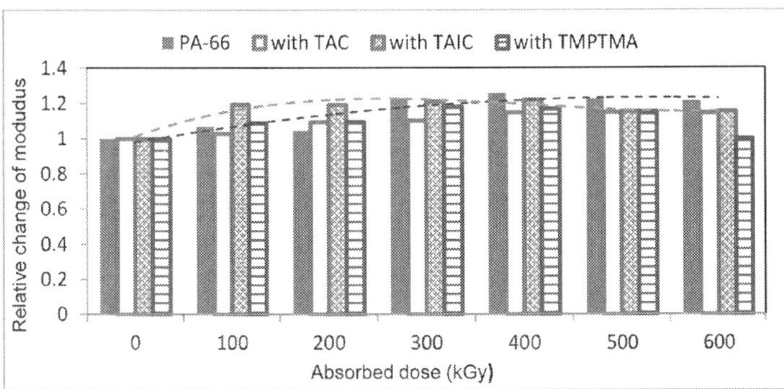

Figure 10 Relative variation of tensile modulus of PA-66 with and without crosslinking agents with dose of e-beam. Adapted from Pramanik et al. [12].

The addition of the crosslinking agents increases the modulus over the virgin PA-66 in the order of TAIC>TAC>TMPTMA, although the dependences follow individual courses. The crosslinking efficacy of the agents in PA-66 is of the same order as well.

The effect of another agent, glycidyl methacrylate (GMA), is similar [21]. PA-6 doped and then irradiated with a dose in the range of 0 to 200 kGy in nitrogen to condensate mutually shows a modulus at 100 and 200 kGy, which is 36% and 45%, respectively, above the unexposed virgin PA-6. At the same time, the corresponding tensile strength is determined to be approximately 15% over the virgin PA-6.

Radiation technology is used in an effort to improve the deficient compatibility of different polymers in blends or composites and to model the resulting properties to a certain extent due to the binding of each other through free radicals generated in the components. PA-6 and linear low-density polyethylene (LLDPE) are typical immiscible polymers. The morphologic examination of the mixture of the PA-6/LLDPE/GMA exposed to electron beam under nitrogen atmosphere within the range of 5 to 200 kGy displays a reduced diameter of the dispersion particles and an increase in the interfacial adhesion. The elongation at break of the blend irradiated at 100 kGy is approximately four times higher than that of the virgin PA-6. This parameter is strongly reduced when the blend is irradiated at 200 kGy, which is supposed to be the crosslinking of LLDPE [20]. The tensile strength and modulus increase with rising irradiation dose nonlinearly.

Electron beam irradiation of multilayer film LDPE/PA-6/LDPE with doses up to 125 kGy in nitrogen results in the increase of the tensile strength and the decrease of the elongation [18].The growth is more rapid up to 50 kGy and then is more moderate. In contrast, the elongation decreases presenting a mirror curve to the strength. This phenomenon is explained by the formation of carbonyl groups in LDPE in the case of the low doses, which facilitate the miscibility of the LDPE with PA-6. On the contrary, at high doses, the presence of crosslinked LDPE would most likely introduce microregions of immiscibility with the PA-6.

The preparation of a thermoplastic elastomer from an immiscible blend consisting of the EPDM, maleated EPDM, and PA-12 is another demonstration of the utility of electron beam irradiation [22]. The irradiation was conducted up to 100 kGy in nitrogen so that PA-12 chain scission occurred generating free radicals but no crosslinking and led to a mutual link of the components and from disastrous mechanical properties to desired ones. The increase is quoted for Young's modulus, tensile strength, and elongation; in addition, the recyclability of the thermoplastic elastomer is gained for three cycles at least.

Waste polymers can be used again when they are combined with other appropriate components. Therefore, waste PA copolymer PA-66/PA-6 blended with acrylonitrile butadiene rubber and subjected to electron beam irradiation shows tensile strength and elongation, depending on dose and composition, at levels between the parameters of those individual components. The obtained compatibility matches with scanning electron microscopy (SEM) images [26].

A composite consisting of ethylene-vinyl acetate copolymer (EVA) flame retarded by a combination of cellulose acetate butyrate microencapsulated ammonium polyphosphate, PA-6, and TAIC reveals a drastic increase in tensile strength by 62% over the initial value at 160 kGy. Beyond this dose, the tensile strength falls, whereas the elongation at break decreases continually from the beginning [27].

Adhesive joints in composite items often involve PAs. Polycarbonate (PC) sheet covered with PA-6 being irradiated with dose of 43 to 432 kGy [28] displays variation in the fracture stress and elasticity of the joint as dependence on dose differing from the neat components. Whereas the fracture stress for the composite increases up to approximately 120 kGy and levels beyond this, the strain falls with rising dose up to 220 kGy and then increases. Some optimum in the characteristics could be found according to topical design requirements.

Young's modulus of composite (PP/PA-6+talc) with or without a compatibilizer or crosslinking agent TAIC, applying a dose of 0 to 200 kGy [25], gives the dependence on dose with composition. Whereas the modulus of the mixture containing all components accounts for only 73% (PP/PA-6+talc) before irradiation, it exceeds the other mixtures after 200 kGy irradiation. Qualitatively, the same tendency is observed for tensile strength, too. These results indicate that, with a combination of suitable composition and electron irradiation, materials could be designed with the desired properties.

Further research in this field is open and not all efforts lead to desired results. For example, experiments with composites of monomer casting PA-6 containing 2% nanofillers as particle carbon or silicon carbide or carbon shortcut fibers with no crosslinking agent give only 6% increase in tensile strength and Young's modulus when irradiated by electron beam with 20 kGy dose [29].

The weakness of PAs as semicrystalline materials is quantified for a low impact strength or toughness. After virgin PA-6 is electron irradiated, a considerable deterioration of these parameters is found. The Izod impact strength falls sharply with the rising dose, and after absorbing 600 kGy, the impact strength retains 31% of the initial level only. The addition of the crosslinking agents changes this uniformly decreasing behavior. According to the agent type, the dependence of Izod impact strength on dose is of variable character, reaching a maximum at 400 kGy followed by a sharp decline beyond this dose [11]. However, the net result is a decrease when compared to the nonirradiated sample. A rather different situation occurs for PA-66 [12]. The Izod impact strength of virgin PA-66 is reduced at 100 kGy to 68% of the initial value and then displays a leveling effect up to 300 kGy, and beyond this dose, a sharp decrease is followed by no significant change at higher doses. Again, the crosslinking agents have a variable influence and result in a dependence including smaller fluctuations in comparison to the virgin PA-66. An observation of the fracture surface leads to the finding that both virgin and crosslinker-doped PA-66 irradiated up to 300 kGy show ductile failure, whereas the materials irradiated with doses of 400 to 600 kGy indicate brittle failure. One of the several reasons why electron irradiation lowers the impact strength may be a certain obstruction in the dissipation of the impact energy when the initial amorphous phase is progressively crosslinked and restricted in the chain motion and relaxation. The contribution of oxidative degradation becomes more evident in the reduction at higher doses.

The flexural characteristics for PA-6 reveal an analogous behavior with the corresponding tensile strength and tensile modulus (i.e., they increase with dose) [11]. The increase in the flexural modulus at 500 kGy is 13% over the initial level and the flexural strength 20%, respectively. The addition of effective crosslinking agents TAIC and TAC raises these values. In the case of PA-66, the flexural modulus increases up to 400 kGy by 40% and then it falls sharply, remaining still at an increase of 20% over the initial level [12]. The same crosslinking agents as in the PA-6 increase the flexural modulus, almost copying the behavior of virgin PA-66 at a higher level.

The irradiation of composite (PP/PA-6+talc) with 0 to 200 kGy doses leads to an increase in flexural strength by 18%; when the composite contains also the compatibilizer and crosslinking agent TAIC, the increase is higher by 137%. Both corresponding flexural moduli are increased by 24% and 200%, respectively [25].

It can be concluded that, besides irradiation conditions, the mechanical properties in blends of PA with other polymers or modifiers are dependent on the character of all components and the interactions between them induced by the irradiation.

2.4. Thermal resistance

The thermal resistance of polymers is usually measured by heat deflection temperature (HDT) and Vicat softening temperature. There are other specific tests adjusted to the factual requirements of manufacturers. However, they are used by narrow groups of design engineers and in the framework of quality testing.

The HDT (or heat distortion temperature) is the temperature at which a polymer sample deforms under a specified load. It might be expected that parameter HDT will increase with rising dose. In fact, for PA-6 irradiated within 0 to 500 kGy, it is found that there is a progressive growth up by 9°C at 500 kGy. All the related values appear near 50°C, the T_g of PA-6 [7]. The HDT increase with increasing dose reflects the progressive restriction of the chain mobility in the amorphous phase as a consequence of the network structure formation and the lower deformability (**Figure 11**).

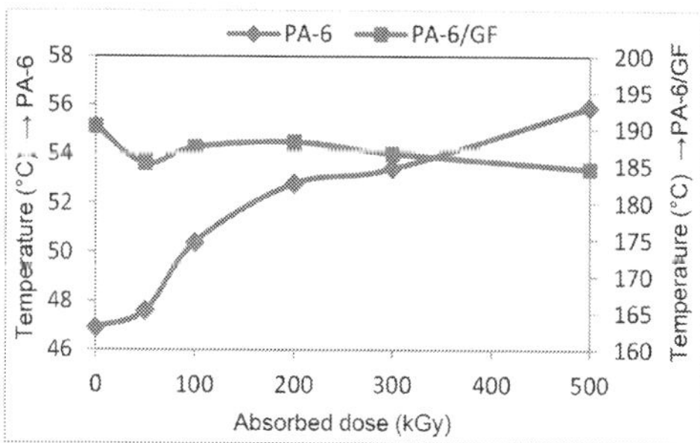

Figure 11. Variation of HDT with absorbed electron beam dose for virgin PA-6 and PA-6/GF (30%) composite. Adapted from Porubská et al. [7].

A different situation is in GF-reinforced PA-6. The presence of GF restrains the segmental motion of the polymer chains, and therefore, the initial HDT (190°C) observed for the compositePA-6/GF is much higher than virgin PA-6 (47°C). Unlike the virgin PA, the HDT values for PA-6/GF decreases in response to irradiation, with a decrease of 6°C at the highest dose [7]. Analogous results can be assumed for other composites as well.

How much crosslinking can affect HDT for PA-6 is seen from the electron beam irradiation of PA-6 containing 2% TAC. Whereas the HDT before exposure is 120°C, after absorbing 80 kGy dose, this figure increases up to 170°C [6] due to ~96% gel content (**Figure 12**).

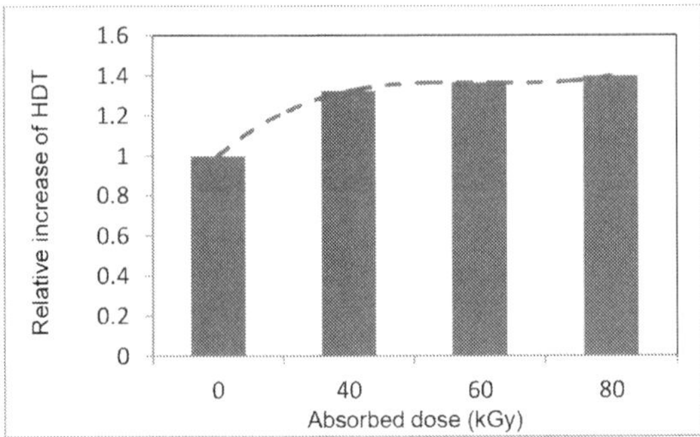

Figure 12. Variation of HDT with absorbed dose for PA-6 at 2% TAC level. Adapted from Dadbin et al. [6].

The irradiation of composite (PP/PA-6+talc) in the range of 0 to 200 kGy displays the rise in HDT depending on the dose from 63°C for the nonirradiated sample to 72°C for that irradiated with 200 kGy. However, when compatibilizer and crosslinking agent TAIC is incorporated in the composite, the HDT gives value of 84°C [25].

The Vicat softening temperature (or Vicat hardness) is taken as the temperature at which the specimen is penetrated to a depth of 1 mm by a flat-ended needle with a 1 mm^2 cross-section. There is not much information on the effect of electron irradiation on the Vicat softening temperature for PAs in the scientific sources. Unlike the HDT, the Vicat temperature for virgin PA-6 (186°C) reaches a maximum (190°C) at the lowest dose of 50 kGy and then decreases with increasing dose nearly returning to its original level (187°C) [7]. The mentioned increase is possibly related to the release of physical entanglements due to the supplied energy from the electron radiation. That enables the involvement of some released segments in amorphous phase into the crystalline phase and its enlargement. A moderate increase in the melting enthalpy/crystallinity at 50 kGy dose corresponds to this. The decrease in the Vicat temperature at a higher dose is assigned to the progressive disruption of the surface structure and the thinning of the lamellae what is obvious also from the lower melting onset (**Figure 3**). The softening temperature for PA-6/GF composite is less pronounced, with a total decrease of only 2°C (**Figure 6**). This observation corresponds to the different crystallinities of both materials (PA-6/GF > PA-6).

2.5. Water absorption

Generally, PAs tend to absorb water due to the nonbinding interactions of water molecules with polar groups in the matrix. The water present in PA acts as a plasticizer. The water absorption affects the dimensional stability and mechanical properties of the polymers, whereas humidity in PA matrix can support hydrolytic destruction and can also be a source of oxygen when PAs are processed or irradiated accelerating the polymer oxidative degradation. Therefore, water absorptivity is one of the crucial parameters affecting the PA properties. That is why the comparison of some characteristic values for various PAs requires a standard conditioning (temperature, humidity, and time) before the PA is tested.

The entry of water molecules into a PA matrix is controlled by diffusion. The penetration of the matrix by water molecules requires enough space inside the matrix for the translation motion of water molecules. The space is formed by vacancies due to the rotating movement of segments of macromolecule chains. The fewer the restrictions of motion in the polymer segments, the more vacancies are created. Because the crosslinking shortens the chain sequences between the cross-points and thus restricts the segmental rotation, water molecule mobility in the polymer should be more difficult suppressing water diffusion. This supposition is confirmed empirically. Water absorption in PA-6 irradiated in range of 0 to 600 kGy decreases throughout all doses from 2.3% to 1.7% at the highest dose [11]. As expected, the decline is greater with the incorporation of the crosslinking agents TAIC or TAC (**Figure 13**).

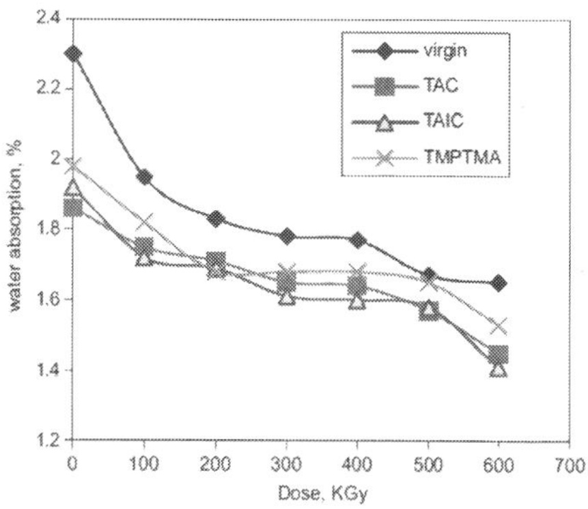

Figure 13. Variation of water absorption for PA-6 without and with 1% crosslinking agents with dose of e-beam radiation [11]. With permission of Elsevier.

Because the concentration of the effective crosslinking agent affects the gel content, it is expected to influence water absorption in the PA correspondingly. Such an example is shown in PA-6 doped with 1%, 2%, and 3% TAC [6]. The measured data confirmed this supposition and the largest decrease at 80 kGy is read for PA-6 with 3%>2%>1% TAC. The decline of water absorption compared to virgin PA-6 irradiated within 0 to 200 kGy occurs also after adding approximately 3% GMA under identical conditions [21]. Similarly, injection-molded PA-66 reveals a lowering of water absorption when irradiated in the range of 0 to 500 kGy, but the PA-66 dipped in TAC solution before the irradiation shows a further decrease in this parameter [15].

3. γ-Irradiation of PAs

In polymer chemistry, as for electron irradiation, γ-irradiation is employed to initiate chemical reactions also in the solid phase without the addition of initiator. The purpose can be to initiate the crosslinking of individual polymers or the grafting of various monomers onto PA chains. Basically with γ-irradiation of PAs, the same processes run as those with electron irradiation. However, some differences occur due to variable conditions.

3.1. Variations in chemical structure

γ-Ray radiation differs from electron beam mainly by a much slower rate of dose due to γ-rays being always of less energy than 10 MeV. Therefore, γ-irradiation requires a longer period to supply the same dose as an electron beam. When a polymer is γ-irradiated in air, enough time is available for generated radicals to react with oxygen. Therefore, considerable oxidative degradation can be expected along with crosslinking. In contrast, an inert atmosphere during irradiation as well as after the irradiation suppresses oxidative degradation and supports crosslinking. However, the oxidative degradation cannot be excluded absolutely due to some portion of oxygen (and possibly also humidity) present in the polar PA matrix. These anticipations were confirmed and both chain scission and crosslinking are observed to occur in PA-6 under γ-irradiation in either air or inert atmosphere, with chain scission prevailing over crosslinking if irradiation proceeds in air [30].

A thorough comparison is provided by a recent study [31] examining PA-6 and GF-reinforced PA-6 (30% GF) irradiated with different γ-ray doses in the range of 0 to 500 kGy in either air or inert atmosphere. As displayed in **Figure 14**, it can be seen that the irradiation in air generates a small amount of gel in PA-6 only, whereas no gel is found in the PA-6/GF composite.

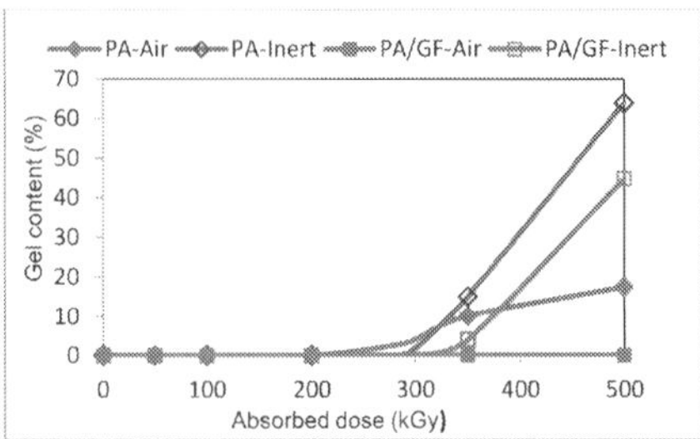

Figure 14. Dependence of gel content in PA-6 and PA-6/GF composite on absorbed dose when γ-irradiated in air or inert atmosphere.

In the PA-6, the crosslinked portion irradiated in air increases slightly with the rising absorbed dose. Concerning the gel point, a calculation following the Charlesby-Pinner equation [14] using the experimental data and taking into consideration the experimental errors gives an estimation that the gel point is to be in the vicinity of a dose of 300 kGy for the PA-6 irradiated in both atmospheres. Such gel point is somewhat higher compared to 200 kGy dose value determined for the same materials irradiated with electron beam in air [7]. The irradiation of PA-6 in argon atmosphere produced considerably more gel in comparison to exposure in air. It demonstrates oxygen influence. The absence of oxygen (or low content of it) affords better facilities to form crosslinks from generated macroradicals because the macroradicals are not attacked. In argon atmosphere, the gel point for PA-6 as well as PA-6/GF is observed to be approximately 300 kGy. However, also in inert atmosphere, PA-6 shows more gel (65%) than PA-6/GF (45%) for the ultimate dose of 500 kGy. The lower gel content in the composite confirms the retarding action of the filler on the networking of the PA matrix. Simultaneously measured solution viscosity for PA-6 increases up to 200 kGy, and above this dose, the viscosity could not already be measured correctly due to the incomplete dissolution of the PA-6 polymer indicating incoming gel point. The increase of the PA-6 solution viscosity below 200 kGy indicates growth in the molecular mass via the recombination of the secondary macroradicals formed on PA-6 chains, which leads to branching as pre-crosslinking stage up to the gel point [5,6]. Whereas the irradiation of PA-6/GF in air did not generate any gel within the applied dose, viscosity increased (see **Table 3**), illustrating some recombination of the macroradicals and relating growth in molecular mass. The solubility of the PA-6/GF matrix was observed within all doses corresponding with no gel content. The reduction of the viscosity beyond 350 kGy is attributed to the continuing branching of already branched chains. A consequence is that the amount of the particles and the corresponding gyration radius of the macromolecules decrease, reducing the viscosity.

Dose (kGy)	Viscosity number (mL/g)	
	PA-6	PA-6/GF
0	143	130
50	146	131
100	154	133
350	–	212
500	–	180

Table 3. Viscosity number of PA-6 and PA-6/GF irradiated with γ-rays in air (solvent formic acid, dose rate of 9.5 kGy/h).

Gupta and Pandey [30] presented a similar fact when, irradiating in air, the chain scission of PA-6 prevailed over crosslinking. Because γ-irradiation took a longer time (from 5 to 52 h), during this period, oxygen could attack the produced macroradicals. The result was the overbalance of scission due to oxidation over crosslinking. In addition, the presence of GF hindered the diffusion of the macroradicals in the matrix, so that the branching was supported and crosslinking was suppressed.

An opposite effect can occur as reported by Aytaç et al. [13] for PA-6 and PA-6,6 tyre cords being calendered and then γ-irradiated within 0 to 75 kGy in air. As already mentioned about the electron exposure of those materials (Section 2.1), also the γ-irradiation of those PAs led to the reduction of the limiting viscosity number with increasing dose. This demonstrates how any pretreatment of material subsequently subjected to the irradiation can be important. In general, calendering itself can start mechano-oxidative degradation leading to the decrease in molecular mass as well as in the corresponding viscosity. In addition, during the calendering, the diffusion of oxygen into the PAs is more facile, as the molecular movement in matrix is increased at higher temperature. From this point of view, the decrease in the limiting viscosity number is expected. The decrease is larger than that under comparable exposure to electron beam, indicating more deteriorative effect of γ-irradiation in comparison to electron beam. Corresponding dependences of the breaking strength on dose show a lower strength, too, in conformity with the viscosity results.

PAs are combined with other polymers in various microfiltration membranes especially to enhance the mechanical properties. The membranes are exposed to γ-irradiation to be sterilized. Such a membrane involving PA-66 as reinforcing part was put in Pyrex glass, purged with argon, and, after adding deionized water again, purged with argon and sealed. The glass with the membrane was γ-irradiated in the range of 0 to 100 kGy and then several characteristics were tested [32]. At first sight, this is different from the PA-6 and PA-6,6 cords mentioned previously [13]. However, also in this case, the reduced viscosity as a function of dose displays a decrease with dose (**Figure 15**).

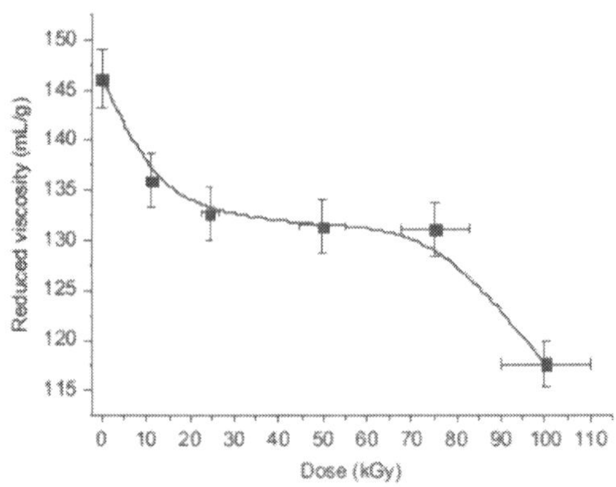

Figure 15. Evolution of reduced viscosity of PA-66 as a function of γ-irradiation doses [32]. With permission of Elsevier.

The downward trend of the reduced viscosity indicates the decrease in molecular mass due to chain scission despite the exposure being conducted in argon purged deionized water. The authors [32] suggested that the viscosity reduction could be due to the chains connecting different lamellae. However, the water medium is a rich source of oxygen and could be the main reason in supporting oxidative degradation. Free radicals generated by γ-rays induce the cleavage of water molecules. In addition, water penetrates PA relatively easy acting as a plasticizer, and the applied dose within 10 to 100 kGy is sufficient to attain the dissociation energy for water producing oxygen (HO-H ~498 kJ mol^{-1} and O-H ~428 kJ mol^{-1}) and that is comparable to the dissociation energy for H-C cleavage (H-C ~339 kJ mol^{-1}, H-CH ~452 kJ mol^{-1}, and H-CH_2 ~473 kJ mol^{-1}) [33]. That is why corresponding tensile properties (Young's modulus, stress at break, elongation, and energy at break) [32] mirrors the course of the decreasing viscosity.

Radicals generated during irradiation are capable of surviving in the polymer matrix for a long time as found by Menchaca et al. [34,35] after 6 years from exposure of PA-6,12 crystalline fibers with applied low dose of 1 to 25 kGy at ambient conditions. The survival of frozen free radicals in the matrix demonstrated itself by the changes in some properties as thermal and morphology characteristics; the melting temperature decreased and the crystallinity increased with the period of storing. It indicates that formed shorter chains generated thinner lamellae and integrated into the crystalline phase.

Applying γ-irradiation can also lead to an improvement in the compatibilization of immiscible polymers in a blend. An example is the blend of PA-6 with LDPE either irradiated or nonirradiated [36]. The structure and properties of the blends with γ-irradiated LDPE differ significantly compared to PA-6 blends with the nonirradiated materials. The difference was ascribed to the formation of functionalized groups on the polyethylene chain during irradiation in air and these interact with PA. However, when analogous blends of LDPE/PA-6 are irradiated in vacuum [37], crosslinking is achieved mainly in the PE component, whereas the main effect on PA-6 is chain branching.

3.2. Thermal properties related to supermolecular structure

A comparison of DSC characteristics for PA-6 and composite PA-6/GF after being γ-irradiated in air or in inert [31] within 0 to 500 kGy can provide some framework observations. Similar to Section 2.2, both first and second heating runs give complementary information. First, the effect of irradiation dose on T_{m1} for PA-6 is qualitatively the same if irradiated in air or inert atmosphere; however, the absolute values for each particular dose indicate lower T_{m1} values for samples irradiated in air (**Figure 16**).

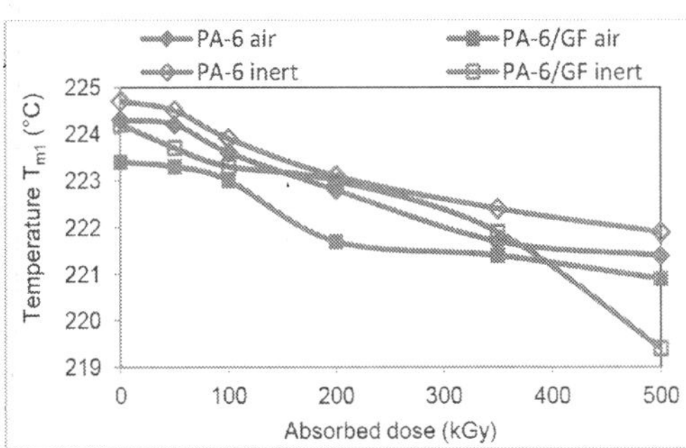

Figure 16. Variation of the first melting temperature of PA-6 and composite PA-6/GF (30%) with absorbed dose of γ-irradiation in air and inert atmosphere. Adapted from Porubská et al. [31].

The effect of GF presence consists of a certain soft increase of sensitivity towards irradiation, which is in certain contradiction with lower gel formation. It may mean that the presence of GF can act as a nucleating agent. If so, the crystallites are formed especially around the fiber surface. However, during exposure, the fiber surface is more heated than the matrix and some destruction in crystalline portion occurs in this area. That is why the noticeable decrease in T_{m1} is observed for the composite irradiated in inert atmosphere.

The figures of melting heat ΔH_1 (**Figure 17**), directly proportional to crystallinity, do not show any tendency with increasing dose in air, changing only within variance of the experimental results.

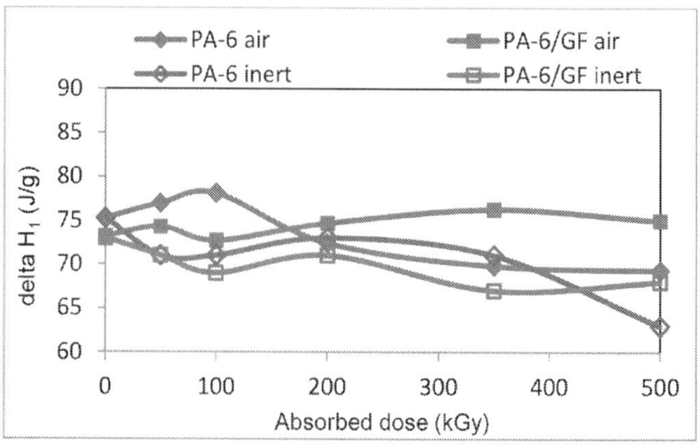

Figure 17. Variation of the first melting enthalpy of PA-6 and composite PA-6/GF (30%) with absorbed dose of γ-irradiation in air and inert atmosphere. Adapted from Porubská et al. [31].

It can be seen that the values of ΔH of both PA-6 and composite PA-6/GF samples irradiated in air are higher compared to those irradiated in inert atmosphere. The course of changes of ΔH measured in the second DSC run, ΔH_2 (**Figure 18**), depending on the absorbed dose is in conformity with expectation, consisting of a gradual decrease of ΔH_2 values with rising dose. Irradiation in air results in higher melting heat values than irradiation in inert atmosphere. In this case, perhaps the important role consists of heating the samples during irradiation and the extent of heating depends on the absorbed dose.

All samples show a decrease in remelting temperature T_{m2} with rising dose (**Figure 19**), which is conformable with expectation. The reason is that the crystallization of polymer remelted after being irradiated is hindered due to the defects generated in polymer within irradiation. However, the same decrease in T_{m1} would not be expected if the irradiation impose structural changes in amorphous phase first. Because T_m and ΔH vary, the crystalline phase also is affected by irradiation, inducing the lamellae thinning. It is evident that the process runs already from the beginning of irradiation.

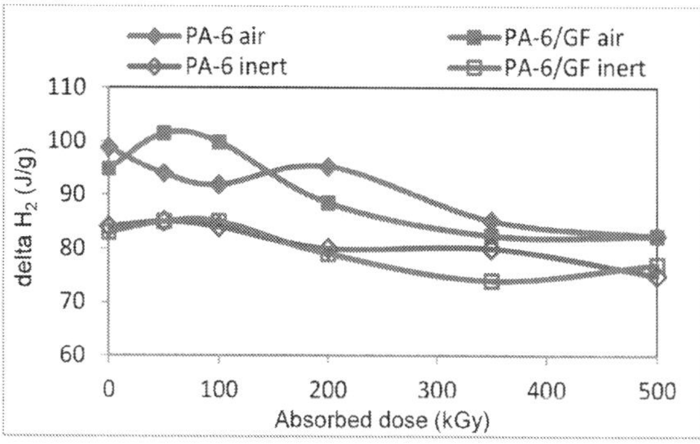

Figure 18. Variation of the second melting enthalpy of PA-6 and composite PA-6/GF (30%) with absorbed dose of γ-irradiation in air and inert atmosphere. Adapted from Porubská et al. [31].

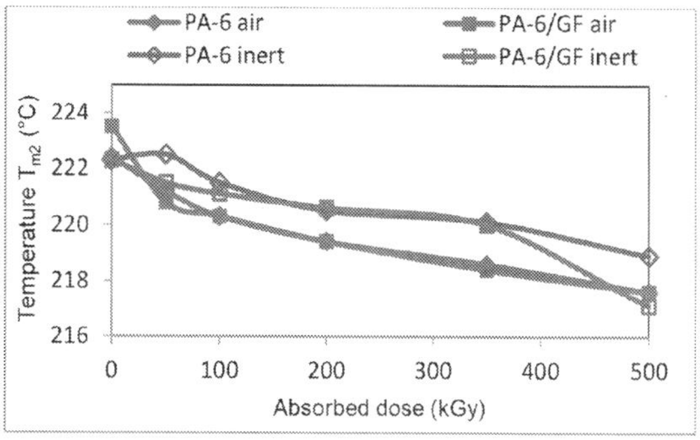

Figure 19. Variation of the second melting temperature of PA-6 and composite PA-6/GF (30%) with absorbed dose of γ-irradiation in air and inert atmosphere. Adapted from Porubská et al. [31]

Similar to electron beam irradiation, the presence of crosslinking agents can modify the crystallinity of γ-irradiated PAs, too. Electrospun PA-66 fibers exposed in nitrogen atmosphere with 20 and 50 kGy dose do not display any change in crystallinity, whereas the addition of TAC raises the crystallinity and this increases with applied dose [38]. This fact is interpreted as the consequence of a TAC-induced crosslinking and formation of a tighter network. Thermal stability measured as weight loss depending on temperature is lower with TAC than without it. After irradiation, the stability with TAC improves, however, without achievement of stability for nonadditive PA-66, whereby the dose of 20 kGy provides better result than 50 kGy.

The exposure of PA-66 to various doses of γ-rays ranging from 100 to 1250 kGy shows an increase in the crystalline nature of the polymer at higher doses as a result of significant decrease in the peak width of X-ray diffraction (XRD) patterns [39]. Higher doses induce more macromolecular fragments of higher mobility and these can integrate easily into crystalline phase modifying supermolecular structure.

Waste PA finds exploitation in various material combinations. Hassan et al. [40] studied the effect of γ-irradiation on blends containing waste PA-6/PA-66 copolymer and ground rubber from tires with various ratios of these incompatible components. The blends irradiated in the range of 0 to 200 kGy give the melting temperature and crystallinity decreasing with increasing dose due to the crosslinking at interphase. The visible side shoulder in the endotherm for 100 kGy is missing in the 200 kGy endotherm and microphotographs show a relatively smooth fracture surface. Thermal stability measured by thermogravimetry is a little worse after irradiation. Montmorillonite clay is then added into the blend to formulate nanocomposite [41]. The composite after being γ-irradiated between 0 and 200 kGy obtains a markedly magnified thermal stability. When 12% montmorillonite is present in the mixture, the DSC data indicate the increase in the melting temperature with dose with reverse order of the onset in melting endotherm. The crystallinity is observed to be highest for the 100 kGy dose and the corresponding endotherm outlines multiplicity. The increase of montmorillonite portion to 18% leads to a decrease in the melting temperature. Also, the temperature onset falls. In this case, the highest crystallinity belongs to the nonirradiated composite and is followed by the 200 kGy dose. The dose of 100 kGy corresponds to the lowest crystallinity with the most structured melting endotherm. The multiplicity of the endotherm indicates a new element in the supermolecular structure as a consequence of γ-irradiation. Another composite consisting of the same polymer components PA-6/PA-6,6 copolymer and ground rubber but with added carbon black was examined by the same authors [42], applying the same doses of 0 to 200 kGy. As reported, the content of carbon black within 6% to 24% improves the thermal stability in both cases without and with γ-irradiation. The melting temperature and crystallinity of the composite with 12% carbon black decrease with rising dose slightly more when compared to 18%. The melting endotherm becomes smoother and the composite irradiated with 200 kGy presents a homogeneous fracture surface. Such studies are useful in optimizing a filler portion regarding other required properties. Usually, some compromise is necessary.

3.3. Mechanical properties

The development of mechanical properties of PAs irradiated with γ-rays is, as usual, linked to changes in the molecular and supermolecular characteristics. Evaluating the behavior of PA-6 and composite PA-6/GF γ-irradiated in air [31] concludes that the Young's modulus of PA-6 is soft decreasing with rising irradiation dose, whereas, for the nonirradiated composite PA-6/GF, the values are almost identical with the material irradiated with the highest dose (**Figure 20**).

For both materials, the dependencies exhibit a shallow minimum around the gel point as a result of the superposition of the two opposite effects, namely, branching, chain scission, and cross-bond formation. These effects are more evident for PA-6 in comparison to the composite PA-6/GF. At first sight, it seems to be a misinterpretation. However, one has to take into consideration generally lower figures for PA-6 modulus and a small variation leads to a larger relative variation than in the case of the composite. Overall marginal variations of modulus are linked to a low gel content for PA-6 and no gel for PA-6/GF (Section 3.1) as well as small changes in crystallinity (Section 3.2). Because the level of crystallinity and crosslinking

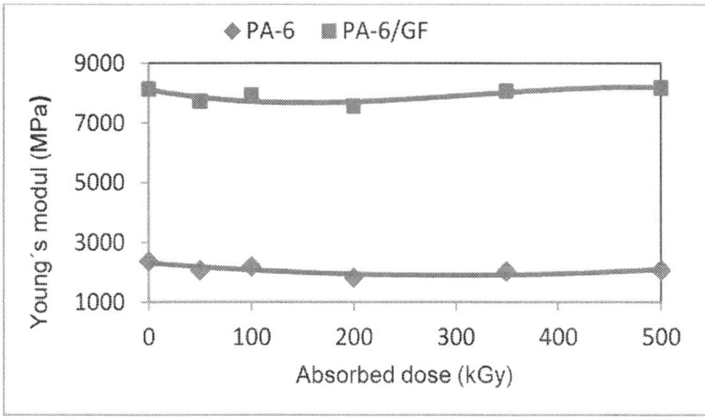

Figure 20 Variations of Young's modulus for virgin PA-6 and composite PA-6/GF γ-irradiated in air.

determines the modulus value, a significant variation of the modulus values should not be expected as demonstrated in **Figure 20**.

Testing tensile strength reveals that yield point can be observed only for PA-6, and the composite PA-6/GF exhibits brittle behavior without signs of yielding. A considerable decrease of yield stress from 71 to 51 MPa is observed already at the lowest dose of 50 kGy and then the curve leveled off up to the highest dose of 500 kGy. The tensile strength at break for PA-6 does not vary considerably with rising irradiation dose. This result is attributed to the gel content (Section 3.1), which contributes to retain the strength at break with no significant variation in the whole dose range compensating the strength decrease induced by degradative influence of γ-irradiation in air. In contrast to PA-6, all exposed PA-6/GF samples showed a little reduction in tensile strength at break. The reason is the different gel formation; whereas no gel was measured in the composite, a certain amount of gel was determined in PA-6 beyond gel point. Therefore, zero gel in PA-6/GF could not compensate the decrease in the strength due to the degradation of the polymer matrix.

Variations in elongation at yield for PA-6 as well as for the composite are negligible and this parameter for PA-6/GF is identical with the elongation at break. For PA-6, the starting elongation at break at ~147% increased to 240% already at 50 kGy without any change for the other doses. Such mild increase in the elongation at break is caused by the minor decrease of crystalline portion as well as the lamellae thinning as a result of oxidative degradation with the consequence of easier plastic deformation. In addition, at low gel content (maximum ~17%), shorter recombined chains act in the matrix as plasticizers and increase its deformability. The effect is not observed for the composite PA-6/GF, as it is overlapped by enhanced brittleness of the material due to the presence of the anisotropic GF.

Concerning irradiation in an inert argon atmosphere, a comparison of effect on the moduli is demonstrated in **Figure 21**.

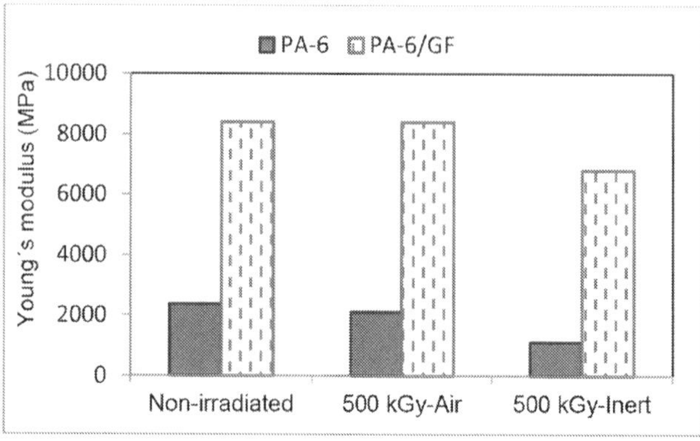

Figure 21. Comparison of variations of Young's modulus for virgin PA-6 and composite PA-6/GF γ-irradiated by 500 kGy dose.

When PA-6 and PA-6/GF were irradiated in inert atmosphere, the relevant moduli showed lower values in comparison to the materials exposed to γ-irradiation in air. In consideration of the marginally changing results of ΔH in the first DSC run and so also the minor changes in crystallinity, the most probable reason consists of the lamellae thinning as a result of irradiation, demonstrated by the lowering in the melting temperature. It is known that the stiffness of the material may decrease if crystallites are smaller even when crystallinity does not change [4].

The reduction of tensile strength for both samples is more enhanced when irradiated in inert atmosphere. As displayed in **Figure 22**, the changes are rather small, except for yield, where the extent of changes is close to the values for Young's modulus.

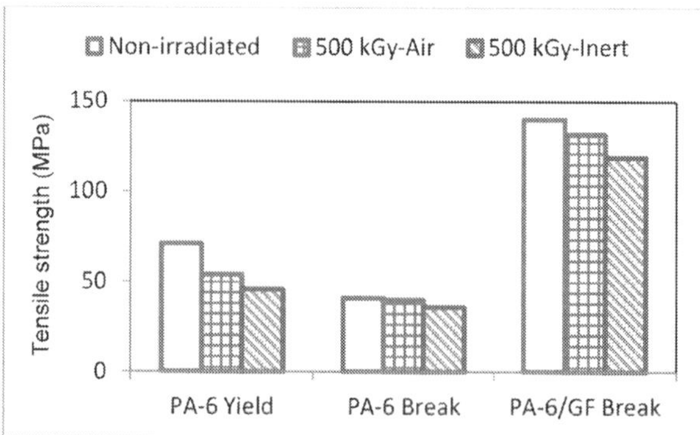

Figure 22. Comparison of variations of tensile strength for virgin PA-6 and composite PA-6/GF γ-irradiated by 500 kGy dose.

The changes in elongation at break and at yield (in the latter case only for PA-6) are shown in **Figure 23**.

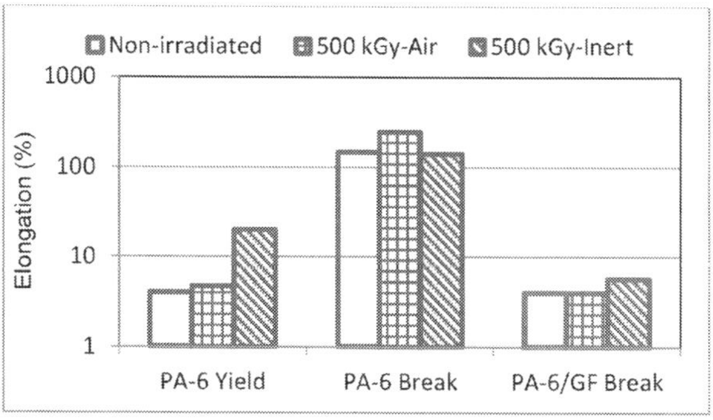

Figure 23. Comparison of variations of elongation for virgin PA-6 and composite PA-6/GF γ-irradiated by 500 kGy dose.

The elongation at yield for PA-6 irradiated with 500 kGy in inert atmosphere increased for up to quintuple value, whereas irradiation in air led to negligible change. In contrast, the elongation at break is higher after irradiation in air when compared to irradiation of PA-6 in inert atmosphere, and this is lower than for the unexposed sample. Irradiation of PA-6/GF in air led to the elongation at break with no change, whereas a little increase is observed when irradiated in inert atmosphere. This indicates some increase of the matrix deformability due to a low crosslinking level.

Concerning exposure in inert atmosphere, similar data are given for electrospun PA-66 fibers irradiated in nitrogen atmosphere with 20 and 50 kGy dose. Tensile stress lowers with dose, whereas an increase is observed after adding the crosslinking agent TAC [38]. The same is reported for Young's modulus.

Notched impact strength of PAs is often unfavorable property. An examination of this aspect was demonstrated by Charpy notched impact test being carried out for both PA-6 and PA-6/GF irradiated within 0 to 500 kGy in air [31]. The dependence of the impact strength on dose reveals a different behavior. The PA-6 curve shows a maximum at 50 kGy. The corresponding impact strength is 80% higher compared to the initial value. This can be attributed to the perturbation of initial physical nodes in amorphous phase and branching, which facilitates more elastic dissipation of impact energy and supports a deceleration of crack propagation. Higher doses form a network, although not a dense one, but it supports the absorbing impact because of its elasticity. Whereas oxidation-degradation occurs simultaneously, the result represents a superposition of the positive contribution of elasticity and a negative oxidation-degradation under irradiation. At 500 kGy, the final impact strength for PA-6 is 1.5 times higher than the starting value. However, an opposite trend is observed for the PA-6/GF composite. In this case, the impact strength decreases mildly with rising dose up to 82% of the initial value linking the zero gel observed (**Figure 14**) and more or less constant crystallinity (Figures **17** and **18**). As evidenced by SEM photos, although easier GF dewetting in nonirradiated composite may act as an obstacle for crack propagation at impact, after irradiation, higher GF adhesion makes the material more compact and less resistant to impact failure.

Blends containing waste PA-6/PA-66 copolymer and ground rubber from tires irradiated within 0 to 200 kGy in air show lower values in tensile strength and elongation compared to the unexposed sample, whereas elastic modulus is changed little. All the variations are a function of the component ratio [40]. When montmorillonite clay is involved in the blend, formulating a nanocomposite, a decrease in tensile strength, and an elongation at break is observed for each dose in comparison to the starting material. However, the courses of both parameters are a function of the montmorillonite portion, indicating an optimum for 12% clay. Concerning modulus, a decrease is measured regardless of dose, but the differences between the compositions are not large [41]. The same doses of 0 to 200 kGy were applied on the mixture of the above-mentioned polymer components PA-6/PA-66 copolymer and ground rubber with the addition of carbon black [42]. The tensile strength of nonirradiated sample is observed to be higher than the irradiated samples, except the dose of 50 kGy being above the initial material. However, irradiation led to a decrease in elongation for all doses compared to that unexposed. For the latter case, the addition of carbon black increased elongation significantly; for others, there was a slower increase. In both tensile characteristics, the differences between doses 100 and 200 kGy are marginal, whereas the variations are more profound for the dose of 50 kGy. The dependences of the parameters on carbon black portion within 6% to 24% are almost flat.

γ-Rays are applied in the field of flame retardation of polymers, too. The recent study of Sonnier et al. [43] demonstrates for blend PP/PA-6/crosslinking agent that, depending on the type of retardation, the best flame-retarded blend before irradiation can become the worst one after irradiation at a higher dose. In this case, a dose up to 100 kGy was used. Testing of the dependence of Young's modulus on temperature provided results similar for all blends at RT for whatever the dose, and these decreased slowly when the temperature increased. Above 150°C, the decrease was faster. Heat distortion of the blends accelerated with rising dose, whereas the distortion was not observed for nonirradiated blend. It is concluded that, if heat shielding effect is applied to provide the flame retardation, the top protective layer can be disrupted and heat release rate will increase to a considerable extent. In such case, the barrier layer is not capable to prevent the subsequent transfer of heat.

3.4. Sterilization of PA materials by γ-radiation

Low doses of γ-radiation are often used for the sterilization of food packaging. Polymers used to package food intended for irradiation must receive relevant approvals. Evaluation of e-beam, γ- and X-ray treatment on the chemistry and safety of polymers used with prepackaged irradiated foods showed that the three forms of irradiation have virtually indistinguishable effects on polymers irradiated in vacuum [44]. However, γ-irradiation in air results in facilitated damage due to slow dose rate providing enough time to oxidative degradation. It is accepted, in general, that the foods in contact with irradiated polymeric materials should not be endangered by radiolytic products with adverse impact on health.

PAs are governed by excellent barrier performance. That is why several works are devoted to the issue of food packaging involving PAs. An examination of various plastic multilayer PA-6 films, used for meat and cheese, after being irradiated (up to 12 kGy) reveals that the release of ε-caprolactam from exposed PA-6 is of a much higher extent compared to that of nonirradiated samples, indicating chain scission [45]. Félix et al. [46] conducted a migration assay at 40°C for 10 days focusing on the effect of γ-irradiation with 12 kGy dose on ε-caprolactam migration from multilayer PA-6 films into food simulants. The results revealed that the irradiation caused almost no changes in ε-caprolactam levels, with the exception of olive oil, which showed an increase in the caprolactam level. However, all the tested films were within the legislation and did not exceed limits for ε-caprolactam migration. Park et al. [47] reported that 5 kGy γ-irradiation significantly increased the formation of ε-caprolactam in PA-6 from 70.76 to 164.10 ppm. The formation of ε-caprolactam ranged between 122 and 164 ppm in the dose range of 5 to 200 kGy.

Barrier five-layer food packaging films, consisting of two outer PA-6 layers (~15%) and a middle LDPE layer (50%), after being irradiated with a larger range of doses of 5 to 60 kGy were analyzed focusing on volatile and nonvolatile radiolytic products and sensory changes [48]. The data show that a large number of radiolytic products are produced such as hydrocarbons, alcohols, carbonyl compounds, and carboxylic acid but also amide type of products. These substances are detected even at the lower doses of 5 and 10 kGy. Most of the substances are assumed to come from LDPE because that is used also as recycled. The type and concentration of radiolytic products increase progressively with the absorbed dose. In addition, irradiation dose appears to influence the sensory properties of table water in contact with the films being classified according to stricter requirements.

In another study [49], the authors analyzed 13 different multilayer polymeric materials for food used before and after their exposure to γ-radiation regarding the profile of volatile compounds released from the polymeric materials. Thermosealed bags of different materials were filled with either air or nitrogen to evaluate the oxygen influence. One third of the samples were analyzed without irradiation, whereas the rest were irradiated at 15 and 25 kGy. Half of the samples were processed just after preparation and the other half was stored for 8 months at RT before analysis. Significant differences between nonirradiated and irradiated bags were found. Sixty to 80 compounds were released and identified per sample. Independent of the filling gas, the results of nonirradiated materials were almost identical. In contrast, the chromatographic profile and the odor of irradiated bags filled with nitrogen were completely different from those filled with air. The migration of compounds from irradiated materials to the vapor phase was much lower than the limits established in the relevant EU Commission Regulation.

4. Proton irradiation of PAs

Compared to electron and γ-irradiation, there is far less information on proton beam irradiation in the scientific sources. One possible reason is that proton irradiation is used mostly in the medical domain for therapy and production of isotope-labeled pharmaceutical preparations.

Currently, proton beam is used as a direct writing method using focused proton beam to decorate resistant polymers at the nanolevel. In comparison to electron beam, the proton writing offers a unique advantage consisting of the fact that proton penetrates materials more deeply, maintains a straight line, and makes three-dimensional structures with vertical and smooth walls as well as low roughness. This advantage is a consequence of a greater proton mass when compared to electron.

Unlike PAs, some works dealing with other polymers (PMMA, PDMS, fluorinated PI, PP, PTFE, PS, LDPE, PP, PET, and PVC) can be found. Semicrystalline PAs are not typical materials for optical or decorative purposes because of their nontransparency. Maybe this is why studies on PAs irradiated with proton beam are rare if not missing.

Concerning polymers, it is known that also protons affect the polymers mainly through scission and crosslinking of macromolecule chains. Solitary data on PA-6 irradiated by proton beam in air can be found in the paper dealing with Fourier transform infrared (FTIR) spectroscopy comparison of electron with proton beam impact [50]. When irradiating the same PA-6 with an equal dose of 500 kGy, the proton beam generates less gel (57.5%) than the electron beam (67.6%). Therefore, the first DSC melting temperature is a little higher for a proton-irradiated sample (217.6°C) than for an electron-irradiated one (216.9°C). Accordingly, the crystallinity for proton-irradiated PA-6 is slightly higher when compared to the corresponding electron-irradiated PA-6, but the difference in the mentioned thermal characteristics is marginal. FTIR spectroscopy shows some differences in the postirradiation species related to nonidentical gel formation. The proton beam irradiation results in a finer structure of some absorption bands, particularly in the range of 400 to 1650 cm^{-1}, indicating the generation of structures that are more varied. The consequence is that the cleavage of PA-6 macromolecules by the proton beam produces fragments containing amine groups and terminal methyl groups, whereas the increase in concentration of these groups in electron-irradiated PA-6 appears to be insignificant in comparison to virgin PA-6. Based on the finding, besides crosslinking [Equation (11)] and oxidation degradation [Equations (13) and (14)], a possible parallel scenario can occur when proton beam interacts with PA:

a) Unsaturated structures formation:

$$[-CO-NH-(CH_2)_5-]_n \xrightarrow{-2H^{\bullet}} [-CO-NH-CH_2-CH=CH-CH_2-CH_2-]_x \tag{15}$$

The double bond can occur between any carbons within the ethylene segments.

b) Amine species formation:

$$[-CO-NH-(CH_2)_5-]_n \rightsquigarrow [-C^{*}O-HN^{*}-(CH_2)_5-]_y \\ \xrightarrow{+2H^{\circ}} y \sim -CHO + y\,H_2N-(CH_2)_5- \sim \quad (16)$$

c) Both methyl-ended and shorter amide chain formation:

$$[-CO-NH-(CH_2)-(CH_2)_4-]_n \rightsquigarrow [-CO-N^{*}H-C^{*}H_2-(CH_2)_4-]_z \\ \xrightarrow{+2H^{\circ}} z \sim -CO-NH_2 + z\,CH_3-(CH_2)_4- \sim \quad (17)$$

The processes reflect some differences also in tensile properties as seen in **Table 4**.

Tensile property	Relative change after 500 kGy dose	
	PA-6/EB	PA-6/PB
Young's modulus	1.256	1.167
Tensile strength at break	1.582	1.398
Elongation at yield	1.075	2.077

Table 4. Relative changes of selected tensile properties for PA-6 irradiated by electron (PA-6/EB) and proton (PA-6/PB) beams in air towards properties of nonirradiated PA 6.

A thorough comparison of the advantages versus disadvantages of the proton beam irradiation in comparison to other radiation technologies in the domain of PAs cannot be concluded unless more data are available.

Author details

Mária Porubská

Address all correspondence to: mporubska@ukf.sk

Constantine the Philosopher University in Nitra, Nitra, Slovakia

References

[1] Anonymous. Polyamide [Internet]. Available at: http://www.plasticseurope.org/what-is-plastic/types-of-plastics-11148/engineering-plastics/pa.aspx. Accessed: 2016-01-10.

[2] Narkis M, Raiter I, Shkolnik A, Eyerer PJ: Structure and tensile behavior of irradiation- and peroxide-crosslinked polyethylenes. Journal of Macromolecular Science Part B Physics. 1987;26:37–58.

[3] Lazár M, Rado R, Rychlý J: Crosslinking of polyolefins. Journal of Applied Polymer Science. 1990;95:149–197.

[4] Chodák I: Properties of crosslinked polyolefin-based materials. Progress in Polymer Science. 1995;20:1165–1199.

[5] Rybnikář F. Chemie a struktura polyamidu. In: Veselý R, Sochor M, editors. Polyamidy —jejich chemie, výroba a použití (In Czech). SNTL: Praha; 1963. p. 55–57.

[6] Dadbin S, Frounchi M, Goudarzi D: Electron beam induced crosslinking of nylon 6 with and without the presence of TAC. Polymer Degradation and Stability. 2005;89:436–441.

[7] Porubská M, Janigová I, Jomová K, Chodák I: The effect of electron beam irradiation on properties of virgin and glass fiber-reinforced polyamide 6. Radiation Physics and Chemistry. 2014;102:159–166.

[8] Sengupta R, Tikku VK, Somani AK, Chaki TK, Bhownick AK: Electron-beam irradiatied polyamide-6,6 films— I: characterization by wide angle X-rays scattering and infrared spectroscopy. Radiation Physics and Chemistry. 2005;72:625–633.

[9] Sengupta R, Tikku VK, Somani AK, Chaki TK, Bhownick AK: Electron-beam irradiated polyamide-6,6 films—II: mechanical and dynamical properties and water absorption behavior. Radiation Physics and Chemistry. 2005;72:751–757.

[10] Burillo G, Adem E, Muñoz E, Vásquez M: Electron beam irradiated polyamide-6 at different temperatures. Radiation Physics and Chemistry. 2013;84:140–144.

[11] Pramanik NK, Haldar RS, Bhardwaj YK, Sabharwal S, Niyogi UK, Khandal RK: Radiation processing of Nylon 6 by e-beam for improved properties and performance. Radiation Physics and Chemistry. 2009;78:199–205.

[12] Pramanik NK, Haldar RS, Bhardwaj YK, Sabharwal S, Niyogi UK, Khandal RK: Modification of nylon 66 by electron beam irradiation for improved properties and superior performances. Journal of Applied Polymer Science. 2011;122:193–202.

[13] Aytaç A, Deniz V, Şen M, Hegazy ES, Güven O: Effects of γ and electron beam irradiation on the properties of calendered cord fabrics. Radiation Physics and Chemistry. 2010;79:297–300.

[14] Charlesby A, Pinner SH: Analysis of the solubility behavior of irradiated polyethylene and other polymers. Proceedings of the Royal Society A. 1959;249:367–377.

[15] Sengupta R, Sabharwal S, Tikku VK, Somani AK, Chaki TK, Bhownick AK: Effect of ambient-temperature and high-temperature electron beam radiation on the structural, thermal, mechanical, and dynamic mechanical properties of injection-molded polyamide-6,6. Journal of Applied Polymer Science. 2006;99:1633–1644.

[16] Adem E, Burillo G, del CastilloL F, Vásquez M, Avalos-Borja M: Polyamide-6: the effect on mechanical and physicochemical properties by electron beam irradiation at different temperatures. Radiation Physics and Chemistry. 2014;97:165–171.

[17] Evora MC, Machado LD, Lourenço VL, Gonçalez OL, Wiebeck H, de Andrade e Silva LGJ: Thermal analysis of ionizing radiation effects on recycled polyamide-6. Journal of Thermal Analysis and Calorimetry. 2002;67:327–333.

[18] Rajaee E, Jahani J, Safarpour MA, Vaseghi MR: Mechanical properties, surface chemistry, and barrier characteristics of electron beam irradiated-annealed LDPE/PA6/LDPE multi-layer films at N_2. Polymers for Advanced Technologies. 2011;22:724–731.

[19] Wang B, Hong N, Shi Y, Wang B, Sheng H, Song L, Tang Q, Hu Y: Comparative study on the effect of electron beam irradiation on the physical properties of ethylene-vinyl acetate copolymer composites. Radiation Physics and Chemistry. 2014;97:284–291.

[20] Shin BY, Han DH: Morphological and mechanical properties of polyamide 6/linear low density polyethylene blend compatibilized by electron-beam initiated mediation process. Radiation Physics and Chemistry. 2014;97:198–207.

[21] Shin BY, Kim JH: Rheological and mechanical properties of polyamide 6 by electron-beam initiated mediation process. Radiation Physics and Chemistry. 2015;112:88–96.

[22] Jung HS, Choi MC, Chang YW, Kang PH, Hong SC: Facile preparation of thermoplastic elastomer with high service temperature from dry selective curing of compatibilized EPDM/polyamide-12 blends. European Polymer Journal. 2015;66:367–375.

[23] Kaji K, Okada T, Sakurada I: Radiation crosslinking of nylons. Nippon Genshiryoku Kenkyusho Nempo. 1973;26:5027.

[24] Lyons BJ, GloverLCJr: Radiolytic crosslinking and chain scission in aliphatic and alkyl-aromatic polyamides. Part I. Radiation Physics and Chemistry. 1990;35:139–147.

[25] Nakamura S, Tokumitsu K, Yamaguchi T: Influence of electron beam irradiation on the mechanical and thermal properties of polypropylene/polyamide6 blends. Journal of Applied Polymer Science. 2013:4318–4326.

[26] El-Nemr KF, Hassan MH, Ali MA: Effect of electron beam radiation on mechanical and thermal properties of waste polyamide copolymer blended with nitrile-butadiene rubber. Polymer Advanced Technologies. 2010; 21:735–741.

[27] Wang B, Hong N, Shi Y, Wang B, Sheng H, Song L, Tang Q, Hu Y: Comparative study on the effect of electron beam irradiation on the physical properties of ethylene-vinyl acetate copolymer composites. Radiation Physics and Chemistry. 2014;97:284–291.

[28] Sato H, Iwata K, Nishi Y: Effects of homogeneous low voltage electron beam irradiation on adhesive strength of polycarbonate (PC) sheet covered with Nylon6 film. Materials Transactions. 2009;50:1859–1863.

[29] Engelman G, Gohs U, Ganster J. Monomer cast polyamide 6 composites and their treatment with high-energy electrons. Journal of Applied Polymer Science. 2012;123:1201–1211.

[30] Gupta MC, Pandey RR: γ-Irradiation of nylon 6. Journal of Polymer Science. Part A-1. 1988;26:491–502.

[31] Porubská A, Babic′ D, Janigová I, Šlouf M, Jomová K, Chodák I: The effect of γ irradiation in air and inert atmosphere on structure and properties of unfilled or glass fibre-reinforced polyamide 6. Polymer Bulletin. 2015. In press.

[32] Fortin N, AlbelaB, Bonneviot L, Rouif S, Sanchez JY, Portinha D, Fleury E: How does γ-irradiation affect the properties of a microfiltration membrane constituted of two polymers with different radiolytic behaviour? Radiation Physics and Chemistry. 2012;81:331–338.

[33] Kerr JA. Strengths of chemical bonds. In: Weast RC, editor. CRC Handbook of Chemistry and Physics. Boca Raton, FL: CRC Press; 1985. p. F1 71–90.

[34] Menchaca C, Martínez-Barrera G, Fainleib A: Nylon 6,12 fibers under low-dose γ irradiation. Journal of Polymer Engineering. 2011;31:457–461.

[35] Menchaca C, Martínez-Barrera G, López-Valdivia H, Carrasco H, Álvarez-Castillo A: Post-irradiation effects on γ-irradiated nylon 6,12 fibers. Journal of Polymer Engineering. 2013;33:823–828.

[36] Spadaro G, Acierno D, Calderaro E, Valenza A: Structural induced modifications in blends of polyamide with γ irradiated low density polyethylene. Radiation Physics and Chemistry. 1992;40:249–254.

[37] Valenza A, Calderaro E, Spadaro G: Molecular modifications-mechanical behaviour relationships for γ irradiated LLDPE/PA6 blends. Radiation Physics and Chemistry. 1994;43:315–322.

[38] Kang HK, Shin HK, Jeun JP, KimHBin, Kang PH: Fabrication and characterization of electrospun polyamide 66 fibers crosslinked by γ irradiation. Macromolecular Research. 2011; 19:364–369.

[39] Gupta SK, Singh P, Kumar R: Modifications induced by γ irradiation upon structural, optical and chemical properties of polyamide nylon-6,6 polymer. Radiation Effects and Defects in Solids. 2014;196:679–685.

[40] Hassan MM, Badway NA, Gamal AM, Elnaggar MY, Hegazy EA: Studies on mechanical, thermal and morphological properties of irradiated recycled polyamide and waste rubber powder blends. Nuclear Instruments and Methods in Physics Research B. 2010;268:1427–1434.

[41] Hassan MM: Synergistic effect of montmorillonite-clay and γ irradiation on the characterizations of waste polyamide copolymer and reclaimed rubber powder nanocomposites. Composites Part B. 2015;79:28–34.

[42] Hassan MM, Badway NA, Gamal AM, Elnaggar MY, Hegazy EA: Effect of carbon black on the properties of irradiated recycled polyamide/rubber waste composites. Nuclear Instruments and Methods in Physics Research B. 2010;268:2517–2534.

[43] Sonnier R, Caro-Bretelle AS, Dumazert L, Longerey M, Otazaghine B: Influence of radiation-crosslinking on flame retarded polymer materials-How crosslinking disrupts the barrier effect. Radiation Physics and Chemistry. 2015;106:278–288.

[44] Sadler G, Chappas W, Pierce DE: Evaluation of e-beam, γ- and X-ray treatment on the chemistry and safety of polymers used with pre-packaged irradiated foods: A review. Food Additives and Contaminants. 2001;18:475–501.

[45] Araújo HP, Félix JS, Manzoli JE, Padula M. Effect of γ-irradiation on caprolactam level from multilayer PA-6 films for food packaging: Development and validation of a gas chromatographic method. Radiation Physics and Chemistry. 2008;77:913–917.

[46] Félix JS, Monteiro M, Manzoli JE, Padula M: Effect of γ irradiation on caprolactam migration from multilayer polyamide 6 films into food simulants: Development and validation of a gas chromatographic method. Journal of AOAC International. 2010;93:1874–1881.

[47] Park GY, Cho SY, Jeon DH, Kwak IS, Lee KH, Park HJ: Formation of monomer residues in PS, PC, PA-6 and PVC upon γ-irradiation. Radiation Physics and Chemistry. 2006;75:1055–1059.

[48] Chytiri S, Goulas AE, Badeka A, Riganakos KA, Petridis D, Kontominas MG: Determination of radiolysis products in γ-irradiated multilayer barrier food packaging films containing a middle layer of recycled LDPE. Radiation Physics and Chemistry. 2008;77:1039–1045.

[49] Salafranca J, Clemente I, Isella F, Nerín C, Bosetti O: Influence of oxygen and long term storage on the profile of volatile compounds released from polymeric multilayer food contact materials sterilized by γ irradiation. Analytica Chimica Acta. 2015; 878:118–130.

[50] Porubská M, Szöllős O, Kóňová A, Janigová I, Jašková M, Jomová K, Chodák I: FTIR spectroscopy study of polyamide-6 irradiated by electron and proton beams. Polymer Degradation and Stability. 2012;97:523–531.

Chapter 11

Ion-Irradiation-Induced Carbon Nanostructures in Optoelectronic Polymer Materials

Taras S. Kavetskyy and Andrey L. Stepanov

Additional information is available at the end of the chapter

http://dx.doi.org/10.5772/62669

Abstract

The recent results obtained on the ion-irradiation-induced carbon nanostructures in optoelectronic polymer materials exemplified by boron-ion-implanted polymethylmethacrylate (B:PMMA) with an energy of 40 keV, ion doses from 6.25×10^{14} to 5.0×10^{16} ions/cm^2, and current density <2 μA/cm^2 are reviewed. The positron annihilation spectroscopy (slow positron beam spectroscopy based on Doppler broadening of positron annihilation gamma rays as a function of incident positron energy and positron annihilation lifetime at a positron energy of 2.15 keV, and temperature-dependent positron annihilation lifetime spectroscopy), optical UV-visible spectroscopy, Raman spectroscopy, electrical (current–voltage) measurements, and nanoindentation test are chosen as the main experimental tools for the investigation of low-energy ion-induced processes in B:PMMA. The formation of carbon nanostructures is confirmed for the samples irradiated with higher ion fluences (>10^{16} ions/cm^2) and the experimental results of the comprehensive study are found to be in a good agreement with SRIM (stopping and range of ions in matter) simulation results.

Keywords: ion irradiation, optoelectronic materials, boron-ion-implanted polymethylmethacrylate, carbon nanostructures

1. Introduction

Ion implantation is a powerful experimental approach for structural modification of materials. In case of organic media, the interest to ion-irradiated polymers is due to the ion implantation being one of the effective technological methods to turn dielectric polymers into semiconductors [1] as well as to improve surface-sensitive mechanical properties of polymers for hard-materials applications [2]. Substantial improvements in hardness, wear resistance, oxidation resistance, resistance to chemicals, and electrical conductivity are the main characteristics attained by polymeric materials after their low-to-medium-energy ion implantation [3–7].

The formation of free radicals at lower ion doses ($<10^{16}$ ions/cm^2) and carbonization at higher ion doses ($>10^{16}$ ions/cm^2) in the most polymeric materials are of general concept [8–10]. In particular, the ion irradiation of polymers leads to the scission and cross-linking of polymer chains, formation of volatile low-molecular fragments, and carbonization of the implanted layer. The carbonization process depends strongly on the implantation dose or/ and ion current density. As ion dose increases, the several stages can be distinguished for carbonization to be occurred. The formation of pre-carbon structures, nucleation, and growth of the carbon-enriched clusters, aggregation of the clusters, formation of network of conjugated bonds, and transition to amorphous carbon or graphite-like material are the main stages predicted in literature [8–10]. Despite numerical studies of carbonization processes in polymers, the problem remains to understand better how the carbonaceous phase or carbon nanostructures formed under high-dose ion implantation could be dependent on the type of polymer matrix. In this respect, a lot of efforts have been made by researchers studying the B^+-ion implantation into various polymer matrix such as polycarbonate, Kapton, polyethylene, polyamide, polyimide, poly(ethylene terephthalate), cellulose, polypropylene, polystyrene, polyethersulfone, and others ([2, 4, 5, 11–25] and references therein). At the same time, the B^+-ion implantation into widely used in practice polymer polymethylmethacrylate (PMMA) was not studied so far.

The interest to B^+-ion implantation of polymers is due to formation of buried carbonaceous layer leading to increasing conductivity [20, 23]. Also, B^+-ion implantation into polymeric matrix results in significant increase of surface-sensitive mechanical properties. So, Lee et al. [4] reported the ion-induced improvement in hardness of Kapton irradiated by three different ion species, He, B, and Si at 200 keV to a dose of 3.5×10^{19} ions/m^2, where boron produced the largest improvement in hardness among the three, not following the increasing trend in atomic number.

The selection of PMMA matrix to be used for B^+-ion implantation is due to an importance of this polymer for construction of many optical components (waveguides, lenses, prisms, etc.), lithography, biomedical applications, etc. ([26–30] and references therein). PMMA was also a subject for implantation with Ag^+-ions [9, 31] to fabricate composite structures with silver nanoparticles for plasmonic applications as well as for implantation with C^+, N^+ and Ar^+-ions [29, 30] that may find an extensive application in fabrication of various optoelectronic devices including organic light-emitting diodes, backlight components in liquid crystal display systems, diffractive elements, solar cells, waveguides, microcomponents for integrated optical circuits, etc. The well-known key performance and important characteristics of PMMA such as a long-term stability in outdoor environments, excellent surface hardness, light weight, outstanding transmittance and optical clarity, optical design flexibility and control, etc. [32], and high stability upon positron irradiation at room temperature [33] are also taken into account at the selection of basic polymeric matrix for low-energy ion implantation.

In this chapter, a review of recent results obtained using proper experimental and simulation techniques on the boron-ion-implanted polymethylmethacrylate (B:PMMA) is presented. As a result, the formation of ion-irradiation-induced carbon nanostructures in optoelectronic polymer materials exemplified by B:PMMA is evidently confirmed.

2. Experimental

2.1. Sample preparation and SRIM simulation

The B^+-ion implantation with the energy of 40 keV, doses from 6.25×10^{14} to 5.0×10^{16} ions/cm^2 and current density < 2 μA/cm^2 into the optically transparent PMMA plates (1.2 mm thickness) was performed under a pressure of 10^{-5} Torr at room temperature by an "ILU-3" ion accelerator at the Kazan Physical-Technical Institute (KPTI, Russia) similar to as it was earlier done for Xe^+ and Ag^+ ions [34].

For comparative analysis of depth profiles of implanted ions and introduced vacancies in respect to ion mass, SRIM (stopping and range of ions in matter) simulations were carried out for 40 keV He^+, B^+, O^+, P^+, Cl^+, Cu^+, Ag^+, Xe^+, and Au^+ ions implanted into PMMA using the free of charge software of version SRIM-2013 [35].

2.2. Positron annihilation spectroscopy measurements

Positron annihilation spectroscopy (PAS) measurements with slow positron beam spectroscopy (SPBS) based on Doppler broadening of positron annihilation gamma-rays as a function of incident positron energy and positron annihilation lifetime at constant positron energy were performed at the National Institute of Advanced Industrial Science and Technology (AIST, Japan) [36]. Doppler broadening spectra (S–E and W–E) were measured using a slow positron beamline in the range of positron incident energy E from 0 to 30 keV. Positron annihilation lifetime spectra at incident positron energy of 2.15 keV were measured by another slow positron beamline using an electron linear accelerator as an intense source of slow positrons. The details of the SPBS measurements are reported elsewhere [36].

PAS measurements with positron annihilation lifetime spectroscopy (PALS) and Doppler broadening of annihilation line (DBAL) techniques in the temperature range of 50–300 K using helium cryostat (Closed Cycle Refrigerator, Janis Research Company, Inc., USA) and vacuum equipment (Pfeiffer Vacuum, HiCUBE, Germany) were carried out at the Institute of Physics, Slovak Academy of Sciences (IPSAS, Slovakia) [37–39]. The samples were measured in the cycles of heating and cooling with step of 20 K and elapsed time of 4–5 hours per point. At the selected temperatures, the elapsed time was extended for better statistics in the lifetime spectra. For these temperatures, the continuous lifetime analysis technique to obtain the distributions of *ortho*-positronium (*o*-Ps) lifetime (free-volume voids) was employed using maximum entropy lifetime (MELT) program [40]. The positron annihilation lifetime spectra were taken by the conventional fast-fast coincidence method using plastic scintillators coupled to photomultiplier tubes as detectors. The radioactive ^{22}Na positron source (1.5 MBq activity) was deposited in an envelope of Kapton foils and then sandwiched between two samples. This source-sample assembly was placed in a vacuum chamber between two detectors to acquire lifetime spectra at different temperatures. The time resolution (FWHM) of positron lifetime spectrometer was 0.32 ns, measured by defect free Al sample as a standard. Analysis of lifetime spectra was carried out using the PATFIT-88/POSITRONFIT [41] software package with proper source corrections. Three component fitting procedure for PALS data treatment was applied and long-lived lifetime component τ_3 and its intensity I_3, ascribing to the *o*-Ps pick-off annihilation in free-volume spaces, was finally taken into account for analysis. Simultaneously with PALS measurements, the DBAL spectra were recorded using a high-purity Ge detector with energy resolution of 1.9 keV FWHM at the energy 1274 keV. The annihilation line was deconvoluted by Gold algorithm that allows eliminating a linear instability of the measuring equipment if the 1274 keV ^{22}Na peak is measured simultaneously with the annihilation peak, for deconvolution of annihilation peak 511 keV; this procedure permits the measurement of small changes of the annihilation peak with high confidence [42]. The Doppler *S* and *W* parameters were used for analysis of a shape of deconvoluted annihilation peak 511 keV. Their numerical values were defined as the ratio of the central area to the total area of the photon peak for *S* parameter and as the ratio of wings area to the total area of the photon peak for *W* parameter.

2.3. Optical spectroscopy measurements

Optical UV-visible spectroscopy measurements were performed using a SHIMADZU-UV 3100PC spectrophotometer in the range of 200–800 nm at AIST, Japan [36].

2.4. Raman spectroscopy measurements

Room temperature Raman spectra were recorded using a Renishaw Raman inVia Reflex spectrometer in the 400–3800 cm^{-1} range with a spectral resolution of ≤1 cm^{-1} [43]. The 514 nm Modu-Laser Stellar-REN Pro 514/50 Argon Laser and 785 nm Near Infrared Diode Laser lines were applied as the excitation source. In order to avoid heat-induced effects, the laser power was set at 10 mW. A standard calibration of wavenumber scale with Si plate was used. The Raman spectroscopy measurements were performed at The John Paul II Catholic University of Lublin (KUL, Poland).

2.5. Electrical measurements

The electrical measuring system was developed at the IPSAS (Slovak Republic) [43]. The measurement of current vs. voltage (*I–V*) by transient method or DC method was used in this experiment. The method was selected to check simply presence of the conductivity layer in the material studied. The measuring system includes a cryostat with sample, resistor heating, thermocouple, CSP (charge sensitive preamplifier), excitation circuit of sample, preamplifier, power supply, DAQ (data acquisition) card, and PC (personal computer). The NI PCI 6229 DAQ card was applied [44]. The electrical DC measurement setup used and detail description of basic parameters are reported elsewhere [43].

http://dx.doi.org/10.5772/62669

2.6. Nanoindentation test

Nanoindentation test was carried out using an ultra nano hardness tester (UNHT) with a diamond Berkovich indenter at the Lublin University of Technology (LUT, Poland) [45, 46]. Advantages of the new UNHT design applied for nanoindentation test compared to the conventional nano indenter (or NHT) design, both developed by CSM Instruments (Switzerland) [47], allow us to make measurements with high performance. The UNHT experiment was done in a progressive multicycle mode. The details of the progressive multicycle mode parameters used are reported elsewhere [45].

3. Results and discussion

3.1. SRIM simulation data

Figure 1 shows the typical SRIM simulation results for B^+-implantation into PMMA at energy of 40 keV [36, 45]. Numerical values of SRIM simulation for 40 keV He^+, B^+, O^+, P^+, Cl^+, Cu^+, Ag^+, Xe^+, and Au^+ ions into PMMA are gathered in **Table 1** [48].

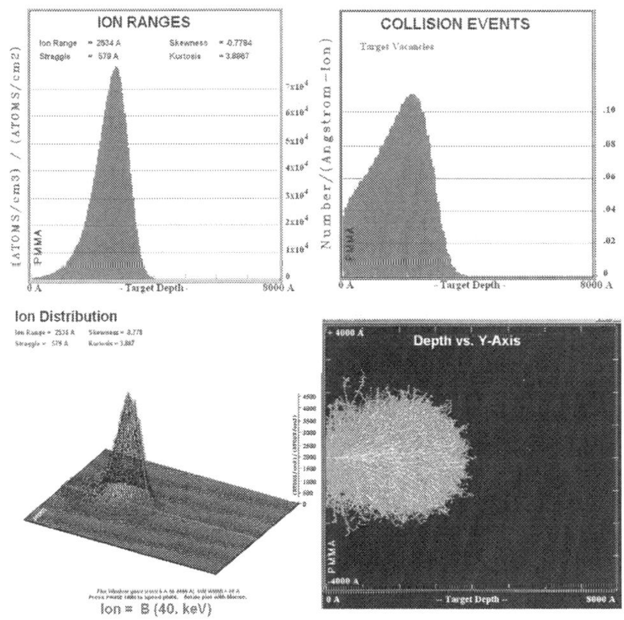

Figure 1. Depth profiles of implanted ions (top, left), introduced vacancies (top, right), 3D image of ion distribution (bottom, left) and depth vs. Y-axis (bottom, right) for B:PMMA. Adapted from [36, 45].

Ion	[b]Atomic weight	[c]R_{max}^{Ion}, nm	R_p^{Ion}, nm	ΔR_p^{Ion}, nm	[c]R_{max}^{V}, nm	[c]R_p^{V}, nm
He^+	4.0026	880	593	102	820	560
[a]B^+	10.81	400	253	58	380	220
[a]O^+	15.999	300	169	45	280	150
P^+	30.974	200	99	29	190	70
Cl^+	35.45	160	85	23	170	60
Cu^+	63.546	130	69	18	140	40
Ag^+	107.87	100	56	11	95	35
Xe^+	131.29	90	56	10	90	30
[a]Au^+	196.97	75	54	7	60	25

Table 1. SRIM simulation results for 40 keV ion implantation into PMMA ([a]data from the work [36]; [b]data from www.webelements.com; [c]data are estimated with error ± 5 nm) [48].

The SRIM data, presented in **Table 1**, were used for estimation a predicted thickness of implanted layer [48]. Let us to explain it on the example of the modified PMMA by 40 keV accelerated B^+ ions. A mean penetration range (R_p^B) is about 253 nm with a longitudinal straggling (ΔR_p^B) of 58 nm in the Gaussian depth distribution. The assumed predicted thickness of the modified PMMA surface layer ($R_p^B + 2\Delta R_p^B$) is about 369 nm, and the maximum penetration depth (R_{max}^B) is about 400 nm. At the same time, the vacancy distribution gives a maximum damage around $R_p^{V,B} = 220$ nm and maximum depth up to around $R_{max}^{V,B} = 380$ nm. Thus, the estimations performed for various ions, listed in **Table 1**, indicate the possible modification of PMMA surface upon low-energy ion implantation in dependence on the ion mass. These values could also be useful in practice for evaluation of geometrical parameters of ion-implanted layers in PMMA matrix.

3.2. Positron annihilation spectroscopy data

The 40 keV B:PMMA polymers with different ion doses have been studied for the first time using PAS techniques such as SPBS [36] and temperature-dependent PALS [37].

SPBS, often called variable-energy PAS, is a powerful experimental tool widely used for evaluation of defects in solids as a function of depth (defect depth profiling) by varying the positron energy in the range of a few eV to tens keV (for review, see Ref. [49]). Thus, SBPS could be a very effective for the detection of defects induced by ion implantation, which are localized near the surface of material. At first, the un-implanted PMMA and implanted B:PMMA were characterized by Doppler broadening of annihilation gamma rays (or DBAL) as a function of incident positron energy in the range of 0–30 keV. And then, using Doppler broadening results, the PAL spectra at incident positron energy of 2.15 keV were measured and analyzed for the investigated samples. **Figure 2** shows the typical variable-energy DBAL and PAL results obtained, the explanation of which has been done in [36].

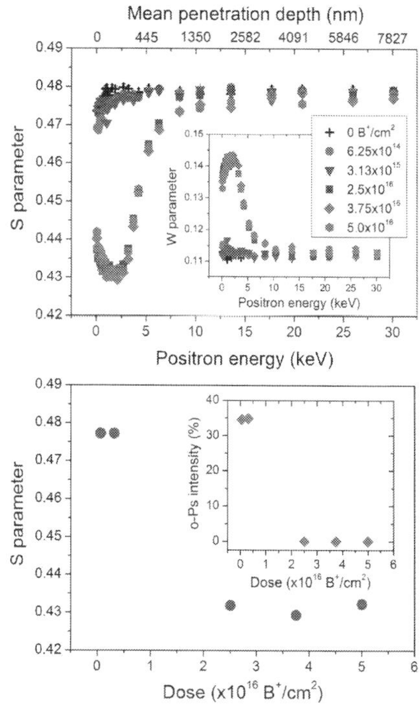

Figure 2. Dose dependence of Doppler S and W parameters as a function of incident positron energy in the range of 0–30 keV (top) and S parameter and *o*-Ps intensity at incident positron energy of 2.15 keV (bottom) for B:PMMA. The error bars are within the size of the symbol. Adapted from [36].

It should only be emphasized here that there are two different processes seen from the S-E and W-E curves in dependence on the ion implantation dose [36]. In particular, (i) $S(E)$ increases and $W(E)$ decreases at lower fluences (6.25×10^{14}–3.13×10^{15} ions/cm^2), while (ii) $S(E)$ decreases and $W(E)$ increases at higher fluences (2.5×10^{16}–5.0×10^{16} ions/cm^2) in the range up to 400 nm in a good agreement with the maximum penetration depth $R_{max}{}^{B}$ = 400 nm after SRIM simulation; positron energy of 2–3 keV, showing the extreme values of $S(E)$ and $W(E)$, corresponds to a mean depth of 100–200 nm in consistent with the maximum damage $R_p{}^{V,B}$ = 220 nm after SRIM simulation (**Figure 2**, top). A mean penetration depth of positrons (z_m) was estimated as $z_m = (40/\rho)E^n$, where z_m is presented in nm, E the positron energy in keV, ρ is the density (1.18 g/cm^3 for PMMA), and n = 1.6 [50]. The results of variable-energy DBAL and PAL measurements at incident positron energy of 2.15 keV were found to be in consistence (**Figure 2**, bottom) [36]. That is, the decreasing suddenly the $S(E)$ values and an absence of any observable *o*-Ps yield (intensity $I_3 \sim 0$) for the implanted samples at higher ion doses were detected to be explained due to carbonization effect, taking into account that no *o*-Ps yield has been observed in carbon-based materials such as, for instance, fullerene C_{60} cage [51] and carbon molecular sieve membranes [52].

Thus, the expected two processes of polymer structure modification upon low-energy ion implantation—formation of free radicals at lower fluences (< 10^{16} ions/cm^2) and carbonization at higher fluences (>10^{16} ions/cm^2)—are plausibly confirmed [36].

The results of temperature dependent PALS measurements of the investigated pristine PMMA and implanted B:PMMA samples at lower ion dose (3.13×10^{15} ions/cm^2) and at higher ion dose (3.75×10^{16} ions/cm^2) are demonstrated in **Figure 3**. The detail description and possible explanation of these data have been presented in [37]. Two structural transitions in the vicinity of ~150 and ~250 K, ascribed to γ and β transitions, respectively, should be noted here, which are observed in the both PMMA and B:PMMA. These structural transitions are found to be in consistent with reference data for PMMA [53, 54].

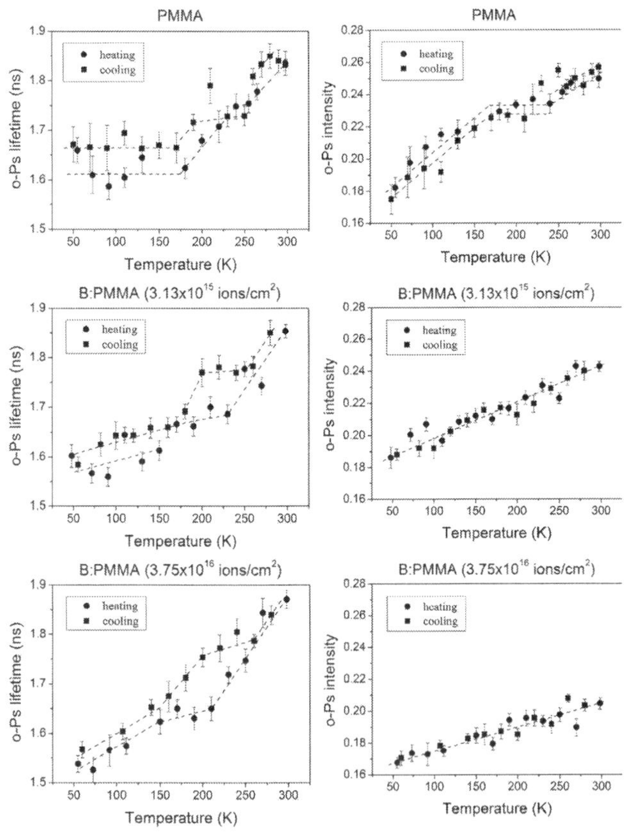

Figure 3. o-Ps lifetime and intensity as a function of temperature in the range of 50–298 K for (top) PMMA, (middle) B:PMMA (3.13×10^{15} ions/cm^2), and (bottom) B:PMMA (3.75×10^{16} ions/cm^2). The dashed lines are drawn as a guide for the eye. Adapted from [37].

The threshold temperatures nearby ~150 and ~250 K for PMMA are also detected on Doppler parameters S/S_0 vs. W/W_0 at different temperatures, normalized to the S_0 and W_0 values at room temperature, using low-temperature DBAL measurements [39] as shown in **Figure 4**. The observed correlation in temperature dependences of S and W parameters and *o*-Ps data (see **Figure 3**) is found to be in a good agreement with literature PALS and DBAL data for PMMA as well [33].

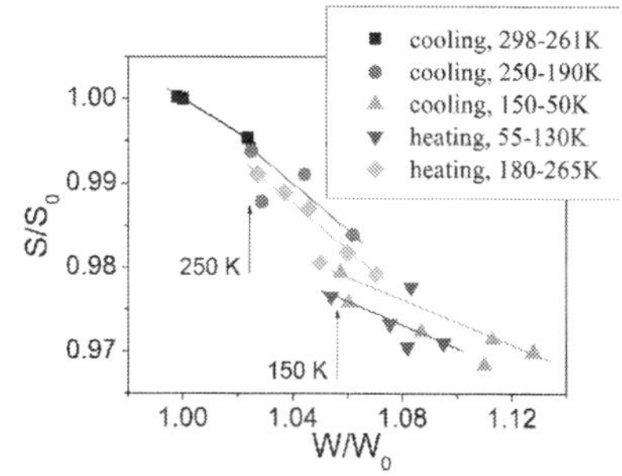

Figure 4. The normalized parameters S/S_0 vs. W/W_0 at different temperatures in the cycles of heating and cooling for PMMA. The solid lines are drawn as a guide for the eye. Adapted from [39].

New results are also obtained from the PALS study of B:PMMA on the free-volume voids distribution at room temperature [37] as shown in **Figure 5**. Namely, estimating the distribution of *o*-Ps lifetime (or the distribution of free volume detected by *o*-Ps) by MELT program similarly as it has been done in work [55], it is found that B^+-ion implantation leads to the shortening the lifetime distribution and decreasing molecular weight in PMMA at lower ion dose (3.13×10^{15} ions/cm^2), while at higher ion dose (3.75×10^{16} ions/cm^2) the broadening the lifetime distribution is probably caused by local destruction of PMMA matrix and generation of additional free volumes.

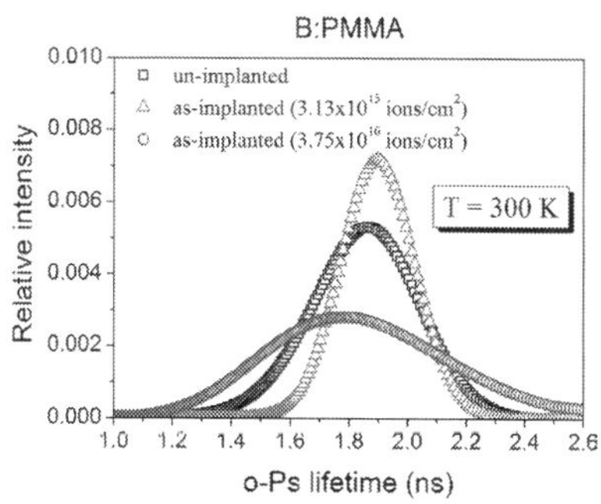

Figure 5. *o*-Ps lifetime distributions in B:PMMA at different ion doses at room temperature. Adapted from [37].

3.3. Optical spectroscopy data

Application of UV-visible optical absorption spectroscopy for investigation of ion-implanted polymeric materials has already been reported for the Ag^+-implanted PMMA, ORMOCER, and Epoxy resin [31, 34, 56–59] as well as for the C^+, N^+, and Ar^+-implanted PMMA [29, 30]. It has been suggested by the authors that ion irradiation creates compact carbonaceous clusters in polymers, which may also be responsible for a narrowing of optical band gap, enhanced electrical conductivity, and increasing optical absorbance (for example, see [31, 34, 56]). In the case of the investigated B:PMMA, it is found the gradual increase of absorbance at lower fluences (<10^{16} ions/cm^2) and saturation of absorbance at higher fluences (>10^{16} ions/cm^2) as shown in **Figure 6** [36].

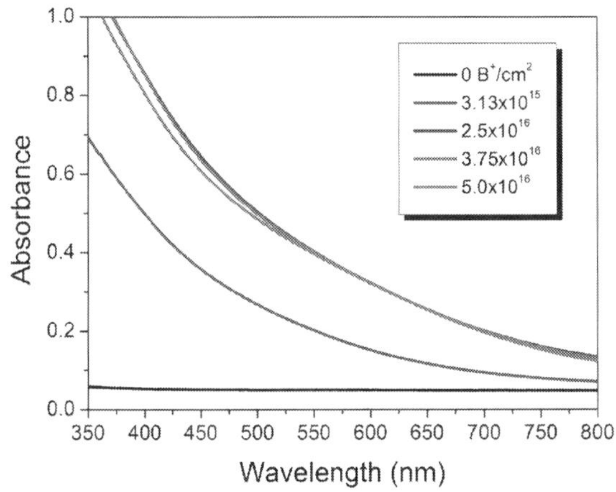

Figure 6. The UV-visible optical absorption spectra of B:PMMA. Adapted from [36].

The observed increasing absorbance for the B:PMMA samples in the course of the ion implantation should be also interpreted as the signature on the formation of carbonaceous clusters which are plausibly confirmed by slow positrons [36].

3.4. Raman spectroscopy data

Figure 7 shows the Raman spectra of the investigated samples excited by 514 nm and 785 nm diode laser lines [43]. Identification of the observed Raman bands has been reported in [43]. The ion-irradiation-induced structural changes as revealed from Raman study can be summarized as follows. New C=C and C–C bands in the vicinity of ~1590 and 1322 cm^{-1}, respectively, are formed for the as-implanted samples at higher fluences (>10^{16} ions/cm^2). At the same time, the decreasing intensity of CH_2 band at ~815 cm^{-1}, C=O band at ~1730 cm^{-1}, O–CH_3 band at ~991 and 2845 cm^{-1}, and C–H band at ~1455, 2955, and 3001 cm^{-1} is detected.

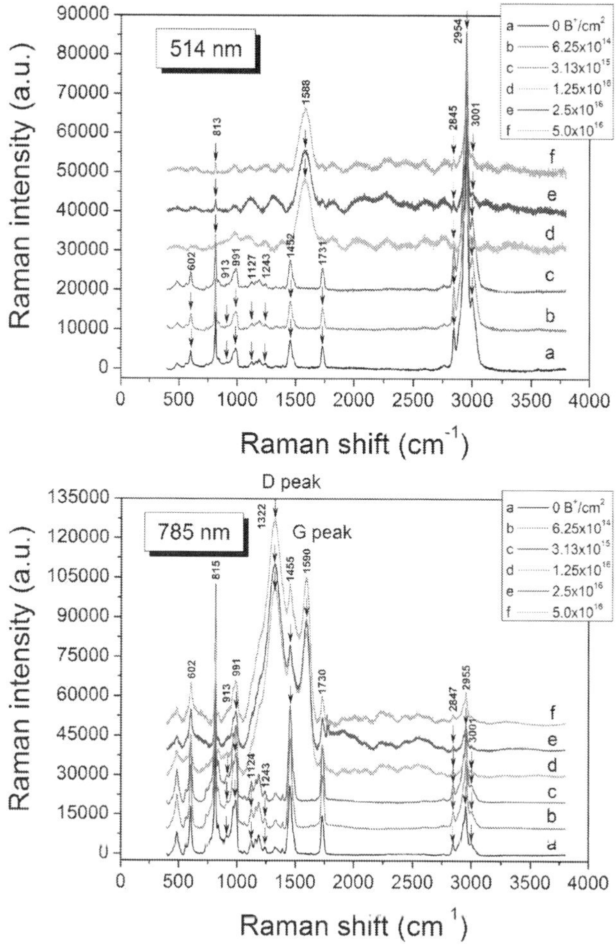

Figure 7. Raman spectra of PMMA and B:PMMA for ion doses from 6.25×10^{14} to 5.0×10^{16} ions/cm^2 excited by (top) 514 nm and (bottom) 785 nm laser lines. Adapted from [43].

In particular, Raman spectroscopy data for excitation by the laser wavelength of 785 nm seem to be additional confirmation for the carbonization processes in the B:PMMA. Indeed, new Raman bands at ~1325 and 1590 cm^{-1}, attributed to C–C and C=C vibrations, respectively, are detected only for the ion-irradiated samples with higher fluences (>10^{16} ions/cm^2). These two new Raman bands may also be attributed to the so-called D and G peaks in the region of 1300-1600 cm^{-1} which are the main peaks characteristic for graphite and graphene structure [60–66]. The intensity ratio of the D and G peaks I_D/I_G is a measure of the size of the sp^2 phase organized in rings [67]. The sp^2 phase is mainly organized in chains, when I_D/I_G is negligible

[63]. The relation between I_D/I_G and the size of the sp^2 phase L_a is given by the equation of Tuinstra and Koenig [60]: $I_D/I_G = C(\lambda)/L_a$. Here, $C(\lambda)$ is a wavelength dependent factor [29, 68]: $C(\lambda) = -126 + 0.033\lambda$, where λ is the excitation wavelength in (Å) at which the Raman spectra were recorded.

A correlation between slow positron beam and Raman spectroscopy results for B:PMMA is mentioned in our recent works [69]. **Table 2** gives the SPBS data, exemplified by *o*-Ps lifetimes and intensities at incident positron energy of 2.15 keV and Raman data, exemplified by I_D/I_G and L_a, for the same B:PMMA samples. A good correlation between these data is clearly observed. Thus, the expected carbonization processes or formation of carbon nanostructures at higher fluences (>10^{16} ions/cm^2) can independently be identified using SPBS and Raman spectroscopy, providing their combination as a powerful experimental tool in the investigation of ion-implanted polymers.

Dose [B$^+$/cm^2]	*o*-Ps lifetime, τ_3 [ns]	*o*-Ps intensity, I_3 [%]	I_D/I_G	L_a [nm]
0	1.765	35.0	No peaks	
6.25×10^{14}	1.752	34.6	No peaks	
3.13×10^{15}	1.752	34.8	No peaks	
2.5×10^{16}		No *o*-Ps or ~0	1.42	9.4
5.0×10^{16}		No *o*-Ps or ~0	1.40	9.5

Table 2. *o*-Ps lifetimes and intensities at incident positron energy of 2.15 keV and the intensity ratio of the D and G peaks, and calculated sizes of sp^2 phase L_a [69].

3.5. Electrical measurements data

The un-implanted PMMA and as-implanted B:PMMA samples were measured at 300 K and 360 K with DC method [43]. It was supposed that the conductive layer in ion-implanted polymer is dependent on the temperature and at a higher temperature the effect of increasing conductivity with temperature will be more pronounced. In order to avoid a possible structural change in polymer matrix, the maximum temperature at 360 K was selected in the electrical measurement experiment, not exceeding the glass transition temperature T_g of PMMA ranging from 358 to 438 K; the range is so wide because of the vast number of commercial compositions which are copolymers with co-monomers other than methyl methacrylate [70]. Similar value $T_g \cong 355 \pm 18$ K of PMMA was obtained using temperature dependent positron annihilation lifetime measurements [53].

Figure 8 shows the obtained *I-V* dependences, fitted by the linear regression line [43]. As an example, the *I-V* characteristic of as-implanted B:PMMA (5.0×10^{16} ions/cm^2) sample is presented with noise (as-implanted + noise (II)) and linear regression line (linear fit of as-implanted (II)). While for other samples only, the linear regression lines of the DC measurements are demonstrated. More details of the electrical measurement experiment have been reported in [43].

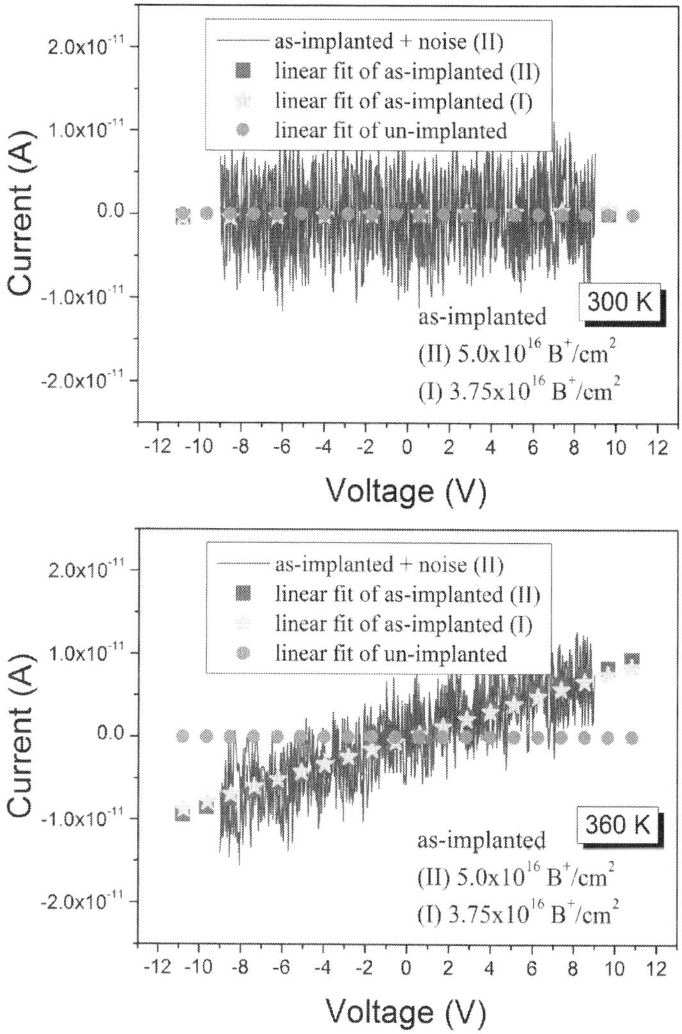

Figure 8. *I-V* dependences for PMMA and B:PMMA (3.75 × 10^{16} and 5.0 × 10^{16} ions/cm^2) at temperatures (top) 300 and (bottom) 360 K. Adapted from [43].

The numerical values of the *I-V* characteristics are presented in **Table 3** [43]. It is evidently proven that the pristine PMMA does not exhibit any conductivity neither at higher temperature applied. At room temperature, the slope of linear regression line of *I-V* dependence remains unchanged and has even negative value. In contrary to unimplanted sample, the results of the *I-V* measurements for the B:PMMA (3.75 × 10^{16} and 5.0 × 10^{16} ions/cm^2) samples revealed the change of slope of the linear regression line of *I-V* dependence at 360 K. It means that the as-implanted samples have created a very thin conductive layer or conductive joints due to carbonization processes or formation of carbon nanostructures in consistent with the results of slow positron beam and Raman spectroscopy measurements [36, 43].

Parameters of linear regression	Temperature 300 K		
	PMMA	B:PMMA (3.75×10^{16} ions/cm^2)	B:PMMA (5.0×10^{16} ions/cm^2)
Slope	−5.85E-15	4.79E-14	2.18E-14
Error	2.64E-14	2.64E-14	3.56E-14
CC	0.008	0.07	0.02
SD	3.59E-12	3.68E-12	4.98E-12
Parameters of linear regression	**Temperature 360 K**		
	PMMA	**B:PMMA (3.75×10^{16} ions/cm^2)**	**B:PMMA (5.0×10^{16} ions/cm^2)**
Slope	2.88E-16	7.96E-13	8.68E-13
Error	1.62E-15	2.71E-14	2.53E-14
CC	0.007	0.73	0.79
SD	2.77E-13	3.77E-12	3.53E-12

Table 3. The evaluated values of *I-V* measurements of PMMA and B:PMMA with higher fluences (3.75×10^{16} and 5.0×10^{16} ions/cm^2) at temperatures 300 and 360 K: Slope is value of slope of line of linear regression, Error is the error of linear regression, CC is correlation coefficient, and SD is standard deviation of the linear regression [43].

3.6. Nanoindentation test data

A first time the results of investigation of the influence of low dose (6.25×10^{14} ions/cm^2) B^+-ion-irradiation on the mechanical properties (hardness and elastic modulus) of PMMA probed by nanoindentation with UNHT in the range of 300–1100 nm indentation depth have been reported in [45]. It has been established that the hardness and elastic modulus versus maximum indentation depth illustrate the main difference between the un-implanted (pristine) and ion-implanted samples in the range up to about 400 nm. The same value of the maximum penetration depth of B^+-ions into PMMA has been found using SPBS and SRIM simulation [36]. As a continuation of the nanoindentation test of the B:PMMA, the averaged values of indentation hardness versus maximum indentation depth for the un-implanted and as-implanted samples with ion doses of 6.25×10^{14}, 1.25×10^{16}, 2.5×10^{16}, and 3.75×10^{16} ions/cm^2 are plotted in **Figure 9** [46].

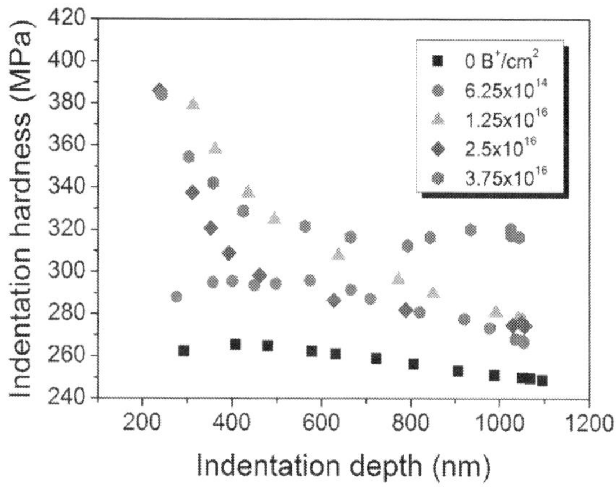

Figure 9. Indentation hardness versus maximum indentation depth for PMMA and B:PMMA (6.25×10^{14}, 1.25×10^{16}, 2.5×10^{16}, and 3.75×10^{16} ions/cm^2). Adapted from [46].

One may see that the hardness dependence on the maximum indentation depth demonstrates the difference between the PMMA and B:PMMA samples in the entire range studied up to 1100 nm with the largest changes in the vicinity of 300-400 nm in consistence with the maximum penetration depth of B^+-ions into PMMA as revealed from SPBS measurements and SRIM simulation [36]. The observed improving of surface-sensitive mechanical properties of B:PMMA by ion beam processing is obviously detected to be more significant as ion dose increases, that may be suitable for hard-materials applications. According to Lee et al. [4] these properties are apparently related to the effectiveness of cross-linking. Besides, the abovementioned formation of carbon nanostructures upon high-dose ion implantation (>10^{16} ions/cm^2) seems to be also important. But further increase of hardness with deeper penetration of indenter up to 1100 nm for higher fluences (3.75×10^{16} B^+/cm^2) was found to be very interesting and not fully understood yet (see **Figure 9**). Actually, similar results that the hardness of B^+-implanted polycarbonate increased with increasing ion dose has also been observed in Ref. [2] but only for penetration of indenter up to 400 nm, that is not so far from the implanted layer as in our case. A deeper understanding the low-energy ion-irradiation-induced processes in polymeric materials exemplified by PMMA irradiated by accelerated light, middle, and heavy ions is still required.

4. Conclusions

The formation of carbon nanostructures has been confirmed for the B:PMMA samples irradiated with higher ion fluences (>10^{16} ions/cm^2) and the experimental results of the comprehensive study have been found to be in a good agreement with SRIM simulation results. It is expected that the results obtained for low-energy B^+-ion implantation into PMMA will have impact on research and development in the fields of nanoscience, nanotechnology, and nanooptoelectronics, in particular, to be potentially of interest for fabrication of organic luminescent devices, backlight components in liquid crystal display systems, diffractive elements and microcomponents for integrated optical circuits, solar cells, waveguides, etc. similarly as it has been foreseen for low-energy C^+, N^+, and Ar^+-ion implantation into PMMA [29, 30].

Acknowledgements

T.S. Kavetskyy acknowledges the SAIA for scholarships in the Institute of Physics of SAS within the National Scholarship Program of the Slovak Republic. This work was also supported in part by the SFFR of Ukraine (Nos. F40.2/019 and F52.2/003) and MES of Ukraine (Nos. 0114U002616 and 0114U002617). A.L. Stepanov thanks the RFBR (No. 15-48-02525) and RSF (No. 14-13-00758) in Russia for financial support.

Author details

Taras S. Kavetskyy[1,2*] and Andrey L. Stepanov[3,4,5]

*Address all correspondence to: kavetskyy@yahoo.com

1 The John Paul II Catholic University of Lublin, Lublin, Poland

2 Drohobych Ivan Franko State Pedagogical University, Drohobych, Ukraine

3 Kazan Physical-Technical Institute, Russian Academy of Sciences, Kazan, Russian Federation

4 Kazan Federal University, Kazan, Russian Federation

5 Kazan National Research Technological University, Kazan, Russian Federation

References

[1] Wasserman B, Braunstein G, Dresselhaus MS, Wnek GE. Implantation-induced conductivity of polymers. Mater. Res. Soc. Symp. Proc. 1983; 27: 423–428. DOI: 10.1557/PROC-27-423

[2] Lee EH, Rao GR, Mansur LK. Improved hardness and wear properties of B-ion implanted polycarbonate. J. Mater. Res. 1992; 7: 1900–1911. DOI: 10.1557/JMR.1992.1900

[3] Bridwell LB. Ion implantation of polymers for electrical conductivity enhancement. Solid State Phenomena. 1992; 27: 163–180. DOI: 10.4028/www.scientific.net/SSP.27.163

[4] Lee EH, Rao GR, Lewis MB, Mansur LK. Ion beam application for improved polymer surface properties. Nucl. Instr. Meth. B. 1993; 74: 326–330. DOI: 10.1016/0168-583X(93)95070-L

[5] Rao GR, Lee EH, Mansur LK. Structure and dose effects on improved wear properties of ion-implanted polymers. Wear 1993; 162–164: 739–747. DOI: 10.1016/0043-1648(93)90074-V

[6] Goyal PK, Kumar V, Gupta R, Kumar S, Kumar P, Kanjilal D. Effect of Ar+ ion implantation on electrical conductivity of polycarbonate. AIP Conf. Proc. 2011; 1349: 543–549. DOI: 10.1063/1.3605973

[7] Goyal PK, Kumar V, Gupta R, Kumar S, Kumar P, Kanjilal D. Study of electrical conductivity of Kr^+ ion implanted polycarbonate in relation to carbon structure. AIP Conf. Proc. 2011; 1393: 147–151. DOI: 10.1063/1.3653652

[8] Sviridov DV. Chemical aspects of implantation of high-energy ions into polymeric materials. Russ. Chem. Rev. 2002; 71: 315–327. DOI: 10.1070/RC2002v071n04ABEH000710

[9] Stepanov AL. Optical properties of metal nanoparticles synthesized in a polymer by ion implantation: A review. Tech. Phys. 2004; 49: 143–153. DOI: 10.1134/1.1648948

[10] Popok VN. Ion implantation of polymers: Formation of nanoparticulate materials. Rev. Adv. Mater. Sci. 2012; 30: 1–26. http://www.ipme.ru/e-journals/RAMS/no_13012/01_popok.pdf

[11] Fink D, Müller M, Stettner U, Behar M, Fichtner PFP, Zawislak FC, Koul S. Non-regular depth profiles of light ions implanted into organic polymer films. Nucl. Instr. Meth. B 1988, 32: 150–154. DOI: 10.1016/0168-583X(88)90199-1

[12] Guimaraes RB, Amaral L, Behar M, Fichtner PFP, Zawislak FC, Fink D. Implanted boron depth profiles in the AZ111 photoresist. J. Appl. Phys. 1988; 63: 2083–2085. DOI: 10.1063/1.341112

[13] Azarko II, Hnatowicz V, Kozlov IP, Kozlova EI, Odzhaev VB, Popok VN. EPR spectroscopy of ion implanted polymer films. Phys. Stat. Sol. A 1994; 146: K23–K27. DOI: 10.1002/pssa.2211460239

[14] Azarko II, Karpovich IA, Kozlov IP, Kozlova EI, Odzhaev VB, Popok VN, Hnatowicz V. Influence of ion implantation on the properties of polymer films. Solid State Commun. 1995; 95: 49–51. DOI: 10.1016/0038-1098(95)00905-1

[15] Odzhaev VB, Azarko II, Karpovich IA, Kozlov IP, Popok VN, Sviridov DV, Hnatowicz V, Jankovskij ON, Rybka V, Švorčík V. The properties of polyethylene and polyamide

implanted with B+ ions to high doses. Mater. Letters 1995; 23: 163–166. DOI: 10.1016/0167-577X(94)00274-6

[16] Popok VN, Odzhaev VB, Kozlov IP, Azarko II, Karpovich IA, Sviridov DV. Ion beam effects in polymer films: Structure evolution of the implanted layer. Nucl. Instr. Meth. B 1997; 129: 60–64. DOI: 10.1016/S0168-583X(97)00208-5

[17] Vacik J, Cervena J, Fink D, Klett R, Hnatowicz V, Popok V, Odzhaev V. High fluence boron ion implantation into polymers. Radiat. Eff. Def. Solids. 1997; 143: 139–156. DOI: 10.1080/10420159708212955

[18] Kozlov IP, Odzhaev VB, Karpovich IA, Popok VN, Sviridov DV. Optical properties of ion-implanted polymer layers. J. Appl. Spectr. 1998; 65: 390–394. DOI: 10.1007/BF02675458

[19] Kozlov IV, Odzhaev VB, Popok VN, Azarko II, Kozlova EI. Paramagnetic properties of ion-implanted polymer layers. J. Appl. Spectr. 1998; 65: 583–588. DOI: 10.1007/BF02675652

[20] Popok VN, Karpovich IA, Odzhaev VB, Sviridov DV. Structure evolution of implanted polymers: Buried conductive layer formation. Nucl. Instr. Meth. B. 1999; 148: 1106–1110. DOI: 10.1016/S0168-583X(98)00799-X

[21] Vacík J, Hnatowicz V, ČervenáJ, Peřina V, Popok V, Odzhaev V, Fink D. High fluence boron implantation into polyimide. Nucl. Instr. Meth. B 1999; 148: 1126–1130. DOI: 10.1016/S0168-583X(98)00814-3

[22] Vacík J, Hnatowicz V, ČervenáJ, Peřina V, Popok V, Odzhaev V, Švorčík V, Rybka V, Arenholz E, Fink D. Annealing behaviour of boron atoms implanted into polyethyleneterephtalate. Nucl. Instr. Meth. B 2000; 166–167: 637–640. DOI: 10.1016/S0168-583X(99)01049-6

[23] Odzhaev VB, Popok VN, Kozlova EI, Jankovskij ON, Karpovich IA. Electrical properties of polyethylene modified by ion implantation and diffusion. Nucl. Instr. Meth. B 2000; 166–167: 655–659. DOI: 10.1016/S0168-583X(99)01212-4

[24] Popok VN, Odzhaev VB, Azarko II, Kozlov IP, Sviridov DV, Hnatowicz V, Vacík J, ČervenáJ. Multistage ion implantation of polyamide-6 films. Nucl. Instr. Meth. B 2000; 166–167: 660–663. DOI: 10.1016/S0168-583X(99)01211-2

[25] Popok VN, Azarko II, Odzhaev VB, Tóth A, Khaibullin RI. High fluence ion beam modification of polymer surfaces: EPR and XPS studies. Nucl. Instr. Meth. B 2001; 178: 305–310. DOI: 10.1016/S0168-583X(00)00491-2

[26] Kuo SW, Kao HC, Chang FC. Thermal behavior and specific interactions in high glass transition temperature PMMA copolymer. Polymer 2003; 44: 6873–6882. DOI: 10.1016/j.polymer.2003.08.026

[27] Teixeira FS, Salvadori MC, Cattani M, Brown IG. Gold-implanted shallow conducting layers in polymethylmethacrylate. J. Appl. Phys. 2009; 105: 064313-1–064313-5. DOI: 10.1063/1.3088874

[28] Tsvetkova T, Balabanov S, Avramov L, Borisova E, Angelov I, Sinning S, Bischoff L. Photoluminescence enhancement in Si^+ implanted PMMA. Vacuum 2009; 83: S252–S255. DOI: 10.1016/j.vacuum.2009.01.075

[29] Wang J, Zhu F, Zhang B, Liu H, Jia G, Liu C. Photoluminescence and reflectivity of polymethylmethacrylate implanted by low-energy carbon ions at high fluences. Appl. Surf. Sci. 2012; 261: 653–658. DOI: 10.1016/j.apsusc.2012.08.076

[30] Gupta R, Kumar V, Goyal PK, Kumar S. Optical characterization of poly(methyl methacrylate) implanted with low energy ions. Appl. Surf. Sci. 2012; 263: 334–338. DOI: 10.1016/j.apsusc.2012.09.056

[31] Stepanov AL. Optical extinction of metal nanoparticles synthesized in polymer by ion implantation. In: Nicolais L, Carotenuto G, editors. Metal-Polymer Nanocomposites. Italy: Wiley; 2005. p. 241–263. DOI: 10.1002/0471695432.ch8

[32] Colburn P. Acrylic materials in PV applications: Making an informed choice. Evonik Industries – NREL PV Reliability Workshop [Internet]; 2011. Available from: http://www1.eere.energy.gov/solar/pdfs/pvmrw2011_23_cpv_colburn.pdf [accessed 2011-02-17].

[33] Qi N, Chen ZQ, Uedono A. Molecular motion and relaxation below glass transition temperature in poly (methyl methacrylate) studied by positron annihilation. Rad. Phys. Chem. 2015; 108: 81–86. DOI: 10.1016/j.radphyschem.2014.11.018

[34] Stepanov AL, Abdullin SN, Petukhov VY, Osin YN, Khaibullin IB. Formation of metal-polymer composites by ion implantation. Phil. Mag. B 2000; 80: 23–28. DOI: 10.1080/13642810008218336

[35] Ziegler JF, Biersak JP, Littmark U. The Stopping and Range of Ions in Solids. New York: Pergamon; 1996. http://www.srim.org/

[36] Kavetskyy T, Tsmots V, Kinomura A, Kobayashi Y, Suzuki R, Mohamed HFM, Šauša O, Nuzhdin V, Valeev V, Stepanov AL. Structural defects and positronium formation in 40 ke V B^+-implanted polymethylmethacrylate. J. Phys. Chem. B 2014; 118: 4194–4200. DOI: 10.1021/jp410763t

[37] Kavetskyy TS, Tsmots VM, Voloshanska SY, Šauša O, Nuzhdin VI, Valeev VF, Osin YN, Stepanov AL. Low-temperature positron annihilation study of B^+-ion implanted PMMA. Low Temp. Phys. 2014; 40: 747–751. DOI: 10.1063/1.4892646

[38] Kavetskyy T, Tsmots V, Šauša O, Stepanov AL. Structural modification of chalcogenide glasses by gamma-irradiation studied with DBAL technique. Phys. Status Solidi C 2012; 9: 2420–2423. DOI: 10.1002/pssc.201200252

[39] Kavetskyy TS, Tsmots VM, Voloshanska SY, Šauša O, Stepanov AL. Low-temperature Doppler broadening of annihilation line study of PMMA. In: Proceedings of the 6th International Scientific and Technical Conference on Sensor Electronics and Microsystem Technologies (SEMST-6); 29 September–3 October 2014; Astroprint; Odessa; 2014. p. 226.

[40] Shukla A, Peter M, Hoffmann L. Analysis of positron lifetime spectra using quantified maximum entropy and a general linear filter. Nucl. Instr. Meth. A 1993; 335: 310–317. DOI: 10.1016/0168-9002(93)90286-Q

[41] Kirkegaard P, Eldrup M, Mogensen OE, Pedersen NJ. Program system for analysing positron lifetime spectra and angular correlation curves. Comp. Phys. Commun. 1981; 23: 307–335. DOI: 10.1016/0010-4655(81)90006-0

[42] Bandžuch P, MorháčM, Krištiak J. Study of the Van Cittert and gold iterative methods of deconvolution and their application in the deconvolution of experimental spectra of positron annihilation. Nucl. Instr. Meth. A 1997; 384: 506–515. DOI: 10.1016/S0168-9002(96)00874-1

[43] Kavetskyy T, Nowak J, Borc J, Rusnák J, Šauša O, Stepanov AL. Carbonization in boron-ion-implanted polymethylmethacrylate as revealed from Raman spectroscopy and electrical measurements. Spectroscop Lett. 2016; 49: 5–10. DOI: 10.1080/00387010.2015.1044113

[44] National Instruments [Internet]. 2014. Available from: http://www.ni.com/datasheet/pdf/en/ds-15 [accessed 2014-11-06].

[45] Kavetskyy TS, Borc J, Kukhazh YY, Stepanov AL. The influence of low dose ion-irradiation on the mechanical properties of PMMA probed by nanoindentation. In: Petkov P, Tsiulyanu D, Kulisch W, Popov C, editors. Nanoscience Advances in CBRN Agents Detection, Information and Energy Security. NATO Science for Peace and Security Series – A: Chemistry and Biology. The Netherlands: Springer; 2015. p. 65–71. DOI: 10.1007/978-94-017-9697-2_7

[46] Kavetskyy TS, Kukhazh YY, Borc J, Stepanov AL. Nanoindentation of boron-ion implanted polymethylmethacrylate up to 1100 nm indentation depth. In: Proceedings of the XXII International Conference on Ion-Surface Interactions (ISI-2015); 20–24 August 2015; NRNU MEPh I; Moscow; 2015. Vol. 2. p. 132–135.

[47] Introduction on Instrumented Indentation [Internet]. 2013. Available from: http://www.csm-instruments.com [accessed 2013-11-15].

[48] Kavetskyy TS, Stepanov AL. Spectroscopic investigations of ion-induced processes in polymethylmethacrylate at low-energy boron-ion implantation. In: Proceedings of the XXII International Conference on Ion-Surface Interactions (ISI-2015); 20–24 August 2015; NRNU MEPh I; Moscow; 2015. Vol. 3. p. 339–342.

[49] Schultz PJ, Lynn KG. Interaction of positron beams with surfaces, thin films, and interfaces. Rev. Mod. Phys. 1988; 60: 701–779. DOI: 10.1103/Rev Mod Phys.60.701

[50] Kobayashi Y, Ito K, Oka T, He C, Mohamed HFM, Suzuki R, Ohdaira T. Application of positron beams to the study of positronium-forming solids. Appl. Surf. Sci. 2008; 255: 174–178. DOI: 10.1016/j.apsusc.2008.05.278

[51] Krištiak J, Krištiakova K, Šauša O. Phase transition in C_{60} observed by positron annihilation. Phys. Rev. B. 1994; 50: 2792–2794. DOI: 10.1103/Phys Rev B.50.2792

[52] Liao KS, Fu YJ, Hu CC, Chen JT, Lin DW, Lee KR, Tung KL, Jean YC, Lai JY. Microstructure of carbon molecular sieve membranes and their application to separation of aqueous bioethanol. Carbon 2012; 50: 4220–4227. DOI: 10.1016/j.carbon.2012.05.003

[53] Wang CL, Hirade T, Maurer FHJ, Eldrup M, Pedersen NJ. Free-volume distribution and positronium formation in amorphous polymers: Temperature and positron-irradiation-time dependence. J. Chem. Phys. 1998; 108: 4654–4661. DOI: 10.1063/1.475876

[54] Pujari PK, Tashiro M, Tseng CY, Honda Y, Nishijima S, Tagawa S. Low-temperature positron studies in polymethylmethacrylate. Mater. Sci. Forum 2001; 363–365: 275–277. DOI: 10.4028/www.scientific.net/MSF.363-365.275

[55] He CQ, Dai YQ, Wang B, Wang SJ. Molecular weight effect on the microstructure and local motion in high molecular weight polymer observed by positrons. Mater. Sci. Forum 2001; 363–365: 309–312. DOI: 10.4028/www.scientific.net/MSF.363-365.309

[56] Stepanov AL. Nanofillers: Synthesis of metal nanoparticles in viscous polymer. In: Nicolais L, Borzacchiello A, editors. Wiley Encyclopedia of Composites. 2nd ed. Italy: Wiley; 2012. p. 1–9. DOI: 10.1002/9781118097298.weoc157

[57] Stepanov AL, Kiyan R, Ovsianikov A, Nuzhdin VI, Valeev VF, Osin YN, Chichkov BN. Synthesis and optical properties of silver nanoparticles in ORMOCER. Appl. Phys. A 2012; 108: 375–378. DOI: 10.1007/s00339-012-6894-6

[58] Stepanov AL. Synthesis of silver nanoparticles in dielectric matrix by ion implantation: A review. Rev. Adv. Mater. Sci. 2010; 26: 1–29. http://www.ipme.ru/e-journals/RAMS/no_12610/stepanov.pdf

[59] Takele H, Greve H, Pochstein C, Zaporojtchenko V, Faupel F. Plasmonic properties of Ag nanoclusters in various polymer matrices. Nanotechnology 2006; 17: 3499–3505. DOI: 10.1088/0957-4484/17/14/023

[60] Tuinstra F, Koenig JL. Raman spectrum of graphite. J. Chem. Phys. 1970; 53: 1126–1130. DOI: 10.1063/1.1674108

[61] Elman BS, Dresselhaus MS, Dresselhaus G, Maby EW, Mazurek H. Raman scattering from ion-implanted graphite. Phys. Rev. B 1981; 24: 1027–1034. DOI: 10.1103/Phys Rev B.24.1027

[62] Lee EH, Hembree DMJr, Rao GR, Mansur LK. Raman scattering from ion-implanted diamond, graphite, and polymers. Phys. Rev. B 1993; 48: 15540–15551. DOI: 10.1103/Phys Rev B.48.15540

[63] Ferrari AC, Robertson J. Interpretation of Raman spectra of disordered and amorphous carbon. Phys. Rev. B 2000; 61: 14095–14107. DOI: 10.1103/Phys Rev B.61.14095

[64] Ferrari AC. Raman spectroscopy of graphene and graphite: Disorder, electron-phonon coupling, doping and nonadiabatic effects. Solid State Commun. 2007; 143: 47–57. DOI: 10.1016/j.ssc.2007.03.052

[65] Morales GM, Schifani P, Ellis G, Ballesteros C, Martinez G, Barbero C, Salavagione HJ. High-quality few layer graphene produced by electrochemical intercalation and microwave-assisted expansion of graphite. Carbon 2011; 49: 2809–2816. DOI: 10.1016/j.carbon.2011.03.008

[66] Petrovski A, Dimitrov A, Grozdanov A, Andonovic B, Paunovic P. Characterization of graphene produced by electrolysis in aqueous electrolytes. In: Petkov P, Tsiulyanu D, Kulisch W, Popov C, editors. Nanoscience Advances in CBRN Agents Detection, Information and Energy Security. NATO Science for Peace and Security Series – A: Chemistry and Biology. The Netherlands: Springer; 2015. p. 103–110. DOI: 10.1007/978-94-017-9697-2_11

[67] Casiraghi C, Ferrari AC, Robertson J. Raman spectroscopy of hydrogenated amorphous carbon. Phys. Rev. B 2005; 72: 085401-1–085401-14. DOI: 10.1103/Phys Rev B.72.085401

[68] Matthews MJ, Pimenta MA, Dresselhaus G, Dresselhaus MS, Endo M. Origin of dispersive effects of the Raman D band in carbon materials. Phys. Rev. B 1999; 59: R6585–R6588. DOI: 10.1103/Phys Rev B.59.R6585

[69] Kavetskyy TS, Tsmots VM, Nowak J, Kuczumow A, Kinomura A, Kobayashi Y, Suzuki R, Mohamed HFM, Šauša O, Nuzhdin V, Valeev V, Stepanov AL. Comprehensive study of carbonization in B^+-implanted PMMA: A correlation between slow positron beam and Raman spectroscopy results. In: Proceedings of the 17th International Conference on Positron Annihilation (ICPA-17); 20–25 September 2015; Wuhan University; Wuhan; 2015. p. 45.

[70] Poly(methyl_methacrylate) [Internet]. 2016. Available from: http://en.wikipedia.org/wiki/Poly(methyl_methacrylate) [accessed 2016-01-28].

Chapter 12

Radiation Effects in Textile Materials

Sheila Shahidi and Jakub Wiener

Additional information is available at the end of the chapter

http://dx.doi.org/10.5772/63731

Abstract

Irradiation processes have several commercial applications, in the coating of metals, plastics, and glass, in printing, wood finishing, film and plastic cross-linking, and in the fields of adhesive and electrical insulations. The advantages of this technology are well known.

Radiation treatment on fabric and garments can add value in coloration, antifelting, printing, and coating. In this chapter, different irradiation methods such as plasma, electron beam, laser, ion implantation, microwave, gamma, and ablation are described, and the effects of these process in textile industry as a finishing method are discussed and compared with conventional methods.

Keywords: Irradiation, Fabric, Textile, Material, Finishing

1. Introduction

The advancement of technology in the textile market has led scientists and researchers to develop novel finishes to add high value on deferent textile materials. It was a new window for researchers to advent and explore new research fields which include geotextiles, flame retardant textiles, insect repellent textiles, aroma textiles, medical textiles, smart textiles, antibacterial textiles, and nanotextiles, etc.

Recent developments in the textile industry are mainly focused on physical and chemical modifications on surface of fibers and fabrics. Different chemical and biological methods have been used electively to improve or impart permanent functional properties on the surface of textile materials. However, some of these chemicals are toxic and sometimes expensive [1].

Alternatively, radiation technology involving low energy use, no chemicals, ease to handling, and high treatment speed can modify the surface of textiles and improve dye uptake, printing, fastness properties, adhesion of coatings, and adsorption of used chemicals.

In this chapter the effect of ultraviolet (UV), gamma, plasma, laser, microwave, electron beam, ion beam, ultrasonic on surface, chemical, and physical and mechanical properties of textile materials is fully discussed.

2. UV irradiation and its application in textile industry

One of the effective and economical methods of surface modification for both natural and synthetic polymers is UV irradiation. Excitation and dissociation of the polymeric molecules take place during exposing the surface to the UV treatment, which is known as a photosensitized oxidation process.

The color depth of the polylactide (PLA) and polyethylene terephthalate (PET) fabrics has been increased by the UV irradiation. This increase in the depth of dyeing is believed to be due to the surface roughening of the fabrics produced by the UV [2].

According to previous research work which has been done, the prepared thin film polyamide nanofiltration membranes were modified by acrylic acid (AA) and UV irradiation. The effect of UV-irradiation time and addition of AA on the performance and morphology of nanofiltration membranes were investigated. The obtained membranes illustrate the ability for rejection of Na_2SO_4 and significant properties for separation of divalent ions from monovalent ions [3].

In recent years, many methods have been reported for modifying PET fabric to be hydrophilic. Different types of chemical auxiliaries can be used for this purpose. However, the hydrophilicity of the treated PET surface is not durable because the coated hydrophilic material easily dissolves during repeated washes. Photo-chemical reactions and photophysical processes are gaining attention as techniques for modifying the surface of PET fabrics [4].

UV irradiation and treatment with nano-TiO_2, H_2O_2, and NaOH increased the hydrophilicity of PET fabric, with an irradiation time of 30 min making the fabric nearly wettable and an irradiation time of 40 min making the fabric superhydrophilic. The greatest and most durable wettability results were obtained after 40 min of UV irradiation combined with 30 g/L nano-TiO_2, 50 g/L H_2O_2, and 30 g/L NaOH treatment. The excellent mechanical and physical properties of PET fabric were retained after modification, although the breaking strength and elongation were slightly reduced. The induced hydrophilicity of the PET fabric can be considered permanent because the surface of the PET fabric was chemically modified with the introduction of hydrophilic groups [4].

In the other research, a new technique has been presented to graft sulfonic groups at the surface of PET fabric, which is based on UVC treatment with SO_3 gaseous molecule. These modifications improved the hydrophilic character of PET fabric surface and consequently its dying

ability. The Schematic of system is shown in **Figure 1**. The air flow containing fuming SO molecules is injected into the vessel between which is composed of an UVC lamp and the fabric to be treated. The UVC lamp emits intense and nearly monochromatic light (254 nm). In fact, the UVC lamp emits 512 kJ mol^{-1} per photon, which allows the PET fibers functionalization without damaging their bulk properties. Sulfonic (SOH) groups can be introduced into aromatic compounds through an electrophilic substitution reaction. Such a reaction is referred as sulfonation [5].

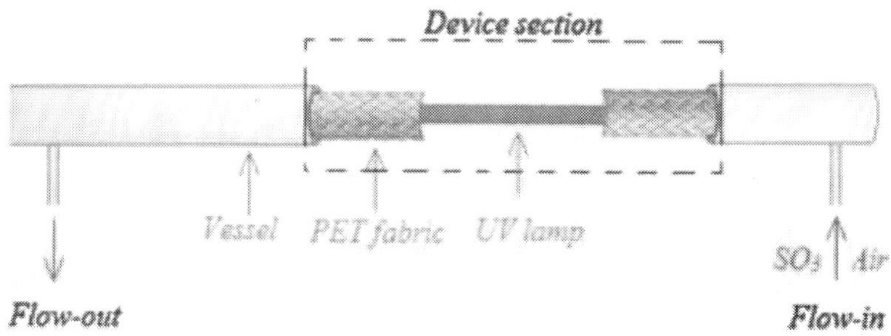

Figure 1. UVC treatment device [5].

The results reported in the other research demonstrate that the UVC (254 nm) irradiation method under a stream of gaseous Cl has the potential to graft PET textile fabrics. This technique has many advantages, such as easy handling, and it is a cheap, solvent-free continuous process without changing the bulk properties. The medication involves only a few nanometres of the surface. Several analytical techniques have confirmed that surface chlorination has occurred. Scanning Electron Microscope (SEM) analyses showed that the chlorination only occurred at the surface of the fibers without changing the bulk of fibers. The Differential scanning calorimetry (DSC) thermograms indicated that the thermal characteristics of the PET fabrics were maintained. X-ray photoelectron spectroscopy (XPS) also showed the presence of Cl atoms in the upper molecular layers of the surface. The affinity of modified fabrics to cationic dyes is more than untreated samples [6].

For a long time, the textile industry has been searching for a rapid way to modify textile polymer surfaces. Obtaining good bonding between two dissimilar materials is critical for several aspects in textile finishes processing. Adhesion between PET and Tuftane Thermoplastic Polyurethane (TPU) Film was improved by grafting NCO groups onto the PET surface using UV irradiation process by Dong Liu and his coworkers [7].

3. Gamma irradiation on textile fabrics

Gamma rays are high-energy electromagnetic radiations having energies above 100 keV and wavelengths less than 10 picometers. Surface modification of textiles using gamma ray is one of the promising methods.

The excellent mechanical properties, heat and oxidation resistance, and environmental stabilities make carbon fiber fabrics ideal reinforcing materials in the advanced composite fields, such as solar panel of space station and electric vehicle body. All these excellent properties depend largely on the interfacial adhesion which finally affects the overall property of the resulting composites. Recently, radiation-induced grafting has been extensively applied as a competitive methodology to develop new functional materials. An easy method to evenly functionalize the fabric surfaces by γ-ray irradiation grafting was reported. This novel technique is simple, green, and versatile. The functionalization was much more uniform compared with the traditional electrochemical method. The interfacial strength of the composites had a dramatic increase [8].

Presently, polymeric membranes are extensively employed in the field of biomedical materials contacting with blood. Sulfonated polypropylene nonwoven fabric (PPNWF) has been successfully prepared via gamma-ray preirradiation-induced graft polymerization of sodium styrenesulfonate (SSS) and acrylamide (AAm) by li et al (**Figure 2**).

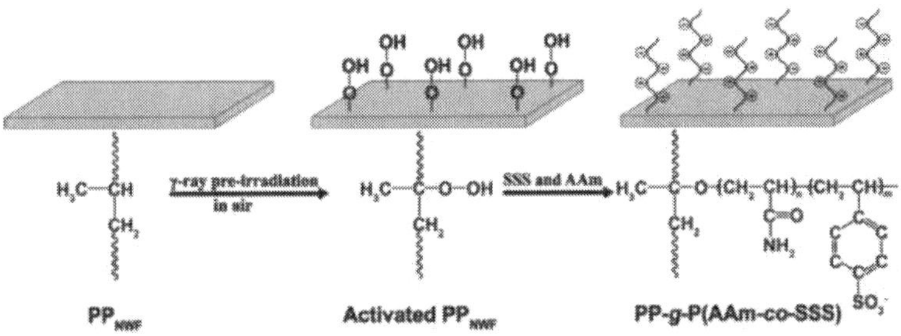

Figure 2. Schematic diagram of preparing sulfonated PP $_{NWF}$ via gamma-ray preirradiation method [9].

Gamma ray preirradiation-induced cograft polymerization has been extensively applied to modified polymer materials because of its simple procedure and even grafting. Acrylonitrile (AN) has been widely used in the modification of materials and an amidoxime group can be obtained by further chemical treatments from the nitrile group to increase adsorption efficiency for heavy ions. In a research that has been done in 2013 by Liu and his coworkers, a preirradiation-induced emulsion cograft polymerization method was used to introduce AN and AA onto a PE nonwoven fabric. The use of AA is meant to improve the hydrophilicity of the modified fabric. The modified nonwoven fabric is ready for further amidoximation to the application of heavy metal ion extraction [10].

In the other research, SSS was grafted onto PPNWF via γ-ray coirradiation method with the existence of *N*-vinyl-2-pyrrolidone. The modified PPNWFs presented good blood compatibility, such as lower hemolysis rate and lower platelet adhesion. Besides, the modified PPNWFs prolonged the clotting time and presented excellent anticoagulant effect [11].

The textile industries are one of the major sources of water pollution in terms of releasing highly colored waste stream in surface water bodies. The wastewater generated in textile processing plants is contaminated with toxic synthetic colorants and various perilous chemicals. The main objectives of the study by Bhuiyan et al. were to degrade the dye molecules and organic pollutants of textile wastewater by using gamma irradiation followed by the investigation of physicochemical parameters of the irradiated water as well as looking into the scope for using treated wastewater for irrigation and dyeing purposes. The wastewater samples were submitted to Cobalt-60 gamma radiation source.

The irradiated wastewater was found to be recyclable in textile wet processing and reusable for irrigation purposes [12, 13].

The presence of toxic metals and pathogenetic microbes in drinking water is a potential health risk. Consequently, numerous investigations have been carried out on the functionalized polymer membrane with AgNPs as effective antimicrobial agents for water treatment. Unfortunately, the AgNPs were commonly inert with polymer surfaces so that silver releases into water filtrate in an overdose compared to the permitted limit of standard at maximum of 0.1 mg/L, according to the US Environmental Protection Agency (EPA) and World Health Organization (WHO).

A new method to immobilize AgNPs onto the acrylic grafted polyethylene nonwoven (PE) fabric by gamma Co-60 irradiation for drinking water treatment has been described in 2013.

The PE fabric pieces were treated by a mixture of acetone/H_2O solution and dried before irradiation. The PE fabric samples were irradiated at the required doses up to 50 kGy by γ ray from Co-60 source. Thereafter, the graft reactions were carried out at ~90°C in a flask containing 1 g preirradiated PE, 0.05 g Mohr's salt and AA with concentrations of 10–50% (v/v) in 100 ml aqueous solution. The PE-g-PAAc samples were soaked overnight at room temperature in Ag NPs colloidal solutions and then squeezed to remove the excess Ag NPs, rinsed with pure water and dried in an oven at 70 C for 2 h. The dried fabrics were then annealed at 120 C for 1 h for esterification of –COOH group of PAAc with –OH group of Polyvinyl alcohol (PVA). The prepared fabrics contained about 10,000 ppm Ag NPs showing strong bactericidal efficiency against *Escherichia coli*. Based on the strong bactericidal efficiency and under permitted limit of silver release into water filtrate, PE-g PAAc/Ag NPs fabrics can be used for the treatment of drinking water. Also, this kind of filters can be used in air cleaners and have other applications [14].

In the other research, the silver ions have been reduced effectively by gamma irradiation and immobilized on the cotton fabrics by in situ synthesis. The Ag NPs content deposited on the fabrics was of 1696 mg/kg, when the fabric sample was irradiated in 1.5 mM Ag NO and 1.0% chitosan solution at the dose of 13.8 kGy at 30°C.

Antibacterial efficacy of the Ag NPs fabrics after washing 40 cycles of washing was about 99.99% for *Staphylococcus aureus* and *E. coli*. The AgNPs/cotton fabrics washing from 1 to 40 cycles were innoxious to skin (k = 0). These results confirm that gamma irradiation of cotton fabrics in the presence of AgNO and chitosan solution is a promising approach for preparation of stable, safe, and efficacious antibacterial fabrics [15].

In the past decades, the materials with high water and oil repellency have attracted much attention from researchers and industries. The perfluoroalkyl phosphate acrylates have been grafted onto a cotton fabric via gamma-ray irradiation to improve the hydrophobic and oleophobic properties.

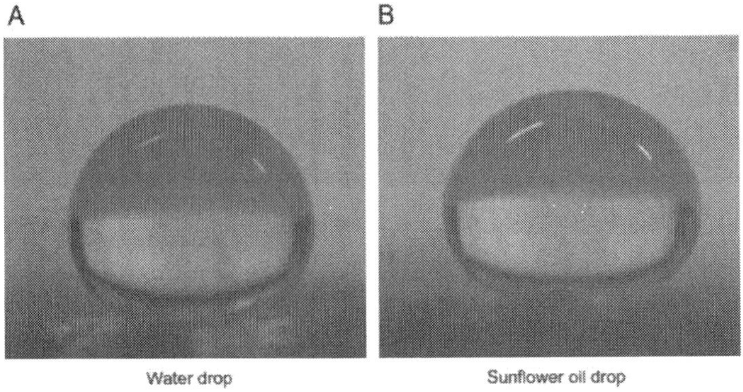

Figure 3. Photographs of water and sunflower oil drops on the grafted sample [16].

The results show that the fabric became highly hydrophobic and oleophobic with the contact angles of above 150° and 140° for water and sunflower oil, respectively (**Figure 3**) [16].

In the other research, a novel coating formulation for improving the UV protection property on cotton, PET, and cotton/PET fabrics was prepared, and gamma rays were applied for surface curing. Aluminum potassium sulfate (Alum) was used individually and in binary coat with Zinc Oxide (ZnO), to induce the UV-blocking properties [17].

4. Microwave applications in textile industry

Microwaves comprise electromagnetic radiation in the frequency range of 300 MHz–300 GHz. As the polar or charged particles in a reaction medium fail to align themselves as fast as the direction of the electric field of microwaves changes, friction is created to heat the medium [18].

They can penetrate into a material and heat the deep layers of the material strongly when they release their energy. Microwave irradiation offers a number of advantages over conventional heating methods, including using less energy, offering a higher heating rate, and offering the ability to more quickly start and stop heating.

Sulfonating a PET fabric by dilute sulfuric acid and microwave irradiation has been found to produce a super-hydrophilic PET fabric, and the fibers sustained minimal damage.

PET fabric has been immersed in H_2SO_4 solutions with different concentrations at room temperature for 5 min.

PET fabric samples were then dried at 50°C for 30 min and then irradiated with microwaves at 2450 MHz using a commercial 700 W microwave oven for 4 min. The SEM images of both

untreated and treated PET are shown in **Figure 4**. The surfaces of the original and the modified PET fibers are all smooth.

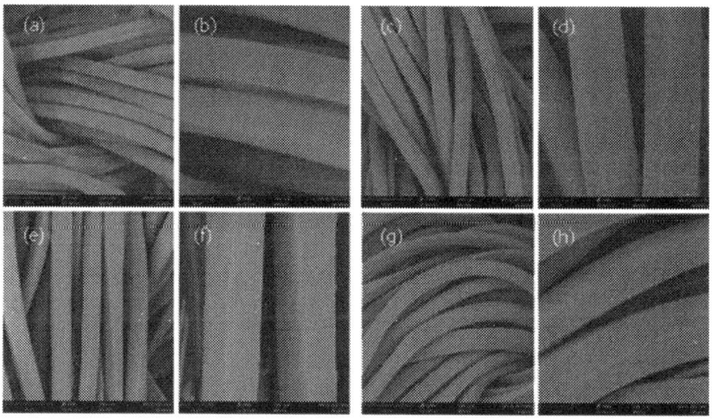

Figure 4. SEM images (×1500 and × 5000) of (a, b) the original PET fibers; (c, d) PET fibers modified by 10 g/L H_2SO_4; (e, f) PET fibers modified by 20 g/L H_2SO_4; and (g, h) PET fibers modified by 25 g/L [19].

Microwave modification process is quick and inexpensive. Therefore, the modification process could be used in a wide range of applications [19].

Polyamide 66 (PA 66) is one of the most abundantly used fabric in many areas attributed to its low cost, resistance to shrinkage, abrasion, etc. However, its combustibility and serious dripping produced during combustion cannot meet industrial and civil requirements in many cases. Zhao et al. researched the efforts on surface modification of PA 66 fabric by microwave-induced grafting with 2-hydroxyethyl methacrylate (HEMA). The grafting reaction was undertaken in water solution to decrease the damage of fabric. The hydrophilicity, dripping tendency, and mechanical properties of grafted samples have been significantly improved. However, the slight change of limiting oxygen index (LOI) values and high damaged length during burning cannot meet the flame retardant requirements in many fields [20].

Microwave (MW) energy which is used in dying of various types of textile staple and fixation processes has brought positive results. Microwave irradiation increases the diffusion of dye molecules in dyeing process when applied to both synthetic and natural fibers. It is also used in prefinishing of silk yarns. In addition MW energy is accepted as more efficient than conventional methods for cotton fabric finishing, drying, and curing processes, durable press finishing, incombustibility, water and oil repellent finishing [21].

5. Ultrasonic in textile industry

Ultrasound technology is among the irradiation technologies whose applications in different branches of industries have soared in recent years.

Ultrasound technology has been extensively used for detecting defects in a large variety of industrial components and materials. Using ultrasound waves through the air is not the first choice due to the high impedance mismatch between air and most of transducer and component materials. As a result, a higher impedance substance is employed as a couplant to optimize the acoustic energy transfer to the sample, for example, water or a coupling gel. Nevertheless, in some processes, it is not allowed to use a wet coupling, which makes room for a specific category of applications usually named as noncontact or airborne ultrasound. An example of this is the inspection of textile goods, where a wet coupling substance could degrade the fabric and/or slow the manufacturing process.

Several previous investigations have shown that ultrasonic technology enhances mass transfer during some textile processing steps such as desizing, scouring, bleaching, mercerizing, and dyeing of natural fabrics.

Pazos-Ospina et al. in 2015 presents a design methodology for half-curved airborne ultrasonic arrays based in cellular ferroelectret film. The geometry of the array proposed allows them focus naturally in the vertical plane and electronically in the horizontal one, obtaining similar spatial resolution in both directions. Theoretical predictions and simulated results were validated with a developed array prototype designed to operate at frequencies between 50 kHz and 300 kHz. The potential of the device was shown by inspecting different textile samples in transmission mode. This multi-transducer design is a low-cost alternative to the use of composite 2D arrays in noncontact ultrasonic inspections [22].

Cleaning of materials is one of the most important applications of ultrasound. However, the use of ultrasonic energy for textile washing has been searched many years without achieving commercial development. The cleaning action of ultrasonic energy is due to cavitations. The implosion of vapor bubbles inside the cleaning auxiliaries and near the surface to be cleaned imposes such stress on the surface that erodes the contaminant and removes the impurities. On the other hand, stable cavitations, may also cause the dispersion of the particles of contaminant removed from the surface.

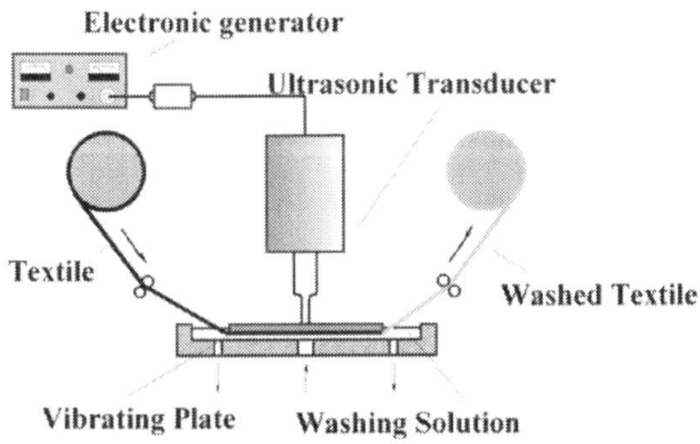

Figure 5. Basic scheme of the ultrasonic process [23].

An ultrasonic system for the continuous washing of textiles in liquid layers based on a procedure has been designed and constructed by Gallego-Juarez et al. in 2010. The system incorporates, as the main part, special plate transducers capable for high-power operation without the interaction of perturbing undesired vibration modes. This system has shown very good washing behavior either with the laboratory or the semiindustrial set-up (**Figure 5**) [23].

Recently, attempts have been made to use acoustic cavitation for washing textiles. Ultrasonic cleaning has been widely employed to remove submicron-sized contaminant particles adhering to solid substrates (e.g., photo masks and wafers) in semiconductor industry. Ultrasonic waves traveling in a liquid result in cavitation and thus produce bubbles. The bubbles exhibit rich dynamic behaviors such as translation, oscillation, growth, and collapse in response to the varying acoustic pressure [24].

In the recent decade, the application of ultrasonic irradiation as an advanced oxidation process has attracted much attention because of the generation of high amounts of •OH radicals due to the ultrasonic cavitation. The ultrasonic waves result in the rapid growth and subsequently, collapse of the cavitation bubbles, which produces extremely high pressure (up to 1800 atm) and temperature (as high as 5000 K) in the bubbles. The high temperature, together with the high pressure, named "hot spots" leads to the generation of •OH within the gas–liquid transition zone near the bubbles and bulk solution as a result of the water dissociation. It has been demonstrated that the catalytically enhanced ultrasonic irradiation based on the application of semiconductors, known as sonocatalysis, has higher degradation efficiency and lower processing time than that of sonication alone.

Darvishi Cheshmeh Soltani et al. in 2016 used a porous clay-like support with unique characteristics for the synthesis and immobilization of ZnO nanostructures to be used as a sonocatalyst for the sonocatalytic decolorization of methylene blue (MB) dye in the aqueous phase. They concluded that the sonocatalytic activity of ZnO–biosilica nanocomposite (77.8%) was higher than that of pure ZnO nanostructures (53.6%). Increasing the initial pH from 3 to 10 led to increasing the color removal from 41.8% to 88.2%, respectively. Increasing the sonocatalyst dosage from 0.5 to 2.5 g/L resulted in increasing the color removal. They also concluded that the ZnO–biosilica nanocomposite can be a suitable sonocatalyst for the sonocatalytic decolorization of colored solutions with high reusability potential and cost-efficiency [25].

As an alternative to the existing finishing technologies, a facile one-step sonochemical route has been suggested for uniform deposition of inorganic nanoparticles on the surface of solid substrates, including textiles.

The antimicrobial finishing is very important for medical textiles, decreasing the risk of hospital-acquired infections.

Petkova et al. in 2016 report a simultaneous sonochemical/enzymatic process for durable antibacterial coating of cotton with zinc oxide nanoparticles (ZnO NPs). The novel technology goes beyond first enzymatic preactivation of the fabrics and subsequent sonochemical nanocoating and is designed to produce "ready-to-use" antibacterial medical textiles in a single step.

A multilayer coating of uniformly dispersed NPs was obtained in the process. The pretreatment with enzymes causes better adhesion of the ZnO NPs on the surface of cotton fabrics. The NPs-coated cotton fabrics inhibited the growth of *S. aureus* and *E. coli*, respectively, by 67% and 100% [26]

Textile dyeing assisted by ultrasonic energy has attained a greater interest in recent years. Ultrasonic-assisted dyeing of cellulosic fibers has already proved to be a better choice among conventional dyeing by many researchers.

Khatri et al. in 2016 reported ultrasonic dyeing of nanofibers. They chose cellulose nanofibers and dyed with two reactive dyes, CI reactive black 5 and CI reactive red 195. Results revealed that the ultrasonic dyeing produced higher color yield (K/S values) than the conventional dyeing. The color fastness test results depicted good dye fixation. Also they have reported that ultrasonic energy during dyeing does not affect surface morphology of nanofibers. The results conclude successful dyeing of cellulose nanofibers using ultrasonic energy with better color yield and color fastness results than conventional dyeing [27].

6. Laser in surface modification of polymer and fabrics

Laser modification on material surface is one of the most studied technologies. It has been shown that various materials modified by laser irradiation often exhibit physical and chemical changes in the material's surface. In general, laser irradiation could not affect the bulk properties of a polymer due to its low penetration depth (**Figure 6**) [28, 29].

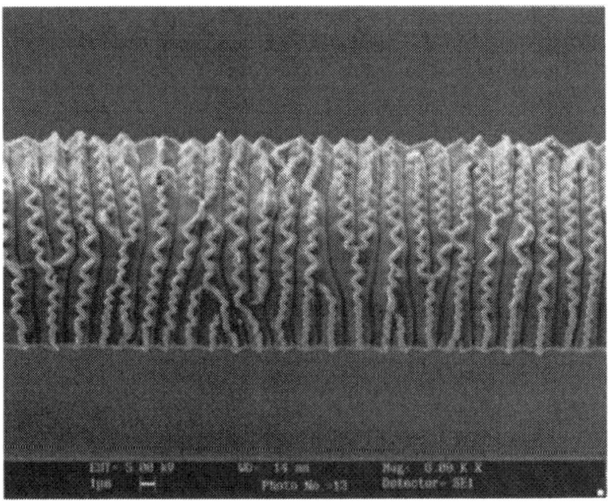

Figure 6. Surface structure of polyester fiber under high fluence (5 pulses at 100 mJ/cm^2) [28].

In the research which has been done by Bahtyari, the effect of CO_2 laser treatment on the dyeability of polyimide fabrics reveals that, following laser treatment, the dyeability of polyamide increased significantly. This is accompanied by a significant bursting strength loss.

It has been observed that, as the laser modification of the fabric was carried out with low intensity, the concentration of free amino groups, which are necessary during dyeing with acid and reactive dyes, increased [30].

The modification induced in PLA by the ArF excimer laser radiation has been investigated by Rytlewski et al. It was found that the surface energy change was affected by surface oxidation as well as by surface roughness. ArF laser surface treatment can be an effective way of improving PLA adhesion properties [31].

Another article focused on the development of a laser pretreatment method for glass-fibre-reinforced polypropylene surfaces for industrial applications. The aim of this research is to create a surface for bonding polypropylene which adheres very poorly to most of the materials and forms to the matrix material for plastic composites [32].

On the other hand, as is known, the laser is a source of energy which can be used for irradiation of different substrates and its power and intensity can be easily controlled. By using laser, it is possible to cut a great variety of material from metal to fabric. Also it would be possible to transfer certain designs onto the surface of textile material by changing the dye molecules in the fabric and changing the color quality values by laser irradiation of fabrics at reduced intensity (**Figure 7**) [33].

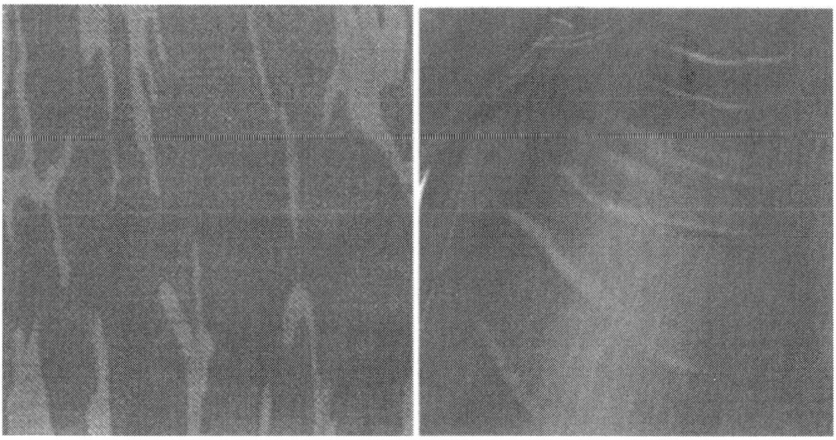

Figure 7. Some examples of denim trousers designed by laser beam method [33].

The CO_2 laser-thinning method has been applied to the PET fiber to prepare the PET nonwoven fabric without using the solvent by Suzuki et al.

The obtained nonwoven fabric was made of continues microfibers with a uniform diameter without a droplet.

The laser-thinning method has been found to be effective for producing other nonwoven fabrics such as poly(L-lactic acid) and poly(glycolic acid). The schematic of setup is shown in **Figure 8** [34].

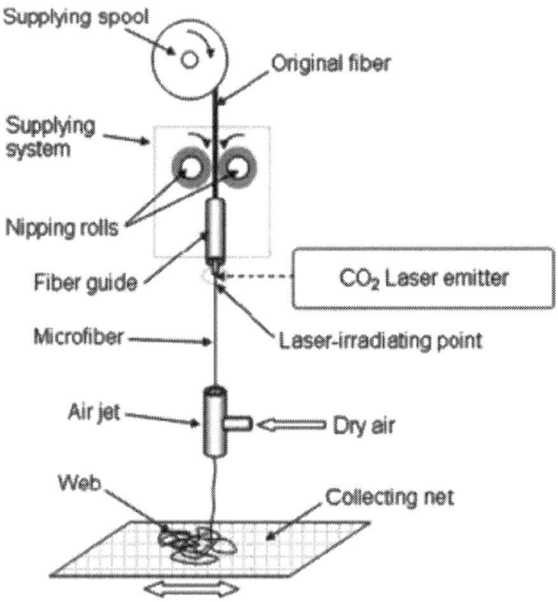

Figure 8. CO2 laser-thinning apparatus used for web formation [34].

CO_2 laser treatment was used as a novel method for creating antibacterial properties on glass mat by Wiener et al. in 2014. Various types of metallic salts such as CuO, ZnO, and $AgNO_3$ were applied on surface of glass mat and irradiated with the laser light beam (100 μs). Metal particles were deposited on the surface of samples. The antibacterial properties of the fabrics were connected with the presence of metal particles on their surface. Wiener et al. concluded that the change in properties induced by laser can effect an improvement in certain textile products [35].

Glass fiber mat surface modifications were carried out using CO_2 laser. The geometry of the experiment is visualized in **Figure 9**. In the laser tube (1) produces IR laser beam (2). In the direction of laser beam, computer-adjusted mirror is located (3) which determined the positron of irradiated place on glass fiber mat (4). The temperature of the glass fiber mat on its irradiated side cannot be measured due to high intensity of IR laser beam. So only the temperature of back side of glass fiber mat was estimated (5) by infrared thermometer (6). Laser light treatment of glass fiber resulted in interesting properties based on the mechanical properties, such as strength, modulus, and elongation of the glass fiber mat, the permeability, morphological properties, and the thickness.

Wiener et al. in 2014 concluded that by increasing the laser intensity, the strength and modulus of the mat decrease. But in the case of laser cycling treatment, the mechanical properties are improved. We observed that laser treatment causes increase of porosity and better air permeability [36].

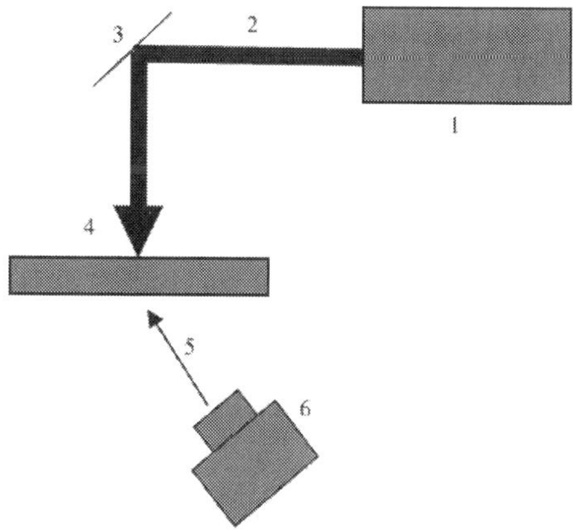

Figure 9. Schematic view of the laser treatment of glass fiber and temperature measurement [36].

7. Plasma and its application in textile industry

Plasma technique can have very important effects on the properties of textile materials. Different types of plasma gases have different effects on the surfaces of textiles. Plasma has many potential for the activation and functionalization of textile materials. Plasma technology is slow but steady in the industrial revolution. Surface modification of textiles cannot replace all wet processes, but it can be a viable pretreatment, which can provide plenty of environmental and economical benefits. Therefore, textile industry should consider the concept of higher initial investments in equipment that will be paid off quickly with respect to environment-related savings and the profit of the sale of high value-added products [37–39].

Improving the fastness properties and antibacterial activity of dyed cotton samples was studied by Shahidi in 2015. In her research, first cotton fabrics were dyed with various types of dyestuffs such as Direct, Vat, and Reactive. Then prepared samples were sputtered using plasma sputtering system for 15 s by silver and copper. For deposition of metal nano layer on the surface of samples, DC magnetron puttering system was used. Samples were placed on the anode. By attacking active ions, radicals, and electrons, the cathode particles were scattered.

Silver or copper particles were deposited on the surface of cotton samples, and through incorporation of metal nano particles on fabric surfaces, the antibacterial has been developed. It can be concluded that sputtering technique can be a novel method for improving the fastness properties of dyed cotton samples [38].

DC magnetron sputtering system for creating antibacterial and ultraviolet protective cotton fabrics has been used by Shahidi et al. in 2016. A silver anode and cathode were used. Silver particles were deposited on the both sides of cotton samples, and the antibacterial property has been developed, through sputtering of silver particles on fabric surfaces. Treated cotton fabrics had an excellent UV-blocking property. According to the standard, the treated cotton fabric can claim to be a "UV Protective product." They concluded that the change in properties induced by plasma can effect an improvement in certain textile products [39].

Conventional wet treatment in textile industry involves high consumption and pollution of water resources. Wastewater processing costs are high, and drying the wetted fibers is energy-, time-, and cost-intensive. So the textile industry has a great interest in alternative dry processes. Low temperature plasma treatment is a dry and ecofriendly technology which has been widely used to modify the chemical and topographical properties of polymers and textiles surface. The application of plasma technologies as a pretreatment and finishing process for textiles has become very popular because this surface modification method changes the outermost layer of the substrate without altering the bulk properties. Low-pressure plasma treatments are known to induce physical and chemical surface changes in textile fibers through several concurrent processes (activation, etching, grafting chemical functional groups, and cross-linking).

Girmoldi et al., in 2015, performed atmospheric pressure plasma treatments of pure cashmere and wool/cashmere textiles with a dielectric barrier discharge (DBD) in humid air (air/water vapor mixtures). Their analyses revealed a surface oxidation of the treated fabrics, which enhances their surface wettability with minor etching effects, an essential feature for the maintenance of the textile softness [40].

Air plasma treatment can modify the physical and chemical properties of the surface, and it causes to increase the hydrophilic character of a material. The production of new textile products for cosmetic applications requires pretreatment on the surface of fabrics. In this kind of application, the textile must have some organoleptic and aesthetic properties which require the use of softeners.

Surface modification of the PA66 fibers by low temperature plasma has been studied by Labay et al. in 2014. Corona plasma treatment has been investigated to achieve surface modification in the first nanometers of polymer fibers surface in order to modulate the incorporation and the release of caffeine. Plasma treatment improved the caffeine release [41].

Polyester fiber is one of the most important materials for textile manufacturing. However, the conventional antistatic finishing processes for polyester involve numerous energies and chemicals, with corresponding environmental pollution. The synergetic effects of low-temperature oxygen plasma (P) and N,O-carboxymethyl chitosan (N) treatments on the antistatic and antibacterial properties of polyester fabrics were investigated by Liu et al. in

2016. They concluded that all the treated polyester fabrics had no obviously antibacterial effect on *E. coli* (Gram-negative). Their findings indicated that the proposed process can provide good antistatic performance for polyester fabrics with a minimum of pollution [42].

In 2016, pure cashmere and wool/nylon textiles were modified by means of an atmospheric pressure plasma treatment with a DBD in humid air followed by a finishing process with a fluorocarbon resin by Zanini et al. Their result indicated a higher amount of fluorocarbon resin on the surface of the fabric which was plasma treated before the finishing process and a more uniform coverage of the fibers of this textile. They concluded that the hydrophilic character of the plasma-activated fabric leads to a higher adsorption of the water-based dispersion that contain the fluorocarbon resin and limit the de-wetting phenomenon of the fibers during the drying step. All these results highlight the importance of the plasma activation step to enhance the hydro- and the oleo-repellent properties of the modified fabrics, as assessed by different analyses (water contact angle, standard tests for hydro- and oleo-repellence, and water adsorption isotherms) [43].

Textile industry wastewater has large amounts of organic dyes that are resistant to the biological methods. Moreover, other physical and chemical processes such as adsorption and coagulation merely transfer contaminants to a secondary phase and require more treatment. Fenton and sonication processes are simple and efficient methods that are applied for the mineralization of various contaminants from polluted water sources. The plasma-treated pyrite (PTP) nanostructures were prepared from natural pyrite (NP) utilizing argon plasma due to its sputtering and cleaning effects resulting in more active surface area by Khataee et al. in 2016. They reported that, environmentally friendly plasma modification of the NP, in situ production of H_2O_2 and OH radicals, low-leached iron concentration and repeated reusability at the milder pH are the significant benefits of the PTP utilization. The significant advantages of the stable PTP are not needed for adding of H_2O_2, low-leached iron amount, and application at the milder pH [44].

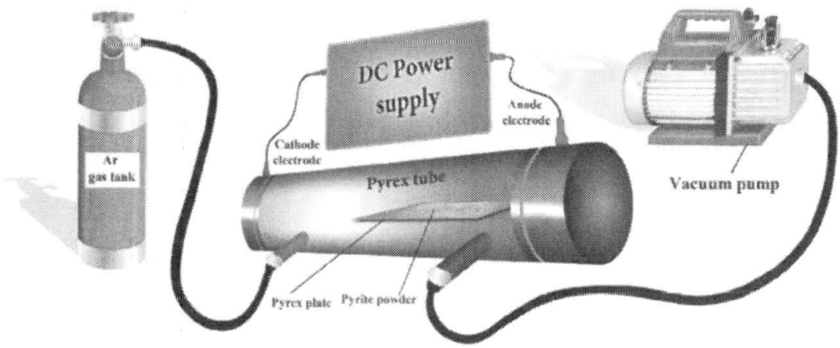

Figure 10. Schematic diagram of the glow discharge plasma system used in Khataee et al. study (2016) [44].

8. Conclusion

In textile industry, the surface modification of fibers is correlated to surface properties such as water repellence and adhesion. In the recent years, there have been many researches in developing surface modification methods to improve the surface properties of textile materials in order to develop new market products. Several surface modification methods are employed. The irradiation treatments have been ahead of recognition as a surface modification technique. Much attention has been paid to treating the surfaces of materials with irradiation methods in the past few decades. In general, treating the surface of a material with irradiation methods can cause physical and chemical changes to occur at the surface but affect the bulk properties of the material little.

Author details

Sheila Shahidi[1*] and Jakub Wiener[2]

*Address all correspondence to: Sh-shahidi@iau-arak.ac.ir

1 Department of Textile, Arak Branch, Islamic Azad University, Arak, Iran

2 Department of Textile Chemistry, Faculty of Textile, Technical University of Liberec, Liberec, Czech Republic

References

[1] High-Energy Radiation Induced Sustainable Coloration and Functional Finishing of Textile Materials, Shahid-ul-Islam and Faqeer Mohammad, Ind. Eng. Chem. Res. (2015), 54, 3727–3745.

[2] Investigation into the Effect of UV/Ozone Irradiation on Dyeing Behaviour of Poly(Lactic Acid) and Poly(Ethylene Terephthalate) Substrates, F. Fattahi, H. Izadan, A. Khoddami, Prog Color Colorants Coat. 5(2012), 15–22.

[3] Preparation and modification of thin film PA membranes with improved antifouling property using acrylic acid and UV irradiation, Y. Mansourpanah, E. Momeni Habili, J Membr Sc 430 (2013) 158–166.

[4] Super hydrophilic surface modification of fabric via coating With nano-TiO2 by UV and alkaline treatment, Mingyu Li, Tingting Deng, Shuxian Liu, Fengxiu Zhang, Guangxian Zhang, Appl Surf Sc, 297 (2014) 147–152.

[5] Sulfonation of polyester fabrics by gaseous sulfur oxide activated by UV irradiation, Bessem Kordoghli, Ramzi Khiari, Mohamed Farouk Mhenni, Mohamed Naceur Belgacem, Faouzi Sakli, Appl Surf Sc 258 (2012) 9737–9741.

[6] UV irradiation-assisted grafting of poly(ethylene terephthalate) Fabrics, Bessem Kordoghli, Ramzi Khiari, Hatem Dhaouadi, Mohamed Naceur Belgacem, Mohamed Farouk Mhenni, Faouzi Sakli, Colloids and Surfaces A: Physicochem. Eng. Aspects 441 (2014) 606–613.

[7] UV-assisted surface modification of PET fiber for adhesion improvement, Xiang-Dong Liu, De-Kun Sheng, Xiu-Mei Gao, Tong-Bing Li, Yu-Ming Yang, Appl Surf Sc 264 (2013) 61–69.

[8] Uniform modification of carbon fibers in high density fabric by X-ray irradiation grafting, Feng Zhao, Yudong Huang, Mat Lett 65 (2011) 3351–3353.

[9] In vitro hemocompatibility of sulfonated polypropylene non-woven fabric prepared via a facile gama-ray pre-irradiation grafting method, Rong Li, Guozhong Wu, Yin Ye, Appl Surf Sc 356 (2015) 1221–1228.

[10] Pre-irradiation induced emulsion co-graft polymerization of acrylonitrile and acrylic acid onto a polyethylene nonwoven fabric, Hanzhou Liu, Ming Yu, Hongjuan Ma, Ziqiang Wang, Linfan Li, Jingye Li, Radiat Phys Chem 94 (2014) 129–132.

[11] Gamma-ray co-irradiation induced graft polymerization of NVP and SSS onto polypropylene non-woven fabric and its blood compatibility, Rong Li, Hengdong Wang, Wenfeng Wang, Yin Ye, Radiat Phys Chem 91 (2013) 132–137.

[12] Scope of reusing and recycling the textile wastewater after treatment with gamma radiation, M.A. Rahman Bhuiyan, M. Mizanur Rahman, Abu Shaid, Mubarak A. Khan, M.M. Bashar, J Clean Prod 112 (2016) 3063–e3071.

[13] Cytotoxicity and mutagenicity evaluation of gamma radiation and Hydrogen peroxide treated textile effluents using bioassays, Munawar Iqbal, Jan Nisar, J Environ Chem Eng 3 (2015) 1912–1917.

[14] Study of incorporation of silver nanoparticles onto PE-g-PAAc nonwoven fabric by gama-irradiation for water treatment, Dang Van Phu, Le Anh Quoc, Nguyen Ngoc Duy, Nguyen Quoc Hien, Radiat Phys Chem 88 (2013) 90–94.

[15] Gamma irradiation of cotton fabrics in AgNO solution for preparation of antibacterial fabrics, Truong Thi Hanh, Dang Van Phu, Nguyen Thi Thu, Le Anh Quoc, Do NuBich Duyen, Nguyen Quoc Hien, Carbohyd Polym 101 (2014) 1243–1248.

[16] Cotton fabric modification for imparting high water and oil repellency using perfluoroalkyl phosphate acrylate via g-ray-induced grafting, Hui Miao, Fenfen Bao, Liangliang Cheng, Wenfang Shi, Radiat Phys Chem 79 (2010) 786–790.

[17] Novel UV-protective formulations for cotton, PET fabrics and their blend utilizing irradiation technique, Maged H. Zohdy, Mamdouh B. El Hossamy, Abdel Wahab M. El-Naggar Abeer I. Fathalla, Nisreen M. Ali, Eur Polym J 45 (2009) 2926–2934.

[18] Use of microwave irradiation in the grafting modification of the polysaccharides.–A review, V. Singh, P. Kumar, R. Sanghi, Prog Polym Sci 37 (2012) 340–364.

[19] Facile preparation of super-hydrophilic poly(ethylene terephthalate) fabric using Dilute sulfuric acid under microwave irradiation, Fang Xu, Guangxian Zhang, Fengxiu Zhang, Yuansong Zhang, Appl Surf Sc 349 (2015) 437–444.

[20] Surface modification of polyamide 66 fabric by microwave induced grafting with 2-hydroxyethyl methacrylate, Qian Zhao, Xiaoyu Gu, Sheng Zhang, Mingzhe Dong, Peng Jiang, Zhongwu Hu, Surf Coat Tech 240 (2014) 197–203.

[21] Usage of Microwave Energy in Turkish Textile Production Sector, Banu Yeşim Büyü-kakinci, Energy Procedia 14 (2012) 424–431.

[22] New dual-focalization ferroelectret-based array for air-coupled ultrasonic inspection of textiles, Pazos-Ospina J.F., EaloJ.L., Camacho J. NDT&E Int 74 (2015) 50–57.

[23] Ultrasonic system for continuous washing of textiles in liquid layers, Juan A. Gallego-Juarez, Enrique Riera, Victor Acosta, GermJn Rodriguez, Alfonso Blanco, Ultrason Sonochem 17 (2010) 234–238.

[24] Ultrasonic washing of textiles, Junhee Choi, Tae-Hong Kim, Ho-Young Kim, Wonjung Kim, Ultrason Sonochem xxx (2015) xxx–xxx.

[25] Ultrasonically induced ZnO–biosilica nanocomposite for degradation of a textile dye in aqueous phase, Reza Darvishi Cheshmeh Soltani, Sahand Jorfi, Hojjatallah Ramezani c, Sudabeh Purfadakari, Ultrason Sonochem 28 (2016) 69–78.

[26] Simultaneous sonochemical-enzymatic coating of medical textiles with antibacterial ZnO nanoparticles, Petya Petkova, Antonio Francesko, Ilana Perelshtein, Aharon Gedanken, Tzanko Tzanov, Ultrason Sonochem 29 (2016) 244–250.

[27] Ultrasonic dyeing of cellulose nanofibers, Muzamil Khatri a, Farooq Ahmed a, Abdul Wahab Jatoi a,b, Rasool Bux Mahar c, Zeeshan Khatri, Ick Soo Kim, Ultrason Sonochem 31 (2016) 350–354.

[28] Impact on textile properties of polyester with laser, C.W. Kan, Opt Laser Technol 40 (2008) 113–119.

[29] Influence of laser irradiation on the optical and structural properties of poly(ethylene terephthalate) fibres, V.N. Wijayathunga, C.A. Lawrence, R.S. Blackburn, M.P.U. Bandara, E.L.V. Lewis, H.M. El-Dessouky, V. Cheung, Opt Laser Technol 39 (2007) 1301–1309.

[30] Laser modification of polyamide fabrics, M.I. Bahtiyari, Opt Laser Technol 43 (2011) 114–118.

[31] Laser induced surface modification of polylactide, Piotr Rytlewski, Waldemar Mrź, Marian Zenkiewicz, Joanna Czwartos, Bogusław Budner, J Mater Process Tech 212 (2012) 1700–1704.

[32] Improving the bond strength at hybrid-yarn textile thermoplastic composites for high-technology applications by laser radiation, Tilo Kckritz, Tom Schiefer, Irene Jansen, Eckhard Beyer, Int J Adhes Adhes 46 (2013) 85–94.

[33] Improving the appearance ofall textile products from clothing to home textile using laser technology Ziynet Ondogan, Oktay Pamuk, Ece Nuket Ondogan, Arif Ozguney, Opt Laser Technol 37 (2005) 631–637.

[34] Preparation of poly(ethylene terephthalate) nonwoven fabric from endless microfibers obtained by CO2 laser-thinning method, Akihiro Suzuki, Mahomi Kishi, Polymer 48 (2007) 2729–e2736.

[35] A novel method for preparing the antibacterial glass fibre mat using laser treatment Jakub Wiener, Sheila Shahidi, Makabongwe Mkhululi Goba, Jana Šašková, Eur Phys J Appl Phys.(2014) 65:20501,http://dx.doi.org/10.1051/epjap/2014130385.

[36] Morphological and mechanical changes of glass fibers mat by CO2 laser, J. Wiener, S. Shahidi, The J Text I, 105(2) 2014, 187–195.

[37] New Advances in Plasma Technology for Textile, Sheila Shahidi, Mahmood Ghoranneviss, Bahareh Moazzenchi, J Fusion Energ, 2014, 33(2), 97–102.

[38] Effect of thymol on the antibacterial efficiency of plasma treated, cotton fabric, Sheila Shahidi, Necdet Aslan, Mahmood Ghoranneviss, May Korachi, Cellulose (2014) 21:1933–1943.

[39] Influence of plasma treatment on CNT absorption of cotton fabric and its electrical conductivity and antibacterial activity, Farzaneh Mojtahed, Sheila Shahidi, Emadaldin Hezavehi, J Exp Nanosci , 11(3), 2016, 215–225.

[40] Plasma sputtering as a novel method for improving fastness and antibacterial properties of dyed cotton fabrics, Sheila Shahidi, 162–172, J Text I, 106(2), 2015.

[41] Plasma Sputtering for Fabrication of Antibacterial and Ultraviolet Protective Fabric, Sheila Shahidi, Mahmood Ghoranneviss, Cloth Text Res J January 2016 34 137–147.

[42] Characterization of atmospheric pressure plasma treatedpure cashmere and wool/ cashmere textiles: Treatment in air/water vapor mixture, Stefano Zanini, Elisa Grimoldi, Attilio Citterio, Claudia Riccardi, Appl Surf Sc 349 (2015) 235–240.

[43] Corona plasma modification of polyamide 66 for the design of textiledelivery systems for cosmetic therapy, C. Labay, J.M. Canal, A. Navarro, C. Canal, Appl Surf Sc 316 (2014) 251–258.

[44] Environmentally friendly surface modification of polyethylene terephthalate (PET) fabric by low-temperature oxygen plasma and carboxymethyl chitosan, Jingchun Lv, Qingqing Zhou, Tian Zhi, Dawei Gao, Chunxia Wang, J Clean Prod xxx (2016) 1e10.

[45] Characterization of hydro- and oleo-repellent pure cashmere and wool/nylon textiles obtained by atmospheric pressure plasma pre-treatment and coating with a fluorocarbon resin, Stefano Zanini, Silvia Freti, Attilio Citterio, Claudia Riccardi, S0257–8972 (16) 30145–1, Surf Coat Tech, doi: 10.1016/j.surfcoat.2016.03.020.

[46] Heterogeneous sono-Fenton-like process using nanostructured pyrite prepared by Ar glow discharge plasma for treatment of a textile dye, Alireza Khataee, Peyman Gholami, Behrouz Vahid, Ultrason Sonochem 29 (2016) 213–225.

Chapter 13

Irradiation Pretreatment of Tropical Biomass and Biofiber for Biofuel Production

Mohd Asyraf Kassim, H.P.S Abdul Khalil,
Noor Aziah Serri,
Mohamad Haafiz Mohamad Kassim,
Muhammad Izzuddin Syakir, N.A. Sri Aprila and
Rudi Dungani

Additional information is available at the end of the chapter

http://dx.doi.org/10.5772/62728

Abstract

Interest on biofuel production from biomass and biofiber has gain great attention globally because these materials are abundant, inexpensive, renewable, and sustainable. Generally, the conversion of biomass and biofiber to biofuel involves several processes including biomass production, pretreatment, hydrolysis, and fermentation. Selecting the most efficient pretreatment is crucial to ensure the success of biofuel production since pretreatment has been reported to contribute substantial portion on the production cost. The main goal of the pretreatment is to enhance digestibility of the biomass and biofiber, and to increase sugar production prior to fermentation process. To date, several pretreatment methods have been introduced to pretreat biomass and biofiber including irradiation. This book chapter reviews and discusses different leading irradiation pretreatment technologies along with their mechanism involved during pretreatment of various tropical biomass and biofiber. This chapter also reviews the effect of irradiation pretreatment on the biomass and biofiber component, which could assist the enzymatic saccharification process.

Keywords: irradiation, pretreatment, biomass, biofiber, biofuel

1. Introduction

Rapid development and increase growth of population has led to global environmental problems. Furthermore, increasing demand on energy source has contributed to a reduction of petroleum reverse. To overcome this problem, production of biofuel from renewable resources such as biomass and biofiber has gain great attention to partially replace fossil fuels in the future. Production of biofuel from these materials is environment friendly and sustainable. Mass production of biomass and biofiber as waste residue from agriculture and forestry industry has created a great concern for environment sustainability globally. Generally, biomass and biofiber can be generated from various origin either direct cultivation, or as residue from agro-waste and forestry industries. The biomass and biofiber residue generated from these industries including sugarcane, bagasse, rice straw, empty fruit bunch (EFB), oil palm trunk (OPT), oil palm frond (OPF), and sago bark. Meanwhile, the examples for the cultivated biofiber are kenaf and hemp.

In order to produce biofuel from these materials, it has to undergo few processes such as biomass production, pretreatment, saccharification, and fermentation. Pretreatment of biomass has been reported to contribute substantial portion of liquid biofuel production cost. There are four established pretreatment methods such as thermal, chemical, biological, and physical pretreatment that have been applied to pretreat biomass and biofiber. However, among of the pretreatments mentioned, physical pretreatment method especially irradiation method is considered as one of the promising approaches applied to reduce the recalcitrant of biomass and biomaterials. Generally, this method utilizes both thermal and non-thermal effect generated by intermolecular collision during the realignment of biomass molecule. Irradiation pretreatment offers great advantages such as having very selective process, and it is energy efficient. Since there are wide ranges of tropical biomass types renewably available in tropical country, thus, exploring the potential of this pretreatment is really much needed.

This chapter discusses comprehensively on irradiation pretreatment of tropical biomass prior to the subsequent enzymatic saccharification and fermentation processes. The emphasis is given on the type of irradiation pretreatments and mechanism that could be beneficial for scientists and researchers to understand the process, which can be applied as an alternative pretreatment approach for biofuel production.

2. Biomass and biofiber

Renewable biomass and biofiber are a carbon based biological material derived from living organism. It is composed of a mixture of organic molecules containing hydrogen, oxygen, nitrogen, and small quantities of other atoms such as alkali, alkaline earth, and heavy metals. These materials are abundant, eco-friendly, low cost, and sustainable biomaterials. The biomass and biofiber produced from various industries and manufacturing can be used in composite, textile, food, and chemical industries. On the other hand, these materials have also gained a great attention as a liquid biofuel feedstock due to low cost feedstock materials and environment friendly conversion process.

The renewable biomass and biofiber materials can be categorized into five major categories based on its origin as presented in **Figure 1**. Five distinct biomass and biofiber categories includes: (1) Wood and non-wood (softwood, hardwood, and residue), (2) animal fiber (wool, silk, hair) (3) aquatic plant (algae and hyacinth) (4) plant fiber (cultivated, residue), and (5) other renewable resource (animal residue, municipal solid waste [MSW], industrial residue, sewage).

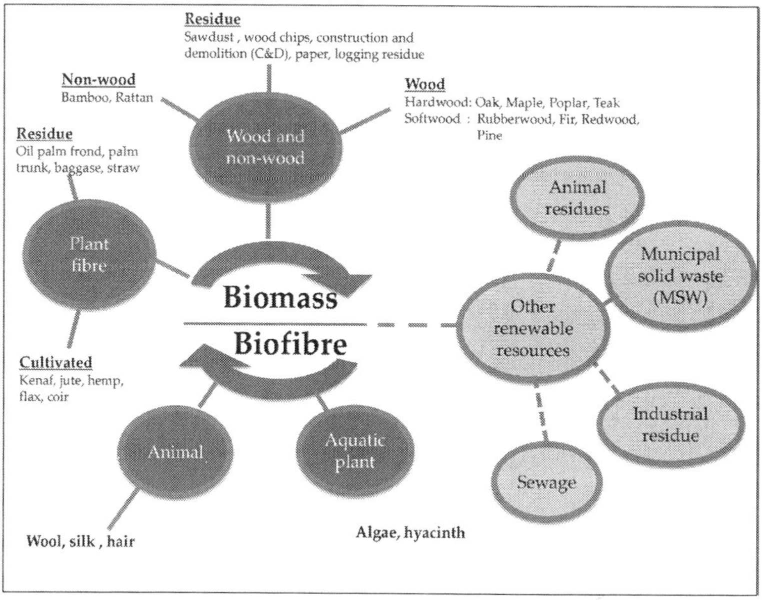

Figure 1. Schematic classification of renewable resources for biofuel production.

These biomass and biofiber are natural biomaterials and can be described as lignocellulosic materials comprised of cellulose, hemicellulose, and lignin. These materials can be further categorized into five categories based on which part it comes from. The five categories are (1) leaf, (2) seed-hair, (3) coir, (4) bark or stem, and (5) other than part mentioned above [1].

Most of the biomass and biofiber generally have a very low economic value. However, due to a broad range of characteristics especially in chemical composition, distribution has provided a variety of applications (**Table 1**).

The biomass and biofiber produced from agro-industry can be used in the plywood, hybrid composite, and animal feed. In any case, the biomass generated from the industry also can be converted into bioenergy and other chemicals, for instance acid and solvent, and liquid biofuel. Due to environmental concern and reduction of fossil fuel reserve, production of liquid biofuel from biomass has gained a great attention because the process is environment friendly. In order to produce biofuel from biomass, it has to go through several processes such as biomass production, pretreatment, followed by enzymatic saccharification, and fermentation (**Figure 2**).

Biomass	Extractive (%)	Hemicellulose (%)	Cellulose (%)	Lignin (%)	Ash (%)	References
Straw	nd	28	40	17	15	[2]
Hemp	nd	18	74	4	4	[2]
Jute	nd	13	72	13	2	[2]
Sugarcane bagasse	nd	24.5	35.2	22.2	20.9	[3]
Corn stover	nd	24.18	37.12	18.20	20.5	[4]
Eucalyptus saligna	nd	48.07	12.69	26.9	12.3	[4]
Montery pine	nd	41.70	20.50	25.90	11.9	[4]
Palm EFB	3.21	29.6	50.49	17.84	3.4	[5]
Palm trunk	5.35	32.04	41.02	24.51	2.2	[5]
Sago hampas	nd	40.5	26.0	7.5	26	[6]
Sago pith	nd	14.5	44.0	4.9	36.6	[6]
Banana stem	10.6	2.0	63.9	18.6	15.5	[7]
Kenaf bast	15.9	9.8	69.8	9.2	1.1	[8]
Kenaf core	7.5	32.3	45.3	19.0	1.4	[8]

nd, Not determined.

Table 1. Chemical composition of various types of renewable biomass and biofiber.

Production of liquid fuel and value-added chemicals from biomass is believed to be one of the approaches to increase the value of biomass and biofiber generated. However, one of the main huddles to ensure the success of this process is the pretreatment process. Pretreatment process has been reported to contribute substantial portion in biofuel production cost. Thus, selecting the most efficient and low cost production could reduce biofuel production cost.

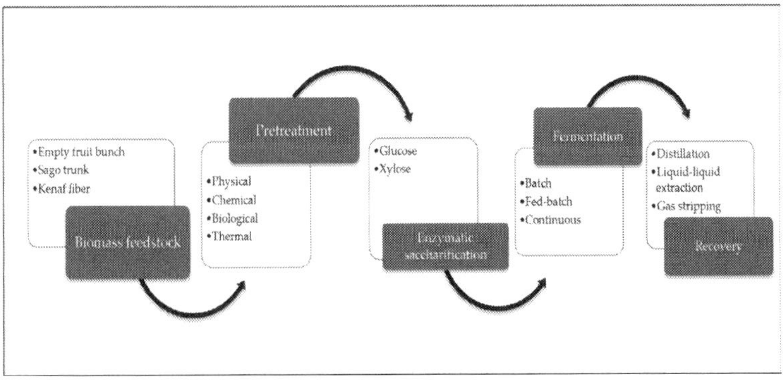

Figure 2. Process flow diagram for biofuel production from biomass and biofiber through biochemical conversion.

3. Pretreatment of biomass and biofiber for liquid biofuel

Pretreatment is one of the most important processed involved in liquid biofuel production through a biochemical conversion pathway. The main goal of the pretreatment is to increase the enzyme accessibility and improve the digestibility of polysaccharides or carbohydrate available in the biomass [9]. The highly organized structure makes plant biomass recalcitrant to physical, chemical, and microbial attack [10]. Thus, the challenge of using lignocellulosic biomass is to have a fast and economical process by integrating variety of pretreatment during the conversion of biofuel. Appropriate selection of pretreatment method must be taken into consideration accordingly to the type of biomass [11]. The pretreatment step involves reduction in biomass size, depolymerization, fractionation, and solubilization of the major components in the biomass, such as hemicellulose, cellulose, lignin, and extractives, making the remaining solid biomass more accessible for further subsequent process. Cellulose is a linear polymer composed of D-anhydroglucopyranose unit which is linked together by β-(1–4) glucosidic bond. This cellulose chains are packed into microfibrils that are attached to each other by hemicelluloses and amorphous polymer. These structures are attached together and covered by lignin. Lignin is an amorphous polymer that provides rigidity to the plant cell wall and protect against microbial attack.

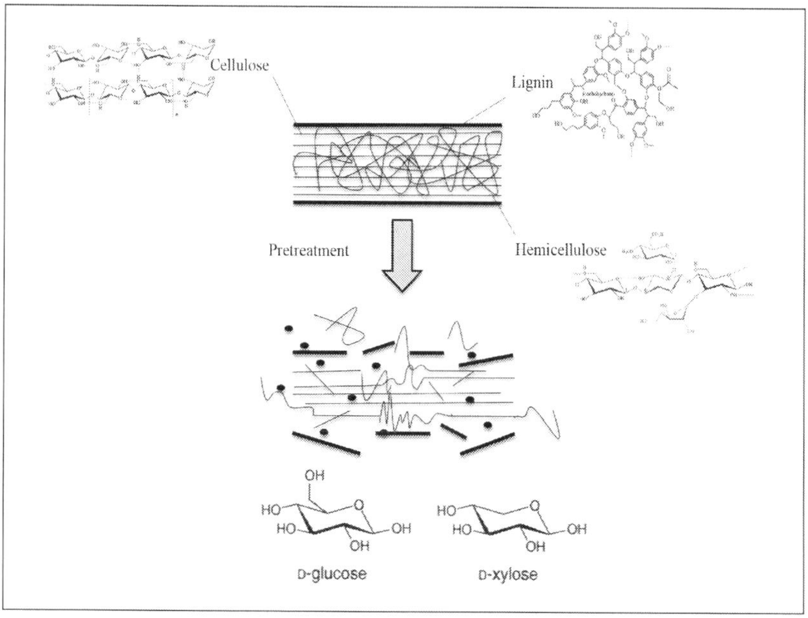

Figure 3. Pretreatment of biomass and biofiber for sugar production.

In order to provide better access for enzymatic saccharification, the lignin, and hemicellulose needs to be separated from the cellulose through pretreatment process (**Figure 3**). Generally, pretreatment of biomass is totally dependent on the chemical composition of the biomass. The key factors for the effectiveness of the pretreatment of biomass and biofiber are: highly digestible, less sugar degradation, and produce less inhibitors that could reduce fermentation performance [12]. Biomass and biofiber that possesses a high recalcitrant component requires a harsh pretreatment condition to disrupt the cell structure.

The pretreatment of biomass and biofiber can be categorized into four different methods, namely thermal, physical, chemical, and biological (**Figure 4**). Thermal pretreatment is a treatment used to solubilize the biomass by applying heat in the pretreatment system. This method is one of the most common method used for the pretreatment of biomass and biofiber. Generally, the thermal pretreatment is sub-divided into three categories: (1) thermal treatment (temperature = < 100°C under atmospheric pressure); (2) hydrothermal treatment (temperature = > 100°C with gradual pressure release after treatment); and (3) thermal treatment with steam explosion (temperature > 100°C with sudden pressure drop after pretreatment). Temperature and reaction time are the most important factor that plays a major role in this pretreatment process [13]. This method proved to display a significant effect on the disruption of biomass and biofiber such as pelletized corn stover, rice hulls, kenaf, Tahoe mix, and switch grass [14–16]. Although this method was reported to display a positive effect on enzymatic saccharification process, this method is not selective and less effective for the biomass with less lignin content. Thermal pretreatment at high temperature would partially degrade hemicellulose and produce more inhibitors that could influence fermentation process [12, 17]. The major fermentation inhibitors such as hydroxymethylfurfural (HMF) and furfural are one of the major products made from the thermal pretreatment process [13].

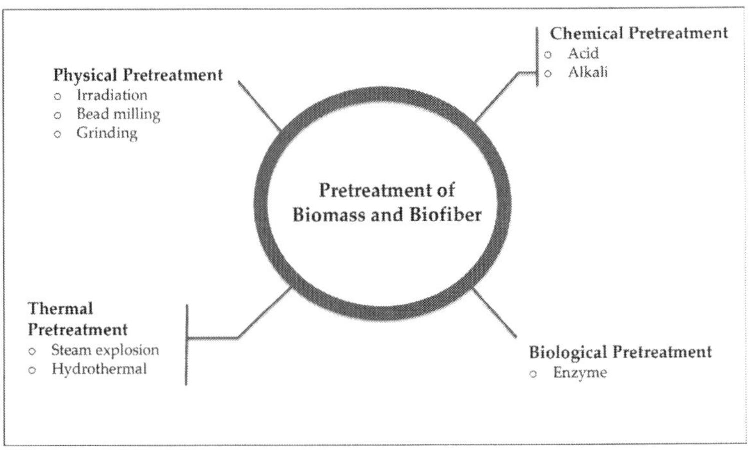

Figure 4. Pretreatment methods of biomass and biofiber for biofuel production.

Chemical pretreatment is one of the most promising methods used to pretreat biomass and biofiber. Generally, this process has been proven successful, particularly when combined with heat [18, 19]. The chemical most commonly applied in this process is either an acid or alkali reagent. The main goal of chemical pretreatment is to solubilize polymers, favoring the availability of carbohydrate in the biomass for enzymatic saccharification. The most common acids used for biomass pretreatment are hydrochloric acid (HCl) and sulfuric acid (H_2SO_4). In this process, the acid will catalyzed the linkage bond and solubilizes hemicellulose. Unlike acid pretreatment, the alkaline pretreatment method is considered very mild and environment friendly as this method uses low concentration of alkali [20]. Pretreatments with alkali such as sodium hydroxide (NaOH), potassium hydroxide (KOH), calcium hydroxide ($Ca(OH)_2$), hydrazine, and anhydrous ammonia cause swelling of biomass, disrupts the lignin structure, and breaks the linkage between lignin and the other carbohydrate fractions. Chemical pretreatment using acid and alkaline has been reported to be a promising approach to pretreat biomass and biofiber prior to enzymatic saccharification due to its capability to remove lignin and hemicellulose from the biomass. However, this pretreatment has few disadvantages such as, generation of inhibitors during the pretreatment and chemical used, which could affect subsequent fermentation process and is not environment friendly [21]. Thus, it has encouraged more exploration on other alternative pretreatment process that is sustainable and could be beneficial for the whole production line.

Another pretreatment that is commonly used to pretreat biomass and biofiber is biological pretreatment. This pretreatment involves microbes and enzymes to degrade the chemical compound and release fermentable sugar from the biomass and biofiber. In this method, microorganisms such as brown-, white-, and soft-rot fungi are used to degrade the biomass and biofiber cell wall. White rot fungi such as *Phanerocheate chrysosprorium, Cleriponopsis subremospera, Phlebia subserialisis,* and *Pleuroisu ostriosis* are commonly used in biological pretreatment [22]. While, brown rot fungi for instance *Gleophylium sepiarum, Fomitopsis pinicola,* and *Laetiporus suiphureus* are among the common brown rot fungi used to pretreat biomass and biofiber via this process [23]. During the biological pretreatment, hydrolytic enzyme such as lignin peroxidase (LiP) is produced by the bacteria or fungi and it will attack biomass and biofiber cell wall to a small compound with a low molecular weight, which subsequently, can be used in anaerobic fermentation for biofuel production. Currently, research on the direct enzymatic saccharification of biomass is still scarce.

This method appears to have a few advantages, for example, it requires low energy input and this process is mildly environment friendly. However, the large diversity of chemical composition among different types of biomass, enzyme production, and low hydrolysis rate are among the drawbacks that needs to be considered before the method is applied in a large-scale biofuel production. In spite of the many pretreatment methods tested, currently available pretreatment techniques can hardly meet the requirements of commercial application due to long processing times, chemical recycle problems, or high operational costs [9, 24]. Therefore, more works are required to understand and generate more information on the pretreatment of biomass and biofiber.

Physical pretreatment is a process that acts directly at breaking the cells through physical force. This method is widely used as a preliminary step for biomass pretreatment process. Physical pretreatment will reduce biomass size and increase the accessible surface area and pore size. Besides, it could also decrease the cellulose crystallinity and polymerization degrees. Various types of physical pretreatment have been introduced to pretreat biomass including comminotium, milling (ball milling, colloid milling, and vibro energy milling), extrusion, and irradiation. The biomass pretreatment using irradiation has been reported to require less energy compared to other approaches mentioned. Furthermore, this approach is selective and easy to control, thus it is more efficient for production of the desired product [25]. The details on the irradiation pretreatment is described in the next section.

4. Irradiation pretreatment and its mechanisms

Among various physical methods, irradiation is considered an attractive method for biomass and biofiber pretreatment. In biomass irradiation process, biomass and biofiber is exposed to high- energy radiations such as ultrasonic waves, microwaves, γ-rays, and electron beam. The irradiation effect on the biomass and process mechanisms varies according to the method applied. Generally, radiation processing technology is defined as: a radiolysis reaction, which uses γ-rays from radioisotopes such as cobalt-60 or cesium-137, or an electron beam produced by an electron accelerator to induce degradation of cellulose. In this process, the high energy radiation generated could change the characteristic of cellulosic biomass including: enhance specific surface area, reduce the degree of polymerization and crystallinity of cellulose, hydrolysis of hemicellulose, and partial depolymerization of lignin [26–29]. Typically, the irradiation pretreatment mechanism mode significantly depends on the technology applied during the pretreatment process. The effect of irradiation pretreatment is assessed base on the reducing sugar production during enzymatic saccharification and the solid residues left after pretreatment. The effectiveness of the treatment depends on several factors such as frequency of radiations, time of exposure, composition of the biomass, and resistance to the radiations by medium between radiations and biomass [30, 31]. Besides, the pretreatment combination used of the irradiation pretreatment and chemical treatment also gives a significant effect on the reducing sugar production during the enzymatic saccharification process [32, 33]. The detailed explanation on the effect of irradiation pretreatment on the biomass and biofiber structure and functional group is described in the next section.

4.1. Type of irradiation pretreatments

There are four different irradiation pretreatment methods that is commonly being used to pretreat biomass prior to enzymatic saccharification process. The irradiation methods are gamma-ray irradiation, electron-beam irradiation, microwave, and ultrasonication. Aforementioned in the previous section, the pretreatment mechanism of each process is different according to the method applied.

4.1.1. Gamma-ray irradiation

Gamma ray is a high-energy ionizing radiation in electromagnetic spectrum that easily penetrates most materials. This irradiation is extremely large high frequency waves and largely depends on the radiation source. This technology is commonly applied in radiotherapy as a tracer in food and medical apparatus sterilization. Recently, the utilization of this technology has gain great attention especially in a biomass and biofiber pretreatment for liquid biofuel production. Radioactive nuclides such as cobalt-60 and cesium-137 are the common radioactive used in this pretreatment [28]. The main goal of this irradiation is to decrease intra and intermolecular order in cellulose due to the breakdown of the intermolecular hydrogen bonds. In this process, the radiation will travel from the seal source and penetrates (bombard) the biomass and biofiber. The energy carried by gamma radiation is transferred to the biomass component by collision of radiation, resulting to the loss of electron by the atom and lead to the ionization. Under exposure to radiation, the biomass component mainly cellulose macromolecules undergo scission, and various short and long-lived radicals are formed [34]. Also, the content of fragments with a low degree of polymerization generated from the process gradually increases, leading to the alteration of biomass structure, thus, providing ease of access for subsequent process such as enzymatic saccharification process.

The potential of gamma irradiation technology in biomass and biofiber pretreatment has been studied on various types of biomass for instance, jute fiber, poplar sawdust, wheat straw, and cotton-cellulose [33, 35]. There were only scanty studies on gamma irradiation pretreatment on tropical biomass and biofiber that has also been reported. A study on gamma irradiation of empty fruit bunches (EFB) indicated that the pretreatment has reduced the lignin and increased the cellulose content in the EFB [36]. Scanning electron microscopy – EDX (SEM-EDX) analysis showed that there is a significant change on the carbon and oxygen content in the EFB biomass. Typically, untreated EFB contains high carbon and low oxygen content, while the study found a decrease of carbon (9% increment) and decrease of oxygen content (16% decrease), indicating the reduction of lignin content in the EFB.

A comparison on gamma ray irradiation pretreatment on soft and hardwood has also been carried out using different level of dosage ranges between 10–100 kGy [37]. The study found that the most suitable condition for softwood was at 40 kGy, while higher dosage is required to pretreat hardwood (90 kGy). The study also concluded that gamma ray pretreatment process is species-dependent, wherein higher dosage is needed to disrupt hardwood cell structure compared to softwood.

4.1.2. Electron-beam irradiation

Electron-beam is one of the irradiation pretreatment used to pretreat biomass prior to enzymatic saccharification. This technology has been widely used in various applications such as welding, drilling, and surface treatment [38]. For commercial use, the most important characteristics of an accelerator are its electron energy and average beam power. Therefore, industrial electron accelerators are usually classified according to their energy ranges, which are divided into low (80–300 keV), medium (300 keV–5 MeV), and high-energy ranges (above 5 MeV). In the electron beam pretreatment, the biomass and biofiber is exposed to a highly charged stream electron. The electron is emitted from an electron beam gun and accelerated by accelerator (**Figure 5**). In this pretreatment process, the electron energy can be controlled and modulated by varying the irradiation dose. The high-energy electrons emitted travel into biomass and biofiber component and transfer the energy within the materials. The heating process initiates chemical and thermal reaction in the biomass including cellulose depolymerization, and production of carbonyl group, resulting from the oxidation of the biomass. Crosslinking of biomass component has also been reported to occur when the biomass is exposed to irradiation beam [39]. Also, reduction of the biomass mechanical strength has been observed from the biomass exposed to electron beam. This could be due to the disruption of hydrogen bond between cellulose chains making it less crystalline and more amorphous [40].

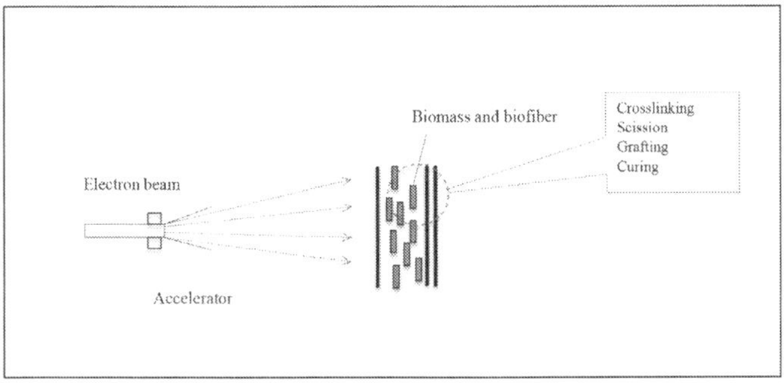

Figure 5. Experimental set-up for electron beam irradiation.

Recently various research groups have studied the potential of electron beam radiation on various type of biomass including tropical biomass and biofiber such as bamboo, rice straw, oil palm, fruit bunch, and kenaf [30, 41, 42]. Overall, most of the EBI pretreatment indicated that a significant cellulose degradation was observed after the process [39]. Moreover, the study also showed that this pretreatment has enhanced enzymatic saccharification and reduce sugar production from biomass [31, 41]. A study on EBI pretreatment of hybrid grass biomass indicated that the pretreatment could enhance 59% of glucose yield from the biomass compared to untreated sample. This is similar to a study by Bak et al. [43] who reported that EBI pretreatment on rice straw could increase enzyme digestibility and energy during the pretreatment process.

Similar to other pretreatment process, EBI pretreatment process could be influenced by several factors. EBI dosage is one of the factors that play a major role in the EBI pretreatment of biomass process [39, 42]. A study on the EBI pretreatment of bamboo chips at various EBI dosage range 0.5–50 kGy, indicated that significant cellulose degradation was attained from the pretreatment dosage between 0–50 kGy. Furthermore, the study showed no significant changes on the hemicellulose content. This indicates that EBI pretreatment process is a selective process and the degradation level can be controlled by the EBI dosage [39].

http://dx.doi.org/10.5772/62728

4.1.3. Microwave irradiation

Microwave is electromagnetic waves between the frequency range of 0.3–300 GHz, and most of the microwave systems used for industrial and domestic purposes range between 0.9 GHz to 2.45 GHz [44]. Microwave radiation is a radiating wave movement and takes a straight-line path type of energy. This radiation do not require any medium to travel through and could penetrate non-metal materials such as plastic and glass. Microwaves can affect the material thermally and non-thermally. Thermally, microwaves heat the material by the interaction of the molecules of material with electromagnetic field produced by microwave energy (**Figure 6**). Non-thermally, microwaves affect and interact with the polar molecules and ions in the materials causing physical, chemical, and biological reactions [45].

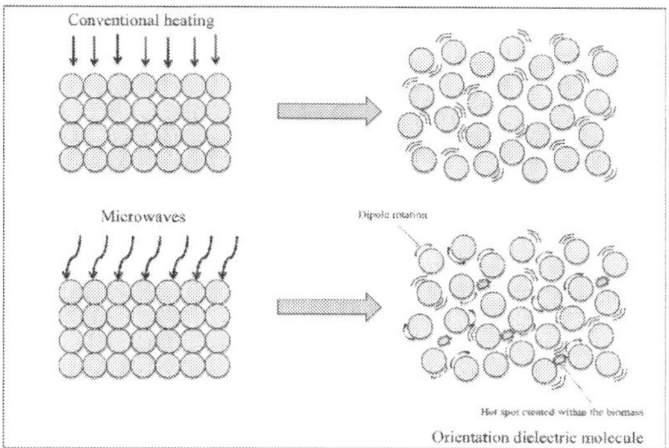

Figure 6. Conventional and microwave heating mechanisms of biomolecule.

Many studies on the potential of microwave pretreatment towards various types of biomass and biofiber digestibilities such as switchgrass, sweet sorghum bagasse and mischantus have been reported [46–48]. Generally, the microwave pretreatment can be carried out through three different approaches:

a. Combination of mechanical and microwave pretreatment

b. Combination of microwave and chemical pretreatment

c. Combination of microwave and steam explosion pretreatment

Mechanical pretreatment is used to reduce biomass particle size and provide more surface area for further microwave pretreatment. Pretreatment through combination of microwave and chemical approaches will generally involve either acid or alkaline as catalyst. In this process, alkaline is used to swell the biomass structure and remove lignin component from the biomass [49]. On the other hand, acid catalyst used in this process will convert hemicellulose and cellulose component into a small monomer sugar such as glucose and xylose, which is the main platform for biofuel production [49]. In contrast to the combination of microwave and steam explosion approach, the high pressure and temperature radically disrupts the lignocellulosic biomass structure and provide better assess for hydrolytic enzyme to degrade cellulose and hemicellulose.

Study on microwaves pretreatment on biomass such as palm biomass has been widely reported [50–52]. Akhbar et al. [53] compared the microwave assisted chemical pretreatment of empty fruit bunches (EFB) with conventional method and found higher lignin removal of up to 72% using microwave assisted chemical treatment. The presence of chemical such as alkaline or acid in microwave pretreatment could assist fractionation of the biomass and biofiber.

The microwave pretreatment has also been applied on other types of palm biomass. Lai and Idris [54] in their study on the microwave pretreatment of oil palm trunk (OPT) and frond (OPF), found that this pretreatment was able to disrupt the OPT and OPF. In this study, the biomass was pretreated at 700 W at 80°C for 60 min, and approximately 41.6% and 64.42% of cellulose was released from the OPT and OPF respectively. They also suggested that pretreatment at this condition is more effective in extracting hemicellulose and cellulose component compared to lignin in both OPT and OPF. In their other study on the determination of optimum condition for lignin extraction from OPT indicated that the highest lignin reduction (22.38%) was attained when the pretreatment was performed at 100°C for 80 min at 900 W. This study is in agreement to the conclusion that microwave pretreatment is significantly influenced by the temperature, reaction time, and microwave power [52].

Apart from oil palm biomass, several studies on the microwave pretreatment on other biomass such as kenaf, sago pith, sago bark waste, banana trunk, and mischantus have also been reported elsewhere [55, 56]. Study by Ooi et al. [57] on the microwave alkali-assisted pretreatment of kenaf pulp showed that the pretreatment at 50°C is the suitable temperature to convert crystalline cellulose to amorphous form, and produce higher sugar yield compared to untreated sample. In another study on microwave pretreatment of sago pith, a starch-based crop that contain substantial amount of starch and fiber, indicated that direct heating of sago pith in water by microwave treatment can swell and gelatinize the starch, resulting to a more amorphous and more susceptible fiber for subsequent enzyme reaction [55]. In a study on microwave chemical assisted pretreatment of miscanthus under different temperature range of 130–200°C, found that the suitable condition for miscanthus pretreatment is at 180°C for 20 min [58]. This study concluded that temperature plays an important role in microwave pretreatment process. Pretreatment at high temperature increases biomass solubility, shorten the pretreatment reaction period, and reduce recalcitrant characteristic of the biomass. However, the pretreatment process at high temperature also produced a substantial amount of inhibitor that is harmful to the subsequent enzymatic saccharification and fermentation.

4.1.4. Ultrasonication

Another irradiation pretreatment that is widely used to pretreat biomass and biofiber for biofuel production is ultrasonication. This process can be performed either using probe-type ultrasonication or an ultrasonic bath. In this process, ultrasonic waves can be generated via piezoelectric or magnetostrictive transducers in the frequency range of 20–1000 kHz, in which the waves induced provide pressure difference in the medium. The pressure wave that travels through the liquid medium has high pressure (compression) and low pressure (rarefaction) regions. The rarefaction of the cycle can stretch the liquid molecules apart and create cavities also known as bubbles. As the wave cycles through the liquid, the bubbles expand and contract with the rarefaction and compression of the wave, respectively, drawing more liquid molecules into the bubbles as they grow. The bubbles that either continue to expand and then float to the surface, are subjected to coalescence due to the forces or collapse during compression of the wave (**Figure 7**). This collapse is almost adiabatic and can result in localized temperatures of around 5000 K and pressures of 1000 atm [59]. The collapse results in the formation of radicals through dissociation of the molecules within and around the bubbles, luminescence due to excited molecules formed losing energy, and microjets shooting out of the bubbles of speeds in the realms of hundreds of km per hour.

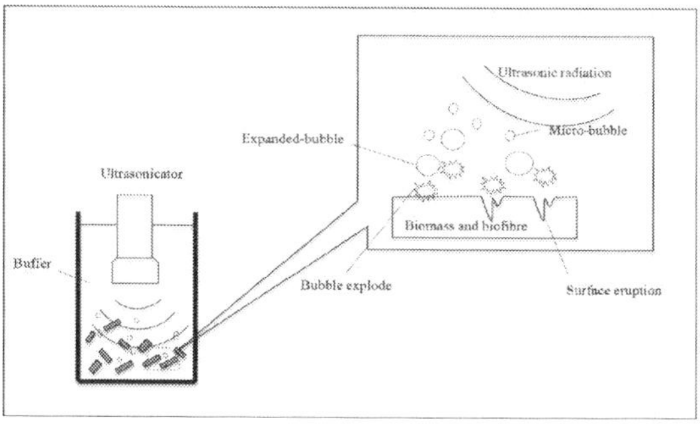

Figure 7. Ultrasonication pretreatment of biomass and biofiber mechanisms.

Ultrasonic pretreatment has been performed on a great variety of lignocellulosic biomass and biofiber including kenaf powder, kenaf bast fiber, corn meal, and corn stover [60–63]. This approach has also been performed on tropical biomass such as EFB and kenaf fiber. Most of the study concluded that ultrasonication pretreatment is capable to enhance conversion of biomass to biofuel. A study on ultrasonic pretreatment of EFB at low temperature indicated that this pretreatment could assist the acid hydrolysis performed at low temperature and pressure [64]. The study showed that xylose production from the pretreated EFB was two times higher than that of un-pretreated sample when the pretreatment was performed at 100°C for 40 min. Similar to the study on the ultrasonic pretreatment of kenaf powder in ionic liquid indicated that higher reducing sugar was attained from pretreated sample [60]. In this study, a significant change on the hemicellulose content was observed in the pretreated biomass.

5. Irradiation effect on biomass and biofiber

The main goal of the pretreatment process in liquid biofuel production is to modify the surface morphology structure and properties aiming to improve digestibility in the subsequent enzymatic saccharification. The pretreatment has also been reported to affect the chemical composition of the biomass (**Table 2**). A significant reduction of lignin and hemicellulose were observed from the EFB after it went through the EBI pretreatment process.

Sample	Total lignin (%)	Cellulose (%)	Hemicellulose (%)
OPEFB untreated	35.94	30.41	20.70
OPEFB-C	10.32	77.5	6.83
OPEFB-CI 100 kGy	8.23	68.87	14.27
OPEFB-CI 200 kGy	5.86	71.96	15.20
OPEFB-CI 300 kGy	7.01	65.64	13.94
OPEFB-CI 400 kGy	7.65	64.92	13.39
OPEFB-CI 500 kGy	7.86	63.81	13.45

Table 2. Chemical composition content of oil palm empty fruit bunches (OPEFB) untreated and gamma irradiated OPEFB. (Adapted from Kristiani et al. [36]).

5.1 Surface morphology and chemical structure

The changes in chemical composition and surface morphology structure of biomass and biofiber are the main obvious effect observed from the irradiation pretreatment process. Typically, biomass or fiber with high crystallinity may consist a significant amount of crystallinity cellulose, and appear to be relatively smooth. The biomass with high degree of crystallinity indicates that it has high tensile strength properties [65].

Various investigations on the effect of irradiation pretreatment on tropical biomass including EFB, kenaf, rubberwood, and bamboo have been reported [29, 42, 64]. The study agreed that the pretreatment applied on these biomass have a significant effect on the biomass structure and chemical properties. For instance, a study on irradiation pretreatment of EFB at 300 kGy indicated a significant change in the surface morphology before and after pretreatment process [36]. The study found that the EFB, which is solid, intact, rough, and rigid structure becomes brittle and flaky after irradiated with gamma ray. Similar observation has been reported on the irradiation pretreatment of rubberwood. Darji et al. [66] compared the rubberwood structure before and after pretreatment and found that most of the fibrous in the rubberwood lost and disappeared after the pretreatment process.

The surface morphology structure change after pretreatment could be attributed to the irradiation process that is able to break the intermolecular hydrogen bond, resulting to the decrease of intra and intermolecular order in cellulose. Furthermore, under high energy and pressure, the cellulose macromolecule will undergo scission and increase the fragment with low degree of polymerization [67]. On the other hand, irradiation has also been reported to influence the biomass pore size. Brunauer–Emmett–Teller (BET) analysis on the kenaf core and cellulose, indicated that a significant increase of pore size was observed for both materials after irradiation pretreatment process [42, 68]. High pore size is a very important characteristic that could provide easy access for subsequent process prior to biofuel production.

http://dx.doi.org/10.5772/62728

Most of the studies reported that the change of the biomass surface morphology is correlated to the chemical structure in the biomass. A change on the degree of crystallinity was found to change surface morphology [69]. Typically, X-ray diffraction (XRD) analysis is applied to evaluate the effect of irradiation on biomass crystallinity. Chen et al. [70] reported that the major diffraction peak for cellulose crystallography can be identified for 2θ ranging between 22° and 23° as a primary peak, whereas a secondary peak is in the range of 16° to 18°. As reported by Liu et al. [28], the I_{002} peak intensity (the maximum intensity of the 002 lattice diffraction) represents the primary peak and is classified as the diffraction intensity of crystalline regions, whereas the secondary peak represents the diffraction intensity of the amorphous zone. XRD analysis of irradiated cellulose at different irradiation dosage between 10–100 kGy indicated that increase of dosage could reduce crystallinity index and crystallite size [68]. In another study on irradiation of OPTT and OPF, it was found that an obvious peak reduction on the primary and second peak, indicates the transformation of cellulose molecular hydrogen bond due to rapid heating during the irradiation pretreatment process [54].

Apart from XRD analysis, the effect of irradiation pretreatment can also be evaluated by Fourier transform infrared spectroscopy (FT-IR). This method is widely used to determine the chemical structure changes after pretreatment of various types of biomass [71, 72]. FT-IR analysis on the irradiated oil palm trunk and oil palm frond indicated that radiation has affected the intensity of all bands in the IR spectra [51, 54]. Obvious changes were observed at absorbance between 3500–3200 cm^{-1}, 2840–2690 cm^{-1}, 1740–1720 cm^{-1}, 1500–1450cm^{-1}, 1300–1000 cm^{-1}, 1315–1318 cm^{-1}, and 900–898 cm^{-1}. These bands represent a specific chemical structure in biomass as summarized in **Table 3**.

Infrared band (cm^{-1})	Functional groups	Infrared band (cm^{-1})	Biomass component assignment
3500–3200	O-H (H-bonded)		
2840–2690	C-H (aldehyde C-H)		
1740–1720	C=O (saturated aldehyde)	1735	C=O in xylan (Hemicellulose)
1695–1630	C=O, C=C	1647	Absorb OH and conjugate C=O
1500–1450	C=C (in ring)	1505	Aromatic skeletal vibration in lignin
		1421	CH deformation in lignin and carbohydrate
1300–1000	C-O	1235	Syringyl ring and CO stretch in lignin and xylan
		1371	CH deformation in cellulose and hemicellulose
		1319	CH vibration in cellulose, CO vibration in syringyl derivative
		1155	C-O-C vibration in cellulose and hemicellulose
		1030	C-O vibration in cellulose and hemicellulose
900	C-H	897	CH deformation in cellulose

Table 3. FT-IR band assignment in biomass [54].

Normally, the structure of lignin consists of guaiacyl propane units (G) and syringyl propane units (S) containing one and two methoxy groups. It is known that the presence of guaiacyl propane could restrict the swelling of biomass [73, 74]. These chemical structures can be identified by FT-IR spectra with frequencies in the region of 1509, 1464, and 1422 cm^{-1}. Reduction of spectra in this region indicated that most of the lignin in OPT and OPF have been removed from the biomass during the pretreatment process. Removal of lignin in the biomass after irradiation gives a better access for enzyme to attack cellulose and hemicellulose.

The FT-IR analysis of the irradiation pretreatment on biomass also indicated that significant changes on absorbance was observed at 1732 cm^{-1} and 3300 cm^{-1}, attributed to the vibration of hydrogen bonded OH-group. Liu et al. [75] reported a shifting and reduction of band 2899 cm $^{-1}$, indicating to the disruption of biomass resulting from the C-H shifting vibration. The study also found that high-energy irradiation pretreatment could interrupt and destroy the intra-molecular and inter molecular hydrogen bond in the cellulose. The degradation of cellulose generated carbonyl group could be determined at band at 1603 cm^{-1}. Apart of this region, the shifting of band region between 1164 cm^{-1}, 1112 cm^{-1}, and 1058 cm^{-1} attributed to the vibration of C-O-C of cellulose.

5.2 Enzymatic saccharification

The pretreatment method aims at facilitating maximum saccharification of cellulose and hemicellulose by enzymatic hydrolysis. In this process, the cellulose and hemicellulose present in the biomass will be hydrolysed by cellulase and hemicellulase enzyme produced from fungi into simple monomer sugar such as glucose and xylose. This monomer sugar is the main chemical platform for biofuel and other chemicals (**Figure 8**). Previous research obtained the cellulose from untreated biomass and biofiber upon enzymatic hydrolysis can yield not more than 15–25% glucose due to the recalcitrance [76]. Most pretreatment methods have some disadvantages in terms of cost, recovery, secondary pollution, and formation of intermediate compounds that will inhibit enzymatic hydrolysis, but implementation of laser, microwave, and electron beam irradiation have become more attractive because of its fast and effective result during experimentation [77].

Biomass pretreated with electron beam irradiation (EBI) enhance enzymatic saccharification by decreasing the crystallinity and molecular weight and simultaneously increase the surface area [36, 68]. Irradiation induces a chain-cleavage mechanism by depolymerizing the polymeric material [78]. Higher cellulose content was found in chemical-irradiated pre-treated oil palm empty fruit bunch (OPEFB) than untreated OPEFB. Higher cellulose content can produce higher glucose, hemicelluloses content can be converted to xylosa, while lignin can produce derivatives compound of phenol [36]. The effectiveness of EBI treatment also depends on the nature of biomass with respect to energy delivered, and sources and concentration of enzymes used [77]. The earliest study by Kumakura and Kaetsu [79] found the pre-irradiation (dosage 107 rad) with presence of chlorine yields six times higher reducing sugar than its absence with subsequent enzymatic hydrolysis on rice straw. Then Ardica et al. [80] used gamma-ray irradiation (doses range from 1 kGy to 1000 kGy) on wood chips, kapok, papers, hays, and grain straw to enhance the enzymatic hydrolysis. Combined pretreatment of gamma-ray and

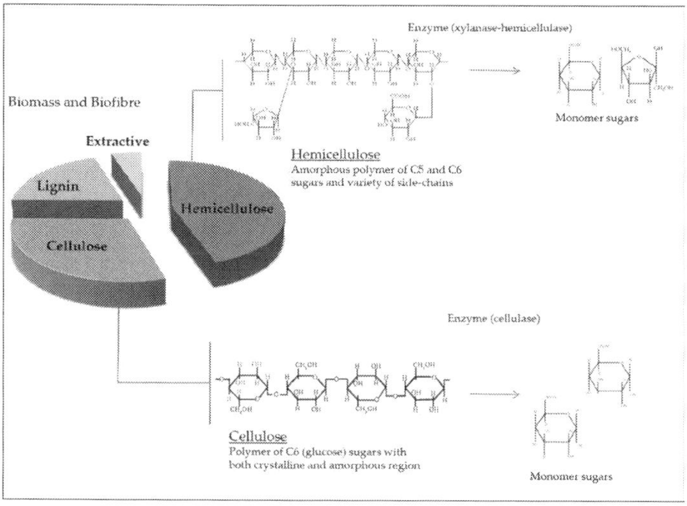

Figure 8. Enzymatic saccharification of cellulose and hemicellulose to monomer sugars by cellulase and hemicellulose enzyme.

diluted acid on poplar bark biomass observed a drastic increased in reducing sugar yield from 56.1 to 83.1% compared to gamma-ray pretreatment alone [81]. **Table 4** shows previous study using electron beam irradiation with various types of biomass.

Zhu et al. [32] reported the rice straw pretreated with microwave-alkali method obtained higher hydrolysis rate than alkaline pretreated alone. The amount of glucose obtained from the enzymatic hydrolysis using *Trichoderma reesei* cellulase was higher (24.8 gl^{-1}), and lower for xylose (2.6 gl^{-1}) concentration for microwave/alkali pretreated rice straw which is more suitable for subsequent fermentation process to produce biofuel. The microwave-alkali assist irradiation has been proven to remove more hemicellulose and lignin, and simultaneously increasing enzyme accessibility [32]. In 2006, they have presented comparison between three techniques for the enzymatic hydrolysis of rice straw by pre-treating them with microwave/alkali, microwave/acid/alkali, and microwave/acid/alkali/H_2O_2 treatment. The result shows that the rice straw pretreated with microwave/acid/alkali/H_2O_2 treatment had the highest hydrolysis rate and glucose content in the hydrolysate. Furthermore, recovery of xylose content could be recovered compared to microwave/alkali treatment [83]. [84] using microwave-assisted alkali treatment. They have presented at optimal condition of 190°C, 50g/l solid content, and 30 min treatment time. The sugar yield from the combined treatment and hydrolysis was 58.7g/100g biomass which is equivalent to 99% of potential maximum sugars. This study was further investigated by using scanning electron microscope, and showed the advantage of microwave over conventional heating was due to the disruption of recalcitrant structures of lignocellulose. In conclusion, microwave/chemical pretreatment is more effective than conventional heating.

Biomass type	EBI dose	Glucose yield (%)		References
		Untreated	Pretreated	
Rice straw (RS)	80 kGy	35	52	[41]
Wheat straw	500 kGy	8.60	10.72	[82]
Poplar bark	0–1000 kGy	44.2	66.7a	[81]
Napier Grass	250 kGy	20	79	[77]
Rice straw (RS)	80 kGy	37	70.4b	[43]
OPEFB	100–500 kGy	na	na	[36]

a, pretreatment added with diluted sulfuric acid.
b, pretreatment added with water soaking-based EBI.
na, not available.

Table 4. Summarization of previous research using EBI pretreatment on various types of biomass and biofiber.

6. Biofuel from biomass and biofiber

In general, sugar substrates from tropical biomass and biofiber are potential sources for biofuel production such as ethanol and butanol because they are abundant, cheap, and renewable [85]. Biomass and biofiber utilization will reduce the dependency on fossil fuel and at the same time it could help in reducing toxic gases emission with an abundant feedstock that can support for a very long period of time. This second generation biofuel does not compete with human food resources which are non-edible in nature [86]. In lignocellulosic biomass conversion for biofuels such as ethanol and butanol, pretreatment plays a major role in separating the major components (lignin, cellulose, and hemicellulose) of the biomass. The conventional chemical and enzymatic pretreatment methods have disadvantages such as producing byproducts and low conversion of biomass components [86]. Numerous numbers of publications reported the potential of bioenergy from biomass wastes through irradiation pretreatment [43, 45, 62, 87, 88]. Most of the studies agreed that irradiation pretreatment could assist the reduction of particle size that provide better access for subsequent process. This pretreatment clearly proved able to enhance enzymatic saccharification and fermentation performance.

Presently, Malaysia is dependent on fossil fuels such as coils, oil, and natural gas as well as renewable energy sources such as hydro, biomass, and solar energy. The demands for energy is increasing by years with some challenges such as the decreasing source of fossil fuels, food versus fuel crisis, and greenhouse gas (GHG) emission that needs to be taken into consideration [11]. In Malaysia, the development of renewable energy is still rather slow. Although, in 10th Malaysia plan, renewable energy usage has to increase >1% in 2009 to 5.5% of total electricity generation in 2010, although several fiscal incentives have been launched by the Malaysian government [89]. On the other hand, Malaysia is geographically located in the tropical and humid climate region which provides easy access to variety of biomass resources. Biomass resources are mainly from palm oil, wood, and agro-industries [90]. Malaysia devotes 11% of the total land area with 62% of the economy agricultural land for planting palm oil. If 20% of palm oil productions are turned into biofuels, it can replace 64% of diesel consumption, and at the same time cutting off 41% of imported crude oil [91]. From these facts, we can estimate the amount of biomass and biofiber waste produced yearly. Thus, there is a potential need to convert the residue into a valuable product by converting them into biomass energy feedstock. Economically, biomass waste from palm oil plantation such as empty fruit bunch (EFB) can be used as resources for conversion of bioethanol, since the production is 6.1 million tons dry EFB and is forecasted to increase to 7.6 million tons in 2025 as shown in **Table 5** [92].

Year	Projected EFB production (Million tons dry matter/year)	Potential bioethanol production (Million/year)	*Energy content in ethanol: 22MJ/litre (*GJoule/year)	Potential Bioethanol (ktoe/year)
2005	6.14	2382	54,793,360	1863
2015	7.59	2945	67,733,160	2303
2025	7.66	2972	68,357,840	2324

Table 5. Potential ethanol and forecasted EFB production by MPOB based on 22% EFB to FFB and moisture at 65% (93).

Currently, biorefineries are increasingly focused on integrated process design for maximum valorization of fractionated biomass components for fuels and a spectrum of co-products. This multi-product "integrated biorefineries concept" is a platform for development of modern biorefineries with economic competitiveness to the current petroleum industry [10]. According to the IEA (International Energy Agency) report from the assessment of available residue in 2030, it was predicted that 10% of global residues could yield around 155 billion lge (5.2 EJ) lignocellulosic ethanol or almost around 4.1% of the projected transport fuel demand in 2030, and 25% of global residues converted to either ethanol, diesel, or syngas that could contribute to 385–554 billion lge (13–23.3 EJ) globally [94].

In conclusion, the viability of biomass and biofiber materials should concentrate more on developing a complete understanding of these materials to form a foundation for significant advancement in sustainable energy. Development in characterization and overcoming the difficulty for enzymatic saccharification of different raw materials is crucial for the development of economically competitive processes based on enzymatic treatment.

7. Important and challenges

Pretreatment process is an important process prior to enzymatic saccharification. Applying the most efficient pretreatment process could reduce production cost, hence reduce the final product price. As per date, the irradiation pretreatment process shows a promising approach and has some advantages compared to other pretreatment processes.

Irradiation pretreatment is an environment friendly process due to less chemical used during the pretreatment process. Also, this process requires less time (<10 minutes) compared to other process, especially biological pretreatment, which requires more than 7 days to remove lignin material from the biomass [95].

The most important advantage of the irradiation process is that this process is very selective to the degradation of the biomass component, unlike chemical pretreatment that could degrade some part of cellulose and hemicellulose during the process [45]. On the other hand, this process also produced less inhibitor that could affect enzymatic saccharification and fermentation process. According to Bak et al. [43], there was no inhibitors produced from the irradiated biomass when the pretreatment was carried out using water as soaking buffer. Hence, could increase the enzymatic saccharification and fermentation process performance for liquid biofuel production.

Even though this irradiation pretreatment procedure is quite simple, it is undeniable that the high-energy consumption associated with it makes the process not preferable for implementation on a commercial scale [40, 96]. Besides, this process requires a special reactor that could affect during large-scale process. For the large-scale pretreatment of biomass, a large microwave irradiator or reactor is required, which is costly, energy consuming, and limits its use in large-scale operations. This drawback hence also could increase the operational cost.

8. Conclusions

There is a wide range of chemical composition distribution in tropical biomass and biofiber making these resources a great potential to be used for biofuel and other value added products. To convert these materials, it has to go through series of processes, and the most environment friendly and efficient method is important to ensure the feasibility of the product produced. Irradiation pretreatment has been reported to have more advantages on the biomass pretreatment because this process is environment friendly, it requires less chemical, and the process can be performed in a short period of time. Irradiation pretreatment such as gamma ray, electron beam, microwave, and ultrasonications proved able to disrupt cell wall structure and provide better access for enzymatic saccharification. Hence, this could increase biofuel production and other chemicals from the tropical biomass and biofiber generated from agro-industry. However, this process still requires high energy and it could give negative impact especially at the large-scale production. Thus, further research to attempt maximum performance using low energy is very crucial to ensure the feasibility of the biofuel production from tropical biomass and biofiber using irradiation pretreatment.

Acknowledgements

The author, would like to thank for the financial support provided by University Sains Malaysia (USM) Short Term Research Grant 304/PTEKIND/6313194

Author details

Mohd Asyraf Kassim[1*], H.P.S Abdul Khalil[1], Noor Aziah Serri[1],
Mohamad Haafiz Mohamad Kassim[1], Muhammad Izzuddin Syakir[1], N.A. Sri Aprila[2] and Rudi Dungani[3]

*Address all correspondence to: asyrafkassim@usm.my

1 School of Industrial Technology, Universiti Sains Malaysia (USM), Penang, Malaysia

2 Department of Chemical Engineering, Engineering Faculty of Syiah Kuala University, Banda Aceh, Indonesia

3 School of Life Sciences and Technology, Institut Teknologi Bandung, Indonesia

References

[1] Abdul Khalil HPS, Bhat AH, Ireana Yusra AF. Green composites from sustainable cellulose nanofibrils: A review. Carbohydrate Polymers. 2012;87(2):963–79.

[2] Mwaikambo LY, Ansell MP. Chemical modification of hemp, sisal, jute, and kapok fibers by alkalization. Journal of Applied Polymer Science. 2002;84(12):2222–34.

[3] Rezende CA, de Lima MA, Maziero P, deAzevedo ER, Garcia W, Polikarpov I. Chemical and morphological characterization of sugarcane bagasse submitted to a delignification process for enhanced enzymatic digestibility. Biotechnology for Biofuels. 2011;4(1):1–19.

[4] Sannigrahi P, Ragauskas AJ, Tuskan GA. Poplar as a feedstock for biofuels: A review of compositional characteristics. Biofuels, Bioproducts and Biorefining. 2010;4(2):209–26.

[5] Abdul Khalil HPS, Siti Alwani M, Ridzuan R, Kamarudin H, Khairul A. Chemical composition, morphological characteristics, and cell wall structure of Malaysian oil palm fibers. Polymer-Plastics Technology and Engineering. 2008;47(3):273–80.

[6] Jenol M, Ibrahim M, Yee P, Salleh M, Aziz S. Sago biomass as a sustainable source for biohydrogen production by Clostridium butyricum A1. BioReseources. 2014;9(1):1007–26.

[7] Abdul Khalil HPS, Siti Alwani M, Mohd Omar A. Chemical composition, anatomy, lignin distribution, and cell wall structure of malaysian plant waste fibers. BioReseources. 2006;1(2):220–32.

[8] Ohtani Y, Mazumder BB, Sameshima K. Influence of the chemical composition of kenaf bast and core on the alkaline pulping response. Journal of Wood Science. 2001;47(1): 30–5.

[9] Agbor VB, Cicek N, Sparling R, Berlin A, Levin DB. Biomass pretreatment: Fundamentals toward application. Biotechnology Advances. 2011;29(6):675–85. Epub 2011/06/01.

[10] Imman S, Arnthong J, Burapatana V, Champreda V, Laosiripojana N. Fractionation of rice straw by a single-step solvothermal process: Effects of solvents, acid promoters, and microwave treatment. Renewable Energy. 2015;83:663–73.

[11] Yusup S, Ahmad M, Ramli A, Zakir K, Mohamad M. Biomass conversion to fuel (solid, liquid and gas fuel). In: Ravindra, P, Rosalam, SH, editors. Advance in Biofuels. Boston: Springer, US. 2013.p. 29-39. Doi: 10.1007/978-1-4614-6249-1_3

[12] Alvira P, Tomás-Pejó E, Ballesteros M, Negro MJ. Pretreatment technologies for an efficient bioethanol production process based on enzymatic hydrolysis: A review. Bioresource Technology. 2010;101(13):4851–61.

[13] Lei H, Cybulska I, Julson J. Hydrothermal pretreatment of lignocellulosic biomass and kinetics. Journal of Sustainable Bioenergy Systems. 2013;03(04):250–9.

[14] Xu J, Wang L, Zhi Z, Qiao Y, Zhao C, Lu X. Enhancement of acidogenic fermentation of corn stover hydrolysates by thermal pretreatment with diluted formic acid as catalyst. Energy & Fuels. 2015;29(12):8157–61.

[15] Kumar S, Kothari U, Kong L, Lee YY, Gupta RB. Hydrothermal pretreatment of switchgrass and corn stover for production of ethanol and carbon microspheres. Biomass and Bioenergy. 2011;35(2):956–68.

[16] Kim SB, Lee SJ, Lee JH, Jung YR, Thapa LP, Kim JS, et al. Pretreatment of rice straw with combined process using dilute sulfuric acid and aqueous ammonia. Biotechnology for Biofuels. 2013;6:109.

[17] Yan W, Acharjee TC, Coronella CJ, Vásquez VR. Thermal pretreatment of lignocellulosic biomass. Environmental Progress & Sustainable Energy. 2009;28(3):435–40.

[18] Mahdy A, Mendez L, Ballesteros M, González-Fernández C. Autohydrolysis and alkaline pretreatment effect on Chlorella vulgaris and Scenedesmus sp. methane production. Energy. 2014;78(0):48–52.

[19] Mendez L, Mahdy A, Ballesteros M, González-Fernández C. Methane production of thermally pretreated Chlorella vulgaris and Scenedesmus sp. biomass at increasing biomass loads. Applied Energy. 2014;129(0):238–42.

[20] Park YC, Kim JS. Comparison of various alkaline pretreatment methods of lignocellulosic biomass. Energy. 2012;47(1):31–5.

[21] Behera S, Arora R, Nandhagopal N, Kumar S. Importance of chemical pretreatment for bioconversion of lignocellulosic biomass. Renewable and Sustainable Energy Reviews. 2014;36:91–106.

[22] Isroi, Ishola MM, Millati R, Syamsiah S, Cahyanto NM, Niklasson C, et al. Structural changes of oil palm empty fruit bunch (OPEFB) after fungal and phosphoric acid pretreatment. Molecules. 2012;17(12): 14996–15012.

[23] Lee J-W, Kim H-Y, Koo B-W, Choi D-H, Kwon M, Choi I-G. Enzymatic saccharification of biologically pretreated Pinus densiflora using enzymes from brown rot fungi. Journal of Bioscience and Bioengineering. 2008;106(2):162–7.

[24] Galbe M, Sassner P, Wingren A, Zacchi G. Process Engineering Economics of Bioethanol Production. In: Olsson L, editor. Biofuels: Springer Berlin Heidelberg; 2007. pp. 303–27.

[25] Saini A, Aggarwal NK, Sharma A, Yadav A. Prospects for irradiation in cellulosic ethanol production. Biotechnology Research International. 2015;2015:13.

[26] Chaturvedi V, Verma P. An overview of key pretreatment processes employed for bioconversion of lignocellulosic biomass into biofuels and value added products. Biotech. 2013;3(5):415–31.

[27] Tang AM, Liang WZ. Studies on the structure changes of fast-growing wood fiber treated by ultrasonic wave. Technical Acoustics. 2000;19: 78–85.

[28] Liu Y, Chen J, Wu X, Wang K, Su X, Chen L, et al. Insights into the effects of [gamma]-irradiation on the microstructure, thermal stability and irradiation-derived degradation components of microcrystalline cellulose (MCC). RSC Advances. 2015;5(43):34353–63.

[29] Sun F, Jiang Z, Sun Q, Lu F. Changes in chemical composition and microstructure of bamboo after gamma ray irradiation. BioReseources. 2014;9(4):5794–800.

[30] Danu S, Darsono H, Kardha M, Oktaviani M. Electron beam degradation of oil palm empty fruit bunch. International Journal of Environment and Bioenergy. 2012;3(4):168–79.

[31] Karthika K, Arun AB, Melo JS, Mittal KC, Kumar M, Rekha PD. Hydrolysis of acid and alkali presoaked lignocellulosic biomass exposed to electron beam irradiation. Bioresource Technology. 2013;129:646–9.

[32] Zhu S, Wu Y, Yu Z, Liao J, Zhang Y. Pre-treatment by microwave/alkali of rice straw and its enzymatic hydrolysis. Process Biochemistry. 2005;40 (9): 3082–3086.

[33] Takács E, Wojnárovits L, Földváry C, Hargittai P, Borsa J, Sajó I. Effect of combined gamma-irradiation and alkali treatment on cotton–cellulose. Radiation Physics and Chemistry. 2000;57(3–6):399–403.

[34] Hon N-S. Formation of free radicals in photo-irradiated cellulose. IV. Effect of ferric ions. Journal of Applied Polymer Science. 1975;19(10):2789–97.

[35] Khan F, Ahmad SR, Kronfli E. γ-Radiation induced changes in the physical and chemical properties of lignocellulose. Biomacromolecules. 2006;7(8):2303–9.

[36] Kristiani A, Effendi N, Aristiawan Y, Aulia F, Sudiyani Y. Effect of combining chemical and irradiation pretreatment process to characteristic of oil palm's empty fruit bunches as raw material for second generation bioethanol. Energy Procedia. 2015;68:195–204.

[37] Betiku E, Adetunji OA, Ojumu TV, Solomon BO. A comparative study of the hydrolysis of gamma irradiated lignocelluloses. Brazilian Journal of Chemical Engineering. 2009;26:251–5.

[38] Kashiwagi M, Hoshi Y. Electron beam processing system and its application. Sei Technical Review. 2012;75:47–54.

[39] Ma X, Zheng X, Zhang M, Yang X, Chen L, Huang L, et al. Electron beam irradiation of bamboo chips: Degradation of cellulose and hemicelluloses. Cellulose. 2014;21(6): 3865–70.

[40] Henniges U, Hasani M, Potthast A, Westman G, Rosenau T. Electron beam irradiation of cellulosic materials-Opportinities and limitations. Materials. 2013;6:1584–98.

[41] Bak JS, Ko JK, Han YH, Lee BC, Choi I-G, Kim KH. Improved enzymatic hydrolysis yield of rice straw using electron beam irradiation pretreatment. Bioresource Technology. 2009;100(3):1285–90.

[42] Jeun J-P, Lee B-M, Lee J-Y, Kang P-H, Park J-K. An irradiation-alkaline pretreatment of kenaf core for improving the sugar yield. Renewable Energy. 2015;79:51–5.

[43] Bak JS. Electron beam irradiation enhances the digestibility and fermentation yield of water-soaked lignocellulosic biomass. Biotechnology Reports. 2014;4:30–3.

[44] Ethaib S, Omar R, Kamal S, Biak D. Microwave-asssisted pretreatment of lignocellulosic biomass: A review. Journal of Engineering Science and Technology. 2015; (21):97–109.

[45] Quitain AT, Sasaki M, Goto M. Microwave-based pretreatment for efficient biomass-to-biofuel conversion. In: Fang Z, editor. Pretreatment Techniques for Biofuels and Biorefineries. Berlin, Heidelberg: Springer Berlin Heidelberg; 2013. pp. 117–30.

[46] Hu Z, Wen Z. Enhancing enzymatic digestibility of switchgrass by microwave-assisted alkali pretreatment. Biochemical Engineering Journal. 2008;38(3):369–78.

[47] Choudhary R, Umagiliyage AL, Liang Y, Siddaramu T, Haddock J, Markevicius G. Microwave pretreatment for enzymatic saccharification of sweet sorghum bagasse. Biomass and Bioenergy. 2012;39:218–26.

[48] Zhu Z, Simister R, Bird S, McQueen-Mason SJ, Gomez LD, Macquarrie DJ. Microwave assisted acid and alkali pretreatment of Miscanthus biomas for biorefineries. AIMS Bioengineering. 2015;2(4):449–68.

[49] Kumar P, Barrett DM, Delwiche MJ, Stroeve P. Methods for pretreatment of lignocellulosic biomass for efficient hydrolysis and biofuel production. Industrial & Engineering Chemistry Research. 2009;48(8):3713–29.

[50] Shahriarinour M, Wahab M, Mustafa S, Rosfarizan M, Arif A. Effect of various pretreatment of oil palm empty fruit bunch fibres for subsequent use as substrate on the performance of cellulase production by Aspergillus terreus. BioReseources. 2011;6(1):291–307.

[51] Normanbhay S, Hussain R, Palamisamy K. Microwave-assisted alkaline pretreatment and microwave-assisted enzymatic saccharification of oil palm empty fruit bunch fiber for enhanced fermentable sugar yield. Journal of Sustainable Bioenergy Systems. 2013;3:7–17.

[52] Akhbar J, Teo C, Lai L, Hassan N, Idris A, Aziz R. Factor affecting delignification of oil palm empty fruit bunch by microwave-assisted dilute acid/alkali pretreatment. BioReseources. 2015;10(1):588–96.

[53] Akhtar J, Idris A, Teo C, Lai L, Hassa N, Khan M. Comparison of delignification of oil palm empty fruit bunch (EFB) by microwave assisted alkali/acid pretreatment and conventional pretreatment method. International Journal of Advances in Chemical Engineering and Biological Sciences. 2014;1(2):155–7.

[54] Lai L, Idris A. Disruption of oil palm trunk and fronds by microwave-alkali pretreatment. BioReseources. 2013;8(2):2792–804.

[55] Sunarti TC, Dwiko M, Derosya V, Meryandini A. Effect of microwave treatment on acid and enzyme susceptibilities of sago pith. Procedia Chemistry. 2012;4:301–7.

[56] Kannan T, Ahmed A, Ani F. Energy efficient microwave irradiation of sago bark waste (SBW) for bioethanol production. Advanced Materials Research. 2013;701:249–53.

[57] Ooi BG, Rambo AL, Hurtado MA. Overcoming the recalcitrance for the conversion of kenaf pulp to glucose via microwave-assisted pre-treatment processes. International Journal of Molecular Sciences. 2011;12(3):1451–63.

[58] Zhu Z, Macquarrie DJ, Simister R, Gomez LD, McQueen-Mason SJ. Microwave assisted chemical pretreatment of Miscanthus under different temperature regimes. Sustainable Chemical Processes. 2015;3(1):1–13.

[59] Pilli S, Bhunia P, Yan S, LeBlanc RJ, Tyagi RD, Surampalli RY. Ultrasonic pretreatment of sludge: A review. Ultrasonics Sonochemistry. 2011;18(1):1–18.

[60] Ninomiya K, Kamide K, Takahashi K, Shimizu N. Enhanced enzymatic saccharification of kenaf powder after ultrasonic pretreatment in ionic liquids at room temperature. Bioresource Technology. 2012;103(1):259–65.

[61] Sosiati H, Harjoso. Effect of combined treatment methods on the crystallinity and surface morphology of kenaf bast fibers. Cellulose Chemistry and Technology. 2013;48(1–2):33–43.

[62] Nikolić S, Mojović L, Rakin M, Pejin D, Pejin J. Ultrasound-assisted production of bioethanol by simultaneous saccharification and fermentation of corn meal. Food Chemistry. 2010;122(1):216–22.

[63] Zhang Y, Fu E, Liang J. Effect of ultrasonic waves on the saccharification processes of lignocellulose. Chemical ENgineering Technology. 2008;31(10):1510–5.

[64] Yunus R, Salleh SF, Abdullah N, Biak DRA. Effect of ultrasonic pre-treatment on low temperature acid hydrolysis of oil palm empty fruit bunch. Bioresource Technology. 2010;101(24):9792–6.

[65] Poletto M, Júnior H, Zattera A. Native cellulose: Structure, characterization and thermal properties. Materials. 2014;7(9):6105–19.

[66] Darji D, Alias Y, Mohd Som F, Abd Razak NH. Microwave heating and hydrolysis of rubber wood biomass in ionic liquids. Journal of Chemical Technology & Biotechnology. 2015;90(11):2050–6.

[67] Ramos LP. The chemistry involved in the steam treatment of lignocellulosic materials. Química Nova. 2003;26:863–71.

[68] Driscoll M, Stipanovic A, Winter W, Cheng K, Manning M, Spiese J, et al. Electron beam irradiation of cellulose. Radiation Physics and Chemistry. 2009;78(7–8):539–42.

[69] Cui T, Li J, Yan Z, Yu M, Li S. The correlation between the enzymatic saccharification and the multidimensional structure of cellulose changed by different pretreatments. Biotechnology for Biofuels. 2014;7:134.

[70] Chen W-H, Tu Y-J, Sheen H-K. Impact of dilute acid pretreatment on the structure of bagasse for bioethanol production. International Journal of Energy Research. 2010;34(3):265–74.

[71] Sabiha-Hanim S, Mohd Noor MA, Rosma A. Fractionation of oil palm frond hemicelluloses by water or alkaline impregnation and steam explosion. Carbohydrate Polymers. 2015;115:533–9.

[72] Kassim MA, Bhattacharya S. Dilute alkaline pretreatment for reducing sugar produc? tion from Tetraselmis suecica and Chlorella sp. biomass. Process Biochemistry.2015: http://dx.doi.org/10.1016/j.procbio.2015.11.027

[73] Taherzadeh MJ, Karimi K. Pretreatment of lignocellulosic wastes to improve ethanol and biogas production: A review. International Journal of Molecular Sciences. 2008;9(9):1621–51.

[74] Pu Y, Hu F, Huang F, Davison BH, Ragauskas AJ. Assessing the molecular structure basis for biomass recalcitrance during dilute acid and hydrothermal pretreatments. Biotechnology for Biofuels. 2013;6(1):1–13.

[75] Liu Q, Wang S, Zheng Y, Luo Z, Cen K. Mechanism study of wood lignin pyrolysis by using TG–FTIR analysis. Journal of Analytical and Applied Pyrolysis. 2008;82(1):170–7.

[76] Zheng Y, Pan Z, Zhang R. Overview of biomass pretreatment for cellulosic ethanol production. International Journal of Agricultural and Biological Engineering. 2009;2(3): 51–68.

[77] Karthika K, Arun AB, Rekha PD. Enzymatic hydrolysis and characterization of lignocellulosic biomass exposed to electron beam irradiation. Carbohydrate Polymers. 2012;90(2):1038–45.

[78] Bak JS, Ko JK, Han YH, Lee BC, Choi IG, Kim KH. Improved enzymatic hydrolysis yield of rice straw using electron beam irradiation. Bioresource Technology. 2009;100(3): 1285–1290.

[79] Minoru K, Isao K. Radiation-induced degradation and subsequent hydrolysis of waste cellulose materials. The International Journal of Applied Radiation and Isotopes. 1979;30(3):139–41.

[80] Ardica S, Calderaro E, Cappadona C. Radiation pretreatments of cellulose materials for the enhancement of enzymatic hydrolysis—II. Wood chips, paper, grain straw, hay, kapok. Radiation Physics and Chemistry (1977). 1985;26(6):701–4.

[81] Chung BY, Lee JT, Bai H-W, Kim U-J, Bae H-J, Gon Wi S, et al. Enhanced enzymatic hydrolysis of poplar bark by combined use of gamma ray and dilute acid for bioethanol production. Radiation Physics and Chemistry. 2012;81(8):1003–7.

[82] Yang C, Shen Z, Yu G, Wang J. Effect and aftereffect of γ radiation pretreatment on enzymatic hydrolysis of wheat straw. Bioresource Technology. 2008;99(14):6240–5.

[83] Zhu S, Wu Y, Yu Z, Wang C, Yu F, Jin S, et al. Comparison of three microwave/chemical pretreatment processes for enzymatic hydrolysis of rice straw. Biosystems Engineering. 2006;93(3):279–83.

[84] Hu Z.,Wen Z. Enhancing enzymatic digestibility of switchgrass by microwave-assisted alkali pretreatment. Biochemical Engineering Journal. 2008; 38 (3):369–378.

[85] Gabhane J, William SPMP, Vaidya AN, Anand D, Wate S. Pretreatment of garden biomass by alkali-assisted ultrasonication: Effects on enzymatic hydrolysis and

ultrastructural changes. Journal of Environmental Health Science and Engineering. 2014;12(1):1–6.

[86] Singh R, Krishna BB, Kumar J, Bhaskar T. Opportunities for utilization of non-conventional energy sources for biomass pretreatment. Bioresource Technology. 2016;199:398–407.

[87] Roda A, De Faveri DM, Giacosa S, Dordoni R, Lambri M. Effect of pre-treatments on the saccharification of pineapple waste as a potential source for vinegar production. Journal of Cleaner Production. 2016;112, Part 5:4477–84.

[88] Chidi E, Oluwatisin S, Deborah K. Microwave-alkaline assisted pretreatment of banana trunk for bioethanol production. Journal of Energy and Power Engineering. 2015;9:705–13.

[89] Mekhilef S, Barimani M, Safari A, Salam Z. Malaysia's renewable energy policies and programs with green aspects. Renewable and Sustainable Energy Reviews. 2014;40:497–504.

[90] Peer review on energy efficiency in Malaysia. APEC Energy Working Group 2011.

[91] Milbrandt A, Overend R. Survey of biomass resources assessments and assessment capabilities in APEC economics. APEC Energy Working Group. 2008.

[92] Hon L, Joseph. A case study on palm empty fruit bunch as energy feedstock. SEGi Review. 2010;3(2):3–15.

[93] Malaysian Palm Oil Board. Retrieved November 13, 2006 from http://www.mpob.gov.my.

[94] Menon V, Rao M. Trends in bioconversion of lignocellulose: Biofuels, platform chemicals & biorefinery concept. Progress in Energy and Combustion Science. 2012;38(4):522–50.

[95] Sindhu R, Binod P, Pandey A. Biological pretreatment of lignocellulosic biomass – An overview. Bioresource Technology. 2016;199:76–82.

[96] Mosier N, Wyman C, Dale B, Elander R, Lee YY, Holtzapple M, et al. Features of promising technologies for pretreatment of lignocellulosic biomass. Bioresource Technology. 2005;96(6): 673–686.

Section 4

Metallic Materials

Chapter 14

Ion Bombardment-Induced Surface Effects in Materials

Farid F. Umarov and Abdiravuf A. Dzhurakhalov

Additional information is available at the end of the chapter

http://dx.doi.org/10.5772/62731

Abstract

This chapter deals with the experimental research and computer simulation of low- and medium-energy (E_0 = 1-30 keV) ion collisions on the surface of a solid and of the accompanying effects, namely scattering, sputtering, and surface implantation. Experimental and computer simulation studies of low-energy (E_0 = 80–500 eV) Cs^+ ions scattering on Ta, W, Re target surfaces and K^+ ions scattering on Ti, V, Cr target surfaces have been performed for more accurate definition of mechanism of scattering, with a purpose of evaluation of use of slow ions scattering as a tool for surface layer analysis. The peculiarities of the process of correlated small angle scattering of 5–15 keV He, Ne, Ar, Kr, Xe, and Rn ions by the Cu(100), Ni(100), and V(100) single-crystal surfaces have been investigated by computer simulation. It has been shown that under these conditions the inelastic energy losses become predominant over the elastic ones. The anomalous energy losses observed experimentally at the grazing ion scattering by the single-crystal surface were explained. It has been shown by computer simulation that the peculiarities of the chain effect at direct and reverse relation of masses of colliding particles and rainbow effect at quasi-single and quasi-double scattering of ions, heavier than adatoms, lead to the appearance of characteristic peaks in the energy and angular distributions of scattered ions. Analysis of these peaks and comparison with experiment give an opportunity to control the initial stages of adsorption and identification of adsorption structures with the help of low-energy ion scattering. It has been shown that from the correlation of the experimental and calculated energy distributions of the scattered particles, one may determine a spatial extension of the isolated atomic steps on the single-crystal surface damaged by the ion bombardment. Results obtained can be also used to study short-range order in alloys undergoing ordering. Grazing ion-sputtering processes of Si(001), SiC(001), and $Cu_3Au(001)$ surfaces at 0.5–5 keV Ne^+ bombardment have been studied by computer simulations. A preferential emission of Cu atoms in the case of Cu_3Au (001) surface sputtering is observed. It was shown that in the case of grazing ion bombardment, the layer-by-layer sputtering is possible, and its optimum is observed within the small angle range of the glancing angles near the threshold sputtering angle. The peculiarities of trajectories, ranges, and energy losses of low-energy different-mass ions channeling in thin single crystals of metals and semiconductors have been thoroughly studied by computer simulation. It has been found that in the case of light ions, even at low energy, the main contribution to energy loss is made by inelastic energy losses, whereas for heavy ions, already at $E < 10$ keV, elastic energy losses exceed inelastic ones. Profiles of the distribution of channeled ions have been calculated depending on the crystal lattice type, kind of ions, and their energy. It has been shown that the channeling of low-energy ions through thin single-crystal metal films can be used to determine the sort and adsorption site of light atoms adsorbed on a clean rear surface.

Keywords: ion scattering, sputtering yield, layer-by-layer sputtering, surface implantation, surface channeling, computer simulation, grazing ion bombardment, elastic and inelastic energy losses, vacancy and atomic steps defects on the surface, method of layer-by-layer analysis of single-crystal surface, ***PACS codes:*** 79.20.Rf; 79.20.Ap

1. Introduction

The ion scattering, sputtering, and implantation processes have been the subject of both scientific investigations for a long time and recent rapid developing thin-film technologies and nanotechnologies. These processes underlie such well-known methods of surface science as Ion Scattering Spectroscopy (ISS), Ion Beam Analysis (IBA), Secondary Ion Mass Spectrometry (SIMS), and Ion Beam Modification of Materials (IBMM). Physically, the energy range under consideration is characterized by the dominance of elastic over inelastic energy losses, and by the possibility of considering classical binary collisions using single-center potentials and disregarding the binding energy of the scattering ion in the crystal lattice. The first of these factors determines the upper, and the second, the lower boundary of the energy range. Speaking about the surface, one should bear in mind that the bulk of the solid also participates in the process of ion scattering. The scattering depth is greater, as the ions are lighter and their energy is higher. In this energy range, heavy low-energy ions are scattered practically by one or two atomic layers. This is a great asset since it offers a possibility of using simple models of single and double scattering, and under grazing incidence, of calculating scattering produced only by surface atomic rows and the semichannels, formed by them. The possibility of probing only one surface atomic layer by heavy-ion scattering is also unique and does not have analogs in the other methods of the surface diagnostics of solids. Particle bombardment of a clean and adsorption-covered solid surfaces leads to radiation-induced vacancy defects, atomic steps, and defect clusters, as well as to an atomic scale relief (<100 Å) formation. The concentration and the type of the radiation defects being formed depend upon the experimental conditions and significantly influence the particles' trajectories and their angular and energy distributions, as well as the number of scattered particles. Moreover, there is a correlation between the defect type, the blocking angles of the reflected beam and the energy distributions of the scattered particles, which allows the determination of the defect type and its surface concentration [1–5]. For the analysis of the first one or two atomic top layers of a solid, noble gas ions with primary energies between about 0.5 and 10 keV are very well suited. This is due to their comparatively large scattering cross sections (of the order of 10^{18} cm^2 /sr) and due to the effective neutralization of ions that penetrate into the sample. Thus, the detection of scattered ions provides a powerful tool for surface analysis that is exclusively sensitive to the outermost atomic layers. The method is known in the literature as "ion scattering spectroscopy" (ISS) [6].

The sputtering process has been the subject of both scientific investigations for a long time and recent rapid developing micro- and nanotechnologies. Processes such as plasma etching and sputter deposition that involve ion bombardment at relatively low (~100 eV) ion energies are widely used in semiconductor processing [7]. Though sputtering and surface modifications of single crystals are widely studied, there are not sufficient data in the case of grazing incidence. However, using glancing-angle ion bombardment for surface modification rather than conventional near-normal incidence ions allows expanding the energy range up to ~10 keV, and has the advantages of reducing damage (such as crater formation) and preferentially removing surface asperities [8] leading to flat surfaces. This is due to the peculiarities of sputtering processes at grazing incidence [9]. Si and SiC crystals have a great importance because of their use in semiconductor technologies. Especially, silicon carbide exhibits a large band gap, a higher breakdown field, a higher thermal conductivity, and a higher saturation velocity, compared to widely used silicon. Besides, SiC is a promising shielding material in nuclear fusion systems such as limiters in Tokamak devices, where the surface erosion is also important [10, 11]. In ref. [12], atomically clean and flat Si(100) surfaces suitable for nanoscale device fabrication were prepared by wet-chemical etching followed by 0.3–1.5 keV Ar ion sputtering. It was found that wet-chemical etching alone cannot produce a clean and flat Si(100) surface which can be achieved by subsequent 300 eV Ar ion sputtering at room temperature followed by a 700°C annealing. Application of grazing angles of incidence of ions on the solid surface opens new perspectives in the investigation of composition, structure, and topography of real surfaces and their modification and polishing by ion beams. Sputtering yields of crystalline silicon carbide and silicon have been experimentally determined, and the results have been compared with Monte Carlo simulations for Ne^+, Ar^+, and Xe^+ ion bombardment in the energy range of 0.5–5 keV under 60° sputtering with respect to the surface normal [13]. The simulation results depend strongly on the input parameters which are not well known, especially for SiC. The TRIM simulation fits the experimental results very well. The evolution of surface morphology during ion beam erosion of Si(111) at 500 eV Ar^+ ion bombardment (60° from normal, 0.75 mA/cm^2 collimated beam current) was studied over a temperature range of 500–730°C [14]. Keeping ion flux, incident angle, and energy fixed, it was found that one-dimensional sputter ripples with wave vector oriented perpendicular to the projected ion beam direction formed during sputtering at the lower end of the temperature range. For temperatures above approximately 690°C, growth modes both parallel and perpendicular to the projected ion beam direction contribute to the surface morphological evolution. Thus, though sputtering and surface modifications of single crystals are widely studied, there are not sufficient data in the case of grazing incidence.

Ion implantation has become a very important technique for modifying surface and impurity doping of semiconductors [15–17]. The ion implantation processes lead to the change of a profile of composition and structure of the subsurface layers. Using glancing-angle ion implantation for surface modification rather than conventional near-normal incidence ions allows expanding the energy range up to ~10 keV and has the advantages of reducing damage (such as crater formation) and preferentially removing surface asperities leading to flat surfaces. Channeling of low-energy ions in metal and semiconductor single crystals offers the opportunity to create the method of local ion implantation in ultrathin film nanotechnology and surface nano-engineering. Therefore, ranges, energy losses, and profiles of distribution of low-energy ions channeling in crystals have received considerable experimental and theoretical interest [18–20]. For small crystal depths, the approaches which are used in the analytical theory of orientation effects on the large depths become unacceptable, and a computer simulation method for the channeling process modeling appears to be the most preferable [21, 22]. So, the theoretical investigation of atomic collision processes in crystals caused by particle irradiation and deposition is usually done using computer simulation, because real physical conditions (e.g., complicated interatomic interaction potential, surfaces, interfaces, defects, etc.) can be taken into account much easier than it is possible by using analytical methods [17, 21–23].

2. Experimental

The measurements of differential energy spectra and angular distributions of scattered ions were performed in experimental equipment (energy analyzer of the spherical deflector type) with high angular ($\Delta\psi \cong 0.6^{\circ}$) and energetic ($\Delta E/E \cong 1/125$) resolutions and with the capability to analyze the secondary ion masses by means of time-of-flight technique [24, 25]. The experimental setup includes a UHV scattering chamber with the oil-free pumping system and a base pressure in the 10^{-9} Torr range. During the measurements, the working pressure rises to about 5×10^{-9} Torr. Ions of alkaline metals were obtained in a thermal ion source with a target density of current $J_0 = 5 \cdot 10^{-7}$ A•cm^{-2} under operating conditions. The repeated cycles of an electron bombardment was used for cleaning the target surface. Beams of E_0 = 50–500 eV of Cs^+ and K^+ ions, with a current density $J_0 = 5 \cdot 10^{-7}$ A•cm^{-2}, scattered from clean Ta, W, Re and Ti, V, Cr polycrystal surfaces under an incidence angle of $\psi = 55°$. At the registration of spectra from contaminated targets, the peak of the straight flight was constantly observed. This peak corresponded to the energy of primary ions E_0, that is, the reflection from oxide film which behaved as a screen. After cleaning by electron bombardment, this peak disappeared and bell-shaped spectra were observed. The incident and scattering beams were laid in the same plane, perpendicular to the surface of the target in the point of incidence of ions. The size of ion spot on the sample at normal incidence of ions on the surface was 2 mm, and the scatter of angles of incidence of the ion beam did not exceed $\Delta\psi = \pm 2°$. The backscattered ions are collected at a scattering angle $\theta = 70°$. The angular resolution of the device is $\Delta\theta = \pm 2°$. Usually, within the method of ion-scattering spectroscopy, the interpretation of both angular and energy distribution of the scattered heavy alkali ions (K, Cs) is based on the differential cross section of scattering only, the ionization degree η^+ being 100% [26]. The measured energy spectra were processed by computers for averaging of statistical fluctuations of impulse registration with the use of low-frequency digital filter of Spencer. Repeating the deflection of the voltage on the plate of energy analyzer with alteration of cleaning of the target, the impulse analyzer was functioning in the mode of accumulation, and summed spectra were registered on the display and digitally printed [24, 25].

3. Computer simulation method

The theoretical investigation of atomic collision processes in crystals caused by ion irradiation is usually done using computer simulation, because real physical conditions (e.g., complicated interatomic interaction potential, surfaces, interfaces, and defects) can be taken into account much easier than it is possible by using analytical methods [4, 5]. The simulation used in our calculations to construct the trajectories of the ions or projectile scattered by target atoms is based on the binary collision approximation [5] with two main assumptions: (1) only binary collisions of ions within target atoms or between two target atoms are considered, and (2) the path in which a projectile goes between collisions is represented by straight-line segments (**Figure 1**). In the binary collision model, particles move along straight-line segments, representing asymptotes to their trajectories in laboratory system, and one determines not a particle trajectory but rather the difference between the angles characterizing the initial and final directions of motion. While this approach permits one to cut the required computer time (compared with direct integration of the equations of motion), it also entails a systematic error due to the fact that over short segments of path, the real ion trajectory differs from the asymptotes used to replace the former. This error was estimated in ref. [27] for the Cu–Cu pair, for a number of potentials and three values of energy. It was established that the deviation of an asymptote from the real trajectory is essential only for head-on collisions and high energies.

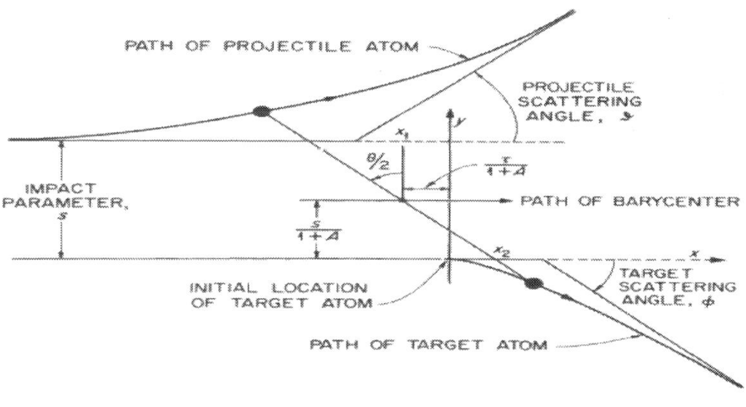

Figure 1. Scheme of the binary collision approximation [27].

For the description of the particle interactions, the repulsive Biersack–Ziegler–Littmark (BZL) potential [28] with regard to the time integral was used. The BZL approximation for the screening function in the Thomas–Fermi potential takes into account the exchange and correlation energies, and the so-called "universal" potential obtained in this way shows good agreement with experiment over a wide range of interatomic separations. Elastic and inelastic energy losses have been summed along trajectories of scattered ions. The inelastic energy losses $\varepsilon\ (E_0, p)$ were regarded as local depending on the impact parameter p and included into the scattering kinematics. These losses have been calculated on the basis of Firsov model modified by Kishinevsky [5] and contain direct dependence on the impact parameter:

$$\varepsilon\ (E_0, p) = 0.3 \times 10^{-7} v\, Z_1 \left(Z_1^{1/2} + Z_2^{1/2}\right)\left(Z_1^{1/6} + Z_2^{1/6}\right)$$
$$\left[1 - 0.68V(r_0)/E_r\right] / \left[1 + 0.67\sqrt{Z_1 r_0} / a_{TF}\left(Z_1^{1/6} + Z_2^{1/6}\right)\right] \quad (1)$$

where v and E_r are the velocity and energy of relative atomic motion, Z_1 is a greater, and Z_2 the smaller of the atomic numbers, and r_0 is in units of Å.

The expressions for the ion E_i and recoil E_r energies after binary collision, taking into account inelastic losses, can be written as follows [29]:

$$E_i = (1+\mu)^{-2} E_0 \left(\cos\theta_i \pm \sqrt{(f\mu)^2 - \sin^2\theta_i}\right)^2,$$

$$E_r = \mu(1+\mu)^{-2} E_0 \left(\cos\theta_r \pm \sqrt{(f)^2 - \sin^2\theta_r}\right)^2 \quad (2)$$

where $f = [1 - (1+\mu)/\mu\varepsilon(E_0, p)/E_0]$; θ_i and θ_r are the angles of ion and recoil scattering in the laboratory system of coordinate; E_0 is initial energy of impinging ion; p is impact parameter, and $\mu = m_2/m_1$. In ref. [29] the dependencies of $\varepsilon(E_0, p)$ on the basis of the Firsov, Kishinevsky [5], and Oen-Robinson [21] models for $Ne^+ \rightarrow Ni$ pair and low-energy $E_0 = 1$–10 keV have been calculated. Estimating the accuracy of models for various values of the impact parameter in the low-energy range, it is necessary to notice that in a small impact parameter region ($p < 0.5$ Å), it is more preferable to use the Kishinevsky model; however, in the region of large impact parameters, all three models give approximately the same results, and they are useful even when the energy $E_0 \sim 100$ eV. The above-mentioned models were checked experimentally more than once. On the whole, experimental data agree well with these theories; however, in some cases, the calculated values exhibit discrepancies from the measurement which reach 30–40% [21, 27]. In order to consider simultaneous collisions of a particle with the atoms of the adjacent chains, the procedure proposed in ref. [27] was used. The inclusion of the thermal vibrations assumed that the target atoms oscillated independently of one another, and their deflections from the equilibrium position are subject to the normal Gaussian distribution.

Sputtering has been simulated in the primary knock-on regime. Only the primary knock-on recoil (PKR) atoms ejected from first, second, and third layers have been considered. The presence of planar potential energy barrier on the surface was taken into account. The number of incident ions is 4×10^4. Each new particle is incident on a reset, pure surface. The incident ions and the recoil atoms were followed throughout their slowing down process until their energy falls below a predetermined energy: 25 eV was used for the incident ions, and the surface binding energy was used for the knock-on atoms. The calculations were performed on the crystals comprising up to 120 atomic layers. The simulations were run with the crystal atoms placed stationary at equilibrium lattice sites.

The channeling simulation program used in the present work is similar by structure to the well-known MARLOWE program and based on the binary collision approximation. But in the case of the solid phase, the binary interaction gets distorted by the influence of neighboring atoms and multiple collisions. It is impossible to calculate inelastic energy losses in this case without exact knowledge of the trajectory of scattering ions. For their calculation, it is necessary to perform computer simulation of ion scattering and channeling in a single crystal. A parallel, uniform, mono-energetic ion beam impinges on an impact area on the surface of a crystal. The angle of incidence of primary ions ψ was counted from a target surface. It is assumed that the incident beam is of small density; so, the ions of the beam do not hit twice at the same place. The impact area covers an elementary cell in the transverse plane of channel axis. The number of incident particles is 4×10^4. The shape of the target area is chosen such that by translating it, one could cover the entire surface of the crystal. Successive multiple scattering of ions from atoms in the rows lying along the principal crystallographic axes is followed in a special search procedure to find the next lattice atom or atoms with which the projectile will interact, with impact parameters for all target atoms forming the walls of a channel calculated for each layer in the crystallite. Around the colliding target atom, the coordinates of the nearest neighbor atoms are consistently set according to the crystal structure of the target. For each set of atoms, the following conditions are checked: (i) it should be at the front part of the ion movement, relatively to a crossing point of asymptotes of the projectile movement directions before and after collision; (ii) the ion impact parameter should be less than p_{lim} (p_{lim} is the impact parameter corresponding to the scattering angle of 0.05°); (iii) among colliding atoms, it should be the first in turn $p_1, p_2, p_3 \ldots$ on the consecutive collisions. After each collision, the scattering angle, energy, and the new movement direction of the channeled ion are determined. It is checked if the projectile is still moving in the given channel. The coordinates of dechanneling are used to obtain the dechanneled and channeled fractions as functions of the depth. The incident ions were followed throughout their slowing-down process, until their energy falls below 25 eV.

In order to consider simultaneous and nearly simultaneous collisions of a particle with the atoms of the adjacent chains, the special procedure proposed in ref. [27] was used. So-called simultaneous collisions which occur if a projectile has a symmetrical position, and which can collide with more than one target atom at the same time, are approximated by successive binary collisions. The inclusion of the thermal vibrations assumed that the target atoms oscillated independently of one another, and their deflections from the equilibrium position are subject to the normal Gaussian distribution. The effect of correlation is equivalent to a reduction of the vibration amplitude of about 5–10%, depending on the effect being looked at [5]. The program allowed the consideration of main peculiarities in channeled particle distribution with depth, such as collision-by-collision details of trajectories, flux-peaking, and difference in specific energy losses for random and channeled trajectories. The number of incident ions is 4 × 10^4. Each new particle is incident on a reset, pure surface. The incident ions and the recoil atoms were followed throughout their slowing-down process until their energy fell below a predetermined energy: 25 eV was used for the incident ions, and the surface-binding energy was used for the knock-on atoms. The calculations were performed on the crystals comprising up to 120 atomic layers. The simulations were run with the crystal atoms placed stationary at the equilibrium lattice sites.

The initial energy of incident ions was varied from 0.5 to 10 keV, a grazing angle of incidence ψ counted from the target surface was 3–30°, and an azimuth angle of incidence ξ realized by rotating the target around its normal and counted from the <100> direction was 0–180°. The polar scattering angle θ was counted from the primary beam direction, the polar escape angle δ—from the target surface, and the azimuthal scattering angle φ—from the incidence plane.

In **Figure 2**, the scattering geometry and scheme of a semichannel on the Cu(100) face along the direction <110> and the target area on it are shown. The impact points on the crystal surface filled a rectangle, whose sides were divided into 100 segments in the beam incidence plane (*I*-coordinate) and 1000 segments in the perpendicular direction (*J*-coordinate). The sizes of the target area were 1.28 Å (half-width of the semichannel) on the *J*-coordinate and 2.56 Å (the interatomic distance along the <110> row) on the *I*-coordinate. A substantial part of the calculations presented in this chapter was made by the computer simulation technique. This was required by the complexity of the scattering, sputtering, and channeling trajectories and by a large number of correlated collision events which prohibit the use of statistical stochastic methods of calculation. Mathematical experiment is similar in some extent to the physical one, while permitting us to extract more information from the latter.

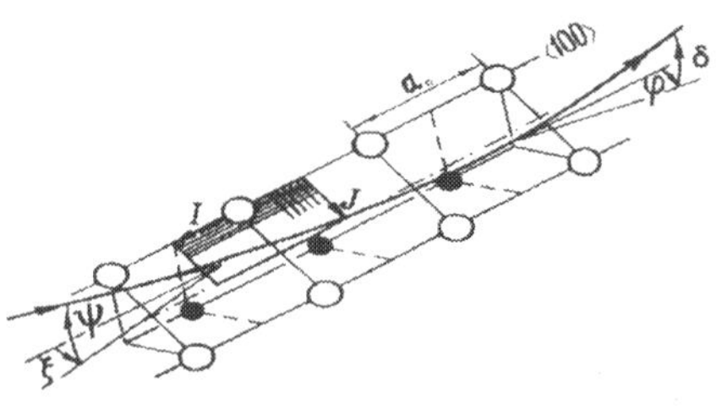

Figure 2. Scheme of ion scattering by a surface semichannel on the Cu(100) face and target area located on it. *I* and *J* are the coordinates of the impact points along and transverse to the semichannel axis, respectively, determining the number of incidence ions [5].

4. Results and discussion

4.1. The low-energy ion scattering by a single and polycrystal surfaces

The peculiarities and mechanism of the low-energy ion scattering by polycrystal targets were experimentally investigated. In refs. [30, 31], the study of low-energy (E_0 = 80–500 eV) Cs^+

and K^+ ions scattering on the surfaces of Ta, W, Re and Ti, V, Cr polycrystal surfaces has been performed experimentally and by means of computer simulation method for more accurate definition of the mechanism of scattering and the evaluation of an opportunity to use heavy ions scattering as a tool for surface layers analysis. The angles of incidence $\psi = 55^0$ and scattering $\theta = 70^0$ correspond to the specular reflection case. In **Figure 3**, the measured differential energy spectra of scattering ions for pairs $Cs^+ \rightarrow$ Ta, $Cs^+ \rightarrow$ W, $Cs^+ \rightarrow$ Re, as well as $K^+ \rightarrow$ Ti, $K^+ \rightarrow$ V, $K^+ \rightarrow$ Cr, in the initial energy range from E_0 = 80 to 500 eV, and the target temperature T = 1250 K are shown. Some physical parameters of investigated samples at $\mu >$ 1 are given in **Table 1**. The same picture can be seen for the case of V, where $E_b{}^V$ = 5.3 eV/atom, but $E_b{}^{Ti}$ and $E_b{}^{Cr}$ are 4.86 and 4.10 eV/atom, respectively.

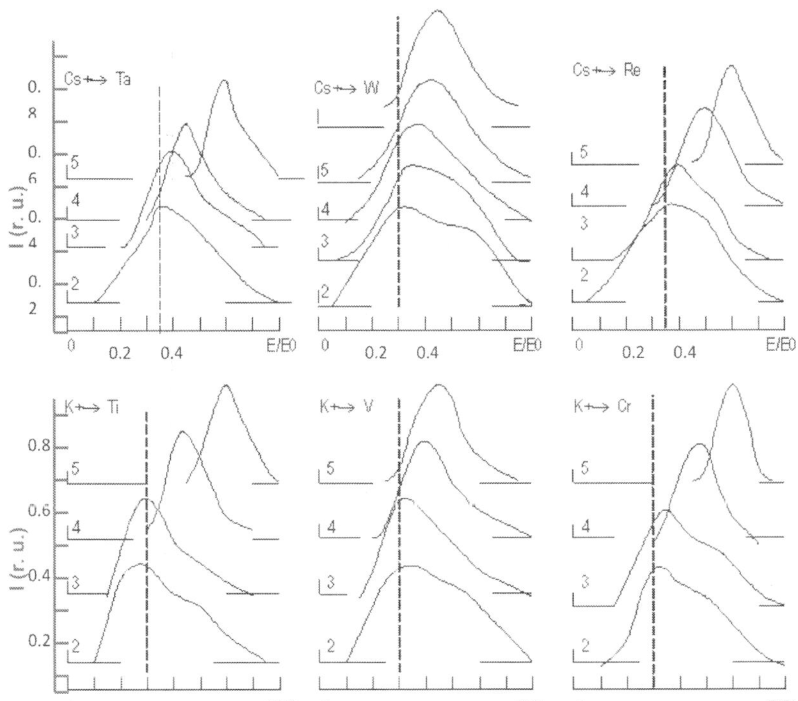

Figure 3. Differential energy spectra of Cs^+ and K^+ ions scattering on corresponding polycrystalline targets at E_0 (eV) = 500 (1); 400 (2); 200 (3); 100 (4); 80 (5). Dashed line shows the calculated results under the formula (3).

Here $\mu = m_2/m_1$ is the mass ratio of target atom and ion, respectively. It is seen that the form of spectra in all the cases is similar. It is characterized by a peak in the low-energy range. When E_0 decreases, the position of the maxima of energy spectra shifts toward greater relative energy retained by ions. This shift is the smallest in the case of W when the binding energy of target atoms is E_b^{W} = 8.66 eV/atom, and we shall note that for Ta the binding energy is E_b^{Ta} = 8.089 eV/atom and for Re, E_b^{Re} = 8.10 eV/atom.

The position of the maxima in the binary collision approximation (energy of scattered ion retained at single collision with target atom) can be defined approximately as follows [32]:

$$E / E_0 = \{ \left[\cos\theta \pm \sqrt{(m_2 / m_1)^2 - \sin^2\theta} \right] / [1 + m_2 / m_1 \}^2 \tag{3}$$

where E is the relative energy of the scattering ions which suffered a single binary collision with a target atom (dashed lines in Figure 3), E_0 is the energy of primary ions, and θ is the angle of scattering. In Figure 3, one can notice the shift of the maxima from the calculated position. As we can see, this shift depends on the binding energy of target atoms and initial energy of scattered ions. Let us note that in case of binary elastic collisions, the energy retained by the scattering ion, at the given angle of scattering, depends on the mass ratio of colliding particles, and the position of the maxima of the energy spectra should be kept constant.

Ions	**Cs^+ (m_1 = 132.90 a.u.)**			**K^+ (m_1 = 39.10 a.u.)**		
Targets	Ta	W	Re	Ti	V	Cr
Mass m_2 (a.u.)	180.95	183.85	186.20	47.90	50.94	51.99
Mass ratio ($\mu = m_2/m_1$)	1.36	1.38	1.40	1.23	1.30	1.33
Binding energy (eV/atom)	8.09	8.66	8.10	4.86	5.3	4.10
Melting temperature (°C)	2996	3387	3180	1660	1900	1857

Table 1. Some physical parameters of investigated samples at direct mass ratio of colliding particles $\mu > 1$.

In **Figure 4a and b**, the dependencies of relative energy retained by scattering ions E_m/E_0 in the maximum energy distributions of Cs^+ and K^+ ions scattered by Ta, W, Re and by Ti, V, Cr polycrystals, respectively, versus the initial energy of impinging ions are presented. It is seen in the whole investigated range of primary energies E_0 that the significant growth of relative energy E_m/E_0 (where E_m is the energy of scattering ions in a maximum energy distribution) with E_0 decrease is observed. The curves for targets with close values of binding energy E_b practically coincide, and it may be observed that this growth of relative energy for vanadium and tungsten is flatter. Thus, the least increase in the relative energy of scattered ions E_m/E_0 with decrease of initial energy E_0 occurs for targets with the largest binding energy E_b (8.66 eV/atom for W and 5.3 eV/atom for V), correspondingly. Violation of the binary character of interaction at initial energy decrease began for these pairs at $E_0 \approx 200$ eV. For other pairs with smaller binding energy of target atoms, this breach began at $E_0 \approx 400$eV.

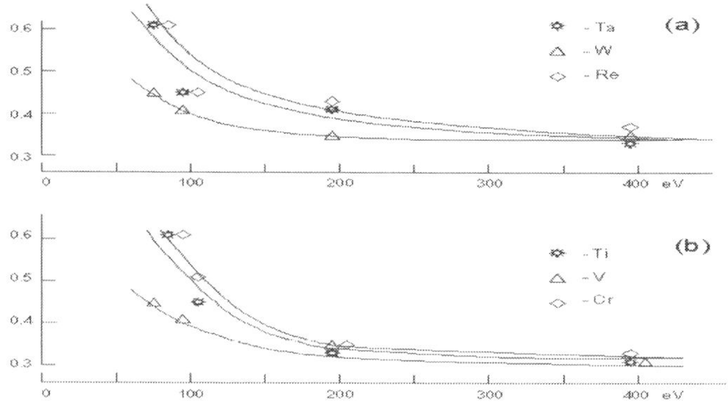

Figure 4. Dependencies of relative energy E_m/E_0 retained by scattering Cs^+ (a) and K^+ (b) ions in a maximum of energy distribution versus the initial energy of impinging ions E_0.

Formerly, it was shown that the picture represented on Figure 4 occurs also for the cases of heavy Cs^+ scattering on the polycrystal surfaces with small atomic masses (Al, Si, Ni) for which the mass ratio of colliding particles is $\mu < 1$ [30]. The angles of incidence $\psi = 45^0$ and scattering $\theta = 90^0$ also correspond to the specular reflection case. In contrast to the results for $\mu > 1$, in cases with $\mu < 1$, the smallest increase of the relative energy of scattered ions E_m/E_0 with decrease of initial energy E_0 occurred for the Al target with the smallest binding energy E_b.

In ref. [30], we showed that at an inverse mass ratio of colliding particles ($\mu < 1$), for example, Cs^+ on Al ($\mu = 0.20$), the possible scattering angles considerably exceed a limiting scattering angle in a single collision ($\theta_{lim} = \arcsin \mu$). For Cs^+ ions scattering on Al target, $\theta_{lim} = 11.7°$. This is possible either in the presence of seven or more successive binary collisions of ion with target atoms. This is not probably at that low energy, or at the interaction of an impinging ion Cs^+ with the several bonded between each other surface Al atoms simultaneously. The determined correlation between relative energy suffered by scattering ions and binding energy of target atoms confirms such many-particle character of interaction. As established in refs. [30, 31], contrary to the scattering results on targets with direct mass ratio $\mu > 1$ (Figure 2), in the cases with inverse mass ratio $\mu < 1$ the smallest increase of the relative energy of Cs^+ scattered ions E_m/E_0 with decrease of initial energy E_0 occurs for the Al target with the smallest binding energy ($E_b = 3.34$ eV/atom; $\mu = 0.20$). For Co, Si и Ni targets, binding energies are 4.387, 4.64, and 4.435 eV/atom, correspondingly. In both cases, the curves E_m/E_0 (E_0) are placed above each other relative to the binding energies of target atoms to show the influence of binding energy on a process of low-energy ions scattering. There is a correlation between the value of energy change maintained by ions for different values of E_0 in the case of scattering by targets with different masses of atoms (for $\mu > 1$ and $\mu < 1$) and its binding energies. From this contrary behavior between the E_m/E_0 (E_0) dependencies and the target atom binding energy value E_b for cases with direct ($\mu > 1$) and inverse ($\mu < 1$) mass ratio of colliding particles, one can explain the different character of many-particle interaction in these cases [30, 31].

Some physical parameters of the investigated samples at direct ($\mu > 1$) and inverse ($\mu < 1$) mass ratio of colliding particles are presented in **Table 2**. The growth of E_m/E_0 with E_0 decrease, coinciding with the curves for Ta and Re cases in contrast to the scattering by W, as well as the similarity of curves for Ti and Cr cases in contrast to the V case can apparently testify to the nonbinary multiparticle nature of interaction of low-energy Cs^+ and K^+ ions during the scattering by the investigated targets for which $\mu > 1$. The contrary behavior of the E_m/E_0 (E_0) dependencies concerning the target atom binding energy value E_b for cases with direct ($\mu > 1$) and inverse ($\mu < 1$) mass ratio of colliding particles testifies to the different character of many-particle interaction in these cases.

Ions	**Cs^+ (m_1 = 132.90 a.u.)**					
Targets	Ta	W	Re	Al	Si	Ni
Mass m_2 (a.u.)	180.95	183.85	186.20	26.98	28.09	58.70
Mass ratio ($\mu = m_2/m_1$)	1.36	1.38	1.40	0.20	0.21	0.44
Binding energy (eV/atom)	8.09	8.66	8.10	3.34	4.64	4.435
Melting temperature (°C)	2996	3387	3180	660	1420	1452

Table 2. Some physical parameters of investigated samples at direct ($\mu > 1$) and inverse ($\mu < 1$) mass ratio of colliding particles.

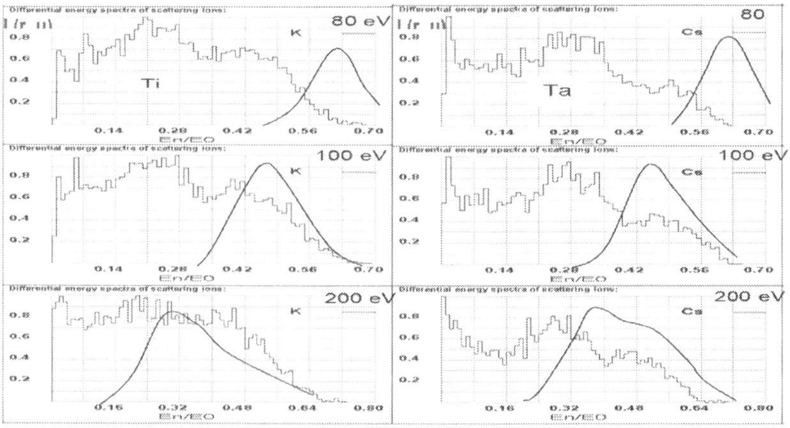

Figure 5. Comparison of experimental energy distribution curves with calculated histograms in the cases of Cs^+ and K^+ ions scattering by Ta and Ti polycrystals, respectively. *I* (r.u)—intensity of the scattered ions in relative units.

The scattering of Cs^+ and K^+ ions on experimentally investigated polycrystals was considered on the basis of binary collision approximation by means of computer simulation to assess the experimentally observed fact of the retaining of abnormally great values of relative energy by scattering ions. The calculations were performed using the visualized CASWIN code [33, 34] based on the Monte Carlo simulation method and treating the interaction of ions with a solid surface to be within the binary collision approximation. The Ziegler–Biersack–Littmark and "Kr-C" potentials were used. The surface potential barrier for ions was assumed to be null, and the values of binding energies of surface atoms were taken as mentioned above. The comparison of calculated results (histograms) with the experiment for K^+–Ti (left) and Cs^+–Ta (right) is given in **Figure 5**. One can see that the agreement became better when E_0 increased, which confirms that the binary collision approximation is fair at sufficiently high energies.

With E_0 decrease, the maximum of the calculated histogram keeps its position while the experimental curve disposes to the greater relative energy range, significantly at 80 eV. To realize what kind of scattering ions contribute to the peaks observed in the calculated histograms, each case of multiple scattering was separated.

On the basis of the conducted research, the correlation between the value of energy change maintained by ion for different values of E_0 in the case of scattering by targets with different masses of atoms and its binding energies is experimentally established. This correlation indicates considerable influence by the binding of surface atoms to each other on the process of scattering in the low-energy range by polycrystal targets. It is established that low-energy scattering with direct ($\mu > 1$) and inverse ($\mu < 1$) mass ratios of colliding particles is not described by the mechanism of binary elastic interactions. Summarizing the effects revealed in the present research, such as (i) the disappearing of a structure of energy spectra; (ii) high values of relative energy of scattering ions E_m/E_0 and its dependence on initial energy E_0; (iii) dependence of scattering ions' energy on binding energy, mass ratio, melting temperature, and packing density of target atoms, one can draw a conclusion about the multiparticle nature of the interaction. The contrary behavior of the E_m/E_0 (E_0) dependencies concerning the target atom binding energy value E_b for cases with direct ($\mu > 1$) and inverse ($\mu < 1$) mass ratio of colliding particles testifies to the different character of many-particle interaction in these cases. In order to use ion-scattering as a tool for diagnostics of disturbances of elemental composition and structure on atomic scale, it is necessary to use much higher initial energies ($E_0 \geq 1$ keV).

4.2. Anomalous inelastic energy losses and trajectory effects at small-angle ion scattering by single-crystal surface

As is known [35], the elastic mechanism of energy losses is dominant in low- and medium-energy range for ions with atomic numbers $Z \geq 10$. However, under the specific conditions of correlated glancing ion scattering from the single-crystal surface, the reverse pattern becomes possible where the inelastic mechanism will predominate. The reasons for this are the large number of collisions involved and the fact that small impact parameters cannot be reached along the scattered ion trajectory [5]. Inelastic processes of ion interaction with crystal exhibit the so-called trajectory effects. Basically, this means that inelastic processes and the associated inelastic energy losses depend on the actual trajectory of the scattered particle [36]. It turned out that the relative magnitude of the anomalous energy losses depends on crystal orientation and increases with decreasing initial ion energy. In ref. [37], this effect is related to the mechanism of surface hyperchanneling (SHC) which dominates for very small grazing angles.

The present section gives the results of a study of elastic and inelastic energy losses and of the specific features of the ion trajectories appearing in scattering from discrete model potentials on an atomic row, a semi-channel and a channel on the surface of a single crystal at small grazing and scattering angles, as well as of the contribution of the various scattering mechanisms to the experimentally observed anomalous energy losses [5, 38]. Trajectories of 15 keV Ar^+ ions suffering grazing scattering at the atomic chains, semi-channels, and channels on the Cu(100) surface were traced in the uppermost 10 atomic layers by computer simulation. They have been simulated in the binary collision approximation using the Biersack–Ziegler–Littmark interaction potential [35] and with regard to the time integral. In order to consider simultaneous collisions of an ion with several target atoms the procedure proposed in ref. [27] was used. Elastic and inelastic energy losses have been summed along trajectories of scattered ions. The latter one has been calculated on the basis of the Firsov model modified by Kishinevsky [5] and included into the scattering kinematics. Figure 2 shows schematically a semi-channel on a Cu(100) face in the <110> direction and the target area on it. It also identifies the angles used in the computation. The calculations have been performed for target points, covering uniformly (on the I and J coordinates) all surfaces of target area, where the total number of filling ions is equal to 5×10^4.

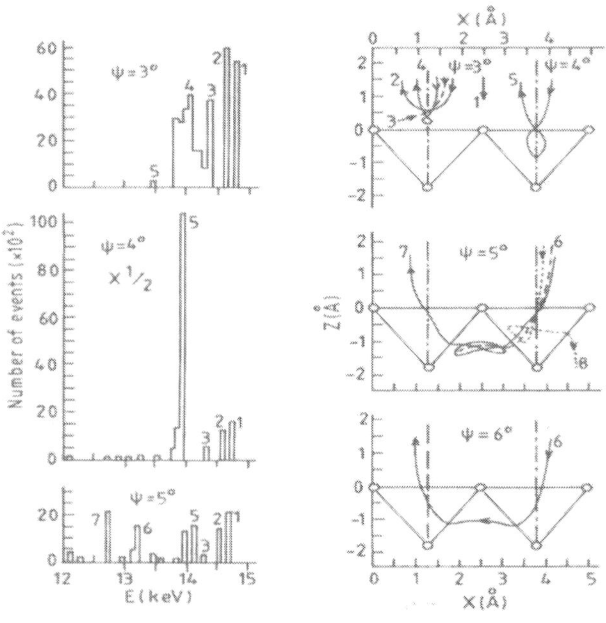

Figure 6. Energy distribution histograms of 15 keV Ar^+ ions scattered specularly ($\theta = 2\psi$) from Cu(100) <110> . On the right-hand side, the most characteristic ion trajectories projected in the transverse plane of the <110> semichannel are depicted.

http://dx.doi.org/10.5772/62731

Figure 6 shows energy distribution histograms for 15 keV Ar^+ ions scattered specularly from Cu(100) <110> into a detector with an angular aperture of ±0.5°. The ions scattered along atomic row ridges contribute to the peaks labeled 1. The peaks located to the left and labeled 2–4 correspond to the ions undergoing surface hyperchanneling (SHC) [37]. In the case of trajectories of type 4, the ions are seen to be focused in the <110> direction, the focus point lying about 0.5 Å above the surface plane. At ψ = 4°, the shape of the spectrum undergoes a substantial change because of a sharp increase in the number of trajectories of the new type 5, and a decrease of the contribution from type 1–4 trajectories. The figure shows that the ions with type 5 trajectories are focused by surface rows with the focus point lying slightly above the surface, and propagate afterwards in a diverging flux toward the walls of the semi-channel.

Trajectories of type 5 differ in shape and character from SHC trajectories. The sharp increase in their number suggests the existence of a peculiar refocusing effect which results in a pronounced narrowing of the spatial distribution of scattered particles. The angular range where this refocusing effect is observed is small (3.5° ≤ ψ < 5°), so that at ψ = 5° it is no more. At ψ = 5°, in addition to peaks 1–5, new, lower energy peaks 6 and 7 appear in the spectrum. The pattern of particle motion along type 7 trajectories is more complex than that along trajectories of type 6. In the latter case, the part of the trajectory within the channel is shorter, the particle actually crossing it. In their shapes and general patterns, these trajectories do not belong to those typical for subsurface hyperchanneling [37], which were also observed in our calculations. There also exist trajectories of type 8 corresponding to ions which, on overcoming the potential barrier of the semi-channel walls, penetrate into deeper lying layers and thus are not backscattered.

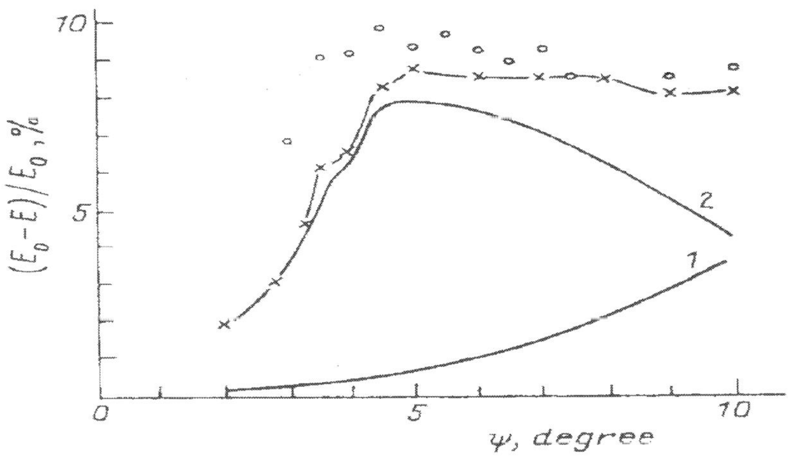

Figure 7. Experimental [39] (circles) and calculated [5, 38] (crosses) dependencies of the relative energy losses versus the grazing angle for the case $\theta = 2\psi$ for 15 keV Ar^+ ions scattered from Cu(100)<110>; elastic (1) and inelastic (2) contributions to the total relative energy losses.

Figure 7 displays experimental [39] and calculated [38] dependencies of the relative energy losses $(E_0 - E)/E_0$ on the grazing angle for 15 keV Ar^+ ions scattered from Cu(100) <110>. Here the circles are the experimental data [39], crosses - calculation [5, 38], 1 - elastic and 2 - inelastic energy losses contributions to the total relative energy losses. The calculated curve was constructed by averaging the losses over the various scattering mechanisms in accordance with their relative contributions to the spectra. As seen from **Figure 7**, the main contribution to the anomalous energy losses comes from inelastic losses. The maximum inelastic energy losses are due to particles with trajectories 5–7, as well as due to those undergoing subsurface hyperchanneling. Thus, the elastic energy losses are considerably smaller than the inelastic ones in a region of glancing scattering. The fact that the inelastic losses exceed the elastic ones for small ψ in the medium energy range is due to an increase in the number of collisions and the particle trajectory length in the surface region, as well as to the absence of small impact parameters in the course of scattering. The predominance of the inelastic energy losses should reveal itself in the efficiency of the various inelastic processes accompanying the glancing ion scattering from a single-crystal surface.

4.3. Low-energy ion scattering by atomic steps on the single-crystal surface

Ion bombardment of a solid surface leads to radiation-induced vacancy defects, atomic steps, and defect clusters, as well as to an atomic scale relief (<100 Å) formation. The concentration and the type of the radiation defects being formed depends upon the experimental conditions

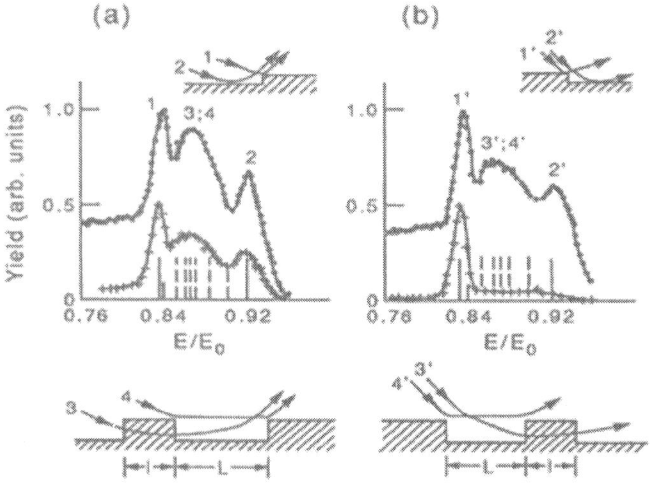

Figure 8. Energy distributions of the total number (.......) and ion component (+++++) of the argon particles scattered at the angle θ = 30° by the surface of Cu(100)<100> at ψ = 7° (a) and 23° (b). The calculated energy positions corresponding to the scattered ions in the spectra are designated by the solid and dotted vertical lines.

and significantly influences the trajectories, angular and energy distributions, as well as the number of the scattered particles. Moreover, there is a correlation between the defect type, the blocking angles of the reflected beam, and the energy distributions of the scattered particles that allows a determination of the defect type and its surface concentration [2–5]. In ref. [40] the number of step atoms formed on the surface of a single crystal Cu(100) at T = 300 K, predamaged by the bombardment with Ar^+ ions, with an initial energy of E_0 = 10 keV and the current density on the target within the range of 10^{-6}–10^{-8} A⊚cm^{-2} has been estimated. Figure 1 shows both the energy distributions of the total number (ions plus neutrals) and the ion component of the argon particles scattered by the Cu(100) surface in the <100> direction for two incidence angles of ψ = 7° (a) and 23° (b), and a constant scattering angle of θ = 30°. The particles having undergone the quasi-single scattering on the step atom of the monoatomic semi-infinite steps contribute to the peaks 1 and 1' of the spectrum, and the particles reflected by the step atom with the previous or subsequent specular scattering on the ordered atomic chain contribute to the peaks 2 and 2' [40]. Schematically, such trajectories are shown in the top of **Figure 8**. The number of step atoms has been estimated according to the intensity of the quasi-single scattering peaks in the energy spectrum of the total number of the scattered particles at ψ = 7° (1) and 23° (1'). In **Figure 9**, these spectra are shown with a solid line.

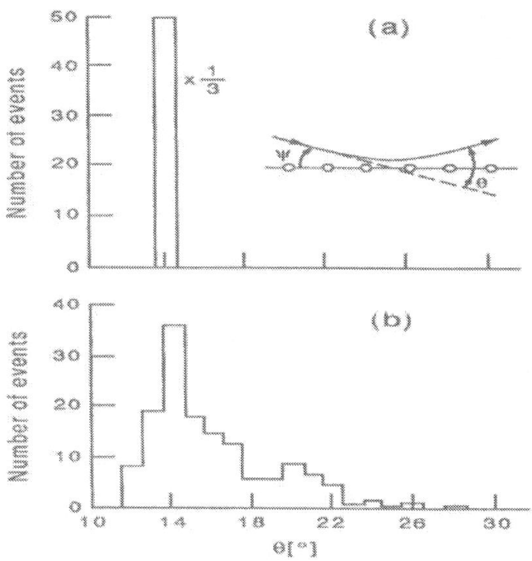

Figure 9. Histograms of the angular distributions of Ar^+ ions with E_0 = 10 keV scattered by the ordered surface of Cu(100)<100> at the incidence angle ψ = 7° for T = 0 (a) and T = 300 K (b).

A possibility of scattering at the angle θ = 30° under small (ψ < 10°) and large (ψ > 20°) incidence angles is conditioned by the fact that the bombardment introduces irregularities into the perfect infinite atomic chain, that is, it becomes finite and is followed with the step-up (a) or down (b).

The number of step atoms formed under ion bombardment , and estimated from the peaks 1 and 1' over the energy spectra, with an accuracy of ~30%, proved to be ~2×10^{14} cm^{-2}.

The aim of the present work is to study the influence of surface atomic steps on the angular and energy distributions of scattered ions, and to determine both the spatial extension of the atomic steps on the damaged surface of a single crystal and the distances between them under experimental conditions [40]. For this purpose, the trajectories of the particles scattered by both the ordered part of the surface and the monoatomic steps of different spatial extensions placed upon it have been studied carefully by means of computer simulation [41, 42]. The computer simulation allowed an investigation of the atomic steps effect on the single-crystal surface upon the trajectory features of the ions being scattered along the ridge of the atomic chains and undergoing the surface hyperchanneling and semichanneling as well.

The calculated energies of the particles scattered along trajectories of the 1, 1', 2, and 2' agreed with the experimentally determined positions of the corresponding peaks in the spectrum, namely: $E/E_0 = 0.84$ and $E/E_0 = 0.92$ (vertical solid lines in Figure1a and b). The ions scattered along the trajectories 2 and 2' before being reflected from the step atom or after it usually underwent 8–10 collisions with the chain atoms. Their range along the surface approximately amounted to 25–30 Å, and the inelastic energy losses were about 20% of the total loss. Broad maxima 3, 4 and 3', 4' between the peaks 1 and 2 as well as 1' and 2' were not explained in ref. [40]. It was suggested in ref. [40] that they could be explained by the particles scattering on the ordered surface, taking into account the thermal vibrations of its atoms. Our calculations have shown, however, that these broad maxima could not be explained in this fashion. In Figure 9, the histograms of the angular distributions of the scattered argon particles at $\psi = 7°$ and the target temperature $T = 0$ (a) and $T = 300$ K (b) are shown. As seen, the scattering at angle $\theta = 30^0$ at $\psi = 7°$ is not observed upon the ordered surface, because of blocking, even taking into account the thermal vibrations of its atoms. Variations in the interaction potential, and in particular the use of the conjugated Firsov and Born–Mayer potential [43] also did not allow an explanation of scattering at $\theta = 30°$ and $\psi = 7°$. In order to explain the maxima 3, 4 and 3', 4' in the spectra, we have calculated the trajectories of the ions scattered on the surface, upon which before or after the semi-infinite monoatomic step and isolated steps (atomic fragments) of different extension l, separated by the parts of the ordered surface of length L, were located. Schematically, such trajectories are shown at the bottom of Figure 8.

The trajectories 3 and 3' contribute to the broad maxima 3, 4 and 3', 4' in the spectra, being formed by two adjacent steps: the particles pass under the first of the steps and then are reflected from the face atom of the second step. The trajectories of quasi-double scattering 4 and 4' on the face atoms of the adjacent steps also contribute to these maxima. In Figure 9, the contributions of the above-mentioned trajectories are designated with vertical dotted lines. It follows from the calculation that the increase of the number of the atoms in the first step (atomic fragment), and consequently, its extension with constant distance between the steps L leads to an energy decrease of the ions being scattered along the trajectories of the 3 and 3' type. A change in the number of atoms in the first step (atomic fragment) from one to four allows 3 and 3' type trajectories of the scattered ions to be obtained with energies covering the whole interval for the relative energies of the broad maxima from $E/E_0 = 0.85$ to $E/E_0 = 0.90$. The steps (atomic fragments) consisting of two and three atoms separated by the parts of the ordered surface of the length L = 15–45 Å turned out to be the most likely ones. The scattering on such steps contributes to the area of the broad peaks 3, 4 and 3', 4' maximum (E/E_0 = 0.86–0.88). The distances between the atomic steps (fragments) vary from a minimum equal to two lattice constants of Cu(100), to ~45 Å. The presence and value of the peaks 2 and 2' in the spectra confirm the existence of the distance between the steps to be within ~25–45Å. The trajectories of the 4 and 4' type have turned out to be weakly sensitive (by energy value) to the distance between the steps, but their probability sharply decreases with the increase of L. It is of interest to note that the disappearance of the 2', 3', and 4' peaks in the ion component of the scattered flux at ψ = 23° agrees well with the character of the corresponding trajectories. The conclusions concerning the spatial extension of the atomic steps and the distances between them obtained by the use of the conjugated potential turned out to be in good agreement with the results described earlier, obtained by the Biersack–Littmark–Ziegler potential.

Thus, from the comparison of the results of computer simulation of the scattered particles' trajectories with the experimental energy distributions, one can draw a conclusion that for bombardment of a Cu(100) surface with 15 keV Ar^+ ions under the experimental conditions [40], isolated monoatomic steps (fragments) consisting of several atoms (from one to four) are formed upon it. The distances between the steps vary from two lattice constants of Cu(100) to ~45 Å. The most likely ones turned out to be the steps (fragments) consisting of two and three atoms separated by the parts of the ordered surface of the length L = 15–45 Å. The estimated value for the number of the atomic steps calculated by us on the basis of the proposed model of the damaged surface agrees with the value ~2×10^{14} cm^{-2} obtained in ref. [40].

4.4. Small angle ion scattering by structures on the single-crystal surface corresponding to the initial stages of adsorption

Determination of equilibrium position of adatoms, initial stages of adsorption, and the structure of submonolayer adsorption coverages on crystal surfaces are fundamental problems of surface physics. Along with low-energy electron diffraction (LEED), alternative methods are being developed for analysis, in particular ion scattering spectroscopy (ISS) [4, 5, 44]. An attractive peculiarity of the ISS method is its high surface sensitivity, enabling monolayer analysis. The structural analysis of surfaces by this method is based on a comparison of experimentally measured energy and angular distributions of scattered particles with calculations based on a chosen surface model. In this section, small-angle scattering of He^+ and Ne^+ ions with initial energies E_0 = 1–5 keV and grazing angles within the interval ψ = 7.5–15° by Ni(100) <100> single-crystal surface with submonolayer coverages by oxygen or deuterium atoms have been studied. The dynamics of changes of angular and energy distributions from scattering on different adsorption structures, corresponding to the initial stages of adsorption have been studied.

Besides the usual atomic chain effect [5] corresponding to the direct mass ratio of colliding particles, that is, to cases where the mass of projectile m_1 is less than that of an atom in the chain m_2, in our case a situation arises where an adatom is lighter than the projectile ion. The atomic chain effect manifests itself in this case in a number of features associated with the existence of a limiting scattering angle $\theta_{lim} = \arcsin(m_2/m_1)$ in quasi-single (QS) and quasi-double (QD) collisions, as well as with the fact that scattering at a given angle $\theta < \theta_{lim}$ is possible for two values of the impact parameter p [5, 45]. Oxygen atoms on a metal surface at submonolayer coverages and $T < 300$ K may be adsorbed in two stages—chemisorption and oxide-state. It has been known for many years from LEED [44] that two ordered surface structures can be obtained during the chemisorption of oxygen on Ni(100). A p(2 × 2)–O pattern is obtained at low coverages (one adsorbed oxygen atom for four nickel atoms on the first substrate plane). The saturation of the adsorption corresponds to one adsorbed atom for two nickel atoms and is associated with a c(2 × 2)–O pattern. For our p(2 × 2)–O and c(2 × 2)–O on Ni(100) calculations, we place the O atoms in the fourfold center sites at a distance 0.9 ± 0.1 Å above the Ni surface plane as found in LEED analysis [44] (**Figure 10a–c**).

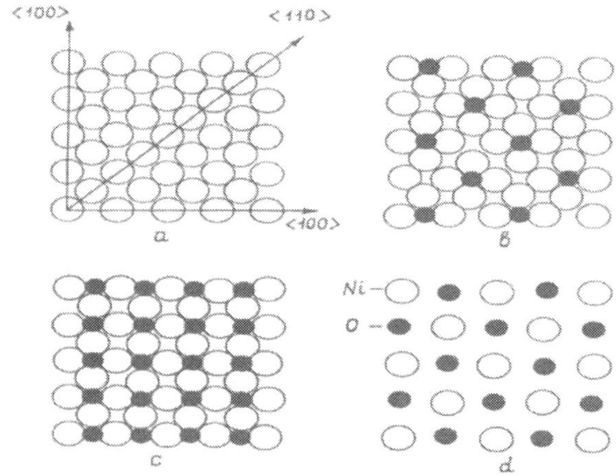

Figure 10. Position of Ni atoms in the first layer of Ni(100) (a) and oxygen atoms in adsorption structures, corresponding to initial stages of adsorption: (b) p(2 × 2); (c) c(2 × 2); (d) NiO(100) (1 × 1).

Adsorption structures p(2 × 2)–O and c(2 × 2)–O on the Ni(100) surface form sequentially and do no lead to substrate reconstruction [44]. At further increase in oxygen exposure and a coverage degree of 0.62, NiO(100)–(1 × 1) oxide islands begin to form with the NaCl type structure (Figure 1(d)). At the initial stages, this process is interpreted as a phase transition of the first order, and unlike chemisorption it is accompanied by a shift of substrate atoms. In Figure 11, histograms of energy distributions of Ne^+ ions, experiencing specular scattering on chains of atoms on a clean Ni(100)<100> surface and on surfaces with different adsorption structures are presented. The acceptance of the analyzer collecting scattered particles, constituted by polar and azimuthal scattering angles is $\Delta\theta = \Delta\varphi = \pm 0.5°$. In a case of a clean surface (**Figure 11a**), the usual peaks of QS- and QD-scattering (1,2) are observed. At specular scattering on the structure p(2 × 2)–O (**Figure 11b**), three peaks, two of which (3 and 4) are of peaks of QS-scattering on oxygen atoms with impact parameters p, smaller and larger p_{lim}

accordingly are present in the spectrum. Here $p_{lim}(E_0)$ is the impact parameter corresponding to scattering at a limiting angle.

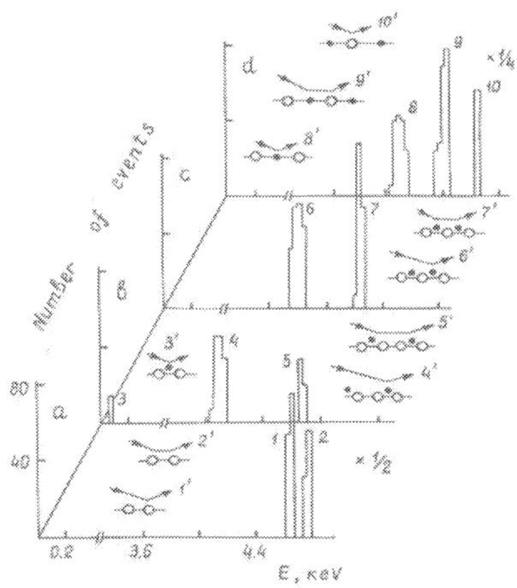

Figure 11. Histograms of energy distributions of 5 keV Ne^+ ions, undergoing specular scattering on a clean Ni(100) surface (a) and one covered by adsorption structures of oxygen p(2 × 2) (b) and c(2 × 2) (c), as well as on nickel oxide NiO (d). The sliding angle is ψ = 12.5°.Typical trajectories of scattered ions numbered 1'–10', contributing to corresponding peaks (1–10) are schematically presented.

Two peaks of the QS-scattering of ions on lighter adatoms were observed experimentally in ref. [46]. The third peak (5) is provided by QD-scattering on oxygen adatoms. In the energy spectrum for structure c(2 × 2)–O (Figure 12c), peaks of QS- and QD-scattering (6,7) on oxygen atoms with impact parameters p larger than p_{lim} are presented. The energy distribution of ions scattered on nickel oxide (Figure 2d) differs greatly from the distributions for chemisorption structures (Figure12b and c): peaks 8 and 10 correspond to QS0 scattering from oxygen and nickel atoms, and peak 9 between them correspond to QD-scattering from oxygen and nickel atoms. The analysis of the calculated spectra follows the fact that a number, intensity, and position of peaks on energy scale, corresponding to scattering on different adsorption structures, are different, which gives an opportunity to interpret them by means of comparison with experimental spectra. In **Figure 12**, calculated angular distributions of Ne^+ ions with E_0 = 5 keV and ψ = 12.5°, scattered from a clean Ni(100)<100> surface and one covered by different adsorption structures of oxygen are presented. In the case of a clean surface, the usual two-peak structure of angular distribution is observed [4, 5]. With the transition to the p(2 × 2)–O structure (**Figure 12b**), angular distribution of scattered ions changes considerably: it broadens to both sides, and in addition to the peaks observed from a clean surface, a peak appears at a limited angle of scattering from oxygen adatoms $\theta = \theta_{lim}$ = 52.5° and a peak of oxygen recoil atoms at θ = 78° (hatched part). The peak at θ_{lim} is provided by rainbow scattering, and its position on the angular scale gives an opportunity to identify the type of adatoms by the formula $m_2 = m_1 + \sin\theta_{lim}$.

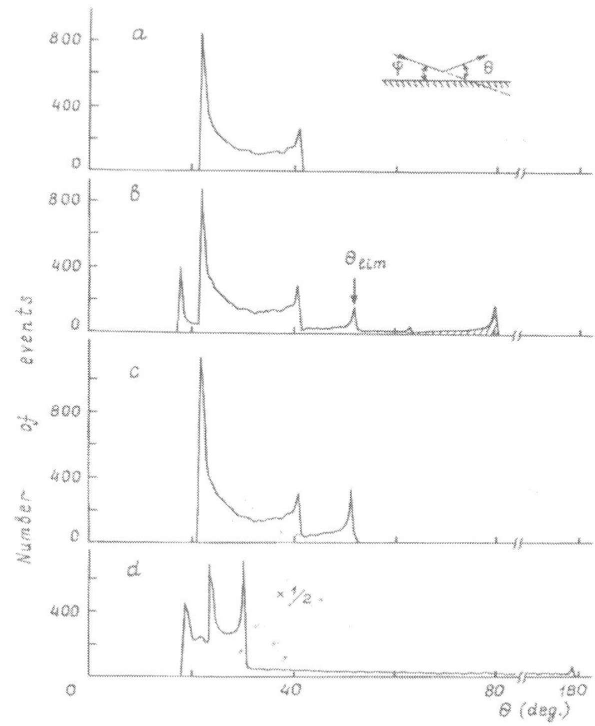

Figure 12. Angular distributions of 5 keV Ne^+ ions and ψ = 12.5°, scattered on a clean Ni(100)<100> surface (a) and covered by adsorption structures of oxygen p(2 × 2) (b) and c(2 × 2) (c), as well as nickel oxide NiO (d). The hatched field in the distribution for the structure p(2 × 2) is the contribution of primary knocked out oxygen adatoms.

Oxygen recoil adatoms are formed by the mechanism of direct knockout, and its detection allows the determination of the type of adsorbate and adsorption stage. At further exposure to oxygen and formation of c(2 × 2)–O chemisorption structure on the Ni(100) surface (**Figure 12c**), peaks at θ_{lim} and oxygen recoil atoms disappear. This is connected with the impossibility of QS-collisions of Ne^+ ions with oxygen adatoms with impact parameters $p < p_{\text{lim}}$ and releasing of recoil adatoms outside. The angular distribution as a whole narrows in comparison with the distribution for previous structures, testifying to the formation of a denser adsorption structure. On the formation on the surface of nickel oxide, the angular distribution of scattered ions (Figure 12d) also changes: the main structure of the distribution (three peaks) is located in a narrow field and has a background, spreading out to 180°. Such sharp broadening of the distribution is the characteristic feature for scattering on chains, consisting of different sort atoms. Thus, peculiarities of the chain effect at direct and inverse mass ratio of colliding particles and rainbow effect at QS- and QD-scattering of ions, heavier than adatoms lead to the appearance of characteristic peaks in the energy and angular distributions of ions, undergoing small-angle correlated scattering on chains of adatoms and target atoms. Analysis of these peaks and comparison with experiment give an opportunity to control the initial stages of adsorption and identification of adsorption structures with the help of low-energy ion scattering.

4.5. Investigation of the dynamics of changes of the $Cu_3Au(100)$ surface in the course of ordering by low-energy ion scattering

$Cu_3Au(100)$ single crystal exhibits an order–disorder phase transition into the bulk at $T_c = 663$ K. This bulk transition is first-order. Theoretically it is predicted that the transition at the surface is of a different type than in the bulk [47, 48]. The transition at the surface has been intensively studied in recent years, including by time-of flight-ion scattering spectroscopy (TOF-ISS) method [49, 50]. In ref. [49] has been shown how the order–disorder transition modifies the atomic sequences in atomic rows, then changing the double-scattering conditions. The atomic row effect exhibits remarkable features in small-angle ion scattering from a single-crystal surface of complex composition when the rows in certain crystallographic directions consist of alternating atoms of different species. In this section $Cu_3Au(100)$ alloy surface has been investigated with low-energy ion scattering at temperatures below and above the bulk order–disorder transition temperature [51].

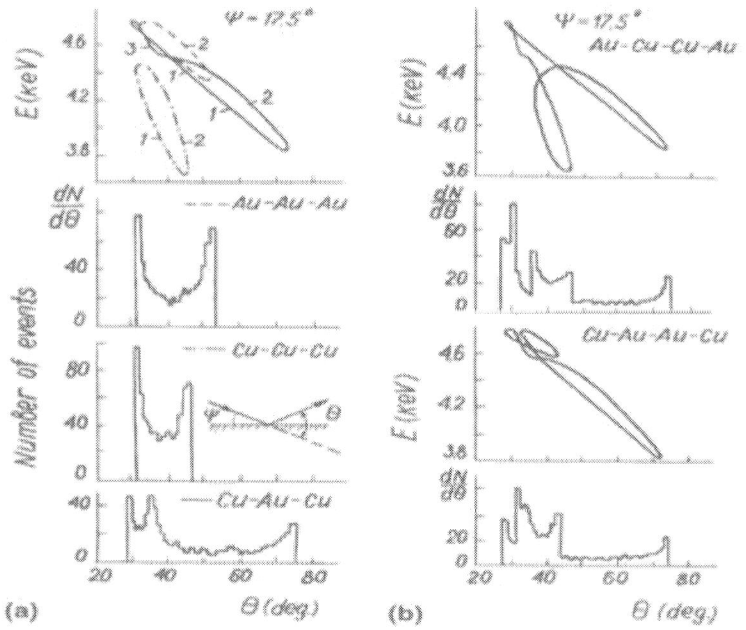

Figure 13. (a) Ovals of Ar^+ ion energy dependence on scattering angle θ and angular distributions at scattering from <110> atomic rows on Cu(100)-chain curve; Au(100)-broken curve, and $Cu_3Au(100)$-full curve; (b) Same dependencies for cases where one of the alloy atoms occurs twice in the row.

At temperatures $T > T_{cr}$ all lattice sites are populated uniformly by atoms of both species, while at lower temperatures nonuniform occupation of sites by atoms of different species is typical. Note that <100> rows contain either only Cu or only Au atoms, whereas in the <110> rows the Cu and Au atoms alternate. **Figure 13a** presents ovals depicting the dependence of the energy of scattered Ar^+ ions with initial energy $E_0 = 5$ keV on the scattering angle θ, which consist of quasi-single (QS)—1 and quasi-double (QD)—2 scattering branches. Shown in the bottom panel are the corresponding angular distributions of ions scattered from <110> rows in the Au(100), Cu(100), and Cu_3Au(100) planes. The dashed ovals correspond to the Au atoms, the chain one to the Cu atoms, and the solid oval with a characteristic break (3) to a row of alternating Cu and Au atoms. In the case of a mixed row, the oval has an unusual shape in that the QD branch is replaced by quasi-triple scattering (3) from a sequence of Au–Cu–Au atoms with the principal deflection from the Cu atom which results in considerable energy losses. The QS- and QD-scattering occur in this case primarily from Au atoms. The angular distribution contains in the region, corresponding to the break in the oval, an additional maximum and exhibits a dramatic broadening. The maximum appears as a shoulder at $\psi = 15°$, grows with increasing grazing angle, and exceeds in height at $\psi = 19°$ the maxima at the limiting scattering angles. As follows from the calculation, the additional maximum comes from the rainbow effect associated with the enhanced scattering close to the shadow-cone boundary of the Au atom toward the nearest Cu atom in a narrow impact parameter region along the row.

Figure 13b shows also the scattering ion energy versus the scattering angle ovals and the corresponding angular distributions; however, in this case, in the alternating sequence of atoms, one of them, either Au or Cu occurs twice in succession. Both the energy ovals and the angular distributions are seen to vary dramatically. It is shown when two neighboring atoms belong to the same species, the characteristic break (3) transforms into an additional oval which lies lower than the principal one and corresponds to QS- and QD-scattering from a pair of Cu atoms. To this oval correspond two additional maxima in the angular distribution. A similar pattern is also observed in the case when two Au atoms adjoin one another in a mixed row. The relative height of the additional maxima was found to depend on the number of neighboring atoms of the same species in a row. Increasing the number of neighboring identical atoms up to three or more results in an increase of the corresponding additional maxima in the angular distribution. Note that the shape of the energy oval and the interval of possible scattering angles are retained, whereas relative height of the maxima at the limiting scattering angles decrease. The characteristic structure of the angular distribution with the number of maxima is also retained when thermal vibrations of atoms in the alloy are included. Thus, a comparison of the angular and energy distributions of ions scattered from the surface of an alloy in the process of ordering with similar distributions for pure targets made up of the alloy components permits a conclusion that two or more neighboring atoms in an alternating sequence are of the same species. The results obtained can be used to study a short-range order in alloys undergoing ordering.

4.6. Ion sputtering of single-crystal surfaces

In this section, the grazing ion sputtering processes of Si(001), SiC(001), and Cu_3Au(001) surfaces at 0.5–5 keV Ne^+ bombardment have been studied by computer simulation [52, 53].

Sputtering has been simulated in the primary knock-on regime. Only the primary knock-on recoil (PKR) atoms ejected from first, second, and third layers have been considered. The presence of planar potential energy barrier on the surface was taken into account. The number of incident ions is 4×10^4. Each new particle is incident on a reset, pure surface. The incident ions and the recoil atoms were followed throughout their slowing-down process until their energy falls below a predetermined energy: 25 eV was used for the incident ions, and the surface-binding energy was used for the knock-on atoms. The calculations were performed on the crystals comprising up to 120 atomic layers. The initial energy of incident ions was varied from 0.5 to 5 keV, an angle of incidence ψ counted from target to the surface was 0–30°, and an azimuthal angle of incidence ξ realized by rotating the target around its normal and counted from the <100> direction was 0–180°.

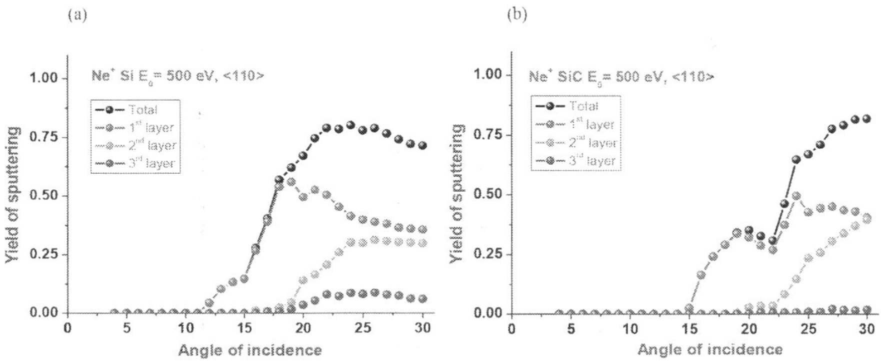

Figure 14. Sputtering yield of Si(001) (a) and SiC(001) (b) versus angle of incidence at Ne^+ ion bombardment.

In **Figure 14a,b** the angular dependences of the sputtering yield for Si(001) and SiC(001) surfaces subdivided into sputtering by the first three surface layers at 0.5 keV Ne^+ ion bombardment are compared. Note the angle of incidence ψ is counted from the surface. It is seen that there is a threshold angle of sputtering in all dependences. At angles of incidence less than the threshold angle, the incident ions cannot penetrate into the crystal and cannot eject target atoms. The threshold angle shifts to the lower values of angle of incidence with increasing the energy of incident ions. At ψ large than a threshold angle, with increasing ψ the number of PKR at first rises and achieves its maximum. There is a plateau (shorter for Si and wider for SiC) near the threshold angle because of insufficient ion energy for both long moving ions within surface semichannels and their penetration to deeper layers. With increasing the initial energy, this plateau disappears, and the sputtering yield decreases at large ψ. This decreasing of PKR yields is explained by partial penetration of ions into deeper layers and domination of the cascade sputtering mechanism. It is clear that the relative contributions of each layer to the total PKR yield strongly depend on the angle of incidence. In the angular range of ψ = 11–20° for Si and 15–23° for SiC, the sputtering occurs only from the first layer.

It is seen that the threshold angle is a bit smaller in the case of Si than for SiC. As results for high initial energy show, in general the sputtering yield is large in the case of SiC. These dependences allow choosing an angle of incidence for an effective sputtering at given initial energy. In **Figure 15a,b**, the sputtering yields of Si(001) and SiC(001) surfaces subdivided into sputtering by the first three surface layers versus the energy of incident Ne^+ ions are shown at ψ =10°. The threshold energy of sputtering is about 1 keV for these cases. There is more drastic increase of sputtering yield in the beginning of dependences for Si than for SiC. It is seen that the main contribution to the total sputtering comes from the sputtering of the first layer. Moreover, in the energy range of 0.5–1.5 keV for Si and 1–3 keV for SiC, the sputtering occurs only from the first layer. Further increasing of the ion energy results in increasing the contribution from second and third layers. The contribution to sputtering from the third layer is larger than the one from the second layer, as the atomic rows in the second layer lies directly under the one of the first layers in the <110> direction. Two local maxima at 2.5 and 4 keV are observed in the total sputtering yield dependence in the case of Si. Sputtering from the first layer gives a basic contribution to the first maximum, while the second maximum is formed by atoms ejected from the second and third layers. In the case of SiC the maximum of total dependence is formed by atoms ejected from the second layer. These results show that by choosing an angle and an energy of incidence, one can produce layer-by-layer sputtering of Si(001) and SiC(001) surfaces.

In **Figure 16**, the sputtering yield of $Cu_3Au(001)$ versus polar (ψ) and azimuth (ξ) angle of incidence at 5 keV Ne^+ ion bombardment is presented. Azimuthal angular dependence is shown only for the range of ξ = 0–90° due to its symmetricity. The main maxima at ξ = 0 and ξ = 90°, two local maxima at ξ = 35° and ξ = 55°, and deep minimum at ξ = 45° are observed in low crystallographic directions and near them. They are caused by the existence of original semichannels and channels in these directions. In these directions the incident ions either penetrate deeper into crystal or undergo the surface semichanneling. Due to ion channeling, the multiple collisions of projectile with target atoms are possible resulting in an intensive surface sputtering. Besides, the sputtering yield of $Cu_3Au(001)$ is two times less than one for SiC(001). These results show that influence of crystal orientation to sputtering yield depends on both crystal structure and its composition.

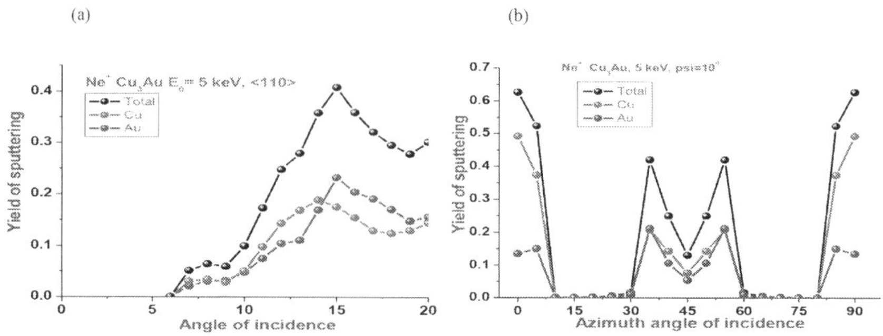

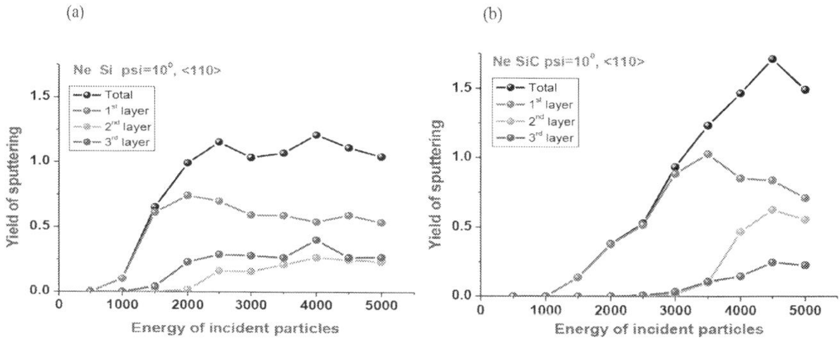

Figure 15. Sputtering yield of Si(001) (a) and SiC(001) (b) versus energy of incident Ne^+ ions at ψ =10°.

Figure 16. Sputtering yield of $Cu_3Au(001)$ versus polar (a) and azimuth (b) angle of incidence at E_0 = 5 keV.

Thus, the peculiarities of formation of PKR atoms at grazing ion incidence beam on an atomically smooth surface of a single crystal promotes its layer-by-layer sputtering. For the realization of the layer-by-layer sputtering mechanism, it is nessesary that a part of the ion energy, corresponding to the normal components of its velocity, will be lower than a threshold of sputtering of atoms of a layer, next to the surface one, that is, $E_i \sin^2 \psi_i < E_d$, where E_i is the ion energy before the ith collision, ψ_i is the angle between the ion movement direction and the semichannel axis before the ith collision, and E_d is the energy of displacement of atoms of a second layer (in the case considered, the bottom chain of semichannel). In these conditions it is possible to achieve successive removal of layers without disturbance of the next layer at removal of the previous one. Ion bombardment at grazing angles reduces considerably the influence of effect of crater walls and ion mixing on the results of layer-by-layer analysis, and increases its accuracy and sensitivity.

Parameters of single crystals (lattice parameter, binding energy, and mass of atoms) influence significantly the angular and energy dependences of sputtering yield. In general, the sputtering yield of the SiC(001) surface is larger than the one of the Si(001) surface. The proposed mechanism of layer-by-layer sputtering of a single-crystal surface at ion bombardment under grazing angles needs detailed checking and investigation by computer simulation as well as on experimental basis.

4.7. Determination of adatom sort and adsorption sites on single-crystal surfaces by ions channeling through thin films

Determination of the equilibrium adsorption sites of atoms adsorbed on crystalline surfaces and initial stages of adsorption is a fundamental problem in surface physics. Stensgaard [54] was the first to propose for the determination of the adsorbed site of D atoms adsorbed on the rear (100) surface of a thin Ni single crystal, 800 keV He^+ ions which channel through the sample. The adatoms were detected by monitoring the yield of the D(^{3}He,p) ^{4}He reaction. In the present section, we propose for the determination of the adsorption site and sort of light adatoms on the rear surface of a thin crystal to use interaction with these adatoms of low energy ions of mass m_1 greater than the adatom mass m_2, channeling through the crystal. For $m_1 > m_2$, there exist a limiting angle $\theta_{lim} = \arcsin(\mu)$, where $\mu = m_2/m_1$, which leads to the rainbow effect in scattering, that is, enhanced reflection that results in the appearance in the angular and energy distributions of transmitted ions of characteristic peaks useful for diagnostics purposes. In order to improve the usefulness of the angular and energy distributions of transmitted particles for diagnostic purposes, it can be recommended to also employ, besides the rainbow effect, the flux peaking phenomenon, that is, a spatial redistribution of the channeling particles in the transverse channel plane, which produces an increase of the flux in the central part of the channel and its sharp decrease on its periphery [56]. The sort and adsorption site of the adatoms are determined from the rainbow effect in the angular distributions of transmitted particles caused by their interaction with adatoms above the rear surface of the crystal [55, 56].

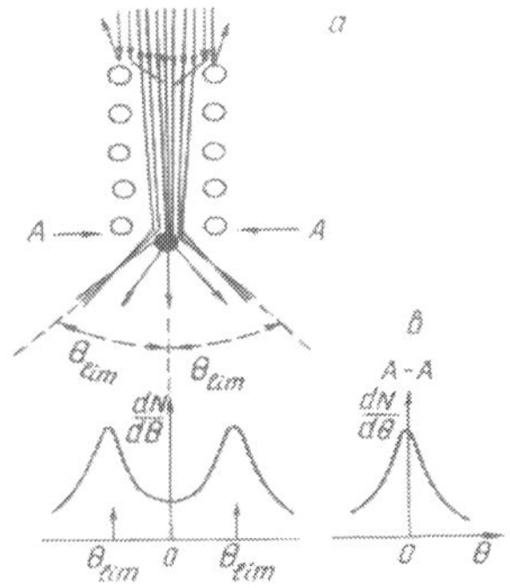

Figure 17. (a) The possibility of using the channeling effect to detect the adatom adsorption site above the rear crystal surface along the principal axes. (b) The changes in the angular distribution of the redistributed transmitted flux caused by the ion interaction with the adatom.

The trajectories of 1–5 keV He^+ and Ne^+ ions channeling in thin (Δz = 100–500 Å) Ni(100) films were computer-simulated in the binary collision approximation using the universal Biersack–Ziegler–Littmark potential and with the inclusion of elastic and inelastic energy losses. For ions emerging out of the rear side of the crystal, one calculates the coordinates of the emergence points projected on a plane perpendicular to the channel axis, the total scattering angle (the angle between the initial and final ion directions), and the final ion energy. Next, one calculates the interaction of the transmitted ions with the adsorbed atoms residing in various positions relative to the channel axis and at given heights h with respect to the rear surface of the crystal. The adatom sort and adsorption site are determined from a comparison of the angular distributions of the transmitted ions with or without adatoms present on the crystal surface (Figure 17a,b). **Figure 17a** illustrates the possibility of using the channeling effect to detect the adatom adsorption site above the rear crystal surface along the principal axes. **Figure 17b** shows schematically the changes in the angular distribution of the redistributed transmitted flux caused by the ion interaction with the adatoms. We readily see that the presence of a peak in the angular distribution at the limiting scattering angle permits the identification of the adatom sort. Thus, the present method is efficient only when $m_2 < m_1$, that is, when the adatoms are lighter than the channeling ions. By choosing properly the ion energy and crystal thickness, one can obtain angular distributions with the main peak corresponding to the limiting scattering angle.

This is illustrated in **Figure 18**, showing angular distributions of 1 keV Ne^+ ions channeling through thin (Δz = 100 Å) Ni(100) films along <100> in the (a) absence and (b) presence of adatoms of carbon and (c) oxygen located at the channel center at a height h = 0.9 Å above the rear crystal surface, which were calculated for T = 0 K. The presence of adatoms affects noticeably the distributions in that they shift toward scattering angles $\theta \approx \theta_{\text{lim}}$, their main peaks corresponding to rainbow scattering (θ_{lim} = 35° for carbon and θ_{lim} = 50° for oxygen). The essential advantage of this method lies in its capability of detecting of very light adatoms, such as hydrogen and deuterium. Thickness of the film, channeling direction, and particle energy have influence only on the magnitude of these maxima, but not on their positions. Thus, this method allows discrimination between the isotopes of an adsorbed element. Similar arguments are valid also for close isotopes of carbon and oxygen. Thus, when $m_1 > m_2$, the angular distributions of transmitted ions allow a correct determination of the adatom sort. The film thickness is not a critical factor in this technique, and it can be taken thicker if one increases correspondingly the initial energy of ions.

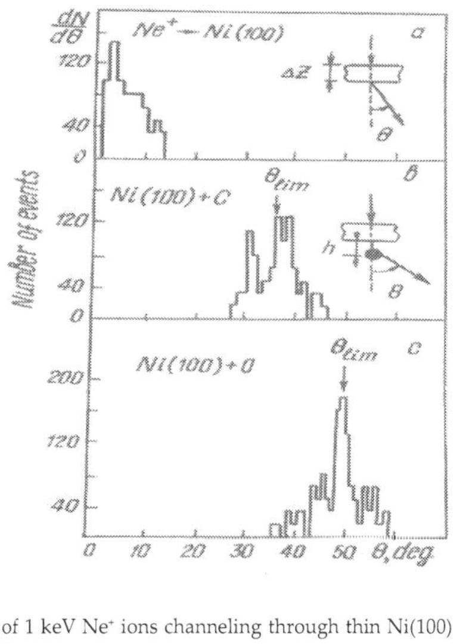

Figure 18. Angular distributions of 1 keV Ne^+ ions channeling through thin Ni(100) films in (a) absence and (b) presence of adatoms of carbon and (c) oxygen.

It is thought that rapid progress of the nanostructures physics and ultra-high vacuum technology allow to obtain in near future single-crystal films only a few hundred Angstroms thick with clean entrance and exit surfaces and well-ordered adsorbate structure on them. It was shown that the magnitude of main maximum at $\theta \approx \theta_{lim}$ depends on the height h of the adatom above the crystal surface. For the case $Ne^+ \rightarrow$ Ni(100) + O, where film thickness (Δz = 100 Å) and E_0 = 1 keV, maximum occurs at h = 0.9 Å, which is in agreement with the experimental value [57]. The sensitivity of the angular distributions of transmitted ions to shadowed adatom positions which do not correspond to the channel center in the given direction can be revealed by varying the direction of particle channeling. The accuracy with which the adatom adsorption site with respect to the unit cell in the transverse plane of the channel can be determined is evaluated as ~0.1 Å. An experimental test of this technique would permit a realistic evaluation of its capabilities.

5. Conclusion

One can draw the following conclusions on the basis of the conducted researches:

- A considerable influence of the binding of surface atoms to each other on the process of ion scattering in the low-energy range by polycrystal targets is experimentally established.
- The elastic energy losses are considerably smaller than the inelastic ones in a region of glancing ion scattering. The predominance of the inelastic energy losses should reveal itself in the efficiency of the various inelastic processes accompanying the glancing ion scattering on a single-crystal surface.
- From the comparison of the results of computer simulation of the scattered particle trajectories with the experimental energy distributions, one can draw a conclusion that for bombardment of a Cu(100) surface with E_0 = 10 keV Ar^+ ions, isolated monoatomic steps consisting of several atoms (from one to four) are formed upon it. The distances between the steps vary from two lattice constants to ~45 Å.
- It was shown that the analysis of the characteristic peaks appearing in energy and angular distributions of ions undergoing small-angle correlated scattering on chains of adatoms and target atoms on the single-crystal surface and comparison with experiment gives an opportunity to control the initial stages of adsorption and identification of adsorption structures with the help of low-energy ion scattering.
- A comparison of the angular and energy distributions of ions scattered from the surface of an alloy in the process of ordering with similar distributions for pure targets made up of the alloy components permits a conclusion that two or more neighboring atoms in an alternating sequence are of the same species. The results obtained can be used to study a short-range order in alloys undergoing ordering.
- Sputtering yields of Cu_3Au(001), Si(001), and SiC(001) surfaces versus the initial energy of incident ions (E_0 = 0.5–5 keV), angle of incidence (ψ = 0–30°), and azimuth angle of incidence

(ξ = 0–180°) have been calculated at Ne ion bombardment. It was shown that effective and layer-by-layer sputtering is possible near threshold angle and energy sputtering. The obtained results allow to select the optimum conditions for obtaining implanted depth distributions with demanded shape in narrow near-surface regions (5–10 atomic layers) of crystals.

- It was shown that the sort and adsorption site of the adatoms one can determined by the instrumentality of the rainbow effect in the angular distributions of transmitted channeled particles provided by their interaction with adatoms above the rear surface of the crystal. The accuracy with which the adatom adsorption site with respect to the unit cell in the transverse plane of the channel can be determined is evaluated as ~0.1 Å.

Author details

Farid F. Umarov[1*] and Abdiravuf A. Dzhurakhalov[2]

*Address all correspondence to: farid1945@yahoo.com

1 Kazakh- British Technical University, Almaty, Kazakhstan

2 University of Antwerp, Middelheimlaan 1, Antwerp, Belgium

References

[1] Van Hove MA *Surf. Sci.*, 603, 1301–1305 (2009).

[2] Begemann, SHA, Boers, AL *Surf. Sci.*, 30, 134 (1972).

[3] Parilis, ES, Turaev, NYu, Umarov, FF *Radiat. Eff.*, 21, 207 (1975).

[4] Mashkova, ES, Molchanov, VA *Medium Energy Ion Reflection from Solids*. North-Holland Publ., Amsterdam (1985).

[5] Parilis, ES, Kishinevsky, LM, Turaev, NYu, Baklitzky, BE, Umarov, FF, Verleger, VKh, Niznaya, S., Bitensky, IS. *Atomic Collisions on Solid Surfaces*. Elsevier, North-Holland Publ., Amsterdam (1993).

[6] Smith, DP *Surf. Sci.*, 25, 171 (1971).

[7] Labanda JGC, Barnett SA, *J. Electron. Mater.*, Sep (1997).

[8] Dzhurakhalov AA *Nucl. Instrum. Methods Phys. Res. B* . , 216, 202 (2004).

[9] Umarov FF, Dzhurakhalov AA, Teshabaeva NA *Appl. Surf. Sci.*, 125, 226 (1998).

[10] Roth J., Bohdansky J., Poschenrieder W., Sinha MK *J. Nucl. Mater.*, 63, 222–229 (1976).

[11] Bischoff L., Teichert J., Heera V. *Appl. Surf. Sci.*, 184, 372 (2001).

[12] Kim JC, Ji J-Y, Kline JS, Tucker JR, Shen T-C *Appl. Surf. Sci.*, 220, 293 (2003).

[13] Ecke G., Kosiba R., Kharlamov V., Trushin Yu., Pezoldt J. *Nucl. Instrum. Methods Phys. Res. B* 196, 39 (2002).

[14] Brown A., George H. B., Aziz M. J., Erlebacher J. *Mater. Res. Soc. Symp. Proc.*, 792, R7.8.1 (2004).

[15] Morgan DV. (Ed.). *Channeling Theory Observation and Applications*. Wiley, London (1973).

[16] Ziegler JF, Biersack JP, Littmark U. *Stopping and Ranges of Ions in Matters*. Pergamon, New York (1983).

[17] Kumakhov MA, Komarov FF *Energy Loss and Ion Ranges in Solids*. Gordon and Breach, New York (1981).

[18] Stepina NP, Kachurin GA *Sov. Phys. Semicond.*, 17, 449 (1983).

[19] Vavilov VS.. *Usp. Fiz. Nauk.*, 145, 9 (1985).

[20] Umarov FF, Rasulov AM, Khaidarov A. Kh. *Radiat. Eff. Defects Solids.*, 158, 481 (2003).

[21] Robinson MT, Oen OS *Phys. Rev.*, 132, 2385 (1963).

[22] Barret JH *Phys. Rev.*, 166, 219 (1968).

[23] Lindhard J. *Usp. Fiz. Nauk.*, 99, 240 (1969).

[24] Umarov FF, Bazarbaev NN, Kudryashova LB, Krylov NM *Nucl. Instrum. Methods Phys. Res. B.* , 196, 155 (2002).

[25] Umarov FF, Bazarbaev NN, Djyrabekova FG *Appl. Surf. Sci.*, 255, 6918 (2009).

[26] Algra AJ, Luitjens SB, Suurmeijer EPThM, Boers AL, *Appl. Surf. Sci.*, 10, 273 (1982).

[27] Robinson, MT, Torrens, IM *Phys. Rev.*, B9, 5008 (1974).

[28] O'Connor DJ, Biersack JP *Nucl. Instrum. Methods Phys. Res. B.*, 15, 14 (1986).

[29] Dzhurakhalov AA, Umarov, FF *Nucl. Instrum. Methods Phys. Res. B.*, 161–163, 377 (2000).

[30] Umarov FF, Bazarbaev NN, Kudryashova LB, Krylov NM *Nucl. Instrum. Methods Phys. Res. B.*, 196, 155 (2002).

[31] Umarov FF, Bazarbaev NN, Djyrabekova FG *Appl. Surf. Sci.*, 255, 6918 (2009).

[32] Landau LD, Lifshiz LM *Mechanica, Nauka* (1998) (in Russian).

[33] Pugacheva TS, Djurabekova FG, Miyagava Y., Valiev S. Kh. *Nucl. Instrum. Methods Phys. Res. B.*, 127–129, 260 (1997).

[34] Djurabekova FG, Umirzakov BE, Umarov FF, Miyagava Y. *Nucl. Instrum. Methods Phys. Res. B.*, 206, 194 (2003).

[35] Ziegler JF, Biersack JP, Littmark U. *The Stopping and Ranges of Ions in Solids*. Pergamon Press, New York (1985).

[36] Nizhnaya SL, Parilis ES, Verleger V.Kh *Radiat. Eff.*, 40, 23 (1979).

[37] Evdokimov IN, Webb RP, Armour DG, Karpuzov DS *Radiat. Eff.*, 42, 83 (1979).

[38] Dzhurakhalov AA, Umarov, FF *Nucl. Instrum. Methods Phys. Res. B.*, 136–138, 1092 (1998).

[39] Evdokimov IN, Mashkova ES, Molchanov VA *Dokl. Akad. Nauk SSSR.*, 186, 549 (1969).

[40] Luitjens, SB, Algra, AJ, Suurmeijer, EPThM, Boers, AL *Surf. Sci.* 100, 315 (1980).

[41] Umarov, FF, Parilis ES, Dzhurakhalov AA *Vacuum*, 44(9), 889–891 (1993).

[42] Dzhurakhalov AA, Kutliev UO, Umarov FF *Radiat. Eff. Defects Solids.*, 159, 293–299 (2004).

[43] Shulga VI *Radiat. Eff.*, 51, 1 (1980).

[44] Woodruff DP, Delchar TA *Modern Techniques of Surface Science*. Cambridge University Press, Cambridge (1986).

[45] Dzhurakhalov AA, Umarov, FF *Surf. Coat. Technol.*, 103, 16–19 (1998).

[46] Bastasz R., Felter TE, Ellis WP *Phys. Rev. Lett.*, 63, 558(1989).

[47] Sanchez JM, Moran-Lopez JL *Surf. Sci. Lett.* 157, 297 (1985).

[48] Lipovsky R. *Phys. Rev. Lett.* 49, 1575 (1982).

[49] Houssian L., Bertrand P. *Nucl. Instrum. Methods Phys. Res. B.*, 118, 467 (1996).

[50] Houssian L., Bertrand P. *Nucl. Instrum. Methods Phys. Res. B.*, 115, 161 (1996).

[51] Dzhurakhalov AA, Umarov, FF *Nucl. Instrum. Methods Phys. Res. B.*, 161–163, 377 (2000).

[52] Umarov FF, Dzhurakhalov AA, Teshabaeva NA *Appl. Surf. Sci.*, 125, 226 (1998).

[53] Umarov FF, Dzhurakhalov AA *Surf. Interf. Anal.*, 45(1), 83 (2013).

[54] Stensgaard I. *Nucl. Instrum. Methods Phys. Res. B.*, 15, 300 (1986).

[55] Dzhurakhalov AA, Parilis ES, Rasulov AM, Umarov FF *Poverkhnost*,6, 148 (1990)(in Russian).

[56] Umarov FF, Rasulov AM *Appl. Surf. Sci.*, 135, 269 (1998).

[57] Demuth JE, Jepsen DW, Marcus PM *Phys. Rev. Lett.*, 31, 540 (1973).

Chapter 15

Neutron Irradiation Effects in 5*xxx* and 6*xxx* Series Aluminum Alloys: A Literature Review

Murthy Kolluri

Additional information is available at the end of the chapter

http://dx.doi.org/10.5772/63294

Abstract

A literature review on highly irradiated 5*xxx* and 6*xxx* series Al alloys is conducted to understand the expected changes in mechanical properties of high flux reactor (HFR) vessel material in relation with microstructural aspects beyond the current surveillance data to support the HFR Surveillance Program (SURP). It was found that the irradiation swelling in 5*xxx* series alloys is not a crucial degradation mechanism. Dislocation damage is expected to reach a saturation limit in both 5*xxx* and 6*xxx* series alloys at relatively low fast-fluence values ($<2 \times 10^{26}$ n/m^2). The damage caused by precipitation of transmutation Si is found to be the dominant mechanism affecting the fracture toughness properties of irradiated 5*xxx* and 6*xxx* series Al alloys at high thermal fluence values. Tensile and fracture toughness data collected from the literature up to very high thermal fluences are analyzed in comparison with the available HFR surveillance data to predict the behavior of the HFR vessel material beyond current surveillance data. The observed changes in mechanical properties are classified into four different regimes. The contribution of various irradiation damage mechanisms, namely the displacement damage and transmutation damage, to the evolution of microstructure and mechanical properties is discussed in all four regimes for 5*xxx* and 6*xxx* series alloys.

Keywords: irradiation defects in aluminum alloys, displacement damage, transmutation damage, Mg_2Si precipitates, fracture toughness, radiation hardening, embrittlement

1. Introduction

Aluminum alloys have a higher tolerance to radiation effects than most other metals when irradiated at ambient temperatures due to its low melting point (T_m). This is because, at room temperature, the homologous temperature of Al alloys is around $0.32T_m$, when compared to, for instance, ~$0.175T_m$ for austenitic steel, ~$0.17T_m$ for ferritic steel, and ~$0.14T_m$ for α-Zr. In metals, it is known that noticeable thermal diffusion of vacancies occurs at homologous temperatures above $0.3T_m$. This thermally induced movement of vacancies at room temperature (RT) promotes mutual recombination of vacancies and interstitials, resulting in a lower density of point defect clusters, which are seeds for the damage microstructure.

In particular, 5*xxx* and 6*xxx* series Al alloys exhibit a good combination of mechanical, thermal, corrosion resistance, and irradiation swelling resistance properties in a research reactor environment, which make these alloys a suitable choice for in-core structures and reactor vessel components of research reactors. The reactor vessel of the high flux reactor (HFR) in Petten has been fabricated from the aluminum alloy ASTM B209 [1], specification Al 5154–O with a restriction on Mg content to a maximum of 3.5 wt.%.

The components of these reactors can experience a large amount of neutron fluences, up to several 10^{27} n/m^2, during their operational life. For the HFR hotspot,[a] a maximum thermal fluence of ~20×10^{26} n/m^2 is expected by the end of 2025.[b] Substantial damage to the material's microstructure and mechanical properties can occur at these high fluence conditions. To this end, a dedicated SURveillance Program (SURP) is executed to understand, predict, and measure the influence of neutron radiation damage on the mechanical properties of the vessel material. As a part of SURP, a literature survey on irradiated Al alloys that are relevant for HFR vessel material is conducted to obtain fundamental understanding on expected mechanical property changes in relation with microstructural damage mechanisms, which forms the goal of this work.

This article is organized as follows. First, a brief review of various irradiation-induced damage mechanisms in Al alloys is presented. Next, the tensile data collected from the literature is analyzed to understand the contributions of various irradiation-induced damage mechanisms to the changes in the mechanical properties of these materials up to high irradiation fluences. Finally, the fracture toughness data from HFR SURP is compared with that of the literature, and the underlying damage mechanisms influencing fracture toughness properties are discussed to explain the suitability of literature data for the prediction of HFR SURP data beyond the current surveillance data.

2. Literature on irradiation effects in Al alloys

A substantial amount of literature is published on the irradiation behavior of Al alloys [2–11]. The available dataset on 6*xxx* series alloys is considerably larger due to their widespread use in several research reactors and cold-neutron sources [2, 4–10]. On the other hand, only limited data were found on 5*xxx* series alloys [3, 9, 11, 12]. The published data from the SURP of the HFR vessel are also included in this review [13]. Although 5*xxx* and 6*xxx* series alloys are

[a] Hotspot is the location on vessel wall where highest neutron fluence is received.

[b] Assuming that the irradiation conditions at the HFR hotspot are kept unchanged as they are in 2015.

fundamentally different in their microstructure and properties in unirradiated condition, the data on irradiated 6*xxx* series Al alloys have great relevance to 5*xxx* series data, because 5*xxx* series alloys slowly transform into 6*xxx* series alloys in the course of neutron irradiation due to the transmutation-produced Si content [9]. Therefore, this study is focused on both the 5*xxx* and 6*xxx* series types of Al alloys. The emphasis is on understanding the influence of neutron irradiation on mechanical and microstructural properties up to very high fluence values and, in particular, the effect of transmutation-produced Si content on the mechanical properties.

To help the discussion on differences in irradiation damage mechanisms in different Al alloys, a brief review on the differences in chemical composition and microstructure of 5*xxx* versus 6*xxx* series Al alloys is presented here. The microstructure of 6*xxx* series alloys is carefully engineered by a suitable age-hardening treatment to form coherent precipitates (GP zones and β″) within the matrix to obtain the required mechanical properties [14]. On the other hand, the as-produced 5*xxx* series has no precipitates within the matrix. The Mg solute atoms present in the solid solution provide the required strength properties. The composition of Al alloys of interest for the current discussion is shown in **Table 1**.

Alloy	Al	Mg	Si	Cu	Cr	Fe	Mn	Ti	Zn	Ni
5154 (HFR vessel)	Balance	3.10–3.50	≤0.25	≤0.05	0.15–0.35	≤0.40	≤0.10	0.10–0.20	≤0.20	–
5052	Balance	2.20	≤0.10	–	0.20	0.18	–	–	–	≤0.30
6061	Balance	0.80–1.20	0.40–0.80	0.10–0.40	0.04–0.35	≤0.70	≤0.15	≤0.15	≤0.25	–

Table 1. Chemical composition of different as-produced Al alloys in wt.%.

3. Irradiation-induced damage mechanisms in Al alloys

The damage caused by neutron irradiation is the major degradation mechanism leading to irradiation hardening and embrittlement of Al alloys used in Materials Test Reactors (MTRs). Both thermal and fast neutrons cause damage in Al alloys. Displacement damage by fast neutrons and transmutation damage by both thermal and fast neutrons are the two major damage mechanisms in irradiated Al alloys [3, 9, 10]. The relative contribution of these different damage mechanisms and the resulting impact on the mechanical properties depend on the alloy composition, thermal-to-fast fluence ratio (TFR), irradiation temperature, and other irradiation conditions.

3.1. Displacement damage

As in other metals, displacement damage is initiated by the production of primary knock-on atoms (PKAs) through elastic collision of fast (high energy) neutrons with the Al matrix. The resulting PKAs trigger displacement cascades leading to the formation of lattice vacancies, self-interstitial atoms, and dislocation loops. With increasing irradiation dose, dislocation loops grow and encounter the other loops or dislocation network. When the loops interact with each other, they coalesce and contribute to the increase in network dislocation density. Interaction between individual dislocations and loops also contribute to the network.

The irradiation-induced dislocation density determines the extent of irradiation hardening and embrittlement resulting from displacement damage. It is known from literature that the dislocation density in irradiated metals evolves toward a saturation value with increasing dose [15]. This occurs when the dislocation annihilation rate reaches the value of the production rate. The resulting contribution of displacement damage to irradiation hardening and embrittlement remains nearly constant above the irradiation dose levels at which dislocation density reaches a saturation value. From that point onward, transmutation-produced Si plays a dominant role in contributing to irradiation hardening of Al alloys as discussed further in the next section. A detailed discussion on the evolution of displacement damage in Al alloys can be found in Refs. [9, 15].

3.2. Transmutation damage

Transmutation damage in aluminum can be caused by both fast and thermal neutrons. Fast neutrons produce gaseous products like He and H through (n, α) and (n, p) transmutation reactions [9]. On the other hand, thermal neutrons cause transmutation of Al into Si through the following sequential reactions,

$$^{27}\mathrm{Al}(n,\gamma)^{28}\mathrm{Al},\ ^{28}\mathrm{Al} \rightarrow {}^{28}\mathrm{Si} + \beta, \tag{1}$$

leading to an increase in Si content with increasing thermal neutron fluence. In most metals, the gaseous transmutation products play a larger role in the development of radiation damage microstructure than nongaseous transmutants. However, Al alloys used in MTRs are different in this respect. Depending upon the thermalization of the neutron spectrum, the solid transmutation product Si can have a stronger effect on radiation damage structure than gaseous transmutation products, as discussed in more detail in the following subsections.

3.2.1. Gaseous transmutation damage

Gaseous transmutation products can have a substantial influence on the radiation damage structure by promoting cavity formation and swelling. Gaseous transmutation products favor cavity nucleation by bubble formation at locations such as grain boundaries and stable particle–matrix interfaces, which otherwise are not suitable for nucleation of pure vacancy clusters.

It should be noted that the resistance to cavity formation and swelling differ between different types of Al alloys even in the presence of similar amounts of gaseous transmutation products. Alloys that promote trapping and recombination of point defects reduce vacancy supersaturation and hence exhibit increased resistance to cavity formation and swelling [3]. For instance, 5052-O and 6061 alloys have an excellent resistance to cavity formation and swelling, when compared to pure Al and grade 1100 alloys. Literature reports [16] show that the incubation dose for cavity formation of 5052-O alloys, ~ 5×10^{26} n/m^2, is about 1000 times that of pure Al. Such strong resistance to cavity formation is imparted to the solute Mg present in the solid solution, which can act as trapping and recombination sites for vacancies and interstitials to reduce vacancy supersaturation [3]. Once the Mg is drawn from solution to form Mg_2Si precipitates, the trapping and recombination sites are presumably shifted to these Mg_2Si precipitates, whose high spatial density might provide overlapping point-defect capture zones. High concentrations of precipitates are expected to contribute to reduced swelling by trapping gases, making these gases not available for cavity nucleation. Farrell et al. [9] reported that the radiation swelling in 5052-O is only about 1% at a fast fluence of ~18×10^{26} n/m^2. The corresponding thermal fluence value is 31×10^{26} n/m^2 with about 7% of transmutation-produced Si. Only sparsely distributed voids are found in 5052-O microstructure at these high fluence values [9]. The contribution of voids to the increase in strength and decrease in ductility of this alloy is found to be negligible at this small amount of swelling [3]. No swelling data was published for 5154-O alloy in these conditions. However, due to the similarity in microstructures of both 5052-O and 5154-O alloys and matching irradiation conditions, a comparable swelling behavior can be predicted in 5154-O alloy at HFR vessel hotspot. Using the swelling data of 5052-O from Farrell et al. [9], the estimated swelling in 5154-O alloy will be ~0.3% for the projected HFR hotspot fluence values by the end of 2025. From these arguments, it can be concluded that the creation of voids and bubbles in 5*xxx* series alloys is not a crucial degradation mechanism for the expected hotspot thermal fluence values of HFR vessel by the end of 2025.

3.2.2. *Solid transmutation damage*

Transmutation-produced Si by thermal neutrons causes substantial radiation damage in Al alloys. Kapusta et al. [11] confirmed that the Si-content is a major indicator for the neutron irradiation effects on the basis of postirradiation testing of Al alloys containing 2.12% Si from transmutation. A quick estimate of the production rate of transmutation-produced Si (~0.084 wt.%/year of 270 effective full power days at HFR hotspot) can be obtained by multiplying the thermal fluence with the standard thermal neutron absorption cross section for Al (=230 milli barn (mb)) [9]. The solubility of Si in the Al matrix below 373 K is negligible. Hence, the transmutation-produced Si will either precipitate in elemental form as in pure Al, grade 1100 and 6061 alloys or forms Mg_2Si precipitates as in 5*xxx* series alloys until all the Mg in solid solution is consumed.

The structure, size, and distribution of these precipitates (Si and Mg_2Si) in the microstructure will determine the resulting mechanical properties of irradiated alloys. For a given volume fraction of precipitates in the microstructure, finer precipitates result in higher strength, but lower ductility and fracture toughness properties. The structure of the Mg_2Si precipitates in 5052 alloy irradiated to 9.7×10^{26} n/m^2 thermal fluence is found to be similar to the thermally aged Mg_2Si precipitates in 6*xxx* alloys [3]. However, irradiation-assisted Mg_2Si precipitates in 5*xxx* alloys are observed to be fine compared to precipitates in thermally aged 6061 alloy [9], probably because irradiation-assisted precipitation occurs at temperatures much lower than the thermal aging temperature of 433 K, thereby favoring a large number of nucleation sites.

In case of control rod drive follower (CRDF) A-2 tubes of the High Flux Beam Reactor (HFBR) in Brookhaven National Laboratory, USA, produced from 6061-T6 alloy, irradiated at 338 K up to a very high thermal fluence of 42 × 10^{26} n/m^2, a high concentration of very fine (8 nm) amorphous Si-rich particles are observed in the microstructure in place of original Mg_2Si precipitates [5]. The corresponding fast fluence is 2 × 10^{26} n/m^2, which gives a high TFR of 21 compared to the HFR hotspot TFR value of maximum 1.4. The total measured Si at this fluence was found to be ~8 wt.%, including 0.6% of the initial Si content.

The location of this transmutation-produced Si precipitates in the microstructure will have substantial impact on the mechanical properties of the alloys. In 1100 and 6061 alloys, it was identified that the transmutation-produced Si will precipitate as elemental Si particles, which are uniformly distributed in the matrix and associated with voids [9]. Farrell et al. [17] reported a noncrystalline Si-coating inside the voids of 1100-O Al alloy at a high thermal fluence ($E <$ 0.025 eV) of ~2.3 × 10^{27} n/m^2. The 6061 alloy irradiated to ~10^{27} n/m^2 at ~328 K has shown a decoration of original Mg_2Si precipitates with transmutation-produced Si in addition to the association of Si particles with voids [9]. Precipitation of this Si along the grain boundary can lower the fracture toughness. For example, CRDF A-2 tubes of HFBR produced from 6061-T6 alloy have shown a drop in fracture toughness to ~8 (MPa)·m$^{1/2}$ from an unirradiated value of 21.75 (MPa)·m$^{1/2}$ after irradiation to a thermal neutron fluence of ~42 × 10^{26} n/m^2 at 338 K (see **Figure 4**). The microstructure of this alloy, with a very high transmutation-produced Si content of 8 wt.%, has shown large silicon flakes occupying less than one-fifth of the grain boundary area [5]. Similarly, heavy discontinuous precipitation at grain boundaries is observed in 5052 alloy irradiated up to a thermal fluence of ~31 × 10^{26} n/m^2 [3].

From the above discussion, it can be concluded that the transmutation-produced Si is the dominant irradiation damage mechanism in 5*xxx* and 6*xxx* series Al alloys irradiated at temperatures <373 K. Consequently, transmutation-produced Si is taken as the measure of the irradiation damage in HFR vessel wall. There are differences in how this transmutation-produced Si will influence the mechanical properties of 5*xxx* and 6*xxx* Al alloys, which will be discussed in the next section.

4. Discussion on irradiation-induced damage effects on mechanical properties of 5*xxx* and 6*xxx* series Al alloys

Although fracture toughness data on irradiated Al alloys is scarce, significant data on tensile properties is available in the literature. In this section, tensile data on irradiated Al alloys collected from literature is plotted as a function of thermal fluence to understand the changes in tensile properties with the evolution of irradiation-induced microstructural damage (or transmutation-produced Si content). Once this relation is established, then one can make a bridge to correlate these changes to corresponding changes in fracture toughness properties, where only limited data is published in the literature.

http://dx.doi.org/10.5772/63294

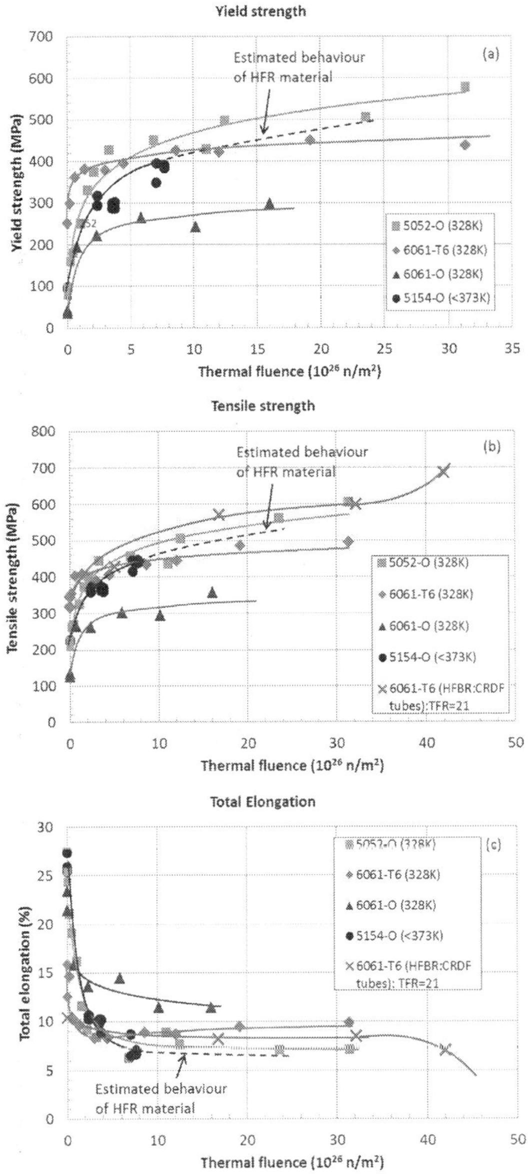

Figure 1. Literature tensile data of irradiated Al alloys in comparison with HFR SURP data. (a) Yield strength versus thermal fluence, (b) tensile strength versus thermal fluence and (c) total elongation versus thermal fluence.

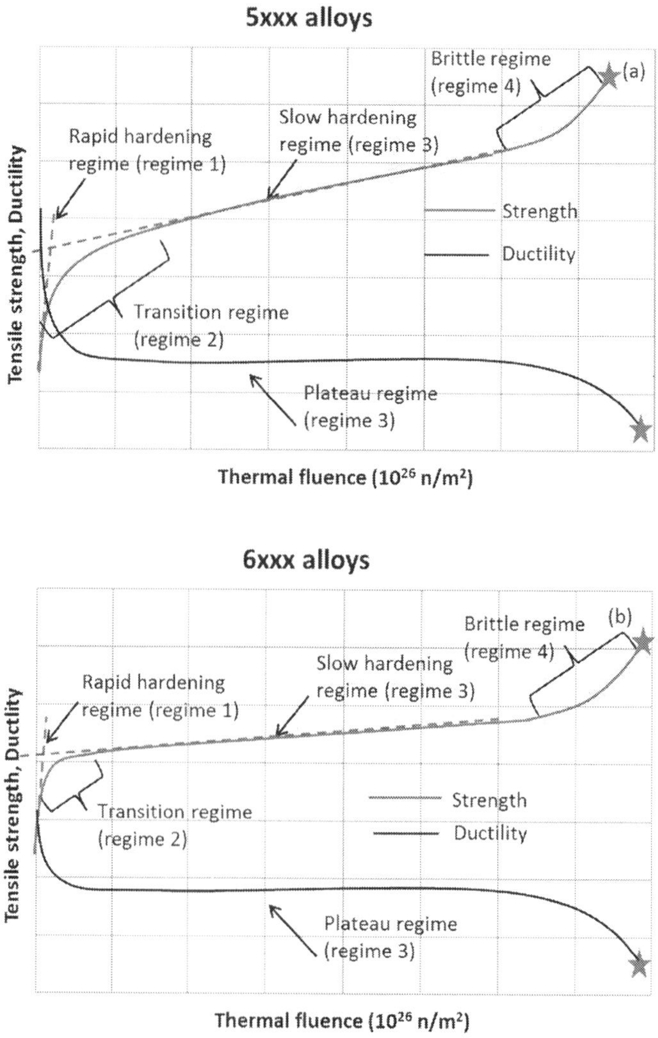

Figure 2. Schematic diagram showing various regimes in irradiation hardening behavior of (a) 5xxx and (b) 6xxx series Al alloys.

Farrell et al. [3] published data on tensile behavior of 5052-O aluminum alloy (Al–2.2% Mg) heavily irradiated in HFIR to fluences greater than 10^{27} n/m^2 in contact with cooling water at 328 K (see **Figure 1**). HFIR is predominantly a thermal reactor with a strong fast neutron component. The thermal neutron fluence (E < 0.0025 eV) of the samples ranged up to ~31 × 10^{26} n/m^2 to produce 7.15 wt.% Si. The fast neutron fluence (E > 0.1 MeV) is a factor of 1.7 (=TFR) lower than the thermal fluence, and results in a damage value of 260 displacements per atom (dpa). Associated gas generation due to fast neutrons was estimated to be 8.5 × 10^{-5} atomic fraction He and 5 × 10^{-4} atomic fraction H. It is important to notice that this data is very relevant to the HFR SURP program because of the (i) similar chemical composition of 5052 and 5154 alloys, except that Mg in 5154-HFR alloy is 3.2% instead of 2.2% in 5052 and (ii) similar irradiation conditions, including temperature, TFR, and high fluence values. Farrell et al. [2] also published tensile data on heavily irradiated 6061-O and 6061-T6 alloys irradiated at 328 K up to ~31 × 10^{26} n/m^2 and tested at three different temperatures (323, 373 and 423 K). **Figure 1** shows only results at 323 K due to their relevance to HFR operating conditions. Additionally, tensile strength data from a 6061-T6 type alloy tested from CRDF A-2 tubes of HFBR, published by Weeks et al. [5] are also shown in **Figure 1**.

Comparison of yield and tensile strength properties of all these alloys including HFR-SURP tensile data, shown with added trend lines in **Figure 1 (a, b)**, reveals specific trends in irradiation hardening and embrittlement behavior. Each of these alloys showed a rapid hardening regime (and corresponding drop in ductility) at the beginning, followed by a transition regime toward a relatively slow hardening (and stable ductility) regime. A brittle regime is observed in some alloys at the end, as shown schematically in **Figure 2**. Depending on whether an alloy is of 5*xxx* or 6*xxx* series, a single or multiple irradiation damage mechanism can be active, determining the hardening rates in each of these regimes.

4.1. Tensile behavior of 6*xxx* series alloys

In case of 6*xxx* series Al alloys, rapid hardening observed at the onset of irradiation (regime 1), for example, curves of 6061-O and 6061-T6 alloys (**Figure 1 (a, b)**), can be attributed to irradiation-induced dislocation damage. It is known that the irradiation-induced dislocation density rapidly increases at the beginning and reaches a saturation value (see Section 3.1) at relatively low fluence values. For steels irradiated at 603 K (~0.35T_m), this saturation dislocation density is expected to reach around a fast fluence of ~2 × 10^{26} n/m^2 (15). For Al alloys irradiated at temperatures about 323 K (~0.35T_m), it is expected that saturation is reached at lower fluences due to the low effective displacement energy of Al (~25 eV) compared to Fe (~40 eV) [9].

In the transition regime (regime 2), precipitation of transmutation-produced Si takes over as the major contributing mechanism, while the dislocation density reaches a saturation limit. It is known from literature that the transmutation Si in 6*xxx* alloys nucleate as amorphous Si particles in the matrix, associate with irradiation-induced voids, and decorate existing Mg_2Si precipitates (see Section 3.2). The contribution of this mechanism to irradiation hardening of 6*xxx* series alloys is low because deformation which occurs by the shearing of soft Si particles produces little strain hardening [18]. At the same time, the hardening contribution from an increase in size of existing Mg_2Si precipitates (due to Si decoration) is also low. This is because for a given particle density, the increase in precipitate size (r) and the decrease in planar spacing (λ) of the precipitates resulting from precipitate growth due to Si decoration have a minimal effect on strength based on the Orowan–Ashby equation [19] (Eq. (2)).

$$\Delta\sigma = \frac{0.13Gb}{\lambda}\ln\frac{r}{b} \tag{2}$$

Assuming a saturation density of ~ 6 × 10^{14} m^{-2} in Al alloys (same as in steel), a rough estimate of the total contribution of dislocation hardening can be made using the following equation [20]:

$$\Delta\sigma_{dis} = \sigma - \sigma_o = Gb\rho^{1/2}, \tag{3}$$

where σ is the strength of the material after introducing dislocation structure, σ_o is the intrinsic strength of the material with low dislocation density, G is shear modulus (=2.648 × 10^4 MPa), b is Burgers vector (0.286 nm) and ρ is dislocation density. The G and b values of pure Al taken from reference [21] are used here. Substituting ρ = 6 × 10^{14} m^{-2} into Eq. (3) gives $\Delta\sigma$ = ~186 MPa. The magnitude of hardening observed at the beginning of 6061-O and 6061-T6 alloys is in good agreement with this value. An estimation of the total irradiation hardening using the microstructure data (r = 4 nm, λ = 4 nm) of CRDF A-2 alloy at 42 × 10^{26} n/m^2 thermal fluence taken from [5] resulted in an estimated yield strength of 605 MPa, which is comparable to the available tensile data for this alloy.

A low hardening rate observed in regime 3 of these alloys can be solely attributed to the growth of existing precipitates. No further increase in the precipitate density occurs in this regime leading to a stable ductility. The final brittle regime (regime 4) with an increasing hardening rate and a decreasing ductility is observed only in 6061-T6 alloy (from CRD A-2 tubes of HFBR) at very high fluences. Although this alloy is the same as 6061-T6 alloy and irradiated at similar temperatures, a difference in behavior is observed due to irradiation at very high TFR, as explained in Section 4.4.

4.2. Tensile behavior of 5*xxx* series alloys

The differences in irradiation hardening trends in all four regimes of 5*xxx* and 6*xxx* series alloys are depicted schematically in **Figure 2**. Both 5052-O and 5154-O alloys show similar irradiation hardening and embrittlement behavior (**Figure 1**) due to similar alloy microstructure and irradiation conditions. The unirradiated strength values of 5052-O and 5154-O alloys are lower, and ductility is higher than 6061-T6 alloy (**Figure 1**) due to the absence of Mg_2Si precipitates before irradiation. In the rapid hardening regime, the magnitude of irradiation hardening and embrittlement in 5*xxx* series alloys is observed to be much higher than 6*xxx* series, because both dislocation hardening (due to displacement damage) and precipitation hardening (due to the formation of Mg_2Si precipitates from transmutation Si) occur simultaneously in 5*xxx* alloys (see Section 3.1).

The contribution of both mechanisms continues in the transition regime until dislocation damage reaches a saturation value (at <2 × 10^{26} n/m^2 of fast fluence or <4 × 10^{26} n/m^2 of thermal fluence). Simultaneously, a saturation in the density of precipitates is expected to occur in this regime, leading to the formation of no new Mg_2Si precipitates.

With further irradiation, the hardening continues with a decreasing rate as Mg_2Si precipitates continue to grow until all the Mg is pulled out from the Al solid solution in the final slow hardening regime. Based on the stoichiometric analysis, production of 0.58 wt.% transmutation Si will consume 1% Mg in the alloy. That means all the Mg in 5154-0 alloy is consumed at ~1.85% transmutation Si (~8.66 × 10^{26} n/m^2 of thermal fluence) and in 5052-O alloy at ~1.27% transmutation Si (~5.95 × 10^{26} n/m^2 of thermal fluence).

With continued irradiation, the newly formed transmutation-produced Si either decorates existing precipitates (like in *6xxx* series) or associates with voids, which are expected to form only at very high fast fluences in *5xxx* alloys as explained in Section 3.2. The higher hardening rate observed in regime 3 of *5xxx* alloys can be attributed to the higher density and volume fraction of Mg_2Si precipitates than that observed in *6xxx* series alloys. A stable ductility is observed in this stage similar to *6xxx* series alloys due to no further increase in precipitation density. In fact, a small decrease in particle density may occur in this regime due to particle coalescence during their growth. The opposite effects of a small decrease in particle density and a slow hardening due to precipitate growth on ductility could be compensating each other, leading to a plateau in the ductility behavior in regime 3. No brittle regime is observed in the available data of *5xxx* alloys until a thermal fluence of 31 × 10^{26} n/m^2.

Similar to *6xxx* alloys, a rough estimation of the total irradiation hardening contribution is performed for 5052-O and 5154-O alloys using Eq. (2) for the dislocation-hardening contribution and Eq. (3) for contribution from precipitation hardening. An average precipitate size of 10 nm and linear planar spacing between precipitates of 10 nm is used from the microstructure data presented in [3] on 5052-O alloy irradiated to a thermal fluence of 31 × 10^{26} n/m^2. This resulted in a total yield strength of 547 MPa (including unirradiated yield strength value of 85 MPa) for 5052-O, matching very well with the data presented in **Figure 1 (a)**. Similar analysis for 5154-O at a thermal fluence of 9.81 × 10^{26} n/m^2, with an average precipitate size of 8 nm and linear planar spacing between precipitates of 20 nm (obtained from **Figure 5**) [13] resulted in a total yield strength of 410 MPa, again matching with the trends seen in **Figure 1 (a)**.

4.3. Fracture toughness behavior of 5*xxx* and 6*xxx* series alloys

In this section, first the fracture toughness data from HFR SURP is plotted against the available literature data on highly irradiated Al alloys. The evolution of fracture toughness behavior of *5xxx* and *6xxx* alloys during neutron irradiation is discussed, and a connection is established between tensile and fracture toughness behavior in various regimes observed in the previous section.

4.3.1. Literature fracture toughness data of 5xxx series Al alloys

No additional data on fracture toughness properties of 5*xxx* Al alloys was found in the literature except the data from 5154-O alloy from HFR SURP [13]. The fracture toughness properties of the HFR surveillance specimens are periodically measured to assess and predict the hotspot behavior of HFR vessel in comparison with the vessel's fracture toughness design limit of 6 (MPa)·m$^{1/2}$ [22]. Fracture toughness properties of surveillance specimens from the current HFR vessel tested until 2010 are plotted as a function of thermal neutron fluence in **Figure 3 (a)**. The thermal neutron fluence is taken as an indicator for the measure of irradiation damage in HFR vessel material, because thermal neutrons are the major cause of damage by producing Si through transmutation, as explained in the previous sections. **Figure 3 (b)** shows the relation between the thermal fluence and transmutation Si in fracture toughness specimens tested in SURP.

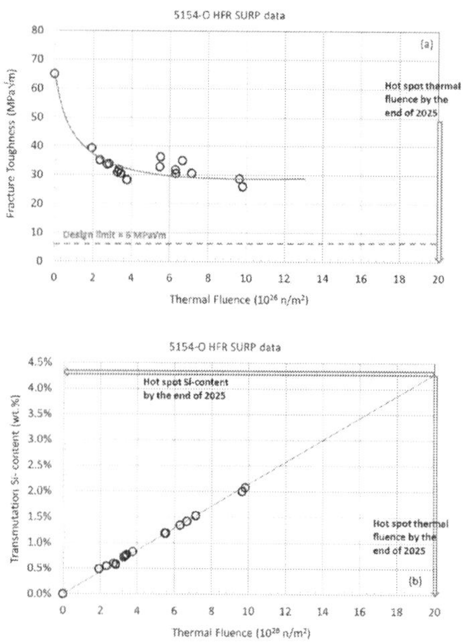

Figure 3. (a) HFR SURP fracture toughness data as a function of thermal neutron fluence. (b) Transmutation Si values as a function of neutron thermal fluence of fracture toughness samples tested in HFR SURP program. Note that the projected hotspot thermal fluence and Si content are based on the assumption that the irradiation conditions at the HFR hotspot are kept unchanged as they are in 2015.

4.3.2. Literature fracture toughness data of 6xxx series Al alloys in comparison with HFR SURP data

Only limited data was published on fracture toughness properties of irradiated Al alloys [5, 8, 10]. The most relevant data for the HFR (irradiation temperatures < 373 K) is plotted in

Figure 4 in comparison with HFR SURP data. Data from 6061-T6 alloy irradiated at < 373 K in the High Flux Isotope Reactor (HFIR) in Oak Ridge National Laboratory, USA matches quite well with the HFR SURP data. As it can be seen from **Figure 4**, there is one high fluence data point published by Weeks et al. [5] beyond the current surveillance data of the HFR vessel. This data is from the CRDF A-2 tubes of the HFBR in Brookhaven National Laboratory, USA, produced from 6061-T6 alloy, irradiated at 338 K up to a thermal fluence of 42 × 10^{26} n/m^2. The corresponding fast fluence of this data point is 2 × 10^{26} n/m^2, which gives a high TFR of 21, compared to the HFR hotspot TFR value of maximum 1.4. The total measured Si at this fluence was found to be ~8 wt.%, including 0.6% of initial Si content. The reported thermal fluence and Si content of this data point are approximately two times the estimated thermal fluence (~20 × 10^{26} n/m^2) and Si (~4.3 %) content of the HFR hotspot by the end of 2025. Note that this data is from the same material and at the same irradiation conditions for which the tensile data at very high thermal fluences (~42 × 10^{26} n/m^2) is also available (see **Figure 1**).

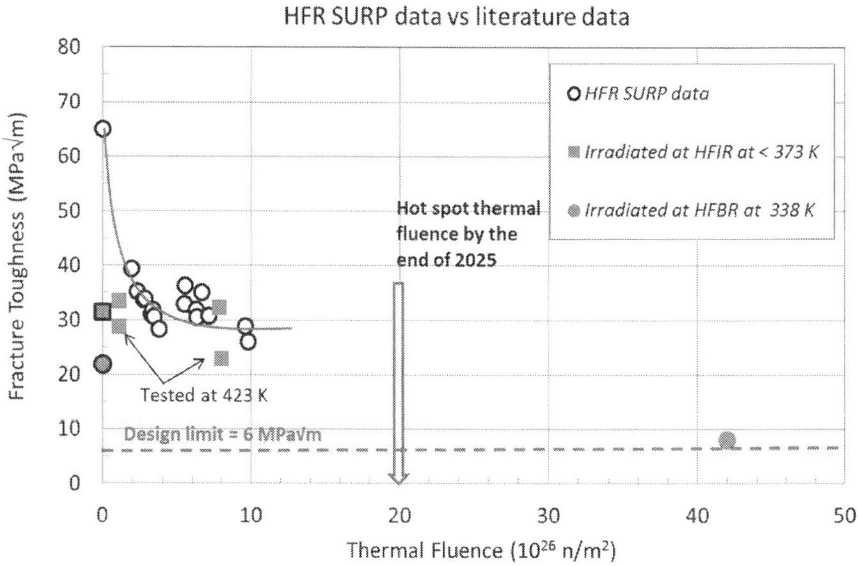

Figure 4. HFR SURP fracture toughness data in comparison with literature data [5,10].

4.3.3. *Fracture toughness behavior of 5xxx and 6xxx series alloys*

In the rapid hardening regime (regime 1), the fracture toughness value drops rapidly in line with the observed hardening and ductility behavior of 5*xxx* alloys (**Figure 3** vs. **Figure 1 (c)**).

This is because the hardening-induced embrittlement causes the decrease in both ductility and fracture toughness properties. In this regime, the magnitude of the drop in 5*xxx* alloys is high compared to 6*xxx* alloys, because both dislocation damage and precipitation damage mechanisms are active in 5*xxx* alloys, whereas only dislocation damage dominates in 6*xxx* alloys. This explains the sharp drop observed in fracture toughness of 5154-O alloy at the onset of irradiation compared to a shallow decrease in fracture toughness properties of 6061-T6 alloy in this regime (see **Figure 4**).

As the irradiation continues, both the dislocation density and the Mg_2Si precipitate density evolve toward a saturation limit describing the slow decrease of fracture toughness toward a plateau in the transition regime. Transmission electron microscopy results of precipitate microstructure reported in [13] are shown in **Figures 5** and **6**. From these results it can be seen that the saturation density is achieved at ~3 × 10^{26} n/m^2 of thermal fluence for 5154-O alloy of HFR vessel. Note that these pictures were taken using a "JEOL JEM-1200ex STEM/TEM" machine operating at 120 keV, located in JGL laboratory at NRG.

After that, the fracture toughness of 5154-O reaches a plateau at a thermal fluence of ~ 4 × 10^{26} n/m^2 after which no further increase in dislocation and precipitate density is expected (**Figure 3**). In fact, a small decrease in particle density may occur later in this regime due to particle coalescence during their growth. The opposite effects of a small decrease in particle density and a slow hardening due to precipitate growth on embrittlement could be compensating each other leading to a plateau in the fracture toughness behavior (similar to ductility) in regime 3. The behavior of 5154-O alloy is expected to be similar to 6*xxx* series alloys in this regime at similar irradiation conditions. This is because within this regime the irradiation hardening in both alloy types occurs primarily due to the growth of existing precipitates by newly produced transmutation Si. A close agreement between the fracture toughness data of 6061-T6 alloy irradiated in HFIR at <373K and HFR SURP data in the plateau regime (see **Figure 4**) confirms this theory.

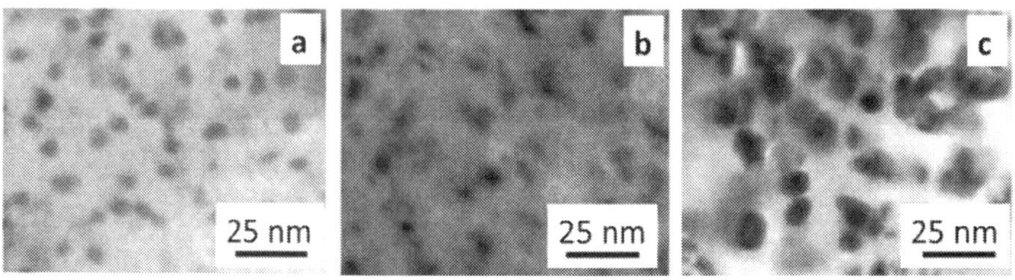

Figure 5. TEM images showing the evolution of precipitate size with thermal fluence in 5154-O Al alloy (HFR SURP specimens). (a) 2.73x1026 n/m2, 0.68 wt.% Si, (b) 3.76x1026 n/m2, 0.88 wt.% Si and (c) 9.81x1026 n/m2, 2.21 wt.% Si. (Photograph courtesy N.V. Luzginova et. al. [13])

http://dx.doi.org/10.5772/63294

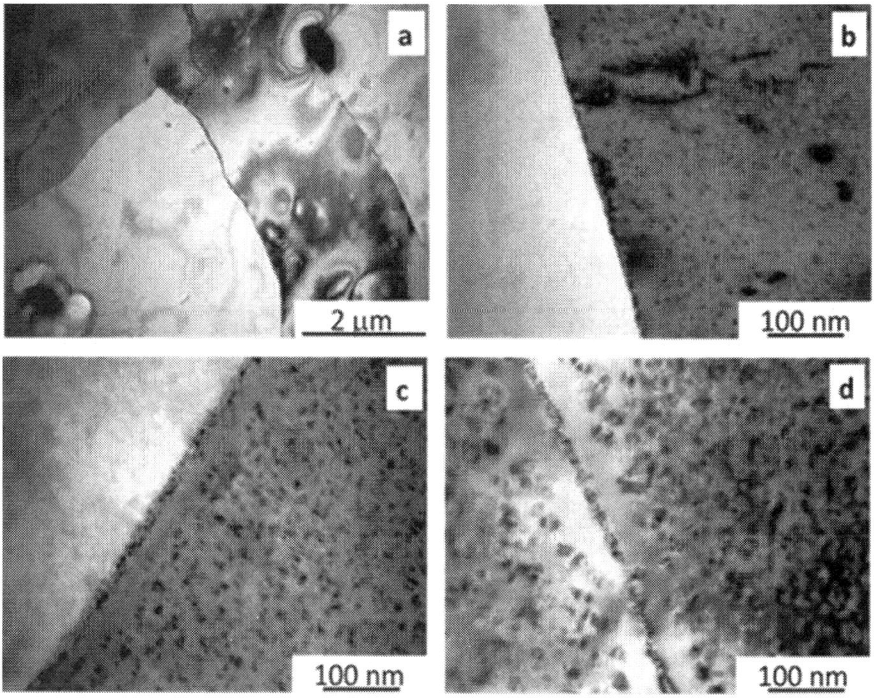

Figure 6. TEM images showing the evolution of precipitate microstructure with thermal fluence in 5154-O Al alloy (HFR SURP specimens). (a) Unirradiated, 0.04 wt.% Si, (b) 2.73 × 10^{26} n/m^2, 0.68 wt.% Si, (c) 3.76 × 10^{26} n/m^2, 0.88 wt.% Si, and (d) 9.81 × 10^{26} n/m^2, 2.21 wt.% Si. (Photograph courtesy N.V. Luzginova et al. [13].)

It is important to understand how long the plateau in the fracture toughness (or regime 3) will continue. This depends on the location of the precipitation of the transmutation Si. As already mentioned in Section 3.2.2, further increase in Si production to high values can lead to Si precipitation at grain boundaries. Fracture toughness value drops when the precipitation of Si at the grain boundaries cumulates to an extent that the dominant deformation and fracture mechanisms shift from the bulk microstructure to the grain boundaries. A heavy discontinuous precipitation observed at the grain boundary in 5052-O alloy at a thermal fluence of 31 × 10^{26} n/m^2 [3] has resulted in no substantial effects on ductility of this alloy. This suggests that the nature of fracture at these high fluence is still controlled by bulk deformation mechanisms (instead of mechanisms controlled by grain boundaries). Due to the similarity in 5154-O and 5052-O alloys (and irradiation conditions), the ductility and fracture toughness properties of the 5154-O alloy are also expected to show a plateau until such high fluences. Indeed, the observation of significant amount of micron-scale dimples on the fracture surface of 5154-O alloy irradiated to a thermal fluence of 9.81× 10^{26} n/m^2 (**Figure 7**) proves that similar behavior can be expected from the 5154-O alloy [13].

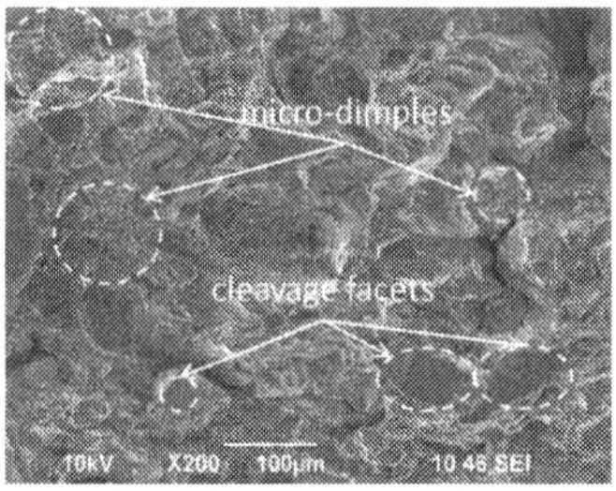

Figure 7. Fracture details of 5154-O alloy with a crack-tip thermal fluence of 9.81 × 10^{26} n/m^2. Figure shows fracture surface characterized by dominant microdimples and some cleavage facets [13].

CRDF A-2 tubes of HFBR produced from 6061-T6 alloy have shown a fracture toughness value of ~8 (MPa)·m$^{1/2}$ after irradiation to a much higher thermal neutron fluence of ~42 × 10^{26} n/m^2 at 338 K (see **Figure 4**). The decrease in fracture toughness from an unirradiated value of 21.75 (MPa)·m$^{1/2}$ for this alloy is primarily attributed to the following: (i) formation of very fine (~8 nm) Si-rich precipitates in the grains due to high TFR of 21 (as explained in Section 4.4) and (ii) large silicon flakes occupying about one-fifth of the grain boundary area at this high transmutation-produced Si content of 8 wt.% [5]. Fracture surface of this alloy revealed substantial intergranular separation with some residual ductility indicating that the contribution of grain boundary fracture mechanisms is increased at such high fluence values to enter into the brittle regime (regime 4).

From the above discussion, no differences in the evolution of irradiation damage at high fluences (in regime 3 and 4) are expected between 5*xxx* and 6*xxx* alloys due to differences in their initial chemical composition and microstructure. Moreover, the difference in the TFR of HFR SURP data and literature data is conservative as explained in the next section. This allows the use of published fracture toughness literature data of 6061-T6 Al alloys to predict the fracture toughness behavior of 5154-O Al alloy of HFR vessel at high fluences.

4.4. Effect of thermal-to-fast flux ratio (TFR)

It is known from the literature that a high difference in TFR can have substantial effect on irradiation hardening and embrittlement behavior of the same material [5]. It was highlighted in [3] that both the thermal and fast neutrons play independent and important roles leading to microstructural damage and corresponding property changes. A very high TFR, ranging from 80 to 500, could explain the observed craze-cracking in AG3-NET alloy (Al–3% Mg) beam tubes in the Reactor Haut Flux (RHF) at Grenoble [23]. Lijbrink et al. [12] pointed out that fast neutron flux reduces the effectiveness of the Si precipitation hardening process. A possible explanation for this behavior (as given in [5, 12]) is as follows. Fast flux has two opposite effects on precipitation:

i. The kinetic energy supplied by fast flux temporarily increases the solubility limit of Si in the matrix and opposes the condensation requirements for the precipitation.

http://dx.doi.org/10.5772/63294

ii. Local energy needed for jumping the nucleation barrier can be readily supplied by the fast flux.

However, the fast flux can be destructive when a freshly formed nucleus is hit by fast neutron collision. That means, at equal thermal fluence values, Si precipitation hardening is more effective at higher TFR. This leads to finer precipitate distribution, causing higher irradiation hardening, lower ductility, and eventually lower fracture toughness values at higher TFR. Indeed, the higher hardening rate observed in 6061-T6 alloy from CRDF A-2 tubes of HFBR irradiated at TFR of 21 compared to the similar 6061-T6 alloy irradiated in HFIR at TFR of 1.7 explains this behavior (**Figure 1 (b)**). Consequently, the high fluence data point from CRDF A-2 of HFBR at TFR = 21 (>>0.8–1.4 for HFR hotspot), shown in **Figure 4**, is likely to give a conservative estimation of the fracture toughness value under HFR conditions.

5. Summary and conclusions

A literature review on highly irradiated 5*xxx* and 6*xxx* series Al alloys is conducted to understand the expected changes in mechanical properties of HFR vessel material in relation with microstructural aspects beyond the current surveillance data to support the HFR SURP program. It was found that the irradiation swelling in 5*xxx* series alloys is not a crucial degradation mechanism for the expected hotspot fluence values of HFR vessel by the end of 2025. Dislocation damage is expected to reach a saturation limit at relatively low fast-fluence values. The damage caused by precipitation of transmutation Si is found to be the dominant mechanism affecting the fracture toughness properties of irradiated 5*xxx* and 6*xxx* series Al alloys at high thermal fluence values. Tensile and fracture toughness data collected from the literature up to very high thermal fluences is analyzed in comparison with the available HFR surveillance data. The observed changes in mechanical properties are classified into four different regimes.

The contribution of various irradiation damage mechanisms to the evolution of microstructure and mechanical properties is discussed in all four regimes for 5*xxx* and 6*xxx* series alloys. A rapid hardening regime characterized by a sharp drop in ductility and fracture toughness is observed at the onset of irradiation. In this regime, a higher degree of embrittlement is observed in 5*xxx* series due to the formation of Mg_2Si precipitates in addition to the contribution from dislocation damage. On the other hand, the hardening and embrittlement observed in regime 1 of 6*xxx* series alloys is primarily due to dislocation damage. The contribution from the precipitation of transmutation Si is estimated to be minor for 6*xxx* alloys in this regime. The contribution of both mechanisms continues in the transition regime (regime 2) for both alloy types, until both dislocation damage and precipitate density evolve toward a saturation value. Due to this, a lowering in irradiation hardening rate and correspondingly a slow decrease in ductility and fracture toughness toward a stable value are observed in this regime. Regime 3 is characterized by a plateau in ductility and fracture toughness values, due to no further increase in dislocation and precipitate density. A slow hardening observed in this regime is primarily due to the growth of existing precipitates. The behavior of 5*xxx* alloys is found to be similar to 6*xxx* series alloys when irradiated under similar conditions in this regime. A final regime (regime 4) with an increasing hardening rate and a decreasing ductility indicates that the contribution of grain boundary fracture mechanisms increases at such high fluence values to enter into the brittle regime (regime 4).

For the 5154-O alloy at the hotspot irradiation conditions, regime 1 ends at ~2×10^{26} n/m^2. Regime 2 is observed between ~2×10^{26} and ~4×10^{26} n/m^2. Finally, the plateau in regime 3 starts at ~4×10^{26} n/m^2 and is expected to continue up to very high thermal fluences, that is, greater than the estimated hotspot thermal fluence by the end of 2025 (~20×10^{26} n/m^2). This is because for a 5052-O alloy, which was irradiated at similar conditions as HFR hotspot and resembles the alloy microstructure and composition of 5154-O, a plateau in ductility was observed from a thermal fluence of ~4×10^{26} n/m^2 until ~31×10^{26} n/m^2. It should be noted that the estimated HFR hotspot thermal fluence by the end of 2025 (~20×10^{26} n/m^2) is only two-thirds of the studied 5052-O alloy.

Additionally, high fluence fracture toughness data is found from the CRDF A-2 tubes of the HFBR in Brookhaven National Laboratory, USA, produced from 6061-T6 alloy, irradiated at 338 K, up to 42×10^{26} n/m^2. The corresponding fast fluence of this data point is 2×10^{26} n/m^2, which gives a high TFR of 21 compared to the HFR hotspot TFR value of maximum 1.4. The reported thermal fluence and Si content of this data point are approximately two times the estimated thermal fluence (~20×10^{26} n/m^2) and Si (~4.3%) content of the HFR hotspot by the end of 2025. Knowing that the transmutation-produced Si induces major damage to the microstructure of irradiated Al alloys, this high fluence data point from CRDF A-2 of HFBR is likely to give a conservative estimation of the fracture toughness value under HFR conditions due to irradiation of this alloy at much higher TFR (leading to high embrittlement) and negligible differences in the embrittlement behavior of 5*xxx* and 6*xxx* series alloys in the plateau regime.

From the above observations of literature tensile and fracture toughness data on irradiated Al alloys, one can conclude that the probability of the fracture toughness of HFR hotspot to fall below the design limit is negligible up until the currently estimated hotspot thermal fluence at the end of 2025.[c]

Acknowledgements

The work presented in this article is performed as a part of SURveillance Program (SURP) of HFR vessel with the financial support of NRG. The author thanks Dr. O. Wouters and Ir. T.O. van Staveren for useful discussion and critical review of this work. The author also acknowledges Dr. C. Li for helping in the extraction of data from the literature.

Export control note

[c] Assuming that the irradiation conditions at the HFR hotspot are kept unchanged as they are in 2015.

http://dx.doi.org/10.5772/63294

The content within this chapter is classified with the code EU DuC = N. EU DuC means European dual use code.

Author details

Murthy Kolluri

Address all correspondence to: kolluri@nrg.eu

Nuclear Research and Consultancy Group (NRG), The Netherlands

References

[1] ASTM-B209. Standard Specification for Aluminum and Aluminum-Alloy Sheet and Plate: ASTM International; West Conshohocken, PA, USA. 2007.

[2] Farrell K, King RT. Tensile Properties of Neutron-Irradiated 6061 Aluminum Alloy in Annealed and Precipitation-Hardened Conditions. ASTM STP 683; West Conshohocken, PA, USA. 1979. pp. 440–449.

[3] Farrell K. Microstructure and tensile properties of heavily irradiated 5052–0 aluminum alloy. J. Nucl. Mater. 1981; 97(1–2): 33–43.

[4] Weeks JR, Czajkowski CJ, Tichler, PR. Effects of High Thermal and High Fast Fluences on the Mechanical Properties of Type 6061 Aluminum in the HFBR. ASTM STP 1046; West Conshohocken, PA, USA. 1990. pp. 441–452.

[5] Weeks JR, Czajkowski CJ, Farrell K. Effects of High Thermal Neutron Fluences on the 6061 Aluminum. ASTM STP 1175; West Conshohocken, PA, USA. 1993.

[6] Munitz A. Mechanical Properties and Microstructure of Neutron Irradiated Cold-Worked Al-1050 and Al-6063 Alloys. Annual Report. IAEC; 1998. Online: http://www.iaea.org/inis/collection/NCLCollectionStore/_Public/31/027/31027808.pdf

[7] Munitz A, Shtechman A, Cotler C, Talianker M, Dahan S. Mechanical properties and microstructure of neutron irradiated cold worked A1-6063 alloy. J. Nucl. Mater. 1998; 252: 79–88.

[8] Yahr GT. Prevention of nonductile fracture in 6061-T6 aluminum nuclear pressure vessel. J. Press. Vessel Technol. 1997; 119(2): 150–156.

[9] Farrell K. Performance of aluminum in research reactors. In: Comprehensive Nuclear Materials. Editor: Konings RJM, Elsevier; Oxford. 2012. pp. 143–157.

[10] Farrell K. Assessment of Aluminum Structural Materials for Service within the ANS Reflector Vessel. ORNL Report. Report No. ORNL/TM-13049; DOE TN (United States) 1995.

[11] Kapusta B, Sainte-Catherine C, Averty X, Campioni G, Ballagny A. Mechanical characteristics pf 5754-NET-O aluminum alloy irradiated up to high fluences: Neutron spectrum and temperature effects. In: Joint Meeting of the National Organization of Test, Research, and Training Reactors and the International Group on Research Reactors. September 12–16, 2005; Gaitherburg.

[12] Lijbrink B, Grol HJV, Dekker F, Witzenburg WV. Effects of neutron irradiation on the mechanical properties of A 5154-O type aluminum alloy. ASTM STP 782; West Conshohocken, PA, USA. 1982. pp. 765–778.

[13] Luzginova NV, Nolles H, van den Berg F, van den Idsert P, van der Schaaf B. Surveillance program results for the high flux reactor vessel material. Effects Rad. Nucl. Mater. 2014; 26: 30–41.

[14] Murayama M, Hono K. Pre-precipitate clusters and precipitation processes in Al-Mg-Si alloys. Acta Mater. 1999; 47(5): 1537–1548.

[15] Was GS. Fundamentals of Radiation Materials Science. Berlin, Heidelberg, New York: Springer; 2007.

[16] Packan NH. Fluence and flux dependence of void formation in pure aluminum. J. Nucl. Mater. 1971; 40(1): 1–16.

[17] Farrell K, Bentley J, Braski DN. Direct observation of radiation-induced coated cavities. Scripta Metal. 1977; 11(3): 243–248.

[18] Dieter GE. Mechanical Metallurgy. In: Bacon D, editor. McGraw-Hill; 1988. Mechanical Metallurgy, McGraw-Hill, Cornell University, USA.

[19] Hart EW. Theory of dispersion hardening in metals. Acta Metal. 1972; 20(2). pp 275–289

[20] Reed-Hill R. Physical Metallurgy Principles. Van Nostrand; the University of Michigan, USA 1973.

[21] Farrell K, Richt AE. Microstructure and tensile properties of heavily irradiated 1100-O aluminum. In: Symposium on Effects of Radiation Structural Materials. STP 683; 1979. pp. 427–439.

[22] Ketema DJ, van der Schaaf B. Principles of HFR Vessel Surveillance Program—revE. Report No. NRG-25146/10,103852; Petten, The Netherlands, Publisher: NRG (internal report) 2011.

[23] Farrell K. Materials Selection for the HFIR Cold Neutron Sources. Report No. ORNL/TM-99-208; TN (United States), DOE 2001.

Chapter 16

A Parallel between Laser Irradiation and Relativistic Electrons Irradiation of Solids

Mihai Oane, Rareş Victor Medianu and Anca Bucă

Additional information is available at the end of the chapter

http://dx.doi.org/10.5772/62353

Abstract

The investigation of the thermal field distribution in a material sample irradiated by a laser beam or an electron beam with the energy of a few MeV appears as a demand for all kinds of experiments that involve irradiation. When investigating the effects of accelerated electrons on a target, it is necessary to figure out the temperature rise in the target. Also during irradiation with laser beams, it is important to know the thermal behavior of the target. A parallel between laser and electron beam irradiation is also made. The results are very interesting. Also, a very interesting case of cluster nano-particles (20–100 nm; inserted in a Cu surface) heated with a laser beam is tacking in to account.

The present chapter is a review article type which comprises the expertise gathers around this domain during the past 15 years within the institute: NILPRP-Romania.

Keywords: laser, electron (beam), irradiation, nano-particle, interaction, W and C

1. Introduction

There are many methods for evaluating the thermal fields in radiation-matter interaction, but most of them require a complex mathematical handling [1–4]. This chapter presents a direct and powerful mathematical approach to compute the thermal field for electron beam-material and laser-sample interaction. The solving procedure is based on applying the integral transform technique which was developed in the 1960s, by the Russian School of Theoretical Physics [5]. As an example, the integral transform technique is used in [3] to solve the heat equation for a sample exposed to an infrared laser beam in order to find the solution for the absorption coefficient, which is then checked experimentally. It should be pointed out straightforwardly that the heat equation has the same form in the case of irradiation with a laser beam or an electron beam, at sufficiently large beam intensities [6, 7]. There is, however, a disadvantage in this model as it cannot take into account simultaneously the variation with temperature of several thermal parameters involved in the interaction like, for example, the thermal conductivity or thermal diffusivity. In consequence, the model should be regarded as a first approximation of the thermal field. The main advantage is that the solution is a series which converges rapidly. It is important to note that the integral transform technique, as it will be shown in the next sections, belongs to the "family" of Eigen functions and Eigen values-based methods.

2. The applicability of the Fourier heat equation for study of laser-nano particles clusters interaction

Light has always played a central role in the study of physics, chemistry and biology. In the past century, a new form of light, laser light, has provided important contributions to medicine, industrial material processing, data storage, printing and defense [8] applications. In all these areas of applications, the laser-solid interaction played a crucial role. The theory of heat conduction was naturally applied to explain this interaction since it was well studied for a long time [9]. For describing this interaction, the classical heat equation was used in a lot of applications. Apart of some criticism [10], the heat equation still remains one of the most powerful tools in describing most thermal effects in laser-solid interactions [11]. In particular, the heat equation can be used for describing both of light interaction with homogeneous and inhomogeneous solids. In the literature, thus a special attention was given to cases of light interaction with multi-layered samples and thin films.

It is undertaken in the following treatment that it has a solid consisting of a layer of a metal such as Au, Ag, Al or Cu, respectively. Assuming that only a photo-thermal interaction takes place, and that all the absorbed energy is transformed into heat, the linear heat flow in the solid is fully described by the heat partial differential equation, Eq. (1):

$$\frac{\partial^2 T}{\partial x^2} + \frac{\partial^2 T}{\partial y^2} + \frac{\partial^2 T}{\partial z^2} - \frac{1}{\gamma}\frac{\partial T}{\partial t} = -\frac{A(x,y,z,t)}{k} \tag{1}$$

where: $T(x, y, z, t)$ is the spatial-temporal temperature function, γ is the thermal diffusivity, k is the thermal conductivity and A is the volume heat source (per unit time). In general, one can consider the linear heat transfer approximation and using the integral transform method assume the following form for the solution of the above heat equation, Eq. (2):

$$\begin{aligned} T(x,y,z,t) = \sum_{i=1}^{\infty}\sum_{j=1}^{\infty}\sum_{k=1}^{\infty} f(\mu_i,v_j,\lambda_k)\cdot g(\mu_i,v_j,\lambda_k,t) \\ \times K_x(\mu_i,x)\cdot K_y(v_j,y)\cdot K_z(\lambda_k,z) \end{aligned} \tag{2}$$

where: $f(\mu_i, v_j, \lambda_k) = \frac{1}{k \cdot C_i \cdot C_j \cdot C_k} \int_0^a \int_{-b}^b \int_{-c}^c A(x, y, z, t) K_x(\mu_i, x) \cdot K_y(v_j, y) \cdot K_z(\lambda_k, z) dxdydz$ and

$$g(\mu_i, v_j, \lambda_k, t) = 1/(\mu_i^2 + v_j^2 + \lambda_k^2)[1 - e^{-\beta_{ijk}^2 t} - (1 - e^{-\beta_{ijk}^2 (t - t_0)}) \cdot h(t - t_0)] \quad (3)$$

with $\beta_{ijk}^2 = \gamma(\mu_i^2 + v_j^2 + \lambda_k^2)$.

Here, t_0 is the light pulse length (assumed rectangular) and h is the step function. The functions $K_x(\mu_i, x)$, $K_y(v_j, y)$ and $K_z(\lambda_k, z)$ are the Eigen functions of the integral operators of the heat equation and μ_i, v_j, λ_k are the Eigen values corresponding to the same operators. Here, for example: $K_x(\mu_i, x) = \cos(\mu_i \cdot x) + (h_{lin}/k \cdot \mu_i) \cdot \sin(\mu_i \cdot x)$, with h_{lin}—the linear heat transfer coefficient of the solid sample along x direction.

The coefficients C_i, C_j and C_k are the normalizing coefficients where, for example: $C_i = \int_{-b}^{b} K_x^2(\mu_i, x) dx$. ($a$, $2b$ and $2c$ are the geometrical target dimensions, which are supposed to be a parallelepiped one).

It has used the thermal parameters of the Cu sample as given in Table 1.

	K [W/cmK]	γ [cm²/s]	α [cm⁻¹]
Cu	3.95	1.14	$7.7 \cdot 10^5$

Table 1. Thermal parameters of Cu.

It has used the heat equation for a configuration where the layers are assumed to have a thickness of 1 mm onto which are included clusters of nano-spheres. The heat term for such a system can be represented by the following equation:

$$A(x, y, z, t) = \sum_{m,n,p} I(x, y, z)((\alpha_1 + r_S \delta(z) + \alpha_{mnp}(\delta(x_m) \cdot \delta(y_n) \cdot \delta(z_p))) \cdot (h(t) - h(t - t_0)) \quad (4)$$

where, m,n,p denote the positions of the nano-particles-clusters, α_1—the optical absorption coefficient, I the incident plane wave radiation intensity incoming from the top –z direction, r_S—the surface absorption coefficient, α_{mnp}—the nano-particles optical absorption coefficients, x,y,t represent the space and time coordinates on the layer surface and h is the step time function.

For the simulation, we have to consider:

$$\alpha_{mnp} >> \alpha_1 + r_S d(z) \tag{5}$$

Inserting groups or clusters of nano-particles-clusters on top of a layer exposed to irradiation gives a detectable increase of temperature in comparison with the bulk material in pure form. This result can be seen in the following simulations.

For m, $n = 1, 2$ and $p = 1$, we have plotted in **Figures 1–3**, the thermal field of 1, 2 and 4 nano-particles-clusters for the case of a Cu layer.

The present chapter continues the numerous ideas developed in the past few years with the integral transform technique applied to classical Fourier heat equation [1, 2].

From practical point of view, consider the formula (2), that: i varies from 1 to 100; j varies from 1 to 100, and k varies from 1 to 100. In consequence, the solutions will be like a sum of 1 million functions. In this way, these solutions become from semi-analytical into analytical one. The solutions are easy to compute in MATHEMATICA, or other package software.

In conclusion, it is considered that the method of integral transform technique is a serious candidate in competition with: Born approximation, Green function method or numerical methods. In **Figure 44** is represented the "geometrical" situation for **Figure 3**.

The nano-particles-clusters should be of the order of magnitude of 20 nm, which is the limit of availability of Fourier model [12].

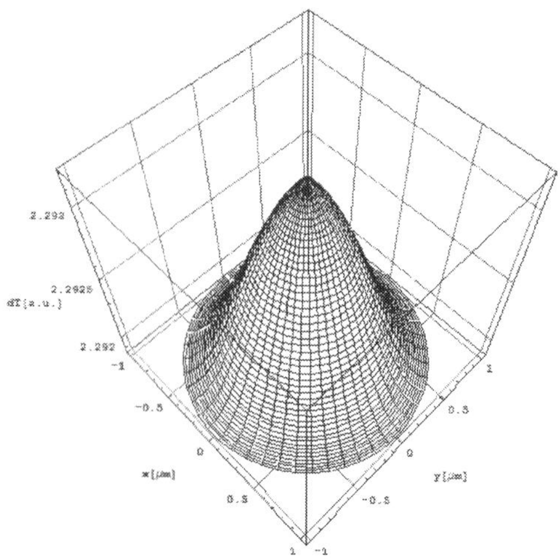

Figure 1. The thermal field produced by one nano-particle-cluster on a Cu substrate. The nano-particle is situated at $x = 0$ and $y = 0$, and 200 nm depth inside Cu sample.

http://dx.doi.org/10.5772/62353

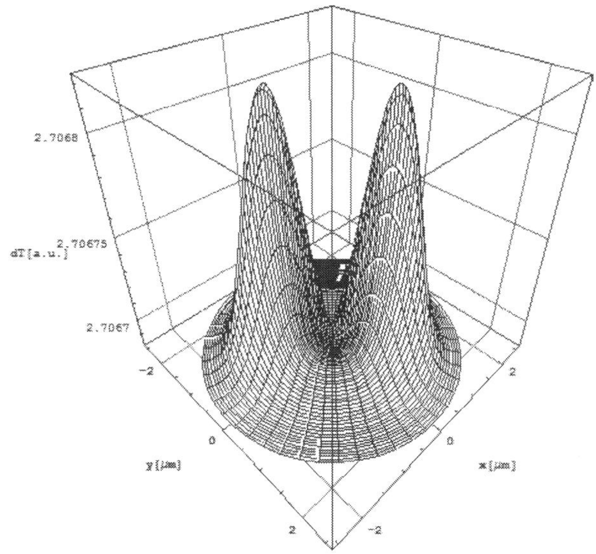

Figure 2. The thermal field produced by two nano-particles-clusters on a Cu substrate. The two nano-particles-clusters have the coordinates symmetric in rapport with the heat source. The two nano-particle-clusters are also 200 nm inside Cu sample.

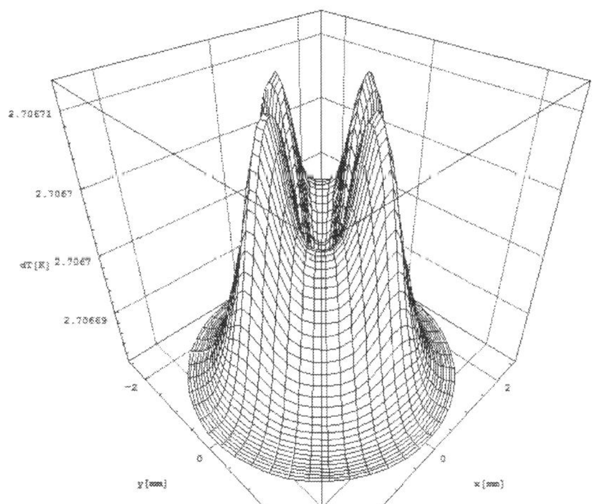

Figure 3. The thermal field produced by four nano-particles-clusters on a Cu substrate. The depth is also 200 nm.

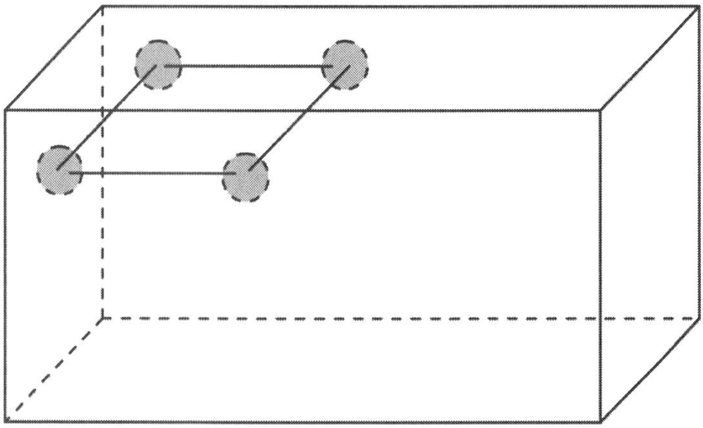

Figure 4. The "geometrical" situation for **Figure 3**.

3. The applicability of the Fourier heat equation for study of relativistic electron-solid interaction

Taking into account the experimental data that were measured at the ALIN-10 linear accelerator from NILPRP [13] can be approximated a power distribution in the electron beam cross section as follows:

$$\mathrm{P}(x, y) \cong P_0 \cdot e^{-(x^2+y^2/169)}. \tag{6}$$

Therefore, it is supposing that the irradiation source emits relativistic electrons with an asymmetric Gaussian distribution [14, 15]. This intensity distribution of the accelerated electron beam is represented in **Figure 5** and is obtained using the experimental data shown in **Figure 6**. The approximation of the ratio between the two planar coordinates of the electron beam spot is obtained from **Figure 6**. This shows the experimental transverse profile of the beam at the exit of the ALIN-10 accelerator. The average beam power is 62 W for a beam current of 10 μA. The measured beam dimensions in the transverse plane are 14 and 2 mm on the x and y coordinates.

The normalization condition is:

$$\int_{-2}^{2}\int_{-8}^{8} e^{-(x^2+y^2/169)} dxdy = 13\pi \cdot \boldsymbol{Erf}[8/13] \cdot \boldsymbol{Erf}[2] \cong 25.3 \tag{7}$$

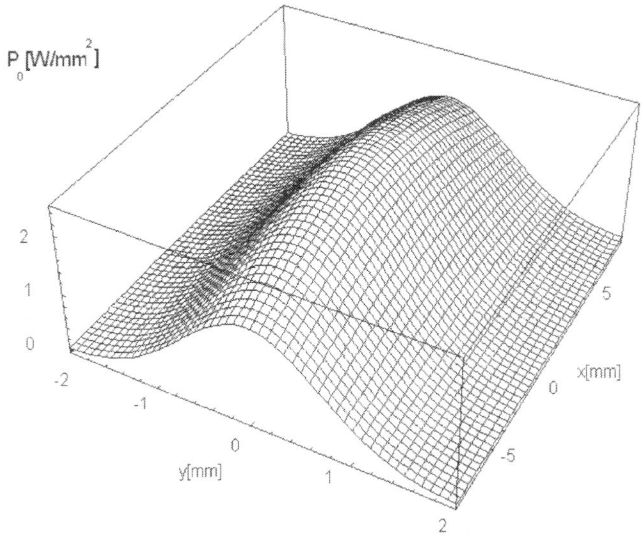

Figure 5. The simulated intensity in the cross section of the beam delivered by ALIN-10.

Taking into account that the average power in time of the electron beam is 62 W, we obtain $P_0 = 2.45\ W$.

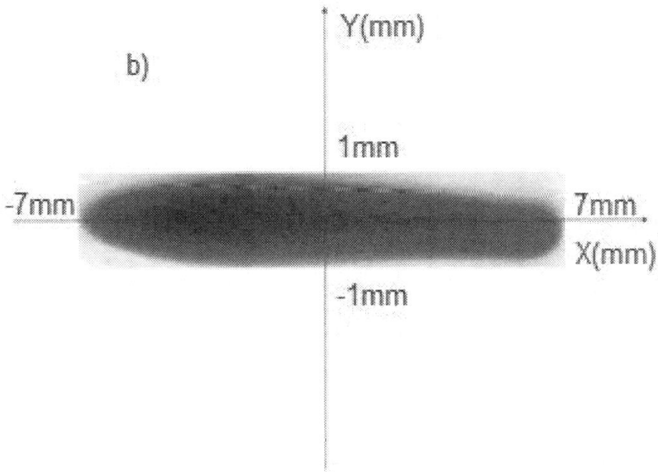

Figure 6. The experimental spot on a plastic sample due to the electron beam after a 10 s exposure.

The geometry of the simulation is shown in **Figure 7** where the electron beam propagates along the z axis and is incident on a graphite sample with dimensions 10 × 10 × 15 mm. The geometry is described in Cartesian coordinates and the transversal plane of the beam is the xy plane.

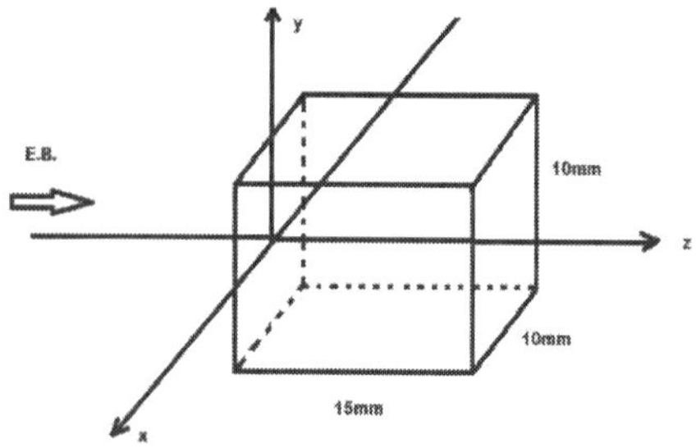

Figure 7. Irradiation geometry for a small graphite sample.

The instantaneous energy loss of electrons passing through the material sample ($\propto \partial E/\partial z = f(E, M_i)$) depends on material constants such as the mass density of the target, the atomic number, the classical radius of the electron and some empirical numerical constants, and does not depend explicitly on the distance travelled, in our case the z direction. When calculating the stopping power, three physical phenomena are taken into account: the secondary electron emission, the polarization of the target and the effect of magnetic field on the incident beam [16, 17].

For electrons with energies greater than 2.5 MeV, their range in the target material is given by the formula put forward by Katz and Penfolds [18]:

$$\boldsymbol{d}_{\max}[\boldsymbol{cm}] = (0.530 \cdot \boldsymbol{E}_{\max}[\boldsymbol{MeV}] - 0.106\,[\boldsymbol{cm}]) / \rho[\boldsymbol{g} / \boldsymbol{cm}^3]. \tag{8}$$

Here, $d_{\max}$ represents the maximum range of electron beam given in cm in a target of density ρ expressed in g/cm^3. The energy E_{max} is introduced in expression of Eq. (8) in MeV and it refers to the maximum energy of the beam which determines the maximum range that the electrons of the beam can go through in a target. Based on this equation, a linear dependency of the energy absorbed in the material with the distance z is considered.

In our study it has used a graphite sample with $\rho = 2.23$ g/cm^3 giving $d_{max} = 1.43$ cm, below the length of our sample. This leads to the following absorption law:

$$\boldsymbol{E}_{abs}(z) = \begin{cases} 6.23 - 4.36 \cdot z,\ \boldsymbol{for}\ z \le \boldsymbol{d}_{\max} \\ 0,\ \boldsymbol{for}\ z > \boldsymbol{d}_{\max} \end{cases}, \tag{9}$$

where: E_{abs} is in MeV and z is expressed in centimeters. Expression of Eq. (9) is the source term for the heat equation as it will be shown in the next section.

4. The Fourier heat equation

The goal to establish the thermal field during electron beam irradiation is not a new issue. For achieving it in the irradiation geometry described in **Figure 7**, the heat equation in Cartesian coordinates is the starting point:

$$\frac{\partial^2 T}{\partial X^2}+\frac{\partial^2 T}{\partial Y^2}+\frac{\partial^2 T}{\partial Z^2}-\frac{1}{\gamma}\frac{\partial T}{\partial t}=-\frac{A(X,Y,Z,T)}{k} \tag{10}$$

Here T represents the temperature variation relative to the initial sample temperature T_0, which occurs during exposure to the electron beam, A is the energy deposited by electrons in the unit volume and unit time, k is the thermal conductivity and γ is the thermal diffusivity of the sample. We have $A(x, y, z, t) = E_{abs}(z)/(V_{sample}\times t_0)$, where $V_{sample}=a\cdot b\cdot c$ and t_0 is the irradiation time. The boundary conditions are:

$$\left[\frac{\partial K_x}{\partial X}+\frac{h}{k}\cdot K_x\right]_{x=-\frac{a}{2}}=0,\quad \left[\frac{\partial K_x}{\partial X}+\frac{h}{k}\cdot K_x\right]_{x=\frac{a}{2}}=0, \tag{11a}$$

$$\left[\frac{\partial K_y}{\partial Y}+\frac{h}{k}\cdot K_y\right]_{y=\frac{b}{2}}=0,\quad \left[\frac{\partial K_y}{\partial Y}+\frac{h}{k}\cdot K_y\right]_{y=-\frac{b}{2}}=0, \tag{11b}$$

$$\left[\frac{\partial K_z}{\partial Z}-\frac{h}{k}\cdot K_z\right]_{z=0}=0,\quad \left[\frac{\partial K_z}{\partial Z}+\frac{h}{k}\cdot K_z\right]_{z=c}=0, \tag{11c}$$

where a, b and c are the geometrical lengths of the sample, along X, Y and Z, respectively, h is the heat transfer coefficient, K_x, K_y, K_z are the Eigen functions and α_i, β_j,χ_o are their corresponding Eigen values, respectively.

The solution for the heat equation is:

$$\Delta T(x,y,z,t)=\sum_{i=1}^{\infty}\sum_{j=1}^{\infty}\sum_{o=1}^{\infty}I_1(\alpha_i,\beta_j,\chi_o)I_2(\alpha_i,\beta_j,\chi_o,t)K_x(\alpha_i,x)K_y(\beta_j,y)K_z(\chi_o,z), \tag{12}$$

where

$$I_1\left(\alpha_i,\beta_j,\chi_o\right)=\frac{1}{C_iC_jC_o}\int_{-\frac{a}{2}}^{\frac{a}{2}}\int_{-\frac{b}{2}}^{\frac{b}{2}}K\left(\alpha_i,x\right)K\left(\beta_j,y\right)\cdot P\left(x,y\right)\cdot dxdy\int_0^c K_o\left(\chi_o,z\right)\left(6.23-4.36\cdot z\right)dz,$$
$$I_2(\alpha_i,\beta_j,\chi_o,t)=\frac{1}{\alpha_i^2+\beta_j^2+\chi_o^2}[1-e^{-\gamma_{ijo}^2 t}-(1-e^{-\gamma_{ijo}^2(t-t_o)})h(t-t_o)], \tag{13}$$

with

$$\gamma_{ijo}^2=\gamma(\alpha_i^2+\beta_j^2+\chi_o^2) \tag{14}$$

Here, C_i, C_j, C_o are normalization constants. The Eigen functions determined for the heat equation (10) with boundary conditions (11a–c) have the following explicit expressions:

$$K_x(\alpha_i,x)=\cos(\alpha_i\cdot x)+(h/k\alpha_i)\cdot\sin(\alpha_i\cdot x) \tag{15a}$$

$$K_y(\beta_j,y)=\cos\left(\beta_j\cdot y\right)+(h/k\beta_j)\sin\left(\beta_j\cdot y\right) \tag{15b}$$

$$K_z(\chi_0,z)=\cos\left(\chi_0\cdot z\right)+(h/k\chi_0)\sin\left(\chi_0\cdot z\right) \tag{15c}$$

The Eigen values can be determined from the boundary equations:

$$2\cot\left(\alpha_i a\right)=\frac{\alpha_i k}{h}-\frac{h}{k\,\alpha_i} \tag{16a}$$

$$2\cot\left(\beta_j b\right)=\frac{\beta_j k}{h}-\frac{h}{k\beta_j}, \tag{16b}$$

$$2\cot\left(\chi_0 c\right)=\frac{\chi_0 k}{h}-\frac{h}{k\,\chi_0}. \tag{16c}$$

5. Experiment and simulations

For small samples the thermal field distribution is determined by two important factors: the energy denoted by the term *A* in Eq. (10) released by the electrons during the time and volume

http://dx.doi.org/10.5772/62353

unit within the target, and the heat transfer constant h, which shows how fast the target loses its heat to the surrounding environment depending on the material of the target, pressure of the surrounding gas and magnitude of the contact surface between the target and the environment. The temperature increases with the absorbed energy A, and with the decrease of h. There are in general three types of heat transfer by: (i) radiation, (ii) convection and (iii) conduction. In the present case, the heat lost by conduction is neglected as the sample is fixed on two Teflon claws. The heat rate lost by radiation may be written as $\sigma \cdot E \cdot (T^4 - T_0^4)$, which in linear approximation is given by $4\sigma \cdot T_0^3 \cdot E \cdot (T - T_0) \equiv h_{rad} \cdot (T - T_0)$. Here, $h_{rad} = 4 \cdot \sigma \cdot T_0^3 \cdot E$, where $T_0 = 298$ K, $\sigma = 5.6 \times 10^{-8}$ $Wm^{-2}K^{-4}$ is the Stephan Boltzmann constant, and E is the thermal emissivity which for polished metallic surfaces can be taken as 0.05. We obtain $h_{rad} = 3 \cdot 10^{-7} W mm^{-2} K^{-1}$. The heat rate loss by convection when the sample is in air obeys a power law given by: $20 \cdot 10^{-9} (T - T_0)^{5/4} [W mm^{-2}]$. This expression can be further made linear: $20 \cdot 10^{-9} (T - T_0)^{1/4} (T - T_0) [W mm^{-2}] = h_{conv} \cdot (T - T_0) [W mm^{-2}]$. In consequence we can conclude: $h_{conv} \cong 0.8 \cdot 10^{-7} W mm^{-2} K^{-1}$, where we have considered: $T - T_0 = 300K$. The total heat transfer coefficient is: $h_{total} = h_{rad} + h_{conv} \cong 3.8 \cdot 10^{-7} W mm^{-2} K^{-1}$, which corresponds to the sample surrounded by air. For a sample in vacuum we neglect cooling by convection $h_{conv} = 0$, and therefore: $h_{total} = h_{rad} \cong 3 \cdot 10^{-7} W mm^{-2} K^{-1}$.

The temperature of a rectangular graphite sample was measured using two thermocouples attached on the lateral and back sides of the sample, respectively. No thermocouple was mounted on the face directly exposed to the incident electron beam to prevent the obstruction of the beam and to protect the sensor. The beam was incident on the square face of the sample with a cross section of 10 × 10 mm and propagated along its length of 15 mm. Each thermocouple consisted in a small size junction with a rounded head of about 1 mm in diameter and was connected to a FLUKA unit which displayed in real time the measured temperature during irradiation. The acquisition of temperature time series was done simultaneously with the two thermocouples. The electron beam exited the vacuum structure of a low-power LINAC through an aluminum window and was incident on the sample placed in air, at normal pressure and temperature. The irradiation time was limited to a few tens of seconds such that no damages would be induced in the sample, its support and the thermocouples. Longer irradiation times of over 50 s could easily induce temperatures well above 500°C.

Figures 8 and **9** present the evolution in time of the temperature on the surface of the graphite sample at two locations (x, y, z) given by (5, 0, 7.5 mm) and (0, 0,15 mm), respectively. Both locations were conveniently chosen to be at the center of the sample faces and coincided with the position of the sensors. In the figures T_{sample} is the temperature of the sample in Celsius. The irradiation time was 36 s, during which T_{sample} increased continuously. When the irradiation stopped, the temperature started to decrease and the sample cooled down. One can observe that T_{sample} dropped relatively fast during the first 20 to 30 s of cooling down process and at a much slower rate after about 100 s.

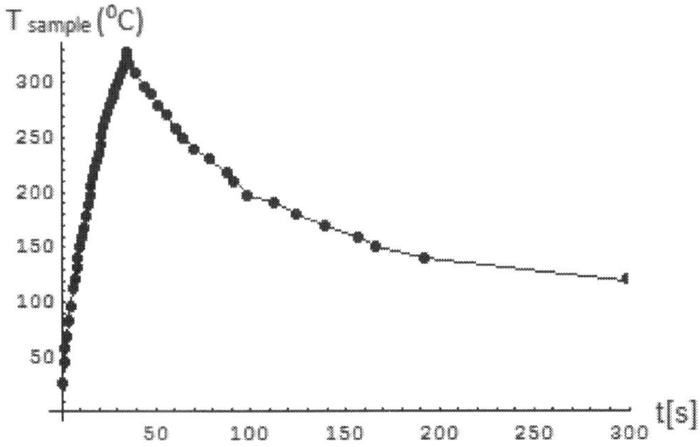

Figure 8. The experimental results for temperature variation in time at the point $(x, y, z) = (5, 0, 7.5 \text{ mm})$.

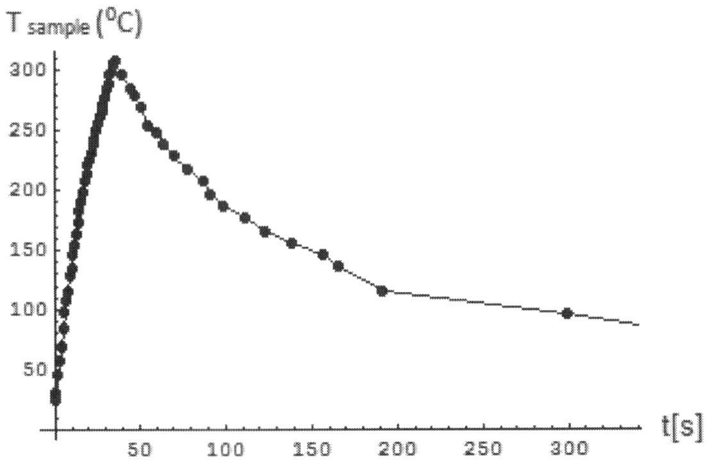

Figure 9. The experimental results for temperature variation in time at the point $(x, y, z) = (0, 0, 15 \text{ mm})$.

In **Figure 8** a slightly higher peak in the temperature with about 30 degrees is observed compared to **Figure 9**. The reason is that the position on the sample surface at which data presented in **Figure 8** has been recorded was closer to the heating source. **Figures 10** and **11** present the comparison of experimental data (dotted line) with our simulations (continuous line) according to integral transform technique. The agreement is quite well, an improving of the future simulations being the consideration of non-Fourier models.

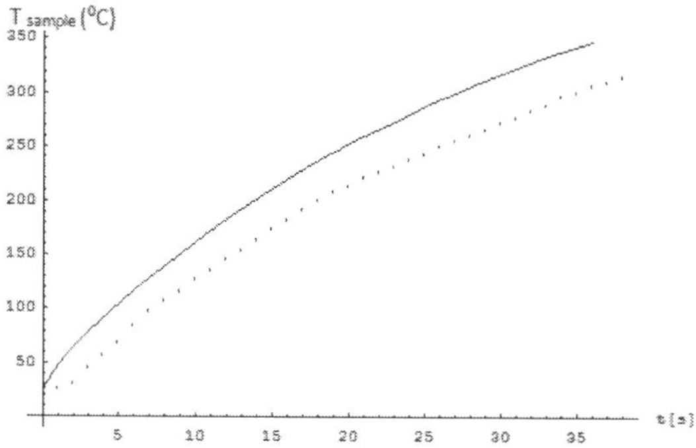

Figure 10. The experimental (dotted line) and simulation (continuous line) results for temperature versus time at the point: $(x, y, z) = (5, 0, 7.5\ \text{mm})$, during 36 s irradiation time.

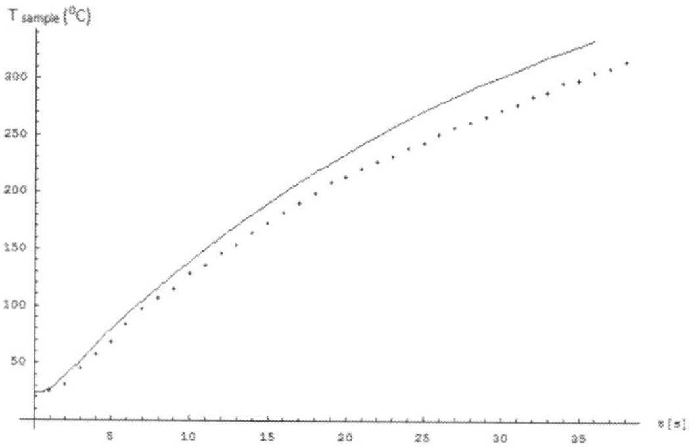

Figure 11. The experimental (dotted line) and simulation (continuous line) results for temperature versus time at the point: $(x, y, z) = (0, 0, 15\ \text{mm})$, during 36 s irradiation time.

6. Laser versus electron interaction in *w* bulk target processing

As it is known, in the case of a Gaussian laser beam having a waist of $w = 1$ mm, the Lambert Beer absorption law reads:

$$I_{mn}(x, y) = I_{0mn}(x, y) \times Exp[-\alpha z] \tag{17}$$

$$I_{mn}(x, y) = I_{0mn}\left[H_m(\tfrac{\sqrt{2}x}{w})H_n(\tfrac{\sqrt{2}y}{w}) \times Exp\left[-\left(\tfrac{x^2+y^2}{w^2}\right)\right]\right]^2 \tag{18}$$

Here, I_{mn}, I_{0mn}, H_m, H_n are lasers intensity in the mode $\{m,n\}$, maximum laser intensity in the mode $\{m,n\}$, the Hermite polynomial of order m, respectively of order n. We are dealing with CO_2 lasers in *cw* mode.

We assumed that one is in the case: $m = 0$ and $n = 0$ in Eq. (18).

For electron irradiation, one should apply, in the particular case of *W*, the empirical absorption Tabata-Ito-Okabe law [19]; it is also considered a Gaussian profile of the electron beam.

The total power of the laser and electron beams is around of 200 W.

The maximum propagation length of an electron beam in cm in targets with high *Z* of density ρ expressed in g/cm³ is:

$$d_{\max} = a_1 \left[(1/a_2) \ \ln \ (1 + a_2\tau) \ - a_3\tau \ / \left(1 + a_4\tau^{a5}\right)\right] /\rho \tag{19}$$

Here, τ is a unitless ratio between the kinetic energy of the electron beam (express in MeV) and the electron rest mass energy.

$$a_1 = b_1 A / Z^{b2}, a_2 = b_3 Z , a_3 = b_4 - b_5 Z , a_4 = b_6 - b_7 Z , a_5 = b_8 / Z^{b9} \tag{20}$$

The constants b_i are given in Table 2:

i	b_i
1.	0.2335
2.	1.209
3.	1.78×10^{-4}
4.	0.9891
5.	3.01×10^{-4}
6.	1.468
7.	1.180×10^{-2}
8.	1.232
9.	0.109

Table 2. The constants b_i in Tabata-Ito-Okabe formula.

Following the formalism from our previous paper, one may write:

$$E_{abs}(z) = \begin{cases} 6.23 - 49.84 \cdot z, for\, z \le 0.125\ cm \\ 0, for\, z > 0.125\ cm \end{cases} \tag{21}$$

where z, E are expressed in cm and MeV, respectively.

In **Figures 12** and **13**, we present the variation of thermal fields under laser and electron irradiation at the same continuous power (200 W) after an exposure time of 25 s. It can be observed that the thermal fields are almost identical, despite the fact that Lambert Beer law (Eq. (17) with: ($\alpha = 8 \times 10^{-2}\text{cm}^{-1}$)) and Tabata-Ito-Okabe (Eq. (21)) law are quite different.

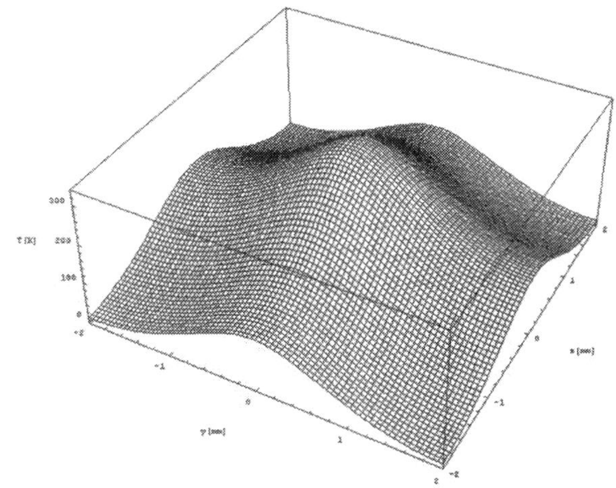

Figure 12. Temperature field on *W* surface after 25 s irradiation with an IR laser beam of 200 W power. In the integral transform technique *T* is the variation temperature rather than the absolute temperature.

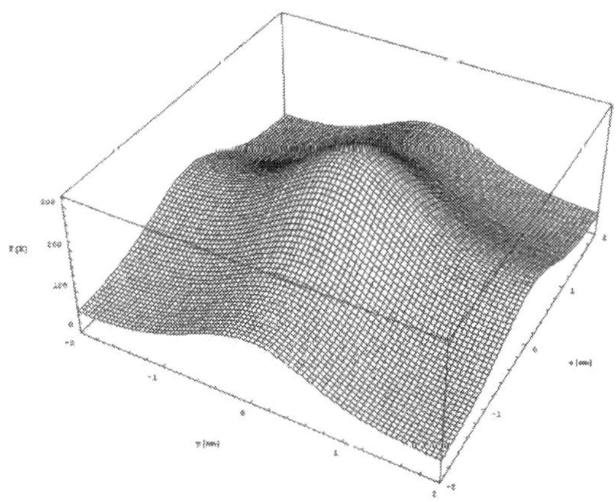

Figure 13. Temperature field on *W* surface after 25 s electron beam irradiation at 200 W power and 6.5 MeV energy.

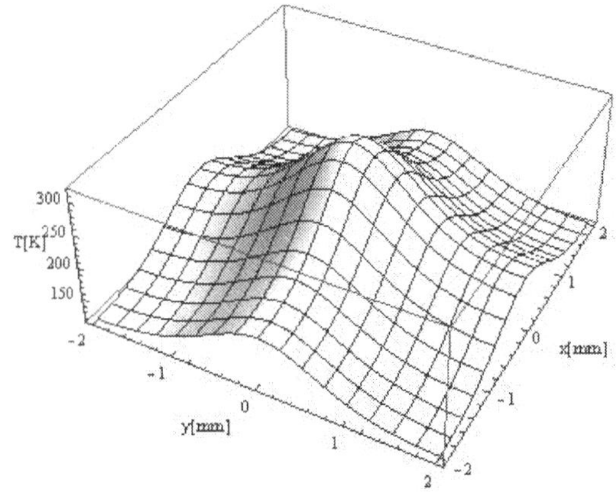

Figure 14. Temperature field on graphite surface after 20 s irradiation with an IR laser beam of 250 W power. In the integral transform technique *T* is the variation temperature rather than the absolute temperature.

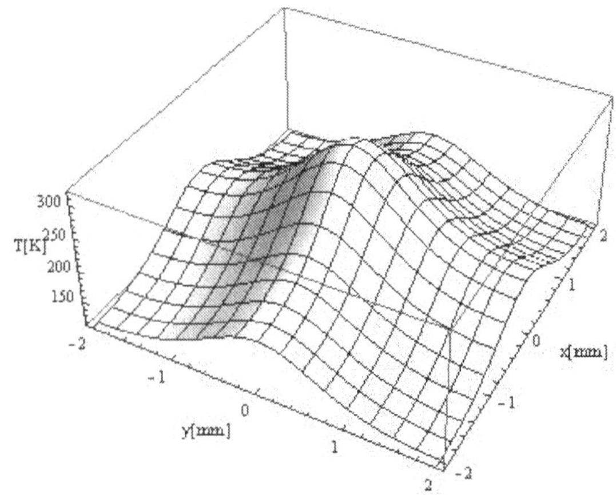

Figure 15. Temperature field on graphite surface after 20 s electron beam irradiation at 250 W power and 6.5 MeV energy.

In **Figures 14** and **15** we present the variation of the thermal fields under laser and electron irradiation at the same continuous power (250 W) after an exposure time of 20 s. Noticeably, the thermal fields are almost identical, also despite the fact that Lambert Beer law (α = $10^{-1}cm^{-1}$) and Katz and Penfolds (Eq. (9)) law are quite different. This implies that initial supposition regarding the similarity between laser and electron irradiation at relatively high power is fully justified. For the near future shall be developed non-Fourier models in order to be more accurate in the tentative to explain the experimental data [20, 21].

Acknowledgements

This work was supported by the NUCLEU project (M. Oane) and M-ERA-NET MAGPHO-GLAS/2013-2015—(R.V. Medianu).

Author details

Mihai Oane[1], Rareş Victor Medianu[2*] and Anca Bucă[1]

*Address all correspondence to: vrmlaser@gmail.com or rares.medianu@nanofizica.ro

1 National Institute for Laser, Plasma Radiation Physics, Bucharest, Romania

2 University of Bucharest, Faculty of Physics, Bucharest, Romania

References

[1] Oane M., Ticoş D., Ticoş C. M., Charged Particle Beams Processing Versus Laser Processing (Monograph), Scholars' Press, Germany, 2015; ISBN: 978-3-639-66753-0.

[2] Oane M., Peled A., Medianu R. V., Notes on Laser Processing (Monograph), Lambert Academic Publishing, Germany, 2013; ISBN: 978-3-659-487-48739-2.

[3] Visan T., Sporea D., Dumitru G., Infrared Physics & Technology, 1998, 39, 335–346.

[4] Bozóki Z., Miklós A., Bicanic D., Applied Physics Letters, 1994, 64, (11), 1362.

[5] Koshlyakov N. S., Smirnov M. M., Gliner E. B., Differential Equations of Mathematical Physics, North Holland Publishing Company, Amsterdam, 1964.

[6] Cline H. E., Anthony T. R., Journal of Applied Physics, 1977, 48(9), 3895–3900.

[7] Yamada M., Nambu K., Yamamoto K., Journal of Applied Physics, 1985, 57(3), 965–968.

[8] Slusher R. E., Reviews of Modern Physics, 1999, 71(2), 471–479.

[9] Carslaw H. S., Jaeger J. C., Conduction of Heat in Solids, Second edition, Oxford University Press, London, 1959.

[10] Korner C., Bergmann H. W., Applied Physics A, 1998, 67,(67),397-401.

[11] Joseph D. D., Preziosi L., Reviews of Modern Physics, 1989, 61(1),41–73.

[12] Bauerle D., Laser Processing and Chemistry, Springer-Verlag, 1996.

[13] Martin D., Fiti M., Radu A., Dragusin M., Cojocaru G., Margarirescu A., Indreas I., Radiation Physics and Chemistry, 1995, 45(4), 615–621.

[14] Oane M., Toader D., Iacob N., Ticoş C. M., Nuclear Instruments and Methods in Physics Research B, 2014, 337, 17–20.

[15] Oane M., Toader D., IacobN., Ticoş C. M., Nuclear Instruments and Methods in Physics Research B, 2014, 318, 232–236.

[16] Yuan X. H., Robinson A. P. L., Quinn M. N., Caroll D. C., Borghesi M., Clarke R. J., Evans R. G., Fuchs J., Gallegos P., Lancia L., Neely D., Quinn K., Romagnani L., Sam G., Wilson P. A., Mc Kenna P., New Journal of Physics, 2010, 12, 063018.

[17] Mahdav M., Ghazizadeh S. F., Journal of Applied Sciences, 2012, 12(2), 145–152.

[18] Katz L., Penfold A. S., Review of Modern Physics, 1952, 24 (1), 28–44.

[19] Tabata T., Ito R., Okabe S., Nuclear Instruments and Methods, 1972, 103, 85–91.

[20] Wang M., Yang N., Guo Z.Y., Journal of Applied Physics, 2011, 110, 064310.

[21] Oane M., Mihăilescu I.N., Ticoş C. M., Banu N., Mitu L. M., Neguţ I., Mihăilescu N., Ticoş D., Journal of Intense Pulsed Laser and Applications in Advanced Physics, 2015, 5(1), 5–8.

Chapter 17

Nanostructuring of Material Surfaces by Laser Ablation

Cinthya Toro Salazar, María Laura Azcárate and
Carlos Alberto Rinaldi

Additional information is available at the end of the chapter

http://dx.doi.org/10.5772/62638

Abstract

Irradiation of materials such as iron and silicon with single nanosecond laser pulses produces nanostructures on its surfaces. Nevertheless, the deposition before irradiation of thin films on the surface of the silicon wafers can modify the shapes of these structures. Upon laser irradiation, different effects are produced on the surfaces of monocrystalline silicon wafers coated with a thin film of Si_3N_4 than on that of bare ones. After irradiation with a Nd:YAG laser pulse of 532 nm, the coated silicon surface presents a nanostructure that, due to its hydrophobic behavior, can be used for biological applications such as cell growth. On the other hand, the nanostructures formed on the surface of metals, such as iron, make them more resistant to oxidation processes by changing their oxidation potentials.

Keywords: nanosecond laser ablation, nanoparticles, surface roughness, nanostructures, laser micromachining

1. Introduction

The surface nanostructuring and the generation of nanoparticles by laser ablation with nanosecond lasers are subjects that have gained importance in the last 15 years [1]. These issues were present since the early experiments with pulsed lasers because they are inherent to pulsed laser interaction with matter, and researchers have made great efforts to get rid of them. In laser ablation micromachining, these effects are known as HAZ (heat affected zone) and are undesirable [1, 3]. It was not until the explosive spring forth of nanotechnology in technological applications in several areas such as medicine and microelectronics among many others that the appearance of different nanostructures in a wide variety of materials started to be reported in the literature.

In this chapter, we will present a review of the nanostructures found by our research group during the quality control of the surfaces of micromachined devices. On the other hand, a method for generating metallic nanoparticles which was developed on the basis of the analysis of the material ejected during the micromachining ablation process will be reported. Nanoparticles were then generated in different media such as air, deionized water, isopropyl alcohol and sodium dodecyl sulfate (SDS) solution. Preliminary results of the generation of Fe@Au nanoparticles (core–shell nanoparticles) will be also discussed.

A brief summary of the fundamentals of laser ablation of solid substrates and of the theory of the resulting ablation plume will be given in the next section. A large variety of studies in this field including basic research and applications have been carried out worldwide by many researchers. However, we will focus on the area of micromachining by laser ablation [1–4].

1.1. Laser ablation

The laser ablation mechanism is one of the most complex phenomena observed when laser radiation interacts with a solid material [1]. Ablation may be produced either with pulsed or with intense continuous wave (cw) lasers [1, 3]. For a given type of material, the onset of ablation takes place around a threshold fluence ($\varnothing_{th}$), which depends on absorption mechanisms and particular properties of the material, such as its microstructure, its morphology, and the presence of defects as well as on laser parameters (intensity, wavelength, and pulse duration) [1, 2, 5, 6]. When irradiation is performed with several consecutive pulses impinging on the same sample's area, $\varnothing_{th}$ can vary due to the accumulation of defects in the previously irradiated zone. At a given wavelength, the amount of material removed per pulse usually shows a logarithmic increase with fluence in accordance with the Lambert–Beer law. On the other hand, Villagran-Muniz et al. [7] reported heuristic equations to describe the fluence dependence of the amount of removed material per pulse of a given substrate with parameters that can be related to physical properties of the substrate.

In particular, for laser pulse durations larger than 10 ps, the laser beam will also interact with the ablation plume as in the case of 10 ns Nd:YAG lasers. In these conditions, the plume will also absorb and scatter radiation. In consequence, the amount of energy reaching the substrate will be less since part of it will be absorbed by the plasma, generating a hot plasma. This excited plasma will then expand and create shock waves in the molten substrate. Frozen shock waves can be observed in the material when the irradiated zone cools [8]. Another effect produced by the excited plume is the explosion of the remnant molten material, which produces splashes. The expelled liquid will inevitably solidify around the irradiated area together with plume's material condensation. In addition, the rapid generation of large thermal gradients may induce excessive thermal stress and thermoelastic excitation of acoustic waves. These stresses may add hardening work, buckling, or cracking [1] to the response of the material, effects which must be taken into account in each particular application.

However, this is not a general feature of all materials. The dynamics of the laser–matter interaction starts with the electronic excitation and relaxation. For weak electric fields, ionization of an atom occurs when the energy of the incident photon exceeds the binding energy of the valence electron. Thus, unlike metals, in a large variety of materials with valence bands, such as dielectrics, electrons are not directly excited into the conduction band. In Chapter 8 of reference [1], an extensive explanation of the differences in the absorption processes of between metals and dielectrics is given.

Only the final results of the ablation process and perhaps some insights of the different phenomena that can possibly occur are experimentally observed. When the material's processing involves a laser, the light–matter interaction mechanisms that simultaneously occur will depend on the pulse duration. Thus, the most relevant mechanisms that take place on a substrate according to the laser pulse duration are as follows: material melting with cw and millisecond pulsed lasers, material vaporizing with nanosecond pulses, and finally, sublimation of material with femtosecond pulses [1]. It is worthy to note that short pulses (1 ns–1 μs) emitted by lasers with Q-switch devices also reduce the thermal impact on the material.

1.1.1. *Applications: Nanostructuring and manufacturing of nanoparticles*

Nanomanufacturing was developed in the last decade driven by the nanotechnology progress. Sub-micrometric structures in the irradiation zones or resulting from the ablation plume were observed since the beginning of pulsed lasers applications to materials' processing. However, at that time, these were a problem and many attempts have been made to mitigate them. It was only in the last decade that they were considered useful [9–12].

The large scope of applications of the sub-micrometric structures' with unique properties that have been lately found [13–15] contributed to this change of perspective. Although the manufacture of nanometric structures and particles with pulsed lasers is quite simple and can be applied to a large variety of materials, the phenomena involved are difficult to analyze and simulate, resulting in a theoretical and experimental research field in continuous growth. In particular, in this chapter, we will present a brief overview of the heuristic functions which describe the laser fluence dependence of the removed material in laser ablation processes of different substrates as well as the nanostructures and nanoparticles found in the analysis with an electronic microscope of the quality of micromachined devices and of the material ejected in the ablation plume, respectively.

1.1.1.1. *Surface modification – nanostructuring*

Nanostructuring by interaction with pulsed lasers is known since the beginning of lasers' applications to machining. However, in recent years, it is an issue that is gaining large importance due to its potential applications in various fields. Different types of nanostructures (NS), named laser-induced periodic surface structures (LIPSS), have been reported. The low spatial frequency LIPSS (LSFL) are characterized by a spatial period Λ of about the laser wavelength (λ), that is, $\Lambda \sim \lambda$. Structures called high spatial frequency LIPSS (HSFL) have a spatial period Λ shorter than the laser wavelength, $\Lambda \ll \lambda$. So far, it is accepted that Λ depends on the laser wavelength and on the radiation incidence angle. These LIPPS are generally described by their period Λ and their orientation with respect to the polarization of the laser [9]. However, the origin of the HSFL is still under debate and some theories have been proposed.

Zhang et al. [10] report NS formed with ultrashort lasers pulses (Ti: Sapphire, 1 kHz, 800 nm, 120 fs). They describe the formation of both types of ripples: LSFL and HSFL. They explain that LSFL arise from optical interference effects due to the coherent interaction between the incident radiation and the electromagnetic wave scattered by the surface. Two main mechanisms are proposed for HSFL formation. One is related to the interaction of the laser pulse with the surface plasma produced by the incident laser. The other mechanism is associated to a combination of interference effects and second harmonic generation. For example, they have experimentally found that the period of the ripples decreases with the increase of the laser fluence on the surface. They report ripples generated perpendicular to the laser polarization in the single-crystal superalloy CMSX-4. According to the experiments of these authors, HSFL ($\Lambda \ll \lambda$) are generated from LSFL ($\Lambda \sim \lambda$) with a period about half of the LSFL spatial period. That is, while the period of the crests of LSFL is about the laser wavelength ($\lambda_L = 800$ nm, Λ_{LSFl} ~ 760 nm), that of the HSFL is $\Lambda_{HSFl} \sim 360$ nm, which is about half the laser wavelength.

NS in silicon were observed in areas up to 100 mm^2 using a versatile laser machining station. Experiences show that the morphology and frequency of the ripples of these structures do not depend either on the focusing geometry, energy, scan rate, polarization, or incident angle of the radiation [11]. It can be thus inferred that laser-based synthesis of nanomaterials is an area that is in its beginnings.

1.1.1.2. Manufacturing of nanoparticles

The manufacture of nanoparticles by pulsed laser ablation is framed within what is known as laser synthesis of nanomaterials (NM). The nanomaterials have extensive applications in electronics, medicine, and energy generation and storage. The physical and chemical properties of these materials strongly depend on their dimensions and shape. The most investigated nanostructures are spheres (nanoparticles (NPs) and "quantum-dots") and cylinders (nanotubes, nanowires, nanorods). In particular, the nanostructures are made by pulsed laser ablation (PLA). In the PLA of a solid, the NPs produced have the same composition of the solid from which they come. If it takes place in air or vacuum, then thin coatings of NPs formed from the ejected clusters can be generated on the surface. When PLA takes place in a liquid, then a colloidal solution is obtained. On the other hand, there is also evidence that the interaction of the laser radiation with nano- and micro-materials in solutions and/or in suspensions in liquids/gases can produce new structures: nanostructures or alloys, as reported in Chapter 7 of reference [1].

Pulsed laser ablation of materials was introduced with the advent of the ruby laser around 1960 [16]. Since then laser for machining research has been directed to obtain new emission wavelengths, increase the repetition rate and output energy and, of course, decrease the pulse duration, nowadays up to attoseconds (10^{-18} s). This research was mainly encouraged by the fact that the laser processing is a powerful tool that does not wear during machining and, in addition, does not produce pollution. On the other hand, in PLA, materials are subjected to high temperatures and pressures, giving rise to a very particular chemistry with the production of various compounds of oxides, carbides, and nitrides. Besides, since cooling rate is very high

http://dx.doi.org/10.5772/62638

due to the rapid expansion of the plasma plume, metastable nanomaterials which are difficult to obtain by other techniques can be produced.

PLA is conceptually simple. However, the material removal mechanisms extend over very long intervals, from fs to ms, at least nine orders of magnitude. When the laser is focused on the surface of a dielectric solid, wavelength dependent non-thermal effects take place. When the solids are metals and semiconductors, thermal processes are also present. The mechanisms and the characteristics of the ejected species strongly depend on the parameters of the laser radiation (wavelength, pulse duration, and fluence).

With a nanosecond laser, PLA is basically a photothermal phenomenon. The incident energy excites the electronic and vibrational levels of the substrate and, in consequence, the material is heated, melted, and evaporated in the first 10–100 ps following the pulse arrival. For relatively low fluence radiation (≤0.3–1 J/cm^2), the ejected material, by desorption and evaporation, is mostly atomic size vapor. Then, the resultant vapor plume expands vertically and is ionized by the photons that keep coming. For fluences near the ablation threshold, the amount of ionized material can be calculated by the Saha equation [17]. If the fluence is larger than the ionization threshold, then optical breakdown occurs and the degree of ionization can be estimated from the Saha-Boltzmann equation.

The plasma role during the laser–matter interaction is not yet fully understood and has been discussed over the past two decades. The plasma is composed of a set of atoms and electrons that remain confined in a high energy field. The laser can produce plasma because during a very short time lapse the photons that interact with the atoms of the solid target can strip off about 15 electrons from each atom. Therefore, the photons that arrive after the solid has been vaporized, contribute to increase the degree of ionization [1]. Ionization strongly influences the dynamics of the plume's condensation.

As the laser fluence increases ($\Phi \geq 10$ J/cm^2), there is a remarkable increase in the amount of ablated material suggesting that a different material ejection mechanism is taking place. Experimentally, it is observed that the removed material consists of a mixture of vapor and micron size droplets (≥10 μm). Different mechanisms have been proposed to explain this behavior. Some suggest that temperature gets close to the critical temperature of the material and then a phase explosion follows [1].

In general, it can be said that in the nanomaterials' synthesis, clusters' ejection and gas to particles condensation, take place. The literature reveals that their final size varies within four orders of magnitude [1]. However, Hubental et al. propose a *tailor*ing method which can be used to homogenize NPs' size [18].

Regardless of the confinement in which the NPs are generated, certain steps taking place in their formation may be outlined:

1. Vaporization (Nd:YAG lasers) and atomization (fs lasers).

2. Phase explosion.

3. Fragmentation.

4. Mechanical exfoliation or chipping.
5. Hydrodynamic spraying.

The high temperature and density of the ejected material near the target's surface develops a pressure exceeding in several orders of magnitude the atmospheric pressure, leading to an expansion of the vapor. During the adiabatic expansion (occurring at low pressure or vacuum) that follows, the thermal energy is converted into kinetic energy, causing a very fast cooling of the plasma (from 10^4 to 10^5 K in 1 μs for a ns laser). The extreme cooling rate leads the plasma to a supersaturation condition in which nucleation becomes energetically favorable [19].

From nucleation theory [20], it is known that the nucleation barrier to form a spherical cluster depends on the cohesive forces between the atoms in the liquid phase, the energetic barrier due to the surface tension, and the plasma ionization (in dielectric materials the electric field leads to polarization). If the radius of the particles with packed atoms is very small, the particle will continue growing, while if it is large, the particle will stop growing and may even break up. The laser beam polarizes the atoms and this effect tends to pack them, so that the critical radius will be even smaller. So, the nucleation energy decreases and a higher amount of nuclei to generate the nanoparticles are produced [21].

Nucleation is said to be homogeneous if clusters are produced from the vaporized material, and the NPs are composed of a few tens of atoms. On the other hand, if clusters are already present during the vapor plume condensation, it is considered a heterogeneous condensation. The already existing clusters are considered nucleation centers and play a predominant role in the condensation stage.

The number of particles is decreased both by collisions and coalescence producing an increase of the NPs' average size. Coalescence in a vapor–liquid medium spontaneously occurs as a reduction of the total surface area during this process, corresponding to a reduction of Gibbs free energy. It takes place up to a few ms after the laser pulse. Then, the NPs cluster due to Van der Waals and electrostatic forces. Clustering is a characteristic feature of NPs' synthesis by laser ablation in gases or liquids that do not contain added stabilizing agents [21–24].

NPs suspended in transparent liquid media can be irradiated, too. Experiments show that when NPs' suspensions are irradiated with pulsed lasers, new structures [25] such as disks, segments, cubes, and pyramids of nanoscale dimensions [26–29] are generated.

2. Research methods

The nanostructures described in this chapter were found when machining micro-devices with a micromachining station based on laser ablation [30]. In particular, the monocrystalline silicon nanostructure was obtained in wafers coated with a sacrificial layer. The nanoparticles were found while analyzing the material ejected during the micro-devices' machining. Later, adequate containers for their generation were developed.

The micromachining station was composed of a laser and a substrates' positioning system. The laser was a frequency doubled Nd:YAG laser (532 nm) emitting pulses of 8 ns FWHM at a repetition rate of 10 Hz. The maximum achievable fluence, for this wavelength, was 250 mJ/cm^2. The substrates' positioning system consisted of eight motors: six stepper motors with micrometric resolution and two piezoelectric motors with nanometric resolution. A diagram of the micromachining station is shown in **Figure 1**.

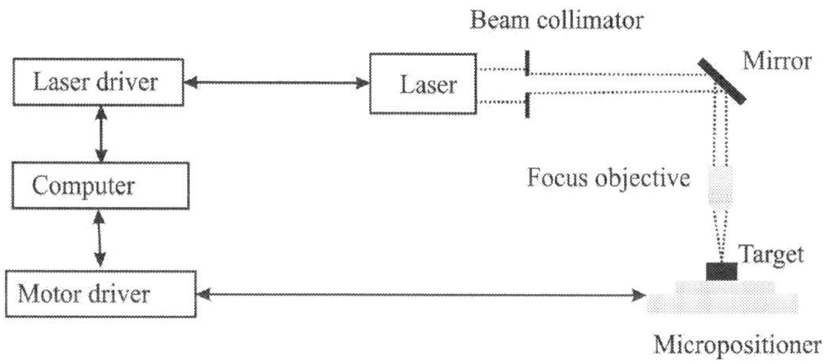

Figure 1. Micromachining system diagram.

Different samples of 2 cm × 2 cm of monocrystalline silicon were cut from c-Si wafers of 4″ diameter and 525 μm thickness. These samples were later coated with different films: Si_2O, Si_3N_4, positive photoresins, resins with dyes, or in some cases, a combination of these films.

The plasma-enhanced chemical vapor deposition (PECVD) technique [30–32] was used to perform Si_2O and Si_3N_4 films coatings. The effect of laser ablation on films of different thicknesses such as 110, 300, 430, 510, 1220, and 2400 nm was studied. The c-Si samples coated with Si_2O were named c-Si + $Si_2O(x)$, where x indicates the film thickness in nm. The thicknesses of the Si_3N_4 films were 180, 420, 500, 570, and 960 nm, and the naming: Si + $Si_3N_4(x)$, x indicating the film thickness in nm. Commercial grade c-Si wafers were also used. These wafers have a 50 nm thick Si_3N_4 film coating on either one or both sides.

Metals such as Fe (99.9%), Ag (99.9%), Au (99.9%), Ta (99.9%), and Cu were other materials processed. Results obtained on Tantalum coated with a 2.5 μm thick AZ1518 resin with the spin-coating technique [33] will be also presented.

2.1. Heuristic equation and its correlation with the physical properties of the irradiated substrate

Since the description of the different regimes observed in the ablation process was found to be a logistic function of the laser fluence, an ad hoc plasma detector was designed based on the works of Bredicce et al. [34] and M. Villagrán et al. [7]. This device, the working principles of which are detailed in reference [30], was used to detect the plasma during the machining. We found that for c-Si, the maximum value of the recorded signals was proportional to the amount of extracted material in each laser shot during the ablation regime according to references [7, 34]. **Figure 2** shows a plot of the induced electric signal maximum amplitude as a function of the incident fluence for c-Si and c-Si + $SiO_2(2400)$.

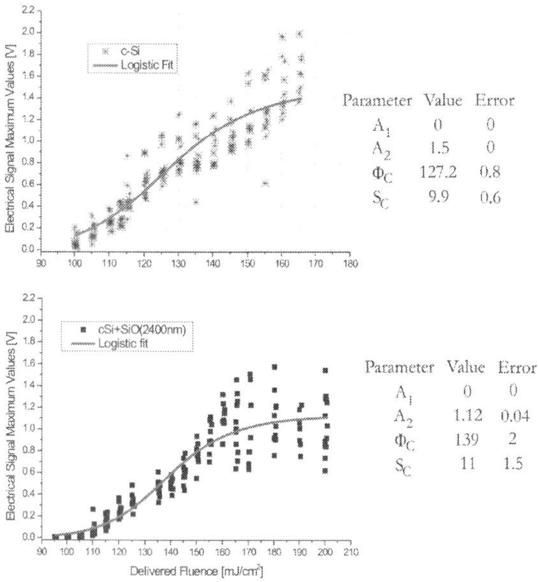

Figure 2. Amplitude of the electric signal versus laser fluence fitted by the logistic function of Eq. (1). (a) For c-Si and (b) for c-Si + SiO_2(2400) along with the values of the fitting functions' parameters.

The results were fitted with the heuristic function described by the following Eq.(1) that relates its parameters to the physical properties of irradiated substrates [34]:

$$M_J(\varnothing) = \left[\frac{A_1(\varnothing) - A_2(\varnothing)}{1 + \left(\frac{\varnothing}{\varnothing_C} \right)^{S_C}} \right] + A_2(\varnothing) \tag{1}$$

where:

$M_J(\varnothing)$ is the amount of material removed per pulse which is proportional to the electric signal. Its units are [mass/(area)(time)], and it is a function of the laser fluence.

$\varnothing$ is in the laser fluence in mJ/cm^2.

A_1 and A_2 represent the lower and higher plateau values related to the onset of the absorption and ablation regimes, respectively. Their units are [mass/(area)(time)].

S_C is a dimensionless fitting parameter related to the surface roughness.

$\varnothing_C$ is the critical fluence. Its value indicates the onset of the true ablation process, that is, the fluence value from which on the material removal occurs mainly through ablation. For example, values of $\varnothing_C$ = 127 mJ/cm² and $\varnothing_C$ = 139 mJ/cm² were found for c-Si and c-Si + Si_2O(2400), respectively (**Figure 2**).

Two plateaus were found for c-Si coated with thin films <1 μm. We postulated that these values indicated whether the laser radiation had reached the substrate or not. **Figures 3** and **4** show the results obtained for two different thicknesses of two kinds of films: c-Si + Si_3N_4(70) and c-Si + Si_3N_4(350), and, c-Si + SiO_2(110) and c-Si + SiO_2(220), respectively, along with the parameters' values.

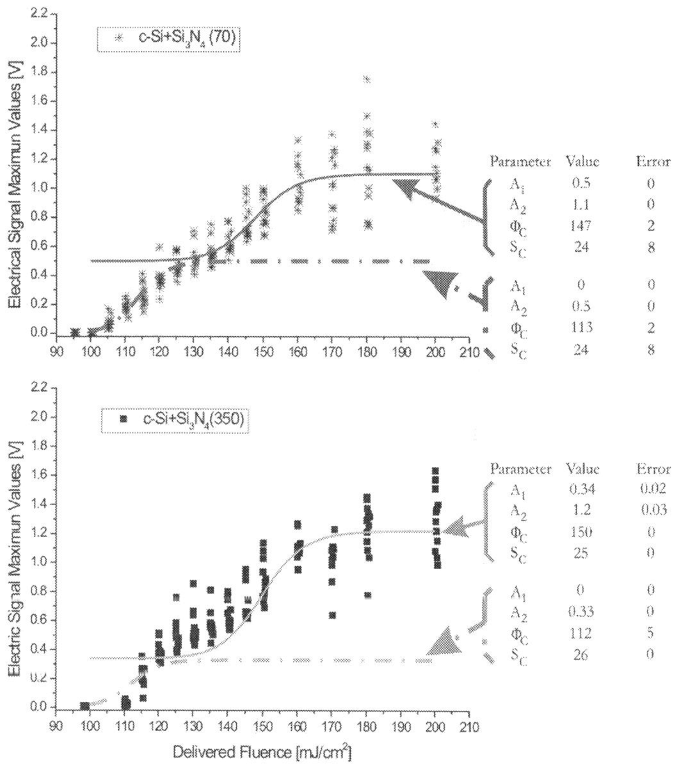

Figure 3. Amplitude of the electric signal versus laser fluence fitted by the logistic function, Eq. (1) for Si_3N_4 thin films thicknesses deposited with the EPCVD technique on monocrystalline silicon: 70 and 350 nm. The parameters of the logistic function are shown too.

The fits show that the effect of the film is to change the surface roughness. This is evidenced by the value of the Sc parameter. The larger the value of this parameter, the lower is the roughness of the surface [7]. Thus, the film acts as a filter until the critical fluence, $\varnothing_C$, is reached. At that moment, the material is ejected from the surface through ablation and generates a nanostructure on the substrate.

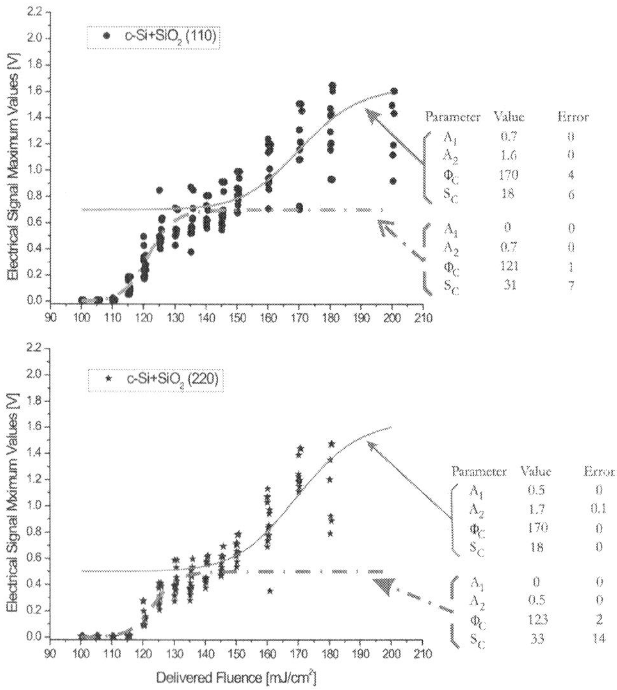

Figure 4. Amplitude of the electric signal versus laser fluence fitted by the logistic function, of Eq. (1) for two silicon dioxide thin films thicknesses deposited with the EPCVD technique on monocrystalline silicon: 110 and 220 nm. The parameters of logistic function are shown too.

2.2. Nanostructuring and nanoparticles manufacturing or generation

2.2.1. Nanostructuring on monocrystalline silicon (c-Si)

Nanostructuring on c-Si was observed during the manufacture of prototypes of nanopores and micro-cavities used as molecules' detectors, drug-delivery control, 2D microstructures, replication matrices, etc. in biological applications. These devices were manufactured using the following techniques in the specified order: (1) laser ablation micromachining (one laser shot cavities), (2) selective chemical attack, and (3) ultrasound cleaning (sonication) [35, 36]. The c-Si nanostructuring was observed after the ns laser pulse impacts on the c-Si + Si_3N_4(370) sample. The cavity formed and its nanostructured base can be observed in the SEM (scanning electron microscope) micrographies shown in **Figure 5**. c-Si nanostructuring has been reported in the literature only with sub-picosecond laser pulses [36–38]. The nanostructures produced in the irradiated zones reduce the reflection of the silicon surface. This effect would be of great interest in the manufacture of silicon photovoltaic cells.

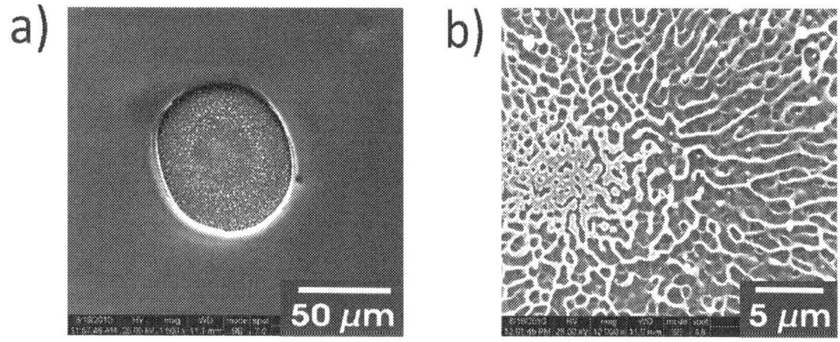

Figure 5. c-Si + Si_3N_4(370) nanostructuring obtained with a ns laser pulse (Nd:YAG, 8 ns, 532 nm). (a) Complete cavity. (b) Magnification of the cavity base.

Nanostructuring was also observed in the machining of a micro-chromatographic column in a c-Si + Si_3N_4(50) commercial wafer [30]. The roughness of the grooves, inherent to the c-Si—laser interaction—was found to be very useful to increase the sample concentration. The micro-column was manufactured as a sequence of straight lines. Part of this micro-chromatographic column and the roughness of the walls of the grooves is shown in **Figure 6**. The larger channels' depth was 50 μm. Devices used for separating substances require roughnesses similar to those of the sample's channels [39].

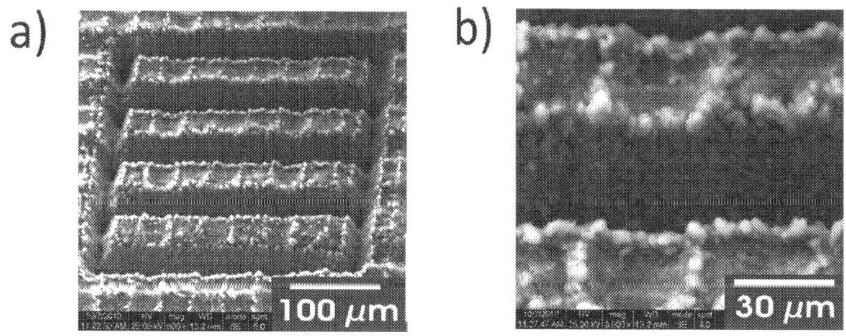

Figure 6. SEM micrographies of (a) the c-Si sample surface after machining. (b) Magnification of part of the machined channels to show the deeper channels. In order to observe the roughness of the channel's wall, the sample's surface was placed forming an angle of 20° with electron beam.

Finally, microstructures and nanostructures were obtained on the walls of cylindrical and conical cavities. For this purpose, the machining station was used with either both percussion and trepanning drilling modes.

In the percussion drilling mode, the laser pulse always impacts in the same sample's area removing material until the desired depth is reached. Grooves of several microns width are obtained as can be seen in **Figure 7a**.

In the trepanning technique, a rotating movement is generated by applying sine wave voltages to the piezoelectric motors with coordinate phases associated to the XY axis of the positioning system. This circular nanometric movement produces a sand-down effect which decreases the drilled cavity walls' roughness. The misalignments of the laser beam focusing system as well as the beam's inhomogeneities are averaged and holes with controlled walls' shapes and roughness can be drilled. As a result, well-controlled circular holes with sub-micrometric structures on the wall can be drilled.

The SEM micrographies of **Figure 7** show cavities drilled on a sample of c-Si + $Si_2O(300)$ with different number of laser pulses of 110 mJ/cm^2. A drilling performed with 2400 pulses with the percussion method can be seen in **Figure 7a**. The grooves on the walls and the debris can be also observed. The debris tower several microns over the silicon surface. A cavity drilled by trepanning with 200 pulses is presented in **Figure 7b**. Marks left by single pulses can be also seen on the sides. In this case, the field of view of electronic microscope allows observing the cavity bottom. Micrography (**Figure 7c**) corresponds to a cavity drilled by trepanning with 2000 laser pulses, the bottom of which can no longer be seen due to its depth. The walls of the cavities of **Figure 7b,c**, show the typical smoothness of the trepanning mode.

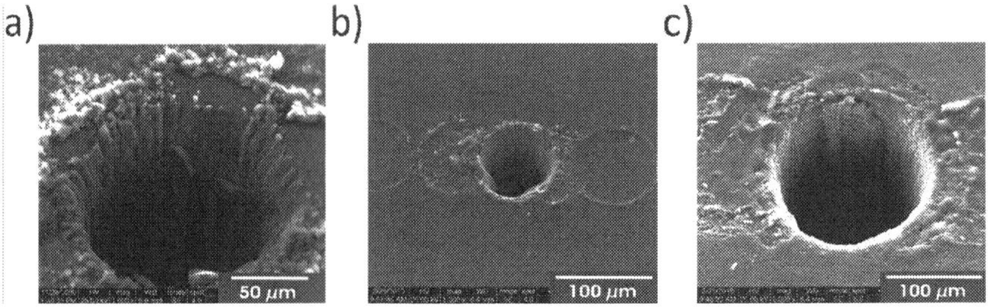

Figure 7. SEM micrographies of (a) c-Si wafer drilled by percussion with 2400 pulses. The sample was inclined 45° to observe the cavity's walls were grooves of ~10 μm are seen. (b, c) Micrographies of cavities drilled by trepanning with 200 and 2000 pulses, respectively. The effect of increasing the laser pulse number for a fixed sample-lens distance and a laser fluence of 110 mJ/cm^2 can be compared. A sine wave voltage of 10 V and 3 Hz was applied to the piezoelectric motors.

2.2.2. *Nanostructuring in metals*

The micromachining experiments in metals were performed with the machining station of **Figure 1**. Nanostructuring on the surface of the devices machined on metallic substrates was observed by electronic microscopy [30]. In this section, the most relevant results obtained are presented.

2.2.2.1. *Cavities in Ta(99.9%)*

Porous tantalum is an alternative metal for the manufacture of total joint arthroplasty components that offer several unique properties since it has excellent biocompatibility and is safe to use in vivo [40]. Upon direct irradiation with several pulses impacting on the same area of a Ta(99.9%) sample, formation of air bubbles of sub-micrometric dimensions gives rise to a porous coating of tantalum and tantalum oxide of the surface. **Figure 8** shows cavities drilled on a Ta sample with 1, 500, and 1000 laser pulses of 200 mJ/cm^2.

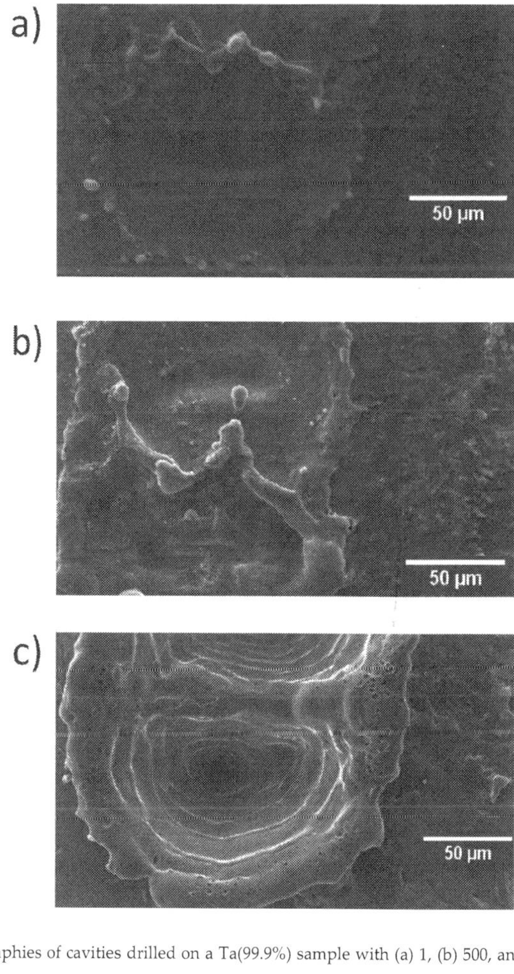

Figure 8. SEM micrographies of cavities drilled on a Ta(99.9%) sample with (a) 1, (b) 500, and (c) 1000 laser pulses of 200 mJ/cm^2.

2.2.2.2. *Nanostructuring in Cu*

Micro-ionizer prototypes to be used in an ion mobility spectrometer [41] were machined on a printed circuit board (PCB). The thickness of the copper coating of these boards is about 300 μm. Direct irradiation of the Cu coating of the PCB was performed by the trepanning method with 200 mJ/cm^2. The SEM micrographies are shown in **Figure 9a**. It shows a sequence of contiguous cavities each drilled with 25 laser pulses each. **Figure 9b**,c shows the details of the cavities at different magnifications.

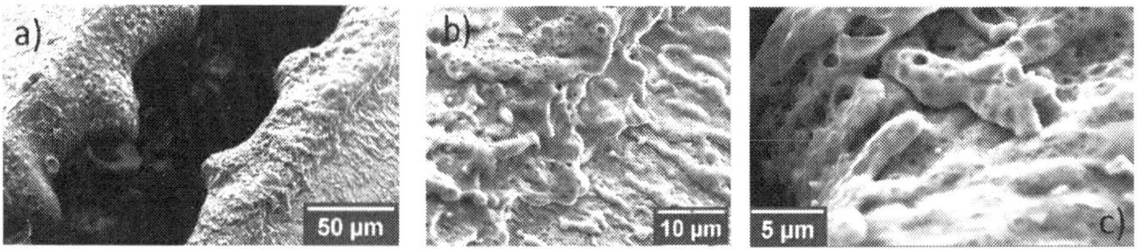

Figure 9. SEM micrographies (a) contiguous cavities machined on the copper coating of a PCB with 25 pulses each. (b, c) Micrographies at larger magnifications to show the porosity.

The appearance of a porous coating of copper and copper oxide is observed as the number of laser pulses impinging on the same sample area is increased. The lower density of porous copper makes it suitable to be used as a lighter material for electrodes and catalysis.

Figure 10 shows the superficial modifications originated by laser irradiation of sub-micrometric structures of copper oxide. **Figures 10a,b** are SEM micrographies of the surface before irradiation; **Figure 10c,d** after irradiation with one square laser shot.

Figure 10. SEM micrographies of a copper oxide surface: (a, b) before irradiation; (c, d) after irradiation with one square laser shot. Cu nanoparticles of lighter color than the rest of the surface due to their larger conductivity can be appreciated.

http://dx.doi.org/10.5772/62638

2.2.2.3. *Nanostructuring in Fe(99.9%)*

The nanostructuring found in Ta and Cu encouraged the irradiation of Fe(99.9%) samples. Only sub-micrometric structures in Fe obtained with sub-picosecond laser pulses have been reported in the literature. The HAZ effect produced on the surface of steel plates is used as anticorrosive. In reference [41], Yang et al. report that the zones in the surroundings of a folding axis are more liable to corrosion due to the larger mechanical strains to which they are subjected. On the other hand, the authors found that steel plates which had been irradiated with laser pulses before being folded became more resistant to oxidation. **Figure 11** shows the surface of a Fe(99.99%) plate after irradiation with laser pulses of 200 mJ/cm^2 in which details of the induced surface nanostructuring can be observed.

Figure 11. Fe (99.99%) plate surface after irradiation laser pulses of 200 mJ/cm^2. The magnification of the electronic microscope (SEM) increases from left to right disclosing details of the surface nanostructuring.

2.2.3. *Fe(99.9%), Au(99.9%), Ag(99.9%) nanoparticles in different media*

The manufacture method of NPs of different metals and the diverse analysis performed will be described in the following. The first NPs were generated in air in an ad hoc chamber suitable for obtaining metallic NPs in gaseous media. The substrate was fixed to the container which was then fixed to the machining system. The laser was focused on the surface of the substrate, and the movement of the positioning system of the substrate was controlled by a PC. To ease the subsequent analysis with an electronic microscope, the NPs formed were stuck on carbon tapes.

Figure 12 shows the Fe NPs generated in air at ambient temperature and pressure with two different irradiation processes. Those shown in **Figure 12a** were generated by drawing lines with a speed of 10 μm/s with pulses of 70 μm spot size. The irradiated zones and the number of pulses are detailed in **Figure 12b**. The NPs shown in **Figure 12c**,d were generated by drilling holes by percussion with 1000 pulses. The difference between both methods is the generation of clusters of NPs. The size of the NPs generated by overlapping pulses on the substrate's surface (**Figure 12a**) is ~200 nm, and, as can be seen, they are grouped in clusters. On the other hand, the size of the NPs generated by drilling is ~50 nm and they are also grouped in clusters. In addition to NPs clusters, the formation of dense spherical microparticles is observed (**Figure 12c**). This effect could be produced by the change of the superficial crystalline structure introduced by the ns 532 nm laser pulse on the metal surface, Fe(99.9%) in this case [43, 44].

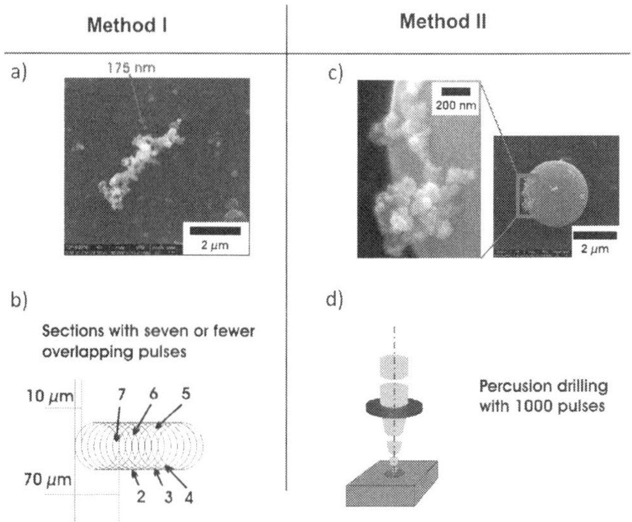

Figure 12. (a) Fe NPs manufactured by drawing lines with up to seven contiguous pulses, (b) Fe NPs generated by percussion (1000 pulses in the same sample area). In both processes, the laser fluence was 220 mJ/cm^2. All micrographies were obtained by SEM.

The generation of NPs in liquid media was performed at ambient temperature and pressure. The substrate was immersed in the liquid contained in a beaker. Laser pulses were focused on the substrate's surface, while the position of the substrate was continuously being changed so that each pulse would impact in a non-irradiated surface area. Metallic NPs dispersions in different liquid media such as ultrapure H_2O, isopropanol, and a solution of sodium dodecyl sulfate (SDS) were obtained following this method.

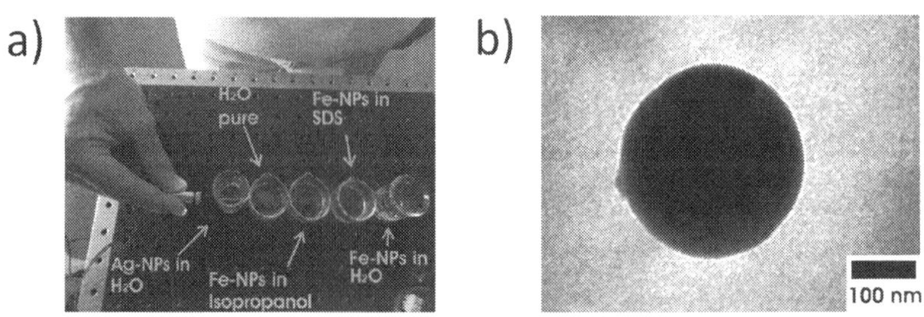

Figure 13. NPs generated in air and in different liquid media. (a) Tyndall effect in dispersions and solutions: (from left to right) Ag NPs in ultrapure H_2O, ultrapure H_2O, Fe NPs in isopropanol, Fe NPs in SDS, and Fe NPs in ultrapure H_2O. (b) Characterization of the Fe NPs in a SDS solution by microscopy (HR-TEM).

Different methods were used for the NPs characterization:

http://dx.doi.org/10.5772/62638

1. Tyndall effect for a rapid verification of the NPs' generation in liquid media. In order to observe this effect, the dispersion is illuminated with a cw laser. For example, in **Figure 13a**, a 98 mW cw laser @405 nm was used. The presence of NPs disperses the laser beam and makes it visible at simple sight. This effect is not observed in solutions or as in the case of ultrapure H_2O in the second beaker from left to right of **Figure 13a**.

2. High-resolution transmission electron microscopy (HR-TEM), which allowed studying the NPs' appearance, form and size (**Figure 13b**).

3. UV–visible light absorption spectrophotometry for the analysis of suspensions.

The advantage of this last method is that it allows preserving the NPs' properties *in situ*, avoiding alterations due to drying effects. Clustering or oxidation effects usually appear when the NPs are separated from the medium in which they were originated (**Figure 14**).

Finally, the size distribution of the NPs in their original medium was estimated from the absorbance spectra of the dispersions and their comparison with those calculated with the MiePlot algorithm (Appendix E, reference [30]).

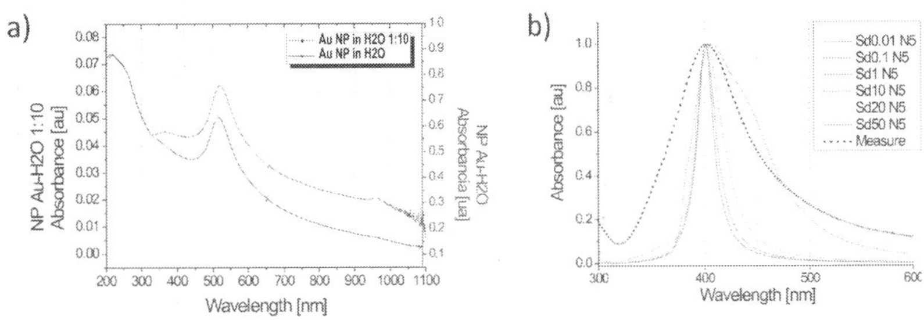

Figure 14. (a) Absorbance spectrum of a dispersion of Au NPs in ultrapure H_2O compared to that of the same dispersion diluted 10 times, (b) comparison of the absorbance spectrum of Ag NPs in H_2O experimentally obtained with that calculated for Ag NPs of 20 nm of radius in H_2O with a log-normal distribution with different standard deviations.

2.2.3.1. *Core–shell nanoparticles (preliminary results)*

Nanoparticles of different nucleus' materials and *core–shell* were produced as prototypes for *drug delivery*. The particles were required to have a ferromagnetic nucleus to be able to be delivered with magnets to the desired zones and a *core–shell* of a material assimilable by the organism so as not to be rejected by it.

The following method was used for the synthesis of *core–shell* NPs. First, a suspension of Fe NPs in ultrapure H_2O was obtained by laser ablation of a Fe(99.9%) plate. An Au(99.9%) plate was subsequently irradiated in the Fe NPs suspension. The Fe NPs acted as nucleation centers for Au crystallization according to the nucleation principles described in Section 1.1.1.2.

The absorbance spectrum of the final suspension obtained, shown in **Figure 15**, resulted in very good agreement with those reported by other authors in references [45, 46].

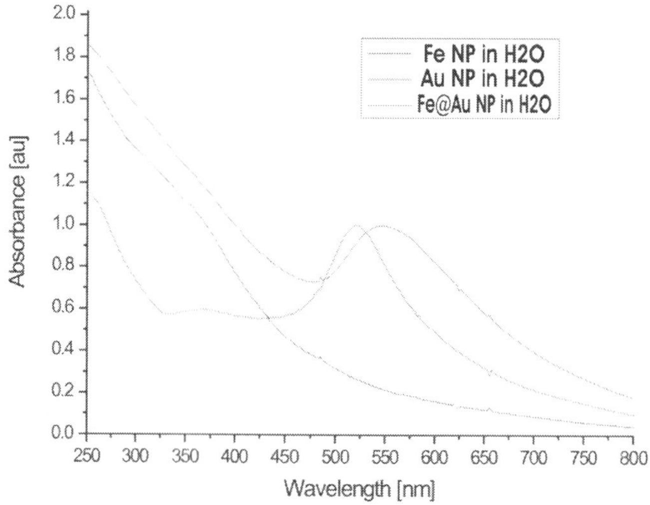

Figure 15. Absorbance spectrum of dispersions of Fe NPs, Au NPs, and Fe@Au NPs in ultrapure H_2O.

3. Conclusions

Laser ablation allows numerous technological developments ranging from laser-induced breakdown spectroscopy (LIBS), pulsed laser deposition (PLD), laser propulsion, to surface modification and generation of nanoparticles, NPs. In this chapter, we have presented a review of the results obtained with these two latter processes. However, it should be noted that these effects or phenomena were found to occur during the implementation of the micromachining processes and have gained relevance since their observation by means of electron microscopy in its different forms SEM and TEM. The possibility of producing nanostructures on surfaces opens up new possibilities for studying their effects on physicochemical surface phenomena. These effects, mentioned in the Section 1.1.1, as laser-induced periodic surface structures and the increase of roughness in the generation of channels for chromatographic columns will allow the production of cheaper systems based on these principles. On the other hand, the heuristic model presented in Section 2.1 to explain the phenomenon of laser ablation allows us to understand the effect of surface modification and use it as a tool to transform surfaces by coating them with thin films. Another very important issue is the possibility of producing core–shell nanoparticles of the type shown in the Section 2.2, using different metals. In our case, the production of Fe@Au NPs resulted a very convenient way to unify in a same material the magnetic properties of Fe with the biological properties of Au. As a final remark, as it was our intention to put forward throughout this chapter, we would like to emphasize that nanosecond lasers are still a very important source for the modification of surfaces for technological uses.

http://dx.doi.org/10.5772/62638

Author details

Cinthya Toro Salazar[1], María Laura Azcárate[1,2] and Carlos Alberto Rinaldi[2,3,4*]

*Address all correspondence to: rinaldi@cnea.gov.ar

1 Centro de Investigaciones en Láseres y Aplicaciones CEILAP (CITEDEF-CONICET), Buenos Aires, Argentina

2 Consejo Nacional de Investigaciones Científicas Técnicas, Buenos Aires, Argentina

3 Comisión Nacional de Energía Atómica, Buenos Aires, Argentina

4 Escuela de Ciencia y Tecnología, UNSAM, Buenos Aires, Argentina

References

[1] K. Sugioka, M. Meunier and Alberto Piqué. Laser Precision Microfabrication. Springer Series in Materials Science. 2010.

[2] N.B. Dahotre and S.P. Harimkar. Laser Fabrication and Machining of Materials. Springer. 2008.

[3] W. M. Steen. Laser Material Processing. Springer-Verlag. 1991.

[4] E. Gentili, L. Tabaglio and F. Aggogeri. Review on Micromachining Techniques. Italy: Department of Mechanical Engineering, University of Brescia. 2005.

[5] M.B. Mason (BP Solar) and J. Fieret (Exited Ltd). Advanced Laser Processing for Industrial Solar Manufacturing. Contractor BP Solar Limited. 2006.

[6] M.M.S. Amera, M.A. El-Ashrya, L.R. Dosserb, K.E. Hixb, J.F. Maguirec and B. Irwin. Femtosecond versus nanosecond laser machining: comparison of induced stresses and structural changes in silicon wafers. Applied Surface Science. 2005;242:162–167.

[7] M. Villagran-Muniz, H. Sobral, C. Rinaldi, I. Cabanillas-Vidosa and J. Ferrero. Optical emission and energy disposal characterization of the laser ablation process of CaF2, BaF2, and NaCl at 1064 nm. Journal of Applied Physics. 2008;104:103112–103117.

[8] C. Toro, C. Lasorsa, C. Sánchez Aké, M. Villagran-Muñiz and C. Rinaldi. New method for nanosecond laser machining. Journal of Laser Micro Nanoengineering. 2012;7:269–273.

[9] J. Eichstädt, G.R.B.E. Römer and A.J. Huisin 't Veld. Determination of irradiation parameters for laser-induced periodic surface structures. Applied Surface Science. 2013;264:79–87.

[10] W. Zhang, G. Cheng, Q. Feng, L. Cao, F. Wang and R. Hui. Abrupt transition from wavelength structure to subwavelength structure in a single-crystal superalloy induced by femtosecond laser. Applied Surface Science. 2011;257:4321–4324.

[11] R. Le Harzic, D. Dörr, D. Sauer, M. Neumeier, M. Epple, H. Zimmermann and F. Stracke. Formation of periodic nanoripples on silicon and germanium induced by femtosecond laser pulses. Physics Procedia. 2011;12:29–36.

[12] T.Q. Jia, H.X. Chen, M. Huang, and F.L. Zhao, J.R. Qiu, R.X. Li and Z.Z. Xu, X. K. He, J. Zhang and H. Kuroda. Formation of nanogratings on the surface of a ZnSe crystal irradiated by femtosecond laser pulses. Physical Review B. 2005;72:125429.

[13] Available from: http://www.nasa.gov/centers/ames/research/technologyonepagers/nanotechnology-landing.html.

[14] Available from: http://www.cnea.gov.ar/institutos/nanociencia_nanotecnologia.php.

[15] Available from: http://www.fan.org.ar.

[16] T.H. Maiman. Nature. 1960;187:493.

[17] G. Bekefi. Principles of Laser Plasmas. 1st ed. New York: Wiley. 1976.

[18] F. Hubenthal. Vol. 1, Sección 1.13. In: Comprehensive Nanoscience and Technology. 2011;375–435.

[19] D. Kashchiev. Nucleation: Basic Theory with Applications. Oxford: Butterworth-Heinemann. 2000.

[20] B.S. Luk'yanchuk, W. Marine, S.I. Anisimov and G.A. Simakina, Condensation of vapor and nanoclusters formation within the vapor plume, produced by ns-laser ablation of Si, Ge and C, Proc. SPIE 3618, 434. 1999.

[21] M.S. Tillack, D.W. Blair and S.S. Harilal. The effect of ionization on cluster formation in laser ablation plumes. Nanotechnology. 2004;15:390–403.

[22] J. Perrière, É. Millon, É. Fogarassy, A.V. Kabashin and M. Meunier. Recent advances in laser processing of materials. Amsterdam: European Materials Research Society: Elsevier. 2006.

[23] R. Hergenroder. Laser-generated aerosols in laser ablation for inductively coupled plasma spectrometry. Spectrochimica Acta Part B: Atomic Spectroscopy. 2006;61(3).

[24] P.S. Liu, W.P. Cai and H.B. Zeng. Journal of Physical Chemistry C. 2008;112:3261.

[25] K.Y. Niu, J. Yang, S.A. Kulinich, J. Sun, H. Li and X.W. Du. Journal of the American Chemical Society. 2010;132:9814.

[26] Z. Yan and D.B. Chrisey. Pulsed laser ablation in liquid for micro-/nanostructure generation. Journal of Photochemistry and Photobiology C: Photochemistry Reviews. 2010;doi:10.1016/j.jphotochemrev.2012.04.004.

[27] D. Werner, S. Hashimoto, T. Tomita, S. Matsuo and Y. Makita. Journal of Physical Chemistry C. 2008;112:1321.

[28] Z.J. Yan, R.Q. Bao and D.B. Chrisey. Langmuir. 2011;27:851.

[29] A.V. Simakin, V.V. Voronov, G.A. Shafeev, R. Brayner and F. Bozon-Verduraz. Chemical Physics Letters. 2001;348:182.

[30] C.T. Salazar. Micromecanizado por ablación láser en silicio, en silicio cubierto por películas y en metales [thesis]. Buenos Aires, Argentina: Facultad de Ingeniería, Universidad de Buenos Aires. 2015;129 p.

[31] C. Toro, C. Lasorsa and C. Rinaldi. Detecting plasma produced by laser in a micromachining system as in-process control. Journal of Laser Micro Nanoengineering. 2011;6:220–224.

[32] A. Rodrigo, et al. Dominant plasma species for TiN film formation by plasma CVD. Journal of Physics D: Applied Physics. 1997;30:2397.

[33] Spin coating theory. Louisville: Micro/Nano Technology Center, University of Louisville. 2013.

[34] H. Sobral, M. Villagrán-Muniz and F. Bredice. Energy balance in laser ablation of metal targets. Journal of Applied Physics. 2005;98:083305.

[35] B. Lerner, M. Perez, C. Toro, C. Lasorsa, C.A. Rinaldi, A. Boselli and A. Lamagna. Generation of cavities in silicon wafers by laser ablation using silicon nitride as sacrificial layer. Applied Surface Science, Amsterdam: Elsevier Science BV. 2012;258:2914–2919.

[36] M. Halbwaxa, T. Sarneta, Ph. Delaportea, M. Sentisa, H. Etienneb, F. Torregrosab, V. Vervischb, I. Perichaudc and S. Martinuzzic. Micro and nano-structuration of silicon by femtosecond laser: application to silicon photovoltaic cells fabrication. doi:10.1016/j.tsf.2007.12.117.

[37] T. Baldacchini, J.E. Carey, M. Zhou and E. Mazur. Superhydrophobic surfaces prepared by microstructuring of silicon using a femtosecond laser. Langmuir. 2006;22:4917–4919.

[38] B.R. Tull, J.E. Carey, E.Mazur, J.P. McDonald and S.M. Yalisove. Silicon surface morphologies after femtosecond laser irradiation. Mrs Bulletin. 2006;31:626–633.

[39] E.H.M. Camara, P. Breuil, D. Briand, L. Guillot, C. Pijolat and N.F. de Rooij. Review: Micro gas preconcentrator in porous silicon filled with a carbon absorbent. Sensors and Actuators B. 2010;148:610–619.

[40] B. Levine, C.J. Della Valle and J.J. Jacobs, Applications of porous tantalum in total hip arthroplasty. Journal of the American Academy of Orthopaedic Surgeons. 2006;14(12): 646–655.

[41] N.G. Boggio, P. Alonso Castillo, J.J. Ortiz, A. Lamagna, A. Boselli and C.A. Rinaldi. Desarrollo de un espectrómetro por movilidad iónica para la detección de compuestos orgánicos (explosivos, drogas y contaminantes). Anales AFA.

[42] L.J. Yang, J. Tang, M.L. Wang, Y. Wang and Y.B. Chen. Surface characteristic of stainless steel sheet after pulsed laser forming. Applied Surface Science. 2010;256:7018–7026.

[43] C. Toro, E.D. Cabanillas and C.A. Rinaldi. New method for micro and nanoparticles fabrication. In: Japan Laser Processing Society, editor. The 12th International Symposium on Laser Precision Microfabrication; Japan, Takamatsu, Kagawa. 2011.

[44] C. Toro, E.D. Cabanillas and C.A. Rinaldi. Generación de nanopartículas y modificación de superficies de Fe con ablación láser. In: Instituto de Química del Noroeste Argentino, editor. XVII Congreso Argentino de Fisicoquímica y Química Inorgánica; Argentina. Córdoba. 2011.

[45] S. Kayal and R.V. Ramanujan. Anti-cancer drug loaded iron–gold core–shell nanoparticles (Fe@Au) for magnetic drug targeting. Journal of Nanoscience and Nanotechnology. 2010;10:1–13.

[46] Journal of Magnetism and Magnetic Materials. 2007;310:2369–2371.